Kinematics and Dynamics of Mechanical Systems Implementation in MATLAB® and Simmechanics®

Kinematics and Dynamics of Mechanical Systems Implementation in MATLAB® and Simmechanics®

Second Edition

Kevin Russell, Qiong Shen, Raj S. Sodhi

CRC Press
Taylor & Francis Group
Boca Raton London New York

CRC Press is an imprint of the
Taylor & Francis Group, an **informa** business

CRC Press
Taylor & Francis Group
6000 Broken Sound Parkway NW, Suite 300
Boca Raton, FL 33487-2742

© 2019 by Taylor & Francis Group, LLC
CRC Press is an imprint of Taylor & Francis Group, an Informa business

No claim to original U.S. Government works

Printed on acid-free paper

International Standard Book Number-13: 978-1-4987-2493-7 (Hardback)

Library of Congress Cataloging-in-Publication Data

Names: Russell, Kevin (Mechanical engineer), author. | Shen, Qiong (Mechanical engineer), author. | Sodhi, R. S. (Raj S.), author.
Title: Kinematics and dynamics of mechanical systems : implementation in MATLAB and SimMechanics / Kevin Russell, Qiong Shen, Raj S. Sodhi.
Description: Boca Raton : CRC Press, Taylor & Francis, 2018. | Includes index.
Identifiers: LCCN 2018022923 | ISBN 9781138584044 (hardback)
Subjects: LCSH: Machinery, Kinematics of--Computer simulation. | Machinery, Dynamics of—Computer simulation. | MATLAB. | SimMechanics.
Classification: LCC TJ175 .R87 2018 | DDC 621.01/5311—dc23
LC record available at https://lccn.loc.gov/2018022923

Visit the Taylor & Francis Web site at
http://www.taylorandfrancis.com

and the CRC Press Web site at
http://www.crcpress.com

Contents

Preface xiii

Authors xvii

1. Introduction to Kinematics 1
   1.1   Kinematics 1
   1.2   Kinematic Chains and Mechanisms 2
   1.3   Mobility, Planar, and Spatial Mechanisms 3
   1.4   Types of Mechanism Motion 5
   1.5   Kinematic Synthesis 7
   1.6   Units and Conversions 8
   1.7   Software Resources 9
   1.8   Summary 9
   References 10
   Additional Reading 11

2. Mathematical Concepts in Kinematics 13
   2.1   Introduction 13
   2.2   Complex Numbers and Operations 13
       2.2.1   Complex Number Forms 13
       2.2.2   Complex Number Addition 15
       2.2.3   Complex Number Multiplication and Differentiation 17
   2.3   Vector and Point Representation 20
   2.4   Linear Simultaneous Equations, Matrices, and Matrix Operations ... 22
       2.4.1   Linear Simultaneous Equation Systems and Matrices 22
       2.4.2   Matrix Transpose, Addition, Subtraction, and Multiplication ... 23
       2.4.3   The Identity Matrix and Matrix Inversion 26
   2.5   Intermediate and Total Spatial Motion 29
   2.6   General Transformation Matrix 33
   2.7   Summary 36
   References 37
   Additional Reading 37
   Problems 37

3. Fundamental Concepts in Kinematics 41
   3.1   Types of Planar and Spatial Mechanisms 41
       3.1.1   Planar Four-Bar Mechanism 41
       3.1.2   Slider-Crank Mechanism 41
       3.1.3   Geared Five-Bar Mechanism 42
       3.1.4   Planar Multiloop Six-Bar Mechanisms 44
       3.1.5   Spatial Four-Bar Mechanisms 45
   3.2   Links, Joints, and Mechanism Mobility 46
   3.3   Number Synthesis 49
   3.4   Grashof's Criteria and Transmission Angle 50
   3.5   Circuit Defect 53

3.6   Mechanism Inversion 54
3.7   Passive Degree of Freedom and Paradoxes 55
3.8   Summary 56
References 57
Problems 58

4. Kinematic Analysis of Planar Mechanisms 63
4.1   Introduction 63
4.2   Numerical Solution Method for Two Simultaneous Equations 64
4.3   Link Velocity and Acceleration Components in Planar Space 64
4.4   Four-Bar Mechanism Analysis 66
        4.4.1   Displacement Equations 66
        4.4.2   Velocity Equations 67
        4.4.3   Acceleration Equations 68
        4.4.4   Kinematics of Coupler Locations of Interest 70
        4.4.5   Instant Center, Centrodes, and Centrode Generation 76
4.5   Slider-Crank Mechanism Analysis 80
        4.5.1   Displacement Equations 80
        4.5.2   Velocity Equations 81
        4.5.3   Acceleration Equations 83
        4.5.4   Centrode Generation 89
4.6   Geared Five-Bar Mechanism Analysis 91
        4.6.1   Displacement Equations 91
        4.6.2   Velocity Equations 92
        4.6.3   Acceleration Equations 93
        4.6.4   Kinematics of Intermediate Link Locations of Interest 95
4.7   Watt II Mechanism Analysis 97
4.8   Stephenson III Mechanism Analysis 100
        4.8.1   Displacement Equations 100
        4.8.2   Velocity Equations 103
        4.8.3   Acceleration Equations 104
        4.8.4   Kinematics of Intermediate Link Locations of Interest 105
4.9   Time and Driver Angular Velocity 108
4.10   Mechanism Configurations 108
4.11   Constructing Cognates 109
4.12   Planar Mechanism Kinematic Analysis and Modeling in Simmechanics® 112
4.13   Summary 117
References 118
Additional Reading 118
Problems 118

5. Dimensional Synthesis 131
5.1   Introduction 131
5.2   Branch and Order Defects 133
5.3   Planar Four-Bar Motion Generation: Three Precision Positions 135
5.4   Order- and Branch-Defect Elimination 140
5.5   Path Generation versus Motion Generation 144
5.6   Stephenson III Motion Generation: Three Precision Positions 145
5.7   Planar Four-Bar Function Generation: Three Precision Points 149

5.8 Planar Four-Bar Function Generation: FSPs and MSPs 153
5.9 Mechanism Dimensions: From Dimensional Synthesis to
     Kinematic Analysis 158
5.10 Summary 163
References 164
Additional Reading 165
Problems 165

6. Static Force Analysis of Planar Mechanisms 175
6.1 Introduction 175
6.2 Static Loading in Planar Space 176
6.3 Four-Bar Mechanism Analysis 177
6.4 Slider-Crank Mechanism Analysis 182
6.5 Geared Five-Bar Mechanism Analysis 185
6.6 Watt II Mechanism Analysis 190
6.7 Stephenson III Mechanism Analysis 196
6.8 Planar Mechanism Static Force Analysis and Modeling in SimMechanics® 201
6.9 Summary 204
References 205
Additional Reading 206
Problems 206

7. Dynamic Force Analysis of Planar Mechanisms 223
7.1 Introduction 223
7.2 Dynamic Loading in Planar Space 224
7.3 Four-Bar Mechanism Analysis 224
7.4 Slider-Crank Mechanism Analysis 230
7.5 Geared Five-Bar Mechanism Analysis 233
7.6 Watt II Mechanism Analysis 238
7.7 Stephenson III Mechanism Analysis 243
7.8 Mass Moment of Inertia and Computer-Aided Design Software 248
7.9 Planar Mechanism Dynamic Force Analysis and Modeling in
     Simmechanics® 250
7.10 Summary 256
References 256
Additional Reading 256
Problems 256

8. Design and Kinematic Analysis of Gears 271
8.1 Introduction 271
8.2 Gear Types 272
8.3 SPUR-Gear Nomenclature and Relationships of Mating Gears 274
     8.3.1 Spur-Gear Nomenclature 274
     8.3.2 Pressure Angle and Involute Tooth Profile 277
     8.3.3 Gear Center Distance and Contact Ratio 280
     8.3.4 Gear-Tooth Interference and Undercutting 282
     8.3.5 Backlash 283
8.4 Helical-Gear Nomenclature 284
8.5 Gear Kinematics 287

8.5.1   Spur Gears and Gear Trains...287
8.5.2   Planetary Gear Trains 291
8.5.3   Rack and Pinion Gears..295
8.5.4   Helical Gears 296
8.5.5   Bevel Gears 298
8.5.6   Worm Gears 301
8.6   Summary...305
References 305
Additional Reading 306
Problems..306

9. **Design and Kinematic Analysis of Disk Cams**..311
9.1   Introduction 311
9.2   Follower Types...312
9.3   Follower Motion 313
9.3.1   Rise, Fall, and Dwell..313
9.3.2   Displacement, Velocity, Acceleration, and Jerk............................315
9.3.3   Constant Velocity Motion 315
9.3.4   Constant Acceleration Motion 317
9.3.5   Simple Harmonic Motion 320
9.3.6   Cycloidal Motion...322
9.3.7   Polynomial Motion 325
9.4   Disk Cam Design and Pressure Angle 331
9.5   Summary...336
References 337
Additional Reading 337
Problems..338

10. **Kinematic Analysis of Spatial Mechanisms**..343
10.1   Introduction 343
10.2   RRSS Mechanism Analysis 344
10.2.1   Displacement Equations 344
10.2.2   Velocity Equations 346
10.2.3   Acceleration Equations...347
10.3   RSSR Mechanism Analysis..350
10.3.1   Displacement Equations 350
10.3.2   Velocity Equations 351
10.3.3   Acceleration Equations...352
10.4   Four-Revolute Spherical Mechanism Analysis....................................355
10.5   Planar Four-Bar Kinematic Analysis Using RRSS and RSSR Kinematic
Equations...359
10.6   Spatial Mechanism Kinematic Analysis and Modeling in Simmechanics®......362
10.7   Summary...364
References 365
Problems..365

11. **Introduction to Robotic Manipulators**...373
11.1   Introduction 373
11.2   Terminology and Nomenclature 374

11.3 Robotic Manipulator Mobility and Types 375

11.4 The General Transformation Matrix 377

11.5 Forward Kinematics 381

    11.5.1 Definition and Application 381

    11.5.2 P-P-P 381

    11.5.3 R-P-P 383

    11.5.4 R-R-P 385

    11.5.5 R-R-R 387

    11.5.6 R-R-C 389

11.6 Inverse Kinematics 392

    11.6.1 Definition and Application 392

    11.6.2 P-P-P 392

    11.6.3 R-P-P 394

    11.6.4 R-R-P 396

    11.6.5 R-R-R 397

    11.6.6 R-R-C 399

11.7 Robotic Manipulator Kinematic Analysis and Modeling in
Simmechanics® 401

11.8 Summary 402

References 402

Additional Reading 403

Problems 403

Appendix A: User Information and Instructions for MATLAB® 409

A.1 Required MATLAB Toolkits 409

A.2 Description of MATLAB Operators and Functions 409

A.3 Preparing and Running Files in MATLAB and Operations in
SimMechanics 409

A.4 Rerunning MATLAB and SimMechanics Files with Existing *.csv* Files 413

A.5 Minimum Precision Requirement for Appendix File User Input 413

Appendix B: User Instructions for Chapter 4 MATLAB® Files 415

B.1 Planar Four-Bar Mechanism 415

B.2 Planar Four-Bar Fixed and Moving Centrode Generation 415

B.3 Slider-Crank Mechanism 417

B.4 Geared Five-Bar Mechanism (Two Gears) 419

B.5 Geared Five-Bar Mechanism (Three Gears) 419

B.6 Watt II Mechanism 420

B.7 Stephenson III Mechanism 422

Appendix C: User Instructions for Chapter 6 MATLAB® Files 425

C.1 Planar Four-Bar Mechanism 425

C.2 Slider-Crank Mechanism 425

C.3 Geared Five-Bar Mechanism (Two Gears) 428

C.4 Geared Five-Bar Mechanism (Three Gears) 429

C.5 Watt II Mechanism 430

C.6 Stephenson III Mechanism 431

Appendix D: User Instructions for Chapter 7 MATLAB® Files 435

D.1 Planar Four-Bar Mechanism 435

D.2 Slider-Crank Mechanism 435

D.3   Geared Five-Bar Mechanism (Two Gears)..437
D.4   Geared Five-Bar Mechanism (Three Gears)...438
D.5   Watt II Mechanism...439
D.6   Stephenson III Mechanism..441

Appendix E: User Instructions for Chapter 9 MATLAB® Files...................................447
E.1   S, V Profile Generation and Cam Design: Constant Velocity Motion 447
E.2   S, V, A Profile Generation and Cam Design: Constant Acceleration
       Motion 447
E.3   S, V, A, J Profile Generation and Cam Design: Simple Harmonic Motion........449
E.4   S, V, A, J Profile Generation and Cam Design: Cycloidal Motion 451
E.5   S, V, A, J Profile Generation and Cam Design: 3-4-5 Polynomial Motion.........451
E.6   S, V, A, J Profile Generation and Cam Design: 4-5-6-7 Polynomial Motion 451

Appendix F: User Instructions for Chapter 10 MATLAB® Files....................................453
F.1   RRSS Mechanism 453
F.2   RSSR Mechanism 453

Appendix G: User Instructions for Chapter 11 MATLAB® Files...................................457
G.1   R-P-P Robotic Manipulator Forward Kinematics.......................................457
G.2   R-R-P Robotic Manipulator Forward Kinematics 457
G.3   R-R-R Robotic Manipulator Forward Kinematics 457
G.4   R-R-C Robotic Manipulator Forward Kinematics 459
G.5   R-P-P Robotic Manipulator Inverse Kinematics.......................................460
G.6   R-R-P Robotic Manipulator Inverse Kinematics..460
G.7   R-R-R Robotic Manipulator Inverse Kinematics.......................................461
G.8   R-R-C Robotic Manipulator Inverse Kinematics 462

**Appendix H: User Instructions for Chapter 4 MATLAB® and
SimMechanics® Files**..465
H.1   Planar Four-Bar Mechanism 465
H.2   Slider-Crank Mechanism 465
H.3   Geared Five-Bar Mechanism (Two Gears)...467
H.4   Geared Five-Bar Mechanism (Three Gears) 469
H.5   Watt II Mechanism...470
H.6   Stephenson III Mechanism..472

Appendix I: User Instructions for Chapter 6 MATLAB® and SimMechanics® Files........475
I.1   Planar Four-Bar Mechanism 475
I.2   Slider-Crank Mechanism 475
I.3   Geared Five-Bar Mechanism (Two Gears)...479
I.4   Geared Five-Bar Mechanism (Three Gears) 479
I.5   Watt II Mechanism...480
I.6   Stephenson III Mechanism..482

Appendix J: User Instructions for Chapter 7 MATLAB® and SimMechanics® Files........487
J.1   Planar Four-Bar Mechanism 487
J.2   Slider-Crank Mechanism 488
J.3   Geared Five-Bar Mechanism (Two Gears)...490
J.4   Geared Five-Bar Mechanism (Three Gears) 494
J.5   Watt II Mechanism...494
J.6   Stephenson III Mechanism..497

**Appendix K: User Instructions for Chapter 10 MATLAB® and
SimMechanics® Files**.. 501
    K.1   RRSS Mechanism..501
    K.2   RSSR Mechanism...501
**Appendix L: User Instructions for Chapter 11 MATLAB® and
SimMechanics® Files**.. 505
    L.1   R-P-P Robotic Manipulator Forward Kinematics...............................505
    L.2   R-R-P Robotic Manipulator Forward Kinematics 505
    L.3   R-R-R Robotic Manipulator Forward Kinematics 506
    L.4   R-R-C Robotic Manipulator Forward Kinematics...............................507
Index...509

Preface

Kinematics is the study of motion without considering forces. In comparison to other engineering design disciplines such as *statics*, where motion and governing loads are considered according to Newton's first law and *dynamics*, where motion and governing loads are considered according to Newton's second law, kinematics is the most fundamental engineering design discipline. Courses pertaining to the kinematics of mechanical systems are core requirements of university undergraduate mechanical engineering curricula.

While a central understanding of classical kinematics will continue to remain relevant in engineering and subsequently, a necessary focus in undergraduate engineering education, it is becoming increasingly important that an undergraduate also acquire a central understanding of static and dynamic mechanism analysis. Such an understanding prepares an undergraduate student to conduct more thorough analyses and produce more relevant solutions in mechanism design. In addition, a central understanding of the design and analysis of *robotic manipulators* has become essential in modern-day undergraduate engineering education due to the expanding use of robotic systems today.

It has become very efficient and extremely practical to utilize mathematical analysis software to conduct engineering analyses in recent years. Of all the mathematical analysis software options available (which are numerous), the authors chose *MATLAB®*. MATLAB is a high-level language and interactive environment for numerical computation, visualization, simulation and programming. Using MATLAB, one can analyze data, develop algorithms, and create models and applications without data type checking, compiling and linking (tasks common in programming languages such as C++ and Java). *SimMechanics®*, a MATLAB toolbox, provides a graphical multi-body simulation environment for 2D and 3D mechanical systems including linkages, robots, cam systems and gear systems. The user models the multi-body system (using blocks representing bodies, joints, constraints, and motion/force actuator elements) and then SimMechanics formulates and solves the governing equations of motion and force for the complete mechanical system. An automatically generated 3D animation lets you visualize the system dynamics. MATLAB and SimMechanics can be used for a broad range of applications including the kinematics, synthesis, statics and dynamics of mechanical systems. Both MATLAB and SimMechanics are well-established (and often the *de facto* standard for mathematical analysis and simulation) in colleges and universities.

There is currently a variety of textbooks available in mechanism kinematics-each book differing from the others primarily in terms of the breadth and depth of kinematics topics presented and the software packages used to implement the concepts and methods presented. In light of the need to go beyond classical kinematics in undergraduate engineering education and fill the gap between theory and the application of theory for real-world problems, this textbook was produced. This textbook introduces the fundamental concepts of mechanism **kinematics**, **synthesis**, **statics** and **dynamics** for *planar* and *spatial linkages, cam systems, gear systems* and **robotic manipulators** by realistic illustrations and practical problems. Also, the commercial software **MATLAB** and its mechanical simulation toolbox **SimMechanics** are thoroughly integrated in the textbook for ease of concept implementation (both during and after one's undergraduate years).

In order to improve both the practicality of the concepts covered as well as the clarity in their presentation, the second edition of *Kinematics and Dynamics of Mechanical Systems: Implementation in MATLAB and SimMechanics* includes the following updates:

- **the inclusion of gravity in static and dynamic force analysis of planar linkages**
  The equation systems, MATLAB files and SimMechanics files for Chapters 6 and 7 have been updated to enable the user to specify gravitational constants in planar linkage static and dynamic force analyses. With this improvement, the user can conduct accurate static and dynamic force analyses of *planar four-bar, slider-crank, geared five-bar, Watt II* and *Stephenson III* mechanisms of any scale.
- **the inclusion of force and torque equations for bevel gears**
  Chapter 8 has been updated to include force and torque equations for straight-toothed bevel gears. With this new content, the user can conduct bevel gear force, torque, work and power analyses of as they have been able to do for *spur, helical* and *worm* gears in the prior textbook version.
- **the inclusion of MATLAB and SimMechanics file input in textbook examples**
  Chapters 4–11 have been updated to include the example problem data input formulated as input in the corresponding MATLAB and SimMechanics files. This new content improves the clarity for the reader regarding example problem interpretation and preparation for use in the library of MATLAB and SimMechanics files that accompany this textbook (these files are available for download at https://www.crcpress.com/product/isbn/9781498724937 and click the *Downloads* tab).

This textbook was written to accommodate students with no working knowledge of MATLAB. In terms of MATLAB knowledge, the ideal user should know how to launch MATLAB and have access to the MATLAB software package itself. Any version of MATLAB post 2013 is suitable to run the MATLAB and SimMechanics files associated with this textbook (provided all of the required toolkits listed in Appendix A.1 are installed).

The intended uses for this textbook are the following:

- as a sole text for an undergraduate course in mechanical system kinematics
- as a sole text for an undergraduate mechanical design course (where mechanisms are then analyzed using *Statics/Dynamics, Stress Analysis, Machine Design, CAE,* etc.).
- as a reference text for mechanical engineering research
- as a reference text for the application of MATLAB and SimMechanics in mechanical engineering

Because our goal is to produce a textbook with sufficient breadth, depth and implementation resources to be an effective resource for 21st century undergraduate engineering education, we look forward to any feedback you may have. For e-mail correspondence, we can be reached at *kevin.russell@njit.edu*. We hope you enjoy utilizing this work as much as we have enjoyed producing it.

The authors would like to acknowledge the contributions of those who assisted in the review and evaluation of this textbook. In particular we thank the students of *ME-231*

(*Kinematics of Machinery*) from the Fall 2014 to the Spring 2018 semesters at New Jersey Institute of Technology.

Kevin Russell
Qiong Shen
Raj S. Sodhi

MATLAB® is a registered trademark of The MathWorks, Inc. For product information, please contact:

The MathWorks, Inc.
3 Apple Hill Drive
Natick, MA 01760-2098 USA
Tel: 508 647 7000
Fax: 508-647-7001
E-mail: info@mathworks.com
Web: www.mathworks.com

Authors

Kevin Russell, Ph.D., P.E. is a member of the teaching faculty in the Department of Mechanical and Industrial Engineering at New Jersey Institute of Technology (NJIT). At NJIT, Dr. Russell teaches courses in kinematics, machine design and mechanical design. Formerly, Dr. Russell was a Senior Mechanical Engineer at the U.S. Army Research, Development and Engineering Center (ARDEC) at Picatinny, New Jersey. Dr. Russell's responsibilities at ARDEC included the utilization of computer-aided design and modeling and simulation tools for small and medium-caliber weapon system improvement, concept development and failure investigations. A fellow of the American Society of Mechanical Engineers (ASME) and a registered Professional Engineer in New Jersey, Dr. Russell holds several patents (and pending patents) for his design contributions relating to small and medium-caliber weapon systems, linkage-based inspection systems and human prosthetics. He has published extensively among mechanical engineering journals in areas such as kinematic synthesis, theoretical kinematics and machine design.

Qiong Shen, Ph.D. (https://www.linkedin.com/in/qiong-shen-57212524/) is the founder of Softalink LLC, a consulting company that applies cloud-computing and big data technologies to help automate and optimize business processes, transform traditional marketing and strategic planning into data-driven manner. Besides business activities, Dr. Shen is also active, as an adjunct professor at New Jersey Institute of Technology (NJIT), in preparing college students for ever-growing challenges in engineering and management. Dr. Shen received Ph.D. degree from a joint program between Mechanical Engineering and Electrical Engineering Departments at NJIT. He has made substantial contributions to researches in Robotics and Mechanism Synthesis by applying technologies from Distributed Parallel Computing, Machine Learning, Visualization, and Simulation.

Raj S. Sodhi, Ph.D., P.E. is a Professor in the Department Mechanical and Industrial Engineering at NJIT. He has over 30 years of experience in research and education related to Mechanical Design, Mechanisms Synthesis and Manufacturing Engineering. Dr. Sodhi is the author or co-author of over one hundred refereed papers in scientific journals and conference proceedings. He was awarded the Society of Manufacturing Engineering's *University Lead Award* in recognition of leadership and excellence in the application and development of computer integrated manufacturing. He also received the *N. Watrous Procter & Gamble Award* from the Society of Applied Mechanisms and Robotics for significant contributions to the science of mechanisms and robotics and the *Ralph R. Teetor New Engineering Educator Award* from the Society of Automotive Engineers. Dr. Sodhi is a registered Professional Engineer in Texas.

1

Introduction to Kinematics

CONCEPT OVERVIEW

In this chapter, the reader will gain a central understanding regarding

1. *Kinematics* and its use in engineering design
2. Distinctions between *kinematic chains* and *mechanisms*
3. Planar and spatial mechanism *mobility*
4. Types of mechanism motion
5. Distinctions between *kinematic analysis* and *kinematic synthesis*
6. Categories of kinematic synthesis

1.1 Kinematics

Kinematics is the study of motion without considering forces. In a *kinematic analysis*, positions, displacements, velocities and accelerations are calculated for mechanical system components without regard to the loads that actually govern them. In comparison to other engineering design studies such as *statics*, where motion and governing loads are considered according to Newton's first law, and *dynamics*, where motion and governing loads are considered according to Newton's second law, kinematics is the most fundamental engineering design study. It is often necessary in the design of a mechanical system to not only consider the motion of its components, but also the following:

- Static or dynamic loads acting on the components (considered in statics and dynamics)
- Component material stress and strain responses to the loads (considered in *stress analysis*)
- Required component dimensions for the working stresses (considered in *machine design*)

Because of this, static, dynamic, stress, and machine design analyses often follow a kinematic analysis.

Figure 1.1 includes kinematics, statics and dynamics, stress analysis and machine design in an ascending order of progression. This order follows the intended order of use of these studies in mechanical design. After a mechanical system has first been determined to

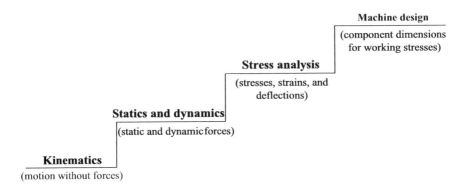

FIGURE 1.1
Kinematics in relation to other associated engineering design studies.

be kinematically feasible, the static or dynamic loads acting on the system components are considered next. After static or dynamic feasibility has been achieved, the stresses and strains produced in the mechanical system components are then considered. Lastly, machine design principles and methodologies are employed to ensure the material and dimensions of the mechanical system components (and subsequently the entire mechanical system) are satisfactory for the known working stresses.*

As illustrated in Figure 1.1, kinematics is the most fundamental of the engineering design study listed. When a design is not kinematically sound, evidence of this will often appear in the other engineering design studies. For example, a discontinuous displacement profile calculated in a kinematic analysis would be revealed as excessive acceleration in a dynamic analysis, which, in turn, could produce excessive dynamic forces. These excessive dynamic forces would likely produce high stresses. These high stresses may require a material selection or component dimensions that make the overall component design impractical for the intended design application. Kinematic feasibility, therefore, must be established first before considering the follow-on engineering design studies in Figure 1.1.

1.2 Kinematic Chains and Mechanisms

This textbook focuses primarily on the kinematic analysis and *kinematic synthesis* of mechanical systems or *mechanisms*, as they are commonly called.† A *kinematic chain*, an overarching classification that includes mechanisms, is an assembly of *links* interconnected by joints where the motion of one link compels the motion of another link (which compels the motion of another link, and so on depending on the number of mechanism links).‡ Complex mechanical systems, such as an automobile engine, for example, can be comprised of multiple kinematic chains, while a single kinematic chain can constitute an entire mechanical system in the case of a simple tool. Figure 1.2 illustrates a commonly

---

* In addition to engineering design factors pertaining to kinematics, statics, dynamics, and machine design— also called *traditional* engineering factors—*nontraditional* or *modern* engineering factors (including *producibility, cost, environmental impact, disposal, aesthetics, ergonomics,* and *human factors*) are often equally important.
† The distinctions between kinematic analysis and kinematic synthesis are first presented in Section 1.5.
‡ Because a mechanism is an assembly of links, it is also called a *linkage*. Links are generally assumed to be nondeforming or *rigid* in kinematics.

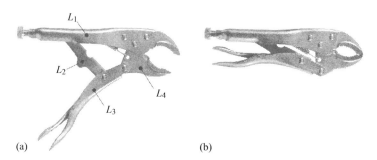

FIGURE 1.2
Pliers in (a) open and (b) closed positions.

used kinematic chain: a pair of pliers. Moving the lower handle (link L_3) toward the upper handle (link L_1) or vice versa compels the motion of the remaining links, including the lower grip (link L_4), which produces a gripping action. Having links compel the motion of each other link in a controlled manner is important because the fundamental objective in the design of a mechanical system is to provide a controlled output motion in response to a supplied input motion.

One characteristic that distinguishes mechanisms from other kinematic chains is that the former has at least one "grounded" link [1]. A grounded link is one that is attached to a particular frame of reference. Some mechanisms have links that are permanently grounded through friction, gravity, or fastening members (e.g., bolts, screws, and welds), whereas with our pliers example, the grounded link can be established according to one's own preferences.

1.3 Mobility, Planar, and Spatial Mechanisms

The *mobility* or the number of *degrees of freedom* of a mechanism is the number of independent parameters required to uniquely define its position in space. Knowing the mobility of a mechanism is particularly important when formulating equation systems for the kinematic analysis or synthesis of the mechanism. This is because the equation systems must include enough parameters to fully define the motion of each mechanism component. To fully define the position of a body in two-dimensional or *planar* space at an instant in time requires three independent parameters. To demonstrate this principle, we will consider the parking automobile example in Figure 1.3a where the X-Y coordinate frame is affixed to the parking space. At any instant in time, the position of the automobile can be measured with respect to the X-Y coordinate frame given three independent parameters. The X and Y coordinates of any point on the automobile are two of the three parameters required to define the planar position of a body. Because the parking automobile also rotates in the coordinate frame, its angular position is also required to fully define its position in the X-Y coordinate frame. Therefore, the three parameters required to define a planar position are the X and Y coordinates of a location on the body and the orientation angle of the body. *Because three independent parameters are required to define the position of the body in the X-Y plane, an individual mechanism link restricted to planar motion can have a mobility of up to three or up to three degrees of freedom.*

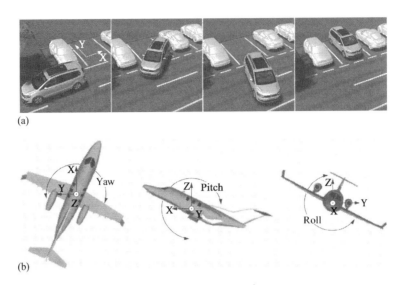

FIGURE 1.3
(a) Parking automobile and (b) aircraft in flight.

To fully define the position of a body in three-dimensional space at an instant in time requires six independent parameters. To demonstrate this principle, we will consider the flying aircraft example in Figure 1.3b where the X-Y-Z coordinate frame is affixed in space. At any instant in time, the position of the aircraft can be measured with respect to the X-Y-Z coordinate frame given six independent parameters. The X, Y, and Z coordinates of any point on the aircraft are three of the six parameters required to define the spatial position of a body. Because the aircraft also rotates about each coordinate frame axis, these three angular positions are also required to fully define its position in the X-Y-Z coordinate frame.* Therefore, the six parameters required to define a spatial position are the X, Y, and Z coordinates of a location on the body and the orientation angles of the body about the X, Y, and Z axes. Because six independent parameters are required to define the position of the body in X-Y-Z space, an individual mechanism link restricted to spatial motion can have a mobility of up to six, or up to six degrees of freedom.

Figure 1.4a illustrates a pair of pliers. As indicated by the overlapping plane, the motion exhibited by this mechanism is restricted to 2D space. A pair of pliers is an example of a commonly used *planar mechanism*. Each link in this particular mechanism has a single degree of freedom—which is consistent with the previously stated condition that a maximum mobility of three is possible with a body in planar motion.

In comparison to the pair of pliers, Figure 1.4b illustrates a particular type of robotic manipulator— the *RPP robotic manipulator* (presented in Chapter 11). As indicated by the overlapping cylindrical volume, this mechanism can exhibit motion in 3D space.† Robotic manipulators are examples of commonly used *spatial mechanisms*. Each link in this particular robotic manipulator has a single degree of freedom—which is also consistent with the previously stated condition that a maximum mobility of six is possible with a body in spatial motion.

---

* As shown in Figure 1.3b, the three aircraft rotation angles are called the *roll*, *pitch*, and *yaw* angles and are about the X, Y, and Z axes, respectively.
† The space (2D or 3D) that encompasses all of the possible positions achieved by a mechanism is called its *workspace*.

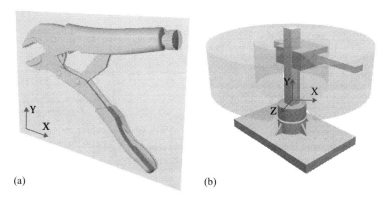

FIGURE 1.4
(a) Pliers in plane and (b) robotic manipulator in spatial workspace.

1.4 Types of Mechanism Motion

The motion exhibited by any mechanism link in 2D or 3D space can be classified as one of four types of motion. These four types are illustrated in Figure 1.5. In *pure rotation* (Figure 1.5a), a link rotates at a constant radius about a fixed axis. A link travels along a linear path in *pure translation* (Figure 1.5b).* *Complex motion* is a type of planar motion that includes both link rotations and translations simultaneously (Figure 1.5c). Any two spatial link positions can be expressed as a rotation about and a translation along a spatial axis (called a *screw axis*). This type of spatial motion is called *screw motion* (Figure 1.5d). Pure rotation and translation can be exhibited by mechanisms in both 2D and 3D space while only mechanisms in 2D space can exhibit complex motion and only mechanisms in 3D space can exhibit screw motion.

Pure rotation and pure translation are commonly called *circular* motion and *linear* motion, respectively. As the names imply, circular motion is exhibited about a circular path and linear motion is exhibited along a linear path. The conversion of circular motion to linear motion (and vice versa) is commonly required for the operation of mechanical systems. Sometimes we are given a circular motion (e.g., from a hand crank, engine, or electric motor) and we desire a linear output motion. On the other hand, we may be given a linear motion and we desire a circular output motion. The circular and linear motion may be constant, oscillatory, or even intermittent.

Several of the noted linear and circular input–output motion combinations appear in the valve train assembly illustrated in Figure 1.6.[†] The valve train assembly is comprised of four major components: the *cam, rod, rocker,* and *valve* (Figure 1.6a). Figure 1.6b includes the motion produced by these four components. The initial input in this assembly is produced by the cam.[‡] The constantly rotating cam produces an oscillating translational rod motion. The oscillating translational rod motion produces an oscillating rotational rocker motion (as the name "rocker" implies). Lastly, the oscillating rotational rocker motion produces an

---

* Pure rotation at a radius of infinity becomes pure translation.
† The valve train assembly is an integral assembly in the internal combustion engine.
‡ The kinematics and design of radial cam systems are introduced in Chapter 9.

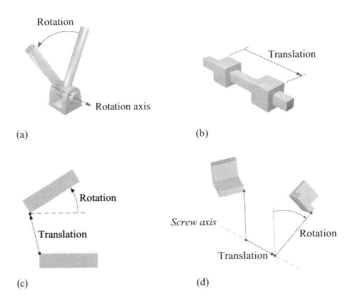

FIGURE 1.5
Links in (a) pure rotation, (b) pure translation, (c) complex, and (d) screw motion.

oscillating translational valve motion. It is the oscillating valve motion that governs the timing in which air and fuel are brought into an internal combustion engine and exhaust products are removed from the engine.

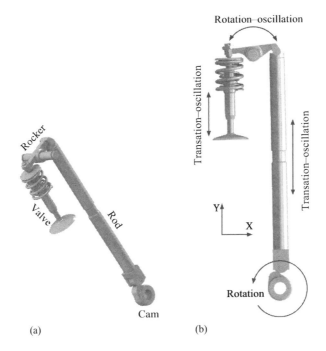

FIGURE 1.6
(a) Valve train assembly and (b) motion of assembly components.

1.5 Kinematic Synthesis

In a kinematic analysis, the mechanism link dimensions are known and the motion characteristics such as positions, displacements, velocities, and accelerations are calculated. Understanding the kinematic behavior of a given mechanism is the scope of a kinematic analysis. In comparison to kinematic analysis, in kinematic synthesis the problem is considered in reverse. Here, the mechanism required to fulfill a predetermined motion is produced.

There are two categories in kinematic synthesis. One category involves the determination of the type of mechanism needed to accomplish a given purpose. This category is called *type synthesis* [2]. *Number synthesis* falls into this category. In number synthesis (presented in Section 3.3), mechanisms are produced that match a given mechanism mobility.

The other category in kinematic synthesis involves the determination of the mechanism dimensions needed to achieve a given motion sequence. This category is called *dimensional synthesis* [3]. *Motion generation* and *function generation* fall into this category. In motion generation, mechanism dimensions required to achieve coupler-link positions are determined, while the achievement of crank and follower-link displacement angles is the objective in function generation. Both motion generation and function generation are presented in Chapter 5.

Example 1.1

Problem Statement: Figure E.1.1 illustrates two positions of a mechanism used to compact trash bundles. By rotating the driving link from its initial position (Figure E.1.1a) to its final position (Figure E.1.1b), the compacting ram is displaced from its initial position to its final position. During compaction, a reaction force is applied to the compacting ram. The mechanism is maintained in a state of static equilibrium by a torque applied to the driving link and the compacting ram reaction force. Describe how the principles of kinematics, statics, stress analysis, and machine design can be used to evaluate the structural integrity of the coupling link (Figure E.1.1) during compaction.

Known Information: Figure E.1.1, background knowledge of kinematics, statics, stress analysis, and machine design principles.

Solution Approach: The mechanism in Figure E.1.1 can be modeled as a slider-crank mechanism.*

Kinematic Analysis: Given the dimensions of this particular slider-crank mechanism, the driving link angular rotation required to achieve the final compacting position can be calculated from the slider-crank mechanism displacement equations.

Static Analysis: A static equilibrium equation can be formulated to calculate the columnar force that acts on the coupling link [4].† Equations for static equilibrium are formulated according to Newton's first law.

Stress Analysis: Given the columnar force on the coupling link, along with its cross-section dimensions and material properties, the normal stress of this link can be calculated. Additionally, the buckling load for this link (which is essentially a column with pinned ends) can also be calculated [5].

---

* Kinematic displacement, velocity, and acceleration equations for the slider-crank mechanism are introduced in Chapter 4.
† Static force analysis for planar mechanisms is introduced in Chapter 6.

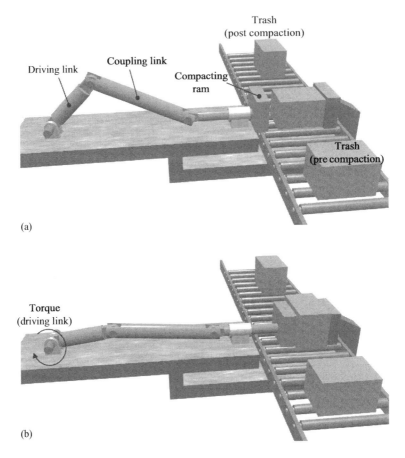

FIGURE E.1.1
(a) Initial and (b) final positions of compacting mechanism.

Machine Design: Based on the values calculated for the normal stress and buckling load for the coupling link, it may be necessary to modify its cross-section dimensions or its material type to achieve an acceptable degree of structural integrity for use in the trash compacting mechanism.

1.6 Units and Conversions

Many of the example problems and end-of-chapter problems presented throughout this textbook are *unitless* because specific dimension units are not required to calculate correct solutions. With such problems, the calculated result can, in fact, be scaled to match the desired unit system. However, there are other problems here, particularly those where force and mass are included, where a particular system of units is not only desired, but required for calculating correct solutions. Among the problems throughout this textbook were units are specified, both the *International System of Units* (SI) and *United States System of Units* (US) are used. Table 1.1 includes common quantities used throughout this textbook and their associated unit conversions between SI and US units.

TABLE 1.1

Common Textbook Quantities and Unit Conversions (from US to SI Units)

Quantity	US Unit	Conversion	SI Unit
Mass	1 pound-mass (lbm)	= 0.4536	kilograms (kg)
Force	1 pound-force (lbf)	= 4.4482	newtons (N)
Length	1 foot (ft)	= 0.3048	meters (m)
Length	1 inch (in)	= 25.4001	millimeters (mm)
Work	1 horsepower (hp)	= 745.6999	watts (W)
Angular velocity	1 revolution/minute (rpm)	= 0.1047	radian/second (rad/s)
Angular velocity	1 degree/second (°/s)	= 0.0174	radian/second (rad/s)
Velocity	1 foot/second (ft/s)	= 0.3048	meters/second (m/s)
Velocity	1 inch/second (in/s)	= 25.4001	millimeters/second (mm/s)
Torque	1 pound-foot (lb-ft)	= 1.3558	newton-meters (N-m)

1.7 Software Resources

The commercial mathematical software package MATLAB® (version 2013b) and its mechanical modeling and simulation toolbox SimMechanics® are thoroughly integrated in this textbook for applied kinematic analysis [6]. Both software resources are well established (and often the *de facto* standard for mathematical analysis and simulation) in many colleges and universities worldwide. A library of MATLAB and SimMechanics files developed for this textbook is available for download at www. crcpress.com/product/ isbn/9781498724937. With these files, the user can calculate solutions for the equation systems presented throughout this textbook and conduct mechanical simulations to independently verify or evaluate equation system results.* The MATLAB and SimMechanics resources that accompany this text provide a *virtual test bed* to utilize the equations systems and methodologies presented in Chapters 2 through 11. This textbook also includes a variety of example problems where solutions are calculated directly through MATLAB's *command window* (see Appendix A.3).

1.8 Summary

Kinematics—the study of motion without considering governing forces—is the most fundamental engineering study in mechanical system design. In mechanical system design, kinematic feasibility should be determined before considering other engineering design studies such as statics, dynamics, stress analysis, and machine design.

Mechanical systems are comprised of kinematic chains—an assembly of interconnected links where the motion of one link compels the motion of another link in a controlled manner. Achieving controlled output motion in response to a supplied input motion is the fundamental objective in mechanical system design. Kinematic chain is an overarching

---

* This textbook also includes example problems where solutions are calculated directly through MATLAB's *command window* (see Appendix A.2).

classification that includes mechanisms (also called *linkages*). The presence of an established ground link is a primary characteristic that distinguishes mechanisms from kinematic chains.

The mobility or the number of degrees of freedom of a mechanism is the number of independent parameters required to uniquely define its position in space. Knowing the mobility of a mechanism is important when formulating equation systems for mechanism kinematic or synthesis. This is because the equation systems must include enough variables to fully define the motion of each mechanism component. An individual link restricted to planar motion can have up to three degrees of freedom and an individual link restricted to spatial motion can have up to six degrees of freedom.

A mechanism link in 2D or 3D space can exhibit *pure rotation, pure translation, complex motion,* or *screw motion*. In pure rotation, a link rotates at a constant radius about a fixed axis. A link travels along a linear path in pure translation. Complex motion is a type of planar motion that includes simultaneous link rotations and translations. Screw motion is a type of spatial motion that includes simultaneous link rotations about and translations along a spatial axis called a *screw axis*.

Circular motion and linear motion are two types of motion often exhibited in mechanical systems. Sometimes we are given a circular motion and desire an output linear motion. On the other hand, we may be given a linear motion and desire a circular output motion. In mechanical systems, the circular and linear motion may be constant, oscillatory, or even intermittent.

In a kinematic analysis, the mechanism link dimensions are known and the motion characteristics such as positions, displacements, velocities, and accelerations are calculated. In comparison to kinematic analysis, the problem is considered in reverse in *kinematic synthesis*. Here, the mechanism required to fulfill specific predetermined motions is produced. One category in kinematic synthesis, called type synthesis, involves the determination of the type of mechanism needed to accomplish a given purpose. *Number synthesis* (presented in Chapter 3) falls into this category. In number synthesis, mechanisms are produced that match a given mechanism mobility.

The other category in kinematic synthesis, called type synthesis, involves the determination of the mechanism dimensions needed to achieve a given motion sequence. Motion generation and function generation (presented in Chapter 5) fall into this category. In motion generation, mechanism dimensions required to achieve coupler-link positions are determined while achieving crank and follower-link displacement angles is the objective in function generation.

The mathematical software packages MATLAB and SimMechanics are fully integrated throughout the remaining textbook chapters for applied kinematic analysis. The library of MATLAB and SimMechanics files for this textbook is available for download at www.crcpress.com/product/ isbn/9781498724937. Example problems are also solved in this textbook through MATLAB's command window.

References

1. Norton, R. L. 2008. *Design of Machinery*, 4th edn. New York: McGraw-Hill.
2. Sandor, G. N. and Erdman, A. G. 1984. *Advanced Mechanism Design: Analysis and Synthesis*, Volume 2. Englewood Cliffs, NJ: Prentice-Hall.

3. Russell, K., Shen, Q., and Sodhi, R. 2014. *Mechanism Design: Visual and Programmable Approaches.* pp. 45–47. Boca Raton: CRC Press.

4. Wilson, C. E. and Sadler, J. P. 2003. *Kinematics and Dynamics of Machinery.* 3rd edn. Saddle River, NJ: Prentice-Hall.

5. Ugural, A. C. and Fenster, S. K. 2009. *Advanced Strength and Applied Elasticity.* 4th edn. Englewood Cliffs, NJ: Prentice-Hall.

6. Mathworks. Products and services. http://www.mathworks.com/products/?s_tid=gn_ps. Accessed July 15, 2015.

7. Mathworks, Products and services, accessed March 12, 2015, http://www.mathworks.com.

Additional Reading

Myszka, D. H. 2005. *Machines and Mechanisms: Applied Kinematic Analysis.* 3rd edn. Saddle River, NJ: Prentice-Hall.

Waldron, K. J. and Kinzel, G. L. 2004. *Kinematics, Dynamics and Design of Machinery.* 2nd edn. Saddle River, NJ: Prentice-Hall.

Wilson, C. E. and Sadler, J. P. 2003. *Kinematics and Dynamics of Machinery.* 3rd edn. Saddle River, NJ: Prentice-Hall.

2

Mathematical Concepts in Kinematics

CONCEPT OVERVIEW

In this chapter, the reader will gain a central understanding regarding

1. Characteristics of vectors, complex vectors, and complex vector forms
2. The formulation of vector-loop displacement, velocity, and acceleration equations using complex vectors
3. Characteristics of point-based vectors and their application in mechanism motion equations
4. Characteristics of linear simultaneous equations and their representation in matrix form
5. Fundamental matrix operations and the identity matrix
6. Matrix inversion and its application in solving linear simultaneous equations
7. Intermediate and total spatial motion and their application in mechanism kinematics
8. The general transformation matrix and its application in the kinematic analysis of robotic manipulators

2.1 Introduction

This chapter introduces the mathematical concepts and methodologies with which the reader should become familiar to gain a solid understanding of the equation systems formulated or presented in Chapters 4 through 11. These mathematical concepts relate to the representation of *complex numbers* and complex number operations, intermediate and total *spatial motion*, and the *general transformation matrix*.

2.2 Complex Numbers and Operations

2.2.1 Complex Number Forms

Vectors are commonly used in the formulation of mechanism equation systems because, being quantities that have both *magnitude* and *direction*, they can appropriately define

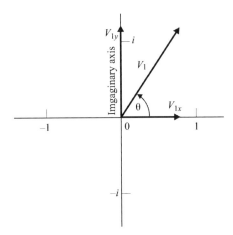

FIGURE 2.1
A vector in two-dimensional complex space.

mechanism motion (specifically, the motion of linkage-based mechanisms). One way to represent a two-dimensional vector (a vector that lies on a plane) is with a *complex number*. A complex number is comprised of a *real component* and an *imaginary component*.

Figure 2.1 illustrates a vector $\mathbf{V}_1$ in two-dimensional complex space. The real and imaginary components of this vector are V_{1x} and V_{1y}, respectively. Vector $\mathbf{V}_1$ has the magnitude V_1, where $V_1 = \sqrt{V_{1x}^2 + V_{1y}^2}$, and its direction angle is θ, where $\theta = \tan^{-1}\left(V_{1y}/V_{1x}\right)$. The real and imaginary components of $\mathbf{V}_1$ can also be expressed as $V_1 \cos\theta_1$ and $V_1 \sin\theta_1$, respectively, due to the vector direction angle.* Vector $\mathbf{V}_1$ can be expressed in the following four forms:

$$\mathbf{V}_1 = V_{1x} + iV_{1y} = V_1\left(\cos\theta_1 + i\sin\theta_1\right) = V_1 e^{i\theta_1} \tag{2.1}$$

where $i = \sqrt{-1}$ and is called the *imaginary unit*.† The second form in Equation 2.1 is the *rectangular form* of the vector $\mathbf{V}_1$—the complex number. The third and last forms in Equation 2.1 are the *polar forms* of $\mathbf{V}_1$ (the last form being the *polar exponential form*).‡

Example 2.1

Problem Statement: Calculate the magnitude and direction angle of vector $\mathbf{V}_1$, where $\mathbf{V}_1 = 1.5 + i2$.

Known Information: Vector $\mathbf{V}_1$ and Section 2.2.1.

Solution Approach 1 (using MATLAB® functions): Figure E.2.1 includes the calculation procedure in MATLAB's command window. After specifying the given values for the real and imaginary components of $\mathbf{V}_1$, this vector is then defined. The magnitude and direction angle of $\mathbf{V}_1$ are produced using MATLAB functions.

Solution Approach 2 (using manual calculations in MATLAB): Figure E.2.2 includes the calculation procedure in MATLAB's command window where the magnitude and direction angle of $\mathbf{V}_1$ were produced using manual calculations.

---

* Therefore, $V_{1x} = V_1 \cos\theta_1$ and $V_{1y} = V_1 \sin\theta_1$.
† Therefore, $i^2 = -1$, $i^3 = -i$ and $i^4 = 1$ (which are also 90° counterclockwise rotations each in the 2D complex space).
‡ The vector polar forms are the result of *Euler's formula*. In this formula, $e^{i\theta} = \cos\theta + i\sin\theta$.

```
>> V1x = 1.5;
>> V1y = 2;
>> V1 = V1x + i*V1y

V1 =

   1.5000 + 2.0000i

>> Magnitude = abs(V1)

Magnitude =

   2.5000

>> Direction = angle(V1)*180/pi

Direction =

   53.1301

>>
```

FIGURE E.2.1
Example 2.1 calculation procedure (for Solution Approach 1) in MATLAB.

```
>> V1x = 1.5;
>> V1y = 2;
>> V1 = V1x + i*V1y

V1 =

   1.5000 + 2.0000i

>> Magnitude = sqrt(V1x^2 + V1y^2)

Magnitude =

   2.5000

>> Direction = atan2(V1y, V1x)*180/pi

Direction =

   53.1301

>>
```

FIGURE E.2.2
Example 2.1 calculation procedure (for Solution Approach 2) in MATLAB.

MATLAB includes a library of functions covering a wide range of conventional calculation methods. A table of the MATLAB functions used in this textbook is included in Appendix A.2. From here on, this textbook will primarily utilize MATLAB functions. There will be occasional examples, however, where manual calculation methods will be utilized to give the reader greater insight into the calculation method or to convey specific techniques in MATLAB.

2.2.2 Complex Number Addition

Equation systems for mechanisms can be formulated by producing vector loops for the mechanisms and taking the sum of the individual vectors in the loops [1]. Figure 2.2 illustrates a loop comprised of vectors V_1, V_2, V_3, and V_4 in two-dimensional complex space. Taking the sum of the vectors in a clockwise loop produces

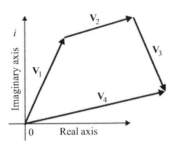

FIGURE 2.2
A vector loop in two-dimensional complex space.

$$\mathbf{V}_1 + \mathbf{V}_2 + \mathbf{V}_3 - \mathbf{V}_4 = 0 \tag{2.2}$$

If we conclude from Figure 2.2 that $\mathbf{V}_1 = V_1 e^{i\theta_1}$, $\mathbf{V}_2 = V_2 e^{i\theta_2}$, $\mathbf{V}_3 = V_3 e^{i\theta_3}$, and $\mathbf{V}_4 = V_4 e^{i\theta_4}$, each vector in the loop can be expressed in polar form as

$$V_1\left(\cos\theta_1 + i\sin\theta_1\right) + V_2\left(\cos\theta_2 + i\sin\theta_2\right) + V_3\left(\cos\theta_3 + i\sin\theta_3\right)$$
$$- V_4\left(\cos\theta_4 + i\sin\theta_4\right) = 0 \tag{2.3}$$

and in rectangular form as

$$\left(V_{1x} + iV_{1y}\right) + \left(V_{2x} + iV_{2y}\right) + \left(V_{3x} + iV_{3y}\right) - \left(V_{4x} + iV_{4y}\right) = 0* \tag{2.4}$$

After grouping and separating the real and imaginary terms in Equation 2.4, the two equations in Equation 2.5 are produced. The imaginary unit in the second equation (being common among all equation terms) can be removed if preferred.

$$V_{1x} + V_{2x} + V_{3x} - V_{4x} = 0$$
$$i\left(V_{1y} + V_{2y} + V_{3y} - V_{4y}\right) = 0 \tag{2.5}$$

Example 2.2

Problem Statement: Determine if the sum $\mathbf{V}_1 + \mathbf{V}_2 + \mathbf{V}_3 + \mathbf{V}_4$ of the following vectors forms a closed loop: $\mathbf{V}_1 = 1.5 + i2$, $\mathbf{V}_2 = -i0.5$, $\mathbf{V}_3 = -1.25 - i2.25$, and $\mathbf{V}_4 = 0.25 - i0.25$.

Known Information: Vectors $\mathbf{V}_1$, $\mathbf{V}_2$, $\mathbf{V}_3$, $\mathbf{V}_4$, and Section 2.2.2.

Solution Approach: Figure E.2.3 includes the calculation procedure in MATLAB's command window. After specifying the given vectors and calculating the vector sum, it can be seen that the vectors do not form a closed loop since the vector sum (called *Sum* in Figure E.2.3) is not zero.

---

* It may be more convenient to use the rectangular form of complex numbers for vector addition.

```
>> V1 = 1.5 + i*2
V1 =
    1.5000 + 2.0000i
>> V2 = - i*0.5
V2 =
    -0.5000i
>> V3 = -1.25 - i*2.25
V3 =
    -1.2500 - 2.2500i
>> V4 = 0.25 - i*0.25
V4 =
    0.2500 - 0.2500i
>> Sum = V1 + V2 + V3 + V4
Sum =
    0.5000 - 1.0000i
>>
```

FIGURE E.2.3
Example 2.2 solution calculation procedure in MATLAB.

2.2.3 Complex Number Multiplication and Differentiation

The product of two complex numbers (e.g., vectors $\mathbf{V}_1 = V_1 e^{i\theta_1}$ and $\mathbf{V}_2 = V_2 e^{i\theta_2}$) in polar exponential form is

$$\mathbf{V}_1\mathbf{V}_2 = V_1 e^{i\theta_1} V_2 e^{i\theta_2} = V_1 V_2 e^{i(\theta_1 + \theta_2)} = V_1 V_2 \left[\cos(\theta_1 + \theta_2) + i\sin(\theta_1 + \theta_2) \right]^* \tag{2.6}$$

The first derivative with respect to time of a complex number (vector $\mathbf{V}_1 = V_1 e^{i\theta_1}$, for example) is

$$\frac{d\mathbf{V}_1}{dt} = i\frac{d\theta_1}{dt} V_1 e^{i\theta_1} = i\dot{\theta}_1 \mathbf{V}_1 \tag{2.7}$$

where vector $\mathbf{V}_1$ represents a *rigid* link (a link having a fixed length).† The second derivative of Equation 2.7 is

$$\frac{d^2\mathbf{V}_1}{dt^2} = \left(\frac{d\theta_1}{dt}\right)^2 V_1 e^{i\theta_1} + i\frac{d^2\theta_1}{dt^2} V_1 e^{i\theta_1} = \mathbf{V}_1\left(i\ddot{\theta}_1 - \dot{\theta}_1^2\right) \tag{2.8}$$

The multiplication and differentiation of complex numbers are used in the formulation of vector-loop equations for higher-order mechanism motion quantities such as velocity and acceleration [2].

---

* It may be more convenient to use the polar exponential form of complex numbers for vector multiplication (or differentiation).
† If $\mathbf{V}_1$ represented a link having a length that changes over time, its derivative would also include the vector length derivative term $d\mathbf{V}_1/dt$ or $\dot{\mathbf{V}}_1$.

Example 2.3

Problem Statement: Formulate an equation system for vector **V** for the vector loop illustrated in Figure E.2.4. In this vector loop, $\mathbf{W} = We^{i\alpha}$ and $\mathbf{X} = Xe^{i\delta}$. Also, formulate equation systems for the first and second derivatives of **V** ($\dot{\mathbf{V}}$ and $\ddot{\mathbf{V}}$, respectively) manually and symbolically in MATLAB. Assume all vectors represent rigid links.

   Known Information: Figure E.2.4, Sections 2.2.2 and 2.2.3.

   Solution Approach: Initial Formulation:

   Taking a clockwise vector-loop sum for the vector loop in Figure E.2.4 and solving for vector **V** produces

$$\mathbf{V} = \mathbf{W} + \mathbf{X} = We^{i\alpha} + Xe^{i\delta} \tag{2.9}$$

After expanding the polar exponential form of **V**, Equation 2.10 is produced, and Equation 2.11 is produced after grouping and separating the real and imaginary terms.

$$\mathbf{V} = W(\cos\alpha + i\sin\alpha) + X(\cos\delta + i\sin\delta) = W_x + iW_y + X_x + iX_y \tag{2.10}$$

$$\begin{aligned} V_x &= W\cos\alpha + X\cos\delta = W_x + X_x \\ V_y &= W\sin\alpha + X\sin\delta = W_y + X_y \end{aligned} \tag{2.11}$$

First Derivative Formulation:

   Taking the first derivative of Equation 2.9 produces

$$\dot{\mathbf{V}} = i\dot{\alpha}We^{i\alpha} + i\dot{\delta}Xe^{i\delta} \tag{2.12}$$

Figure E.2.5 includes this calculation procedure in MATLAB's command window. On observation, it can be seen that Equation 2.12 agrees perfectly with the first derivative produced symbolically in MATLAB. In Figure E.2.5, the terms *alpha(t)* and *diff(alpha(t),t)* represent the terms α, and $\dot{\alpha}$; respectively, in Equation 2.12 and the terms *delta(t)* and *diff(delta(t),t)* represent the terms δ and $\dot{\delta}$, respectively.

   After expanding the polar exponential form of $\dot{\mathbf{V}}$, Equation 2.13 is produced, and Equation 2.14 is produced after grouping and separating the real and imaginary terms.

$$\dot{\mathbf{V}} = \dot{\alpha}W(i\cos\alpha - \sin\alpha) + \dot{\delta}X(i\cos\delta - \sin\delta) = i\dot{\alpha}W_x - \dot{\alpha}W_y + i\dot{\delta}X_x - \dot{\delta}X_y \tag{2.13}$$

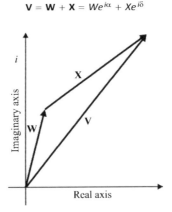

$$\mathbf{V} = \mathbf{W} + \mathbf{X} = We^{i\alpha} + Xe^{i\delta}$$

FIGURE E.2.4
Vector loop in two-dimensional complex space.

```
>> syms W X t alpha(t) delta(t) real;
>> V = W*exp(i*alpha(t)) + X*exp(i*delta(t));
>> dV = diff(V)

dV =

X*exp(delta(t)*i)*diff(delta(t), t)*i +
W*exp(alpha(t)*i)*diff(alpha(t), t)*i

>>
```

FIGURE E.2.5
Example 2.3 vector first-order differentiation procedure in MATLAB.

$$\dot{V}_x = -\dot{\alpha}W\sin\alpha - \dot{\delta}X\sin\delta = -\dot{\alpha}W_y - \dot{\delta}X_y$$

$$\dot{V}_y = \dot{\alpha}W\cos\alpha + \dot{\delta}X\cos\delta = \dot{\alpha}W_x - \dot{\delta}X_x$$

(2.14)

Second Derivative Formulation:
Taking the second derivative of Equation 2.9 produces

$$\ddot{\mathbf{V}} = -\dot{\alpha}^2 We^{i\alpha} + i\ddot{\alpha}We^{i\alpha} - \dot{\delta}^2 Xe^{i\delta} + i\ddot{\delta}Xe^{i\delta}$$

(2.15)

Figure E.2.6 includes this calculation procedure in MATLAB's command window. On observation, it can be seen that Equation 2.15 agrees perfectly with the second derivative produced symbolically in MATLAB. In Figure E.2.6, the terms *diff(alpha(t), t, t)* and *diff(delta(t), t, t)* represent the terms $\ddot{\alpha}$ and $\ddot{\delta}$, respectively, in Equation 2.15.

After expanding the polar exponential form of $\ddot{\mathbf{V}}$, Equation 2.16 is produced, and Equation 2.17 is produced after grouping and separating the real and imaginary terms.

$$\ddot{\mathbf{V}} = -\dot{\alpha}^2 W(\cos\alpha + i\sin\alpha) + \ddot{\alpha}W(i\cos\alpha - \sin\alpha) - \dot{\delta}^2 X(\cos\delta + i\sin\delta)$$

$$+\ddot{\delta}X(i\cos\delta - \sin\delta) = -\dot{\alpha}^2 W_x - i\dot{\alpha}^2 W_y + i\ddot{\alpha}W_x - \ddot{\alpha}W_y - \dot{\delta}^2 X_x - i\dot{\delta}^2 X_y + i\ddot{\delta}X_x - \ddot{\delta}X_y$$

(2.16)

$$\ddot{V}_x = -\dot{\alpha}^2 W\cos\alpha - \ddot{\alpha}W\sin\alpha - \dot{\delta}^2 X\cos\delta - \ddot{\delta}X\sin\delta = -\dot{\alpha}^2 W_x - \ddot{\alpha}W_y - \dot{\delta}^2 X_x - \ddot{\delta}X_y$$

$$\ddot{V}_y = -\dot{\alpha}^2 W\sin\alpha - \ddot{\alpha}W\cos\alpha - \dot{\delta}^2 X\sin\delta - \ddot{\delta}X\cos\delta = -\dot{\alpha}^2 W_y + \ddot{\alpha}W_x - \dot{\delta}^2 X_y + \ddot{\delta}X_x$$

(2.17)

Example 2.4

Problem Statement: Calculate the vector product **VW** where $\mathbf{V} = V_1(\cos\theta_1 + i\sin\theta_1)$ and $\mathbf{W} = W_1(\cos\alpha_1 - i\sin\alpha_1)$. Also formulate **VW** symbolically in MATLAB.

```
>> syms W X t alpha(t) delta(t) real;
>> V = W*exp(i*alpha(t)) + X*exp(i*delta(t));
>> ddV = diff(V, 2)

ddV =

- W*exp(alpha(t)*i)*diff(alpha(t), t)^2 -
X*exp(delta(t)*i)*diff(delta(t), t)^2 +
W*exp(alpha(t)*i)*diff(alpha(t), t, t)*i +
X*exp(delta(t)*i)*diff(delta(t), t, t)*i

>>
```

FIGURE E.2.6
Example 2.3 vector second-order differentiation procedure in MATLAB.

```
>> syms theta1 alpha1 V1 W1 real;
>> V = V1*(cos(theta1) + i*sin(theta1));
>> W = W1*(cos(alpha1) - i*sin(alpha1));
>> Product = simplify(expand(V*W))

Product =

-V1*W1*(sin(alpha1 - theta1)*i - cos(alpha1 - theta1))

>>
```

FIGURE E.2.7
Example 2.4 solution calculation procedure in MATLAB.

Known Information: Vectors **V**, **W**, and Section 2.2.3.
Solution Approach: Equation 2.18 includes the basic form of the product **VW**. Equation 2.19 includes the vector product with the trigonometric identities for cos $(\alpha_1 - \theta_1)$ and sin $(\alpha_1 - \theta_1)$ included.

$$\mathbf{VW} = V_1 W_1 \left(\cos\theta_1 \cos\alpha_1 - i\cos\theta_1 \sin\alpha_1 + i\sin\theta_1 \cos\alpha_1 + \sin\theta_1 \sin\alpha_1 \right) \qquad (2.18)$$

$$\mathbf{VW} = V_1 W_1 \left(\cos(\alpha_1 - \theta_1) - i\sin(\alpha_1 - \theta_1) \right) \qquad (2.19)$$

Figure E.2.7 includes the calculation procedure for **VW** in MATLAB's command window.

2.3 Vector and Point Representation

Vectors as well as *points* can be used to formulate equation systems for mechanisms. Both the magnitude and direction can be calculated for a vector, given the coordinates of its endpoints. Vector **V** and its endpoints $\mathbf{p}_1$ and $\mathbf{p}_2$ in two-dimensional and three-dimensional space are illustrated in Figure 2.3. In terms of point coordinates, the vector magnitude V can be expressed in 2D space as

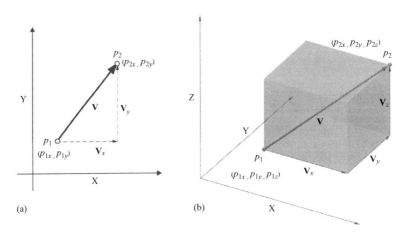

(a)   (b)

FIGURE 2.3
Point and vector representations in (a) 2D and (b) 3D space.

$$V = \|\mathbf{p}_2 - \mathbf{p}_1\| = \sqrt{\left(p_{2x} - p_{1x}\right)^2 + \left(p_{2y} - p_{1y}\right)^2} \tag{2.20}$$

where $\mathbf{p}_1 = [p_{1x}, p_{1y}]^T$ and $\mathbf{p}_2 = [p_{2x}, p_{2y}]^{T}.$* The direction of this vector is expressed in the x- and y-components $p_{2x} - p_{1x}$ and $p_{2y} - p_{1y}$, respectively.

In 3D space, the vector magnitude can be expressed as

$$V = \|\mathbf{p}_2 - \mathbf{p}_1\| = \sqrt{\left(p_{2x} - p_{1x}\right)^2 + \left(p_{2y} - p_{1y}\right)^2 + \left(p_{2z} - p_{1z}\right)^2} \tag{2.21}$$

where $\mathbf{p}_1 = [p_{1x}, p_{1y}, p_{1z}]^T$ and $\mathbf{p}_2 = [p_{2x}, p_{2y}, p_{2z}]^T$. The direction of this vector is expressed in the x-, y-, and z-components as $p_{2x} - p_{1x}$, $p_{2y} - p_{1y}$, and $p_{2z} - p_{1z}$, respectively. Point-based vectors are used in the spatial mechanism equation systems in Chapter 10 [3].

Example 2.5

Problem Statement: Calculate the magnitude and orientation angle of vector **V** where this 2D vector is comprised of points $\mathbf{p}_1 = (1.25, -5)$ and $\mathbf{p}_2 = (-2, 9.65)$. Also calculate the magnitude and orientation angle of vector **V** where this 3D vector is comprised of points $\mathbf{p}_1 = (1.25, -5, 0)$ and $\mathbf{p}_2 = (-2, 9.65, 6)$.

Known Information: Planar and spatial points $\mathbf{p}_1$ and $\mathbf{p}_2$, and Section 2.3.

Solution Approach: Figure E.2.8 includes the calculation procedure in MATLAB's command window.

```
>> p1 = [1.25, -5]';
>> p2 = [-2, 9.65]';
>> V = p2 - p1

V =

    -3.2500
    14.6500

>> Magnitude = norm(V)

Magnitude =

    15.0062

>> p1 = [1.25, -5, 0]';
>> p2 = [-2, 9.65, 6]';
>> V = p2 - p1

V =

    -3.2500
    14.6500
     6.0000

>> Magnitude = norm(V)

Magnitude =

    16.1612

>>
```

FIGURE E.2.8
Example 2.5 solution calculation procedure in MATLAB.

---

* The length of a vector **V** is often called the *norm* of **V** and is denoted by $\|\mathbf{V}\|$.

2.4 Linear Simultaneous Equations, Matrices, and Matrix Operations

2.4.1 Linear Simultaneous Equation Systems and Matrices

A *linear equation* is an equation that includes linear or *first-order* variables.* A system of linear equations (or a *linear system*) is collection of linear equations including the same variables. If a common solution is sought among a system of linear equations, it is called a set of *simultaneous equations*. An arbitrary set of linear simultaneous equations is given in Equation 2.22. Variables x_1, x_2, x_3, and x_4 are included among the four linear equations and true x_1, x_2, x_3, and x_4 solutions for these simultaneous equations must satisfy each equation in the set.

$$
\begin{aligned}
x_1 &- 2x_2 + x_3 - 3x_4 = 1 \\
2x_1 &+ x_2 + 2x_3 - 2x_4 = -2 \\
-x_1 &+ 2x_2 - 4x_3 + x_4 = 1 \\
3x_1 & \qquad\qquad\quad - 3x_4 = 3
\end{aligned}
\tag{2.22}
$$

A set of linear simultaneous equations can be expressed in matrix form. A *matrix* is a rectangular array of numbers, symbols, or expressions arranged in rows and columns. This form can be particularly convenient when solving for the variables in a simultaneous equation set. When expressed in matrix form, Equation 2.20 becomes

$$
\begin{bmatrix}
1 & -2 & 1 & -3 \\
2 & 1 & 2 & -2 \\
-1 & 2 & -4 & 1 \\
3 & 0 & 0 & -3
\end{bmatrix}
\begin{Bmatrix}
x_1 \\ x_2 \\ x_3 \\ x_4
\end{Bmatrix}
=
\begin{bmatrix}
1 \\ -2 \\ 1 \\ 3
\end{bmatrix}
\quad \text{or} \quad [A]\mathbf{x} = \mathbf{b}
\tag{2.23}
$$

where

$$
[A] = [A]_{4\times4} =
\begin{bmatrix}
1 & -2 & 1 & -3 \\
2 & 1 & 2 & -2 \\
-1 & 2 & -4 & 1 \\
3 & 0 & 0 & -3
\end{bmatrix},
\mathbf{x} = \mathbf{x}_{4\times1} =
\begin{Bmatrix}
x_1 \\ x_2 \\ x_3 \\ x_4
\end{Bmatrix}
\quad \text{and} \quad
\mathbf{b} = \mathbf{b}_{4\times1} =
\begin{bmatrix}
1 \\ -2 \\ 1 \\ 3
\end{bmatrix}
$$

Matrix $[A]$ includes the variable coefficients and is subsequently called the *coefficient matrix*. For simultaneous equation sets having n equations and n variables, the coefficient matrix will always be *square*—having n rows and n columns ($[A]_{n \times n}$).

---

* A first-order variable is one that is of degree 1. Considering a first-order variable we will arbitrarily define as x, examples of variables that are not first order include x^n and $x^{1/n}$ (where $n > 1$), cos (x), and log (x).

Matrices **x** and **b**, which include the equation variables and remaining quantities, respectively, are called *column matrices* or *column vectors*. For simultaneous equations sets having n equations and n variables, the column matrices will always have n rows and one column ($\mathbf{x}_{n \times 1}$ and $\mathbf{b}_{n \times 1}$).

Therefore, considering Equation 2.24, a general set of linear simultaneous equations having n equations and n variables,

$$
\begin{array}{ccccccccccc}
a_{11}x_1 & + & a_{12}x_2 & + & a_{13}x_3 & + & \cdots & + & a_{1n}x_n & = & b_1 \\
a_{21}x_1 & + & a_{22}x_2 & + & a_{23}x_3 & + & \cdots & + & a_{2n}x_n & = & b_2 \\
a_{31}x_1 & + & a_{32}x_2 & + & a_{33}x_3 & + & \cdots & + & a_{3n}x_n & = & b_3 \\
& & & & & \vdots & & & & & \\
a_{n1}x_1 & + & a_{n2}x_2 & + & a_{n3}x_3 & + & \cdots & + & a_{nn}x_n & = & b_n
\end{array} \tag{2.24}
$$

the matrix form expression of the this equation set becomes

$$
\begin{bmatrix}
a_{11} & a_{12} & a_{13} & \cdots & a_{1n} \\
a_{21} & a_{22} & a_{23} & \cdots & a_{2n} \\
a_{31} & a_{32} & a_{33} & \cdots & a_{3n} \\
\vdots & \vdots & \vdots & & \vdots \\
a_{n1} & a_{n2} & a_{n3} & \cdots & a_{nn}
\end{bmatrix}
\begin{Bmatrix}
x_1 \\ x_2 \\ x_3 \\ \vdots \\ x_n
\end{Bmatrix}
=
\begin{bmatrix}
b_1 \\ b_2 \\ b_3 \\ \vdots \\ b_n
\end{bmatrix} \tag{2.25}
$$

where

$$
[A] = [A]_{n \times n} =
\begin{bmatrix}
a_{11} & a_{12} & a_{13} & \cdots & a_{1n} \\
a_{21} & a_{22} & a_{23} & \cdots & a_{2n} \\
a_{31} & a_{32} & a_{33} & \cdots & a_{3n} \\
\vdots & \vdots & \vdots & & \vdots \\
a_{n1} & a_{n2} & a_{n3} & \cdots & a_{nn}
\end{bmatrix},
\quad
\mathbf{x} = \mathbf{x}_{n \times 1} =
\begin{Bmatrix}
x_1 \\ x_2 \\ x_3 \\ \vdots \\ x_n
\end{Bmatrix}
\quad \text{and} \quad
\mathbf{b} = \mathbf{b}_{n \times 1} =
\begin{bmatrix}
b_1 \\ b_2 \\ b_3 \\ \vdots \\ b_n
\end{bmatrix}
$$

2.4.2 Matrix Transpose, Addition, Subtraction, and Multiplication

This section presents the most common matrix operations used in this textbook. One operation, which is used in this textbook for column matrices, is called the *transpose*. In this matrix operation, the rows and columns of a matrix are interchanged. Considering the column matrices in Equations 2.25, the transposes of **x** and **b** become

$$
\mathbf{x}^T = \mathbf{x}_{1 \times n} = \begin{bmatrix} x_1 & x_2 & x_3 & \cdots & x_n \end{bmatrix} \quad \text{and} \quad \mathbf{b}^T = \mathbf{b}_{1 \times n} = \begin{bmatrix} b_1 & b_2 & b_3 & \cdots & b_n \end{bmatrix}
$$

where the superscript T in $\mathbf{x}^T$ and $\mathbf{b}^T$ denotes the transpose operation. The transpose of a column matrix produces a *row matrix* or *row vector*.

When adding or subtracting column matrices, the cells in each matching row are added or subtracted. Considering the column matrices in Equations 2.25, the sum $\mathbf{x} + \mathbf{b}$ and difference $\mathbf{x} - \mathbf{b}$ become

$$\mathbf{x} + \mathbf{b} = \mathbf{x}_{n\times1} + \mathbf{b}_{n\times1} = [\mathbf{x} + \mathbf{b}]_{n+1} = \begin{bmatrix} x_1 + b_1 \\ x_2 + b_2 \\ x_3 + b_3 \\ \vdots \\ x_n + b_n \end{bmatrix} \quad \text{and} \quad \mathbf{x} - \mathbf{b} = \mathbf{x}_{n\times1} - \mathbf{b}_{n\times1} = [\mathbf{x} - \mathbf{b}]_{n+1} = \begin{bmatrix} x_1 - b_1 \\ x_2 - b_2 \\ x_3 - b_3 \\ \vdots \\ x_n - b_n \end{bmatrix}$$

When calculating the product of a matrix and a scalar quantity (e.g., $k[A]$, where k is a scalar quantity), the product of each cell in the matrix and the scalar quantity is taken (distributing the constant throughout the matrix). For the product of a square matrix and a column matrix, each row in the square matrix is multiplied by the column matrix. Considering the square and column matrices in Equations 2.25, the product $[A]\mathbf{x}$ becomes

$$[A]\mathbf{x} = [A]_{n\times n}\mathbf{x}_{n\times1} = ([A]\mathbf{x})_{n\times1} = \begin{bmatrix} a_{11}x_1 & + & a_{12}x_2 & + & a_{13}x_3 & + & \cdots & + & a_{1n}x_n \\ a_{21}x_1 & + & a_{22}x_2 & + & a_{23}x_3 & + & \cdots & + & a_{2n}x_n \\ a_{31}x_1 & + & a_{32}x_2 & + & a_{33}x_3 & + & \cdots & + & a_{3n}x_n \\ & & & & \vdots & & & + & \\ a_{n1}x_1 & + & a_{n2}x_2 & + & a_{n3}x_3 & + & \cdots & + & a_{nn}x_n \end{bmatrix}$$

For the product of two column matrices, the first matrix is transposed so that the product of a row matrix and a column matrix is what is actually taken.* Considering the column matrices in Equations 2.25, the product of $\mathbf{x}$ and $\mathbf{b}$ becomes

$$(\mathbf{x})^T(\mathbf{b}) = (\mathbf{x}_{1\times n})(\mathbf{b}_{n\times1}) = \{(\mathbf{x})(\mathbf{b})\}_{1+1} = x_1b_1 + x_2b_2 + x_3b_3 + \cdots + x_nb_n$$

It can be observed that the product of a row vector and a column vector is a single scalar quantity.

Example 2.6

Problem Statement: For the given matrices:

$$\mathbf{v} = \begin{bmatrix} 2.5 & -2 & 1 & 0.5 \end{bmatrix}, \mathbf{w} = \begin{bmatrix} -5 & -1 & 0.75 & -3 \end{bmatrix}, \text{ and } [A] = \begin{bmatrix} 3 & 2 & -1 & 1 \\ 5 & 3 & 2 & 1 \\ 3 & 1 & 3 & 2 \\ -6 & -4 & 2 & -2 \end{bmatrix}$$

calculate $\mathbf{v}^T$, $\mathbf{w}^T$, $\mathbf{v}^T + \mathbf{w}^T$, $[A]\mathbf{v}^T$, and $(\mathbf{v})(\mathbf{w})^T$.

---

* The transpose of the first matrix is taken to make the number of columns in the first matrix equal to the number of rows in the second matrix—a requirement for the multiplication of any two matrices.

Known Information: Given matrices and Section 2.4.2 equations.

Solution Approach: Figure E.2.9 includes the calculation procedure in MATLAB's command window.

Considering the product of two general matrices [A] and [B] (as shown in Matrix (2.27)) where $[A] = [A]_{m \times n}$ and $[B] = [B]_{n \times o}$, the entry in each cell in the product [A][B] (which is labeled AB_{ij} in Matrix (2.27)) is given by

$$AB_{ij} = a_{i1}b_{1j} + a_{i2}b_{2j} + a_{i3}b_{3j} + \cdots a_{in}b_{nj}* \qquad (2.26)$$

```
>> v = [2.5, -2, 1, 0.5];
>> w = [-5, -1, 0.75, -3];
>> A = [3, 2, -1, 1
5, 3, 2, 1
3, 1, 3, 2
-6, -4, 2, -2];
>> v_T = v'

v_T =

    2.5000
   -2.0000
    1.0000
    0.5000

>> w_T = w'

w_T =

   -5.0000
   -1.0000
    0.7500
   -3.0000

>> v_T + w_T

ans =

   -2.5000
   -3.0000
    1.7500
   -2.5000

>> A*v_T

ans =

    3.0000
    9.0000
    9.5000
   -6.0000

>> v*w_T

ans =

  -11.2500

>>
```

FIGURE E.2.9
Example 2.6 solution calculation procedure in MATLAB.

---

* Equation (2.26) is the result of the product of a row in matrix [A] and the corresponding column in matrix [B].

$$[A][B] = \begin{bmatrix} a_{11} & a_{12} & \cdots & a_{1n} \\ a_{21} & a_{22} & \cdots & a_{2n} \\ \vdots & \vdots & & \vdots \\ a_{i1} & a_{i2} & \cdots & a_{in} \\ \vdots & \vdots & & \vdots \\ a_{m1} & a_{m2} & \cdots & a_{mn} \end{bmatrix} \begin{bmatrix} b_{11} & b_{12} & \cdots & b_{1j} & \cdots & b_{1o} \\ b_{21} & b_{22} & \cdots & b_{2j} & \cdots & b_{2o} \\ \vdots & \vdots & & \vdots & & \vdots \\ b_{n1} & b_{n2} & \cdots & b_{nj} & \cdots & b_{no} \end{bmatrix}$$

$$= \begin{bmatrix} AB_{11} & AB_{12} & \cdots & AB_{1j} & \cdots & AB_{1o} \\ AB_{21} & AB_{22} & \cdots & AB_{2j} & \cdots & AB_{2o} \\ \vdots & \vdots & & \vdots & & \vdots \\ AB_{i1} & AB_{i2} & \cdots & AB_{ij} & \cdots & AB_{io} \\ \vdots & \vdots & & \vdots & & \vdots \\ AB_{m1} & AB_{m2} & \cdots & AB_{mj} & \cdots & AB_{mo} \end{bmatrix} \qquad (2.27)$$

The subscript i in Equation (2.26) corresponds to the rows in matrix $[A]$ and the subscript j in the same equation corresponds to the columns in matrix $[B]$. The subscript pair ij in Equation (2.26) corresponds to the cells in the matrix product $[A][B]$. Matrix (2.27) includes matrices $[A]$, $[B]$ and $[A][B]$ with an arbitrary row i, column j and cell ij shaded.

When calculating the product of three or more matrices, the product is calculated from right to left. Therefore to calculate the product of matrices $[A]$, $[B]$ and $[C]$, for example, the product $[B][C]$ is first calculated and the result is multiplied by $[A]$ (or $[A][B][C] = [A]$ $([B][C])$).

Example 2.7

Problem Statement: Calculate the matrix products $[A][B]$ and $[A][B][C]$ where

$$[A] = \begin{bmatrix} a_{11} & a_{12} \\ a_{21} & a_{22} \end{bmatrix}, \quad [B] = \begin{bmatrix} b_{11} & b_{12} \\ b_{21} & b_{22} \end{bmatrix} \text{ and } [C] = \begin{bmatrix} c_{11} & c_{12} \\ c_{21} & c_{22} \end{bmatrix}$$

Known Information: Given matrices
   Solution Approach: Figure E.2.10 includes the calculation procedure in MATLAB's command window.

2.4.3 The Identity Matrix and Matrix Inversion

The *unit matrix* or *identity matrix* (denoted by I) is a square matrix having the number 1 along its main diagonal (with all other cells having the number 0).* The identity matrix is the matrix equivalent of the number 1. The general form of the identity matrix can be expressed as

---

* The main diagonal runs from the top-left matrix corner to the bottom-right matrix corner.

```
>> syms a11 a12 a21 a22 real;
>> syms b11 b12 b21 b22 real;
>> syms c11 c12 c21 c22 real;
>> A = [a11, a12; a21, a22];
>> B = [b11, b12; b21, b22];
>> C = [c11, c12; c21, c22];
>> A*B

ans =

[ a11*b11 + a12*b21, a11*b12 + a12*b22]
[ a21*b11 + a22*b21, a21*b12 + a22*b22]

>> A*B*C

ans =

[ c11*(a11*b11 + a12*b21) + c21*(a11*b12 + a12*b22), c12*(a11*b11
+ a12*b21) + c22*(a11*b12 + a12*b22)]
[ c11*(a21*b11 + a22*b21) + c21*(a21*b12 + a22*b22), c12*(a21*b11
+ a22*b21) + c22*(a21*b12 + a22*b22)]

>>
```

FIGURE E.2.10
Example 2.7 solution calculation procedure in MATLAB.

$$I = \begin{bmatrix} 1 & 0 & 0 & \cdots & 0 \\ 0 & 1 & 0 & \cdots & 0 \\ 0 & 0 & 1 & \cdots & 0 \\ \vdots & \vdots & \vdots & \ddots & \vdots \\ 0 & 0 & 0 & \cdots & 1 \end{bmatrix} \tag{2.28}$$

As noted in Section 2.4.1, expressing a set of linear simultaneous equations in matrix form may be particularly convenient when solving for the variables in the equation set. Considering the matrix form $[A]\mathbf{x} = \mathbf{b}$, the column vector of variables $\mathbf{x}$ is the result of the product $[A]^{-1}\mathbf{b}$ $[A]^{-1}\mathbf{b}$ or $\mathbf{x} = [A]^{-1}\mathbf{b}$ where the superscript −1 in $[A]^{-1}$ represents the *inverse* of $[A]$. Assuming $[A]$ is invertible, its inverse can be defined as

$$[A]^{-1} = \frac{1}{det[A]} adj[A] \tag{2.29}$$

where *det* and *adj* are the *determinant* and *adjoint* (two matrix functions) of $[A]$. Equation 2.29 is used in *Cramer's rule*—a formula for the solution of linear simultaneous equations having n equations and n unknown variables.

This textbook does not include descriptions of the determinant and adjoint functions. This is in part because the procedures for manually calculating the determinant and adjoint of a matrix become increasingly involved for matrices having dimensions beyond 2 × 2. Another reason is that matrix inversion is a simple procedure in the mathematical analysis software MATLAB. For those interested in becoming more familiar with the determinant and adjoint functions (as well as Cramer's rule), we recommended that you refer either to online resources or to textbooks that include the fundamentals of *linear algebra.**

---

* *Linear algebra* is the branch of mathematics concerning vector spaces and linear mappings between such spaces.

The *order of operations* used in computer programming and arithmetic operations for scalar quantities also applies to matrix operations. If we recall, this order is as follows: (1) parentheses, (2) exponents, (3) multiplication, (4) division, (5) addition, and (6) subtraction.

Example 2.8

Problem Statement: Calculate $[A]^{-1}$ where

$$[A] = \begin{bmatrix} a_{11} & a_{12} & a_{13} \\ a_{21} & a_{22} & a_{23} \\ a_{31} & a_{32} & a_{33} \end{bmatrix}$$

Known Information: Given matrix.
  Solution Approach: Figure E.2.11 includes the calculation procedure in MATLAB's command window. It can be observed from scale of $[A]^{-1}$ in Figure E.2.11 that manually producing the inverse of even this 3×3 matrix will be quite involved.

Example 2.9

Problem Statement: Calculate $[A]^{-1}$ and the unknown variables in Equation 2.22.
  Known Information: Equation 2.22.
  Solution Approach: Figure E.2.12 includes the calculation procedure in MATLAB's command window.

```
>> syms a11 a12 a13 a21 a22 a23 a31 a32 a33 real;
>> A = [a11, a12, a13
a21, a22, a23
a31, a32, a33];
>> Inverse_A = inv(A)

Inverse_A =

[ (a22*a33 - a23*a32)/(a11*a22*a33 - a11*a23*a32 - a12*a21*a33 +
a12*a23*a31 + a13*a21*a32 - a13*a22*a31), -(a12*a33 -
a13*a32)/(a11*a22*a33 - a11*a23*a32 - a12*a21*a33 + a12*a23*a31 +
a13*a21*a32 - a13*a22*a31), (a12*a23 - a13*a22)/(a11*a22*a33 -
a11*a23*a32 - a12*a21*a33 + a12*a23*a31 + a13*a21*a32 -
a13*a22*a31)]
[ - (a21*a33 - a23*a31)/(a11*a22*a33 - a11*a23*a32 - a12*a21*a33 +
a12*a23*a31 + a13*a21*a32 - a13*a22*a31), (a11*a33 -
a13*a31)/(a11*a22*a33 - a11*a23*a32 - a12*a21*a33 + a12*a23*a31 +
a13*a21*a32 - a13*a22*a31), - (a11*a23 - a13*a21)/(a11*a22*a33 -
a11*a23*a32 - a12*a21*a33 + a12*a23*a31 + a13*a21*a32
a13*a22*a31)]
[ (a21*a32 - a22*a31)/(a11*a22*a33 - a11*a23*a32 - a12*a21*a33 +
a12*a23*a31 + a13*a21*a32 - a13*a22*a31), -(a11*a32 -
a12*a31)/(a11*a23*a33 - a11*a23*a32 - a12*a21*a33 + a12*a23*a31 +
a13*a21*a32 - a13*a22*a31), (a11*a22 - a12*a21)/(a11*a22*a33 -
a11*a23*a32 - a12*a21*a33 + a12*a23*a31 + a13*a21*a32
a13*a22*a31)]

>>
```

FIGURE E.2.11
Example 2.8 solution calculation procedure in MATLAB.

```
>> b = [1, -2, 1, 3]'

b =

      1
     -2
      1
      3

>> A = [1, -2, 1, -3
2, 1, 2, -2
-1, 2, -4, 1
3, 0, 0, -3]

A =

      1     -2      1     -3
      2      1      2     -2
     -1      2     -4      1
      3      0      0     -3

>> inv(A)

ans =

   -0.5000    -0.3750    -0.3125     0.6458
         0     0.5000     0.2500    -0.2500
         0     0.2500    -0.1250    -0.2083
   -0.5000    -0.3750    -0.3125     0.3125

>> x = A\b

x =

    1.8750
   -1.5000
   -1.2500
    0.8750

>>
```

FIGURE E.2.12
Example 2.9 solution calculation procedure in MATLAB.

2.5 Intermediate and Total Spatial Motion

Matrix 2.30 is a general spatial angular displacement matrix. In this matrix, the rotation angle is represented by the variable δ and the rotation axis is represented by the vector **u** (Figure 2.4).*

$$[R_{\delta,u}] = \begin{bmatrix} u_x^2 v(\delta) + \cos(\delta) & u_x u_y v(\delta) - u_z \sin(\delta) & u_x u_z v(\delta) + u_y \sin(\delta) \\ u_x u_y v(\delta) + u_z \sin(\delta) & u_y^2 v(\delta) + \cos(\delta) & u_y u_z v(\delta) - u_x \sin(\delta) \\ u_x u_z v(\delta) - u_y \sin(\delta) & u_y u_z v(\delta) - u_x \sin(\delta) & u_z^2 v(\delta) + \cos(\delta) \end{bmatrix} \quad (2.30)$$

In Matrix 2.30, $v(\delta) = 1 - \cos(\delta)$.

---

* By specifying a rotation axis of **u** = (0,0,1), Matrix 2.30 is restricted to rotations in two-dimensional space (as well as Matrices 2.35 and 2.39). In this condition, the z-components of the point coordinates could all be specified as zero.

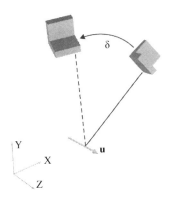

FIGURE 2.4
Spatial rotation of an arbitrary body.

Figure 2.5 illustrates two interconnected members in three-dimensional space. In this system of interconnected members, point $\mathbf{p}_1$ rotates about the axis $\mathbf{u}_0$ by an angle θ and point $\mathbf{q}_1$ rotates about the axis $\mathbf{u}_1$ by an angle β.* Because both members are interconnected, point $\mathbf{q}_1$ and axis $\mathbf{u}_1$ also rotate about the axis $\mathbf{u}_0$. Point $\mathbf{p}_0$, like axis $\mathbf{u}_0$, is fixed or *grounded* in space. An equation for the rotation of $\mathbf{p}_0$ about $\mathbf{u}_0$—or the *total displacement* of $\mathbf{p}_1$ (which will be labeled $\mathbf{p}$)—can be expressed as

$$\mathbf{p} = \left[R_{\theta,\mathbf{u}_0} \right]\left(\mathbf{p}_1 - \mathbf{p}_0 \right) + \mathbf{p}_0 \tag{2.31}$$

and an equation for the rotation of $\mathbf{u}_1$ about $\mathbf{u}_0$—or the total displacement of $\mathbf{u}_1$ (which will be labeled $\mathbf{u}$)—can be expressed as

$$\mathbf{u} = \left[R_{\theta,\mathbf{u}_0} \right]\mathbf{u}_1 \tag{2.32}$$

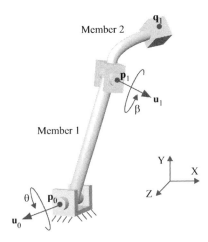

FIGURE 2.5
Spatial motion of a two-body system.

---

* Because Matrix 2.30 is a 3×3 matrix, points $\mathbf{p}_0$, $\mathbf{p}_1$, $\mathbf{q}_1$ and axes $\mathbf{u}_0$ and $\mathbf{u}_1$ are 3×1 *column matrices* (where vector rows 1, 2, and 3 include the *x*-, *y*-, and *z*-components, respectively).

An equation for the *intermediate displacement* of $\mathbf{q}_1$ about $\mathbf{u}_0$ (which will be labeled $\mathbf{q}'$) can be expressed as

$$\mathbf{q}' = \left[R_{\theta, u_0} \right] \left(\mathbf{q}_1 - \mathbf{p}_0 \right) + \mathbf{p}_0 \tag{2.33}$$

Equation 2.33 is included in Equation 2.34 to calculate the total displacement of $\mathbf{q}_1$ about $\mathbf{u}_0$ (which will be labeled $\mathbf{q}$).

$$\mathbf{q} = \left[R_{\beta, u} \right] \left(\mathbf{q}' - \mathbf{p} \right) + \mathbf{p} \tag{2.34}$$

Matrix 2.35 is a general spatial angular velocity matrix. In this matrix, the angular velocity is represented by the variable $\dot{\delta}$ and the rotation axis is represented by the vector $\mathbf{u}$.

$$\left[V_{\dot{\delta}, u} \right] = \begin{bmatrix} 0 & -u_z\dot{\delta} & u_y\dot{\delta} \\ u_z\dot{\delta} & 0 & -u_x\dot{\delta} \\ -u_y\dot{\delta} & u_x\dot{\delta} & 0 \end{bmatrix} \tag{2.35}$$

The velocity of $\mathbf{p}_1$ in reference to $\mathbf{u}_0$—or the *total velocity* of $\mathbf{p}_1$ (which will be labeled $\dot{\mathbf{p}}$)—can be expressed as

$$\dot{\mathbf{p}} = \left[V_{\dot{\theta}, u_0} \right] \left(\mathbf{p}_1 - \mathbf{p}_0 \right) \tag{2.36}$$

An equation for the *intermediate velocity* of $\mathbf{q}_1$ in reference to $\mathbf{u}_0$ (which will be labeled $\dot{\mathbf{q}}'$) can be expressed as

$$\dot{\mathbf{q}}' = \left[V_{\dot{\theta}, u_0} \right] \left(\mathbf{q}_1 - \mathbf{p}_0 \right) \tag{2.37}$$

where the variable $\mathbf{q}$ is included from Equation 2.34. Equation 2.37 is included in Equation 2.38 to calculate the total velocity of $\mathbf{q}_1$ in reference to $\mathbf{u}_0$ (which will be labeled $\dot{\mathbf{q}}$)

$$\dot{\mathbf{q}} = \left[V_{\dot{\beta}, u} \right] \left(\mathbf{q} - \mathbf{p} \right) + \dot{\mathbf{q}}' \tag{2.38}$$

where the variable $\mathbf{p}$ is included from Equation 2.31.

Matrix 2.39 is a general spatial angular acceleration matrix. In this matrix, the angular velocity and acceleration are represented by the variables $\dot{\delta}$ and $\ddot{\delta}$, respectively, and the rotation axis is represented by the vector $\mathbf{u}$ from Equation 2.32. In Matrix 2.39, it is assumed that $\dot{\mathbf{u}} = 0$, since $\mathbf{u}_1$ typically has a fixed orientation with respect to its associated links [4].*

---

* In the RRSS and 4R spherical mechanisms in Chapter 10, the condition $\dot{\mathbf{u}} = 0$ holds true.

$$
\left[A_{\ddot{\delta},\dot{\delta},u} \right] = \begin{bmatrix}
\left(u_x^2 - 1 \right)\dot{\delta}^2 & u_x u_y \dot{\delta}^2 - \dot{u}_z \dot{\delta} - u_z \ddot{\delta} & u_x u_z \dot{\delta}^2 + \dot{u}_y \dot{\delta} + u_y \ddot{\delta} \\[3mm]
u_x u_y \dot{\delta}^2 + \dot{u}_z \dot{\delta} + u_z \ddot{\delta} & \left(u_y^2 - 1 \right)\dot{\delta}^2 & u_y u_z \dot{\delta}^2 - \dot{u}_x \dot{\delta} - u_x \ddot{\delta} \\[3mm]
u_x u_z \dot{\delta}^2 - \dot{u}_y \dot{\delta} - u_y \ddot{\delta} & u_x u_z \dot{\delta}^2 - \dot{u}_x \dot{\delta} + u_x \ddot{\delta} & \left(u_z^2 - 1 \right)\dot{\delta}^2
\end{bmatrix}
\tag{2.39}
$$

The acceleration of $\mathbf{p}_1$ in reference to $\mathbf{u}_0$—or the *total acceleration* of $\mathbf{p}_1$ (which will be labeled $\ddot{\mathbf{p}}$)—can be expressed as

$$
\ddot{\mathbf{p}} = \left[A_{\ddot{\theta},\dot{\theta},\mathbf{u}_0} \right]\left(\mathbf{p} - \mathbf{p}_0 \right)
\tag{2.40}
$$

where the variable $\mathbf{p}$ is included from Equation 2.31. An equation for the *intermediate acceleration* of $\mathbf{q}_1$ in reference to $\mathbf{u}_0$ (which will be labeled $\ddot{\mathbf{q}}'$) can be expressed as

$$
\ddot{\mathbf{q}}' = \left[A_{\ddot{\beta},\dot{\beta},\mathbf{u}_0} \right]\left(\mathbf{q} - \mathbf{p}_0 \right)
\tag{2.41}
$$

where the variable $\mathbf{q}$ is included from Equation 2.34. Equation 2.41 is included in Equation 2.42 to calculate the total acceleration of $\mathbf{q}_1$ in reference to $\mathbf{u}_0$ (which will be labeled $\ddot{\mathbf{q}}$). In Equation 2.42, the variables $\mathbf{p}$ and $\mathbf{q}$ are included from Equations 2.31 and 2.34, respectively.

$$
\ddot{\mathbf{q}} = \ddot{\mathbf{q}}' + \left[A_{\ddot{\beta},\dot{\beta},\mathbf{u}} \right]\left(\mathbf{q} - \mathbf{p} \right) + 2\left[V_{\dot{\theta},\mathbf{u}_0} \right]\left\{ \left[V_{\dot{\beta},\mathbf{u}} \right]\left(\mathbf{q} - \mathbf{p} \right) \right\}
\tag{2.42}
$$

The intermediate and total spatial displacement velocity and acceleration equations appear in the spatial mechanism equation systems in Chapter 10 [3].

Example 2.10

Problem Statement: For the two-body system in Figure 2.5, calculate the displaced values of points $\mathbf{p}_1$ and $\mathbf{q}_1$ and axis $\mathbf{u}_1$. The dimensions and rotation angles in this system are as follows: $\mathbf{p}_0 = (0, 0, 0)$, $\mathbf{p}_1 = (0, 1, 0)$, $\mathbf{u}_0 = (0, 0, 1)$, $\mathbf{u}_1 = (0.7071, 0, 0.7071)$, $\mathbf{q}_1 = (0.25, 1.3536, -0.25)$, $\theta = 30°$, and $\beta = -15°$.

  Known Information: Given dimensions, rotation angles, Matrix 2.30 and Equations 2.31 through 2.34.

  Solution Approach: Figure E.2.13 includes the calculation procedure in MATLAB's command window.

```
>> p0 = [0, 0, 0]';
>> p1 = [0, 1, 0]';
>> q1 = [0.25, 1.3536, -0.25]';
>> u0 = [0, 0, 1];
>> u0x = u0(1); u0y = u0(2); u0z = u0(3);
>> u1 = [0.7071, 0, 0.7071]';
>> theta = 30*pi/180;
>> beta = -15*pi/180;
>> C = cos(theta);
>> S = sin(theta);
>> V = 1 - C;
>> R_theta_u0 = [...
V*u0x^2 + C, V*u0x*u0y - S*u0z, V*u0x*u0z + S*u0y
V*u0x*u0y + S*u0z, V*u0y^2 + C, V*u0y*u0z - S*u0x
V*u0x*u0z - S*u0y, V*u0y*u0z + S*u0x, V*u0z^2 + C];
>> u = R_theta_u0*u1

u =

    0.6124
    0.3535
    0.7071

>> p = R_theta_u0*(p1 - p0) + p0

p =

   -0.5000
    0.8660
         0

>> q_prime = R_theta_u0*(q1 - p0) + p0;
>> C = cos(beta);
>> S = sin(beta);
>> V = 1 - C;
>> ux = u(1); uy = u(2); uz = u(3);
>> R_beta_u = [...
V*ux^2 + C, V*ux*uy - S*uz, V*ux*uz + S*uy
V*ux*uy + S*uz, V*uy^2 + C, V*uy*uz - S*ux
V*ux*uz - S*uy, V*uy*uz + S*ux, V*uz^2 + C];
>> q = R_beta_u*(q_prime - p) + p

q =

   -0.3599
    1.2357
   -0.3062

>>
```

FIGURE E.2.13
Example 2.10 solution calculation procedure in MATLAB.

2.6 General Transformation Matrix

Matrices 2.43 through 2.45 are rotation matrices about the *x*-, *y*-, and *z*-axes of a global coordinate frame, respectively. Given the rotation angle value, the product of any of these matrices and coordinates of a point (in a 3 × 1 column matrix) are the coordinates of the rotated point.

$$[R_{\delta_x}] = \begin{bmatrix} 1 & 0 & 0 \\ 0 & \cos\delta_x & -\sin\delta_x \\ 0 & \sin\delta_x & \cos\delta_x \end{bmatrix} \tag{2.43}$$

$$\left[R_{\delta_y} \right] = \begin{bmatrix} \cos\delta_y & 0 & \sin\delta_y \\ 0 & 1 & 0 \\ -\sin\delta_y & 0 & \cos\delta_y \end{bmatrix} \tag{2.44}$$

$$\left[R_{\delta_z} \right] = \begin{bmatrix} \cos\delta_z & -\sin\delta_z & 0 \\ \sin\delta_z & \cos\delta_z & 0 \\ 0 & 0 & 1 \end{bmatrix} \tag{2.45}$$

The product of the three matrices can be expressed as Matrix 2.46, which can accommodate simultaneous rotations about the x-, y-, and z-axes (by rotation angles δ_x, δ_y, and δ_z, respectively).

$$[R] = \left[R_{\delta_z} \right]\left[R_{\delta_y} \right]\left[R_{\delta_x} \right] = \begin{bmatrix} R_{11} & R_{12} & R_{13} \\ R_{21} & R_{22} & R_{23} \\ R_{31} & R_{32} & R_{33} \end{bmatrix} =$$

$$= \begin{bmatrix} \cos\delta_y \cos\delta_z & \left(\sin\delta_x \sin\delta_y \cos\delta_z - \cos\delta_x \sin\delta_z\right) & \left(\cos\delta_x \sin\delta_y \cos\delta_z + \sin\delta_x \sin\delta_z\right) \\ \cos\delta_y \sin\delta_z & \left(\sin\delta_x \sin\delta_y \sin\delta_z + \cos\delta_x \cos\delta_z\right) & \left(\cos\delta_x \sin\delta_y \sin\delta_z - \sin\delta_x \cos\delta_z\right) \\ -\sin\delta_y & \sin\delta_x \cos\delta_y & \cos\delta_x \cos\delta_y \end{bmatrix}$$

$$\tag{2.46}$$

Including the elements of Matrix 2.46 into a 4×4 matrix that also considers translations along the x-, y-, and z-axes (in the fourth matrix column) produces

$$^i_j[T] = \begin{bmatrix} R_{11} & R_{12} & R_{13} & \Delta_x \\ R_{21} & R_{22} & R_{23} & \Delta_y \\ R_{31} & R_{32} & R_{33} & \Delta_z \\ 0 & 0 & 0 & 1 \end{bmatrix} \tag{2.47}$$

Matrix 2.47 is a *general transformation matrix* for calculating point coordinates given in one coordinate frame (which we will call Frame j) in reference to another coordinate frame (which we will call Frame i) or

$$^i\{\mathbf{p}\} = \,^i_j[T]\,^j\{\mathbf{p}\} \tag{2.48}$$

In Equation 2.48, the coordinates in Frame j or $^j\{\mathbf{p}\}$ are $^j\{\mathbf{p}\} = [p_x \; p_y \; p_z \; 1]^T$. Matrix 2.47 is general because it can accommodate all six possible degrees of freedom (x-y-z rotations and x-y-z translations). The rotation angles δ_x, δ_y, δ_z and translation values Δ_x, Δ_y, and Δ_z in Matrix 2.47 are the angular and linear displacement values required to align Frame i to Frame j.

As an example, Figure 2.6 illustrates a Coordinate Frame i (where $i = 1$) and an arbitrary Coordinate Frame j (where $j = 2$). Using rotation angles $\delta_x = 20°$, $\delta_y = 40°$, $\delta_z = 60°$

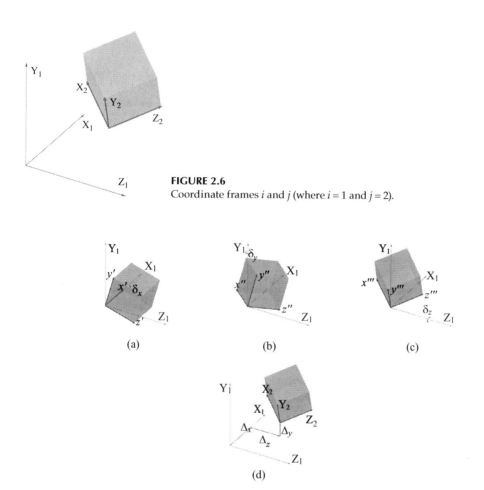

FIGURE 2.6
Coordinate frames i and j (where $i = 1$ and $j = 2$).

(a)   (b)   (c)

(d)

FIGURE 2.7
(a–c) Coordinate frame rotations about and (d) translations along axes X_1, Y_1, and Z_1.

(Figures 2.7a, b, c, respectively) and translation values $\Delta_x = 2$, $\Delta_y = \Delta_z = 1$ (Figure 2.7d), Frame 1 is aligned with Frame 2. As a result, when these rotation and translation values are used in Equation 2.48, any point coordinates given in reference to Frame 2 will be calculated in reference to Frame 1. Transformation matrices are used in in Chapter 11 for the kinematic analysis of robotic manipulators [5].

Example 2.11

Problem Statement: The coordinates of point $\mathbf{p}_1$ in Reference Frame 2 of a robotic system are $^2\{\mathbf{p}_1\} = [2, 5, -1, 1]^T$ (see Figure E.2.14). Calculate the coordinates of this point in Reference Frame 1 of the system ($1\{\mathbf{p}_1\}$). The location of the origin of Frame 2 with respect to Frame 1 is $\Delta = (5, 10, -2)$ and the orientation angles of Frame 2 with respect to Frame 1 are $\delta_x = 0°$, $\delta_y = 15°$, and $\delta_z = 30°$.

Known Information: Given frame rotation and displacement values and Equation 2.48.

Solution Approach: Figure E.2.15 includes the calculation procedure in MATLAB's command window.

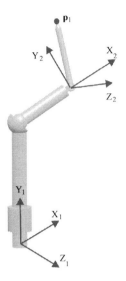

FIGURE E.2.14
Robotic system with reference frames.

```
>> p2 = [2, 5, -1, 1]';
>> Dx = 5;
>> Dy = 10;
>> Dz = -2;
>> delta_x = 0*pi/180;
>> delta_y = 15*pi/180;
>> delta_z = 30*pi/180;
>> Sx = sin(delta_x); Cx = cos(delta_x);
>> Sy = sin(delta_y); Cy = cos(delta_y);
>> Sz = sin(delta_z); Cz = cos(delta_z);
>> T12 = [...
Cy*Cz, Sx*Sy*Cz - Cx*Sz, Cx*Sy*Cz + Sx*Sz, Dx
Cy*Sz, Sx*Sy*Sz + Cx*Cz, Cx*Sy*Sz - Sx*Cz, Dy
-Sy, Sx*Cy, Cx*Cy, Dz
0, 0, 0, 1];
>> p1 = T12*p2

p1 =

    3.9489
   15.1666
   -3.4836
    1.0000

>>
```

FIGURE E.2.15
Example 2.11 solution calculation procedure in MATLAB.

2.7 Summary

Vectors (quantities having both *magnitude* and *direction*) are commonly used in the formulation of equation systems for mechanisms. One way to represent a two-dimensional vector is with a complex number. A complex number is comprised of both a real component and an imaginary component. Equation systems for mechanisms can be formulated by producing vector loops for the mechanisms and taking the sum of the individual vector terms in the loops. First and second derivatives of vector-loop equations are taken to calculate mechanism velocities and accelerations.

In addition to vectors, points can also be used to formulate mechanism displacement, velocity, and acceleration equations. In fact, a vector can be produced from the coordinates of two points in 2D or 3D space. Point-based vectors appear in such spatial mechanism equation systems as intermediate and total displacement, velocity, and acceleration equations.

A linear equation is an equation that includes linear or first-order variables. A system of linear equations (or a linear system) is collection of linear equations including the same variables. If a common solution is sought among a system of linear equations, it is called a set of simultaneous equations.

A set of linear simultaneous equations having n equations and n variables can be expressed in matrix form. A matrix is a rectangular array of numbers, symbols, or expressions arranged in rows and columns. This form can be particularly convenient when solving for the variables in a simultaneous equation set. Cramer's rule is a popular matrix-based formula for the solution of linear simultaneous equations having n equations and n unknown variables.

The transformation matrix, commonly used in the analysis of robotic systems, is used to calculate point coordinates given in one reference frame (Frame j) in terms of another reference frame (Frame i). The general spatial transformation matrix can consider up to all six possible degrees of freedom (x, y, and z rotations and translations).

References

1. Norton, R. L. 2008. *Design of Machinery*, 4th edn, pp. 186–199. New York: McGraw-Hill.
2. Ibid, pp. 309–318.
3. Russell, K., Shen, Q., and Sodhi, R. 2013. *Mechanism Design: Visual and Programmable Approaches*. Chapter 7. Boca Raton: CRC Press.
4. Suh, C. H. and Radcliffe, C. W. 1978. *Kinematics and Mechanisms Design*. p. 85. New York: John Wiley.
5. Wilson, C. E. and Sadler, J. P. 2003. *Kinematics and Dynamics of Machinery*. 3rd edn, Chapter 12. Saddle River, NJ: Prentice-Hall.

Additional Reading

Craig, J. J. 2005. *Introduction to Robotics: Mechanics and Control*. 3rd edn, Chapter 2. Upper Saddle River, NJ: Pearson Prentice-Hall.
Suh, C. H. and Radcliffe, C. W. 1978. *Kinematics and Mechanisms Design*. pp. 67–73. New York: John Wiley.

Problems

1. Formulate an equation system for the vector loop illustrated in Figure P.2.1. Consider that vector $\mathbf{V}_j$ always lies along the real axis.

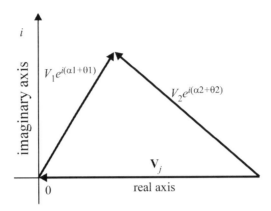

FIGURE P.2.1
Vector loop (three vectors where V_j changes length) in complex space.

2. Formulate an equation system for the vector loop illustrated in Figure P.2.2. Consider that vector $\mathbf{V}_j$ always lies along the real axis and vector $\mathbf{V}_3$ is always perpendicular to the real axis.

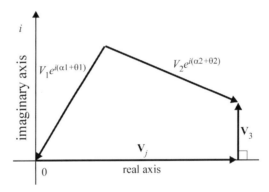

FIGURE P.2.2
Vector loop (four vectors where V_j changes length) in complex space.

3. Calculate the first derivative of the vector-loop equation solution from Problem 2. Consider only angles α_1, α_2 and vector $\mathbf{V}_j$ from Problem 2 to be time-dependent.

4. Calculate the second derivative of the vector-loop equation solution from Problem 2. Consider only angles α_1, α_2 and vector $\mathbf{V}_j$ from Problem 2 to be time-dependent.

5. Formulate an equation system for the vector loop illustrated in Figure P.2.3.

6. Calculate the first derivative of the vector-loop equation solution from Problem 5. Consider only angles α_1, α_2, and α_3 from Problem 5 to be time-dependent.

7. Calculate the second derivative of the vector-loop equation solution from Problem 5. Consider only angles α_1, α_2, and α_3 from Problem 5 to be time-dependent.

8. Formulate an equation system for the vector loop illustrated in Figure P.2.4.

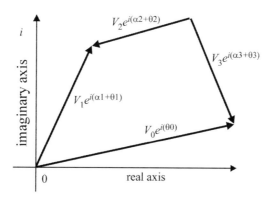

FIGURE P.2.3

Vector loop (four vectors) in complex space.

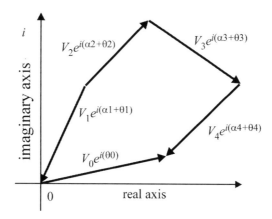

FIGURE P.2.4

Vector loop (five vectors) in complex space.

9. Calculate the first derivative of the vector-loop equation solution from Problem 8. Consider only angles α_1, α_2, α_3, and α_4 from Problem 8 to be time-dependent.

10. Calculate the second derivative of the vector-loop equation solution from Problem 8. Consider only angles α_1, α_2, α_3, and α_4 from Problem 8 to be time-dependent.

3

Fundamental Concepts in Kinematics

CONCEPT OVERVIEW

In this chapter, the reader will gain a central understanding regarding

1. Kinematic and design distinctions among select planar and spatial mechanisms
2. Mechanism components and mechanism construction
3. Mechanism *mobility*, *Gruebler's equation* and *number synthesis*
4. *Grashof criteria* and the *circuit defect*
5. The *transmission angle* and its relationship with follower-link forces
6. *Mechanism inversion*
7. The *passive degree of freedom* and paradoxes to Gruebler's equation

3.1 Types of Planar and Spatial Mechanisms

3.1.1 Planar Four-Bar Mechanism

Figure 3.1a illustrates a *planar four-bar mechanism*. The four interconnected links in this mechanism are the *crank* (the driving link), *coupler, follower*, and *ground*.* A supplied input rotation to the crank link compels the motion of the coupler and follower links. Both the crank and follower links are connected to the ground link (or are *grounded*) and undergo pure rotation (as indicated by the arrows).† The planar four-bar mechanism is one of the most widely utilized kinematic chains in everyday devices (Figure 3.1b, c, and d) including locking pliers, folding chairs, and doorways.

3.1.2 Slider-Crank Mechanism

Figure 3.2a illustrates a *slider-crank mechanism*. The four interconnected links in this mechanism are the *crank* (the driving link), *coupler, slider*, and *ground*. A supplied input rotation to the crank link compels the motion of the coupler and slider links. In this mechanism, the crank link undergoes pure rotation, the slider undergoes pure translation and the coupler undergoes complex motion. The slider-crank mechanism can be theoretically described as a planar four-bar mechanism having a follower link of infinite length (Figure 3.2b) [1]. The

---

* Because mechanisms are comprised of links, they are also called *linkages*.
† The coupler undergoes *complex motion*—a combination of simultaneous rotation and translation (see Section 1.4).

41

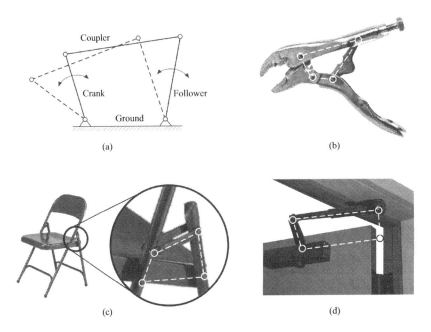

FIGURE 3.1
(a) Planar four-bar mechanism as (b) lock pliers, (c) folding chair, and (d) doorway linkages.

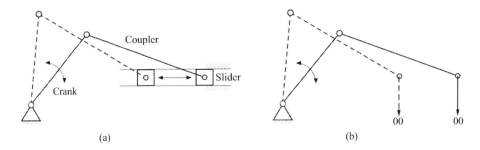

FIGURE 3.2
(a) Slider-crank mechanism and (b) four-bar mechanism as slider-crank mechanism.

slider-crank mechanism is also among the most widely utilized kinematic chains in every-day applications. Among the many everyday applications for the slider-crank mechanism is the crankshaft-connecting rod–piston linkage: a fundamental subsystem of the internal combustion engine (Figure 3.3).

3.1.3 Geared Five-Bar Mechanism

Figure 3.4 illustrates a *geared five-bar mechanism*. This mechanism is comprised of five interconnected links where the crank and output link are generally coupled to each other through a gear pair or gear train.* Because the crank and output links are

---

* Coupling the motion of the crank and output links reduces the resulting mechanism to a single degree of free-dom. In addition to using gears for coupling the crank and output links, other options include using *pulley-belt* systems, *chain-sprocket* systems or *drive motors*.

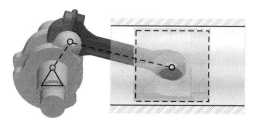

FIGURE 3.3
Slider-crank mechanism as crankshaft-connecting rod–piston linkage.

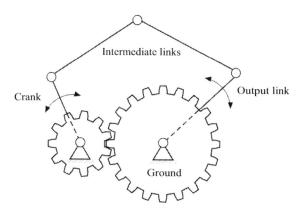

FIGURE 3.4
Geared five-bar mechanism.

indirectly interconnected, the supplied input motion to the crank compels the motion of the output and intermediate links. Both the crank and output links of the geared five-bar mechanism undergo pure rotation while the two intermediate links undergo complex motion.

Though not as commonly utilized in everyday devices as the planar four-bar or slider-crank mechanisms, one advantage the planar five-bar mechanism has over the four-bar mechanism is that it can trace paths of higher orders than the latter. This means that the intermediate links of the geared five-bar mechanism have the capacity to trace paths of more complex curvature than the paths traced by the coupler link of the planar four-bar linkage. Using Equation 3.1 (published by Wunderlich in 1963) to calculate the order (m) of a coupler curve produced by a mechanism having n links connected by revolute joints, we can determine that while a planar four-bar mechanism has a maximum curve order of six, the planar five-bar mechanism has a maximum order of 10 (or 10.392 to be more exact) [2].

$$m = 2\left[3^{\left(\frac{n}{2}-1\right)}\right]$$

(3.1)

3.1.4 Planar Multiloop Six-Bar Mechanisms

Attaching a grounded link pair or a *dyad* (the link pair c-d-e in Figure 3.5) to the coupler of the planar four-bar mechanism produces a *planar multiloop six-bar mechanism*. This particular type of multiloop planar six-bar mechanism is called a *Stephenson type III mechanism* [3]. This mechanism includes three links that undergo pure rotation and two links that undergo complex motion.

The five types of planar multiloop six-bar mechanisms are illustrated in Figure 3.6. These types are grouped into two classifications: *Watt* and *Stephenson*. The supplied input motion to the crank (e.g., Link 2 in Figure 3.6) compels the motion of the remaining links. Though not as commonly utilized in everyday devices as the planar four-bar or slider-crank mechanisms, Watt and Stephenson mechanisms include two to three intermediate links that undergo complex motion. This property enables these mechanisms to deliver

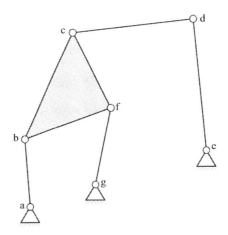

FIGURE 3.5
Planar multiloop six-bar mechanism (Stephenson III six-bar mechanism).

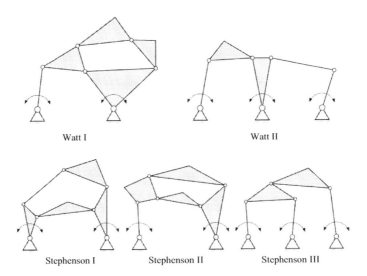

FIGURE 3.6
Watt and Stephenson mechanism types.

both dual and simultaneous *motion* and *path generation* capabilities.* As shown in this figure, both Stephenson and Watt mechanisms include two to three links that undergo pure rotation (as indicated by the arrows in this figure) as well as two to three links that undergo complex motion. The Watt II and Stephenson III mechanisms are the planar multiloop mechanisms of choice for analysis throughout this textbook.

3.1.5 Spatial Four-Bar Mechanisms

Planar mechanisms are restricted to motion in two-dimensional or planar space. *Spatial* mechanisms can exhibit three-dimensional or spatial motion. Spatial mechanism motion is predominantly determined by the degrees of freedom of the mechanism joints used and the spatial orientation of the joints.

Because spatial mechanisms have the capability to exhibit spatial motion, they offer a greater variety of possible motions and are structurally more general than planar mechanisms. However, because the equations for spatial mechanism analysis and synthesis are often much larger in scale and greater in complexity than those for planar mechanisms (not to mention the actual design of spatial mechanisms vs. planar mechanisms), their real-world applications are often limited. In practice, it is not uncommon to find complicated planar mechanism solutions when, in fact, a simpler spatial mechanism solution is also possible. It is, therefore, an ongoing task to devise simple methods of calculation, to produce design aids with diagrams, and to set design standards for spatial mechanisms [4].

Like planar mechanisms, there are also many different types of spatial mechanisms. This textbook considers three types of four-bar spatial mechanisms: the *revolute-revolute-spherical-spherical* or *RRSS*, the *revolute-revolute-revolute-revolute spherical* or *4R spherical*, and the *revolute-spherical-spherical-revolute* or *RSSR* mechanisms (Figure 3.7a, b, and c, respectively) [5,6]. The RRSS, 4R spherical, and RSSR are among the more basic four-bar spatial mechanisms in terms of the types of joints used and the required linkage assembly conditions for motion.

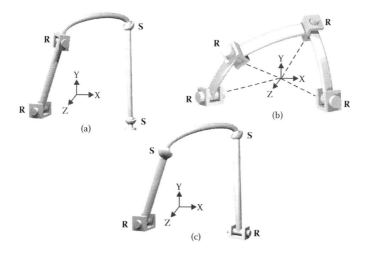

FIGURE 3.7
Spatial (a) RRSS, (b) 4R spherical, and (c) RSSR mechanisms.

---

* *Motion* and *path* generation (mechanism design for prescribed *mechanism link positions* and *path points*, respectively) are introduced in Chapter 5.

3.2 Links, Joints, and Mechanism Mobility

As explained in Chapter 1, a mechanism is an assembly of links and joints. The attachment points to adjacent links are called *nodes*. A link that has two nodes is called a *binary link*, and a *ternary link* has three nodes. The Stephenson III six-bar mechanism in Figure 3.8 includes a ternary link as ground and a movable intermediate ternary link. The remaining links in this mechanism are all binary.

Adjacent mechanism links are interconnected at their nodes by *joints*. Joint types differ by both the number and type of degree(s) of freedom (DOF). Six mechanism joint types are illustrated in Figure 3.9. For planar mechanisms, the revolute joint is more commonly used than any other joint. Although the revolute joint is by far the dominant joint type used in practice, as well as in the mechanisms in this textbook, Figure 3.9 includes other common joint types. The *revolute joint* **(R)** has one rotational DOF, the *prismatic joint* **(P)** has one translational DOF, the *cylindrical joint* **(C)** has 2 DOFs—one rotational and one translational—and the *spherical joint* **(S)** has three rotational DOFs. The term *lower pair* describes joints like the R, P, C, and S joints where *surface* contact occurs (e.g., a ball surrounded by a socket for the S joint or a pin surrounded by a hole for the R joint). Lower pairs are also called *full joints* [7].

Also illustrated in Figure 3.9 are the *cam joint* and the *gear joint* (G). Because the cam joint includes *rolling-sliding* contact (between the cam and follower surfaces), it has two degrees of freedom.* Because the gear joint includes rolling-sliding contact between the surfaces of the gear teeth, it also has two degrees of freedom. The term *higher pair* describes joints like the cam and the G joints where *line* contact occurs (e.g., a convex surface on a flat surface for the cam joint or two curved convex surfaces in contact for the G joint).† Higher pairs are also called *half joints* [7].

Gruebler's equation (Equations 3.2 and 3.3) is used to determine the mobility (the DOF) of a mechanism. Equation 3.2 calculates the mobility of a *planar mechanism*. Since any individual planar mechanism link can have no more than three degrees of freedom, the

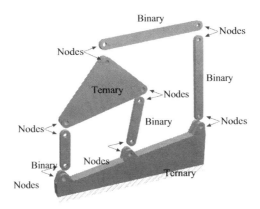

FIGURE 3.8
Link types in the Stephenson III mechanism.

---

* This joint pertains to the classical *radial cam* (or disk cam) and follower type presented in Chapter 9.
† The term "higher pair" also describes joints where *point* contact occurs (e.g., two spheres in contact or a sphere on a flat surface).

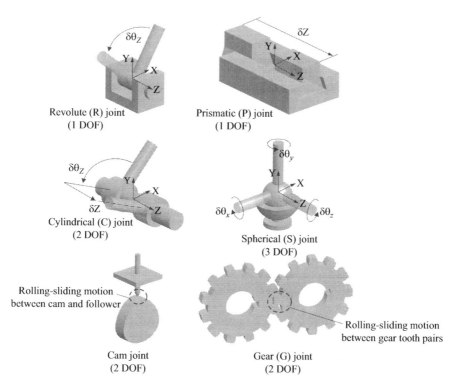

FIGURE 3.9
Six mechanism joint types.

maximum mobility of a planar mechanism with L links is $3L$. Because the ground link is fully constrained, its mobility is subtracted from the maximum mechanism mobility, giving $3(L-1)$. Each 1-DOF joint type removes two degrees of freedom, giving $-2J_1$, and each 2-DOF joint type removes one degree of freedom, giving $-J_2$ (where J_1 and J_2 are the total number of 1-and 2-DOF joints, respectively).

Equation 3.3 calculates the mobility of a *spatial* mechanism. Since any individual spatial mechanism link can have no more than six degrees of freedom, the maximum mobility of a spatial mechanism with L links is $6L$. Because the ground link is fully constrained, its mobility is subtracted from the maximum mobility, giving $6(L-1)$. Each 1-DOF joint type removes five degrees of freedom, giving $-5J_1$. Likewise, each 2- and 3-DOF joint type removes four and three degrees of freedom, respectively. In this textbook, $J_4 = J_5 = 0$, since 4- and 5-DOF joints are not utilized.

$$DOF_{PLANAR} = 3(L-1) - 2J_1 - J_2 \qquad (3.2)$$

$$DOF_{SPATIAL} = 6(L-1) - 5J_1 - 4J_2 - 3J_3 - 2J_4 - J_5 \qquad (3.3)$$

As explained in Section 1.2, the fundamental objective in mechanical system design is to produce specific controlled output motions for supplied input motions. Output motion control is maximized when the mechanical system has a single degree of freedom. Knowing the mobility of a mechanism enables the designer to determine if additional constraints are needed to reach the desired mobility (and if so, how many constraints are needed).

Example 3.1

Problem Statement: Determine the mobility of the planar and spatial mechanisms illustrated in Figure E.3.1.

  Known Information: Figure E.3.1, Equations 3.2 and 3.3.

  Solution Approach: The planar mechanism illustrated in Figure E.3.1 includes eight links, eight revolute joints, and two prismatic joints (therefore, $L = 8$, $J_1 = 10$, and $J_2 = 0$). The spatial linkage illustrated in Figure E.3.1 includes six links, three revolute joints, three spherical joints, and one cylindrical joint (therefore, $L = 6$, $J_1 = 3$, $J_2 = 1$, $J_3 = 3$, and $J_4 J_5 = 0$). Figure E.3.2 includes the calculation procedure in the MATLAB® command window using Gruebler's planar and spatial mechanism equations (Equations 3.2 and 3.3).

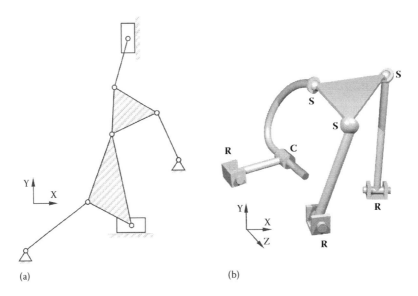

(a)   (b)

FIGURE E.3.1
Planar (a) and spatial (b) mechanisms.

```
>> L = 8;
>> J1 = 10;
>> J2 = 0;
>> DOF_Planar = 3*(L - 1) - 2*J1 - J2

DOF_Planar =

      1

>> L = 6;
>> J1 = 3;
>> J2 = 1;
>> J3 = 3;
>> J4 = 0;
>> J5 = 0;
>> DOF_Spatial = 6*(L - 1) - 5*J1 - 4*J2 - 3*J3 - 2*J4 - J5

DOF_Spatial =

       2

>>
```

FIGURE E.3.2
Example 3.1 solution calculation procedure in MATLAB.

3.3 Number Synthesis

Given the number of mechanism links and the number and order of mechanism joints, mechanism mobility is calculated from Gruebler's equation. An inverse application of Gruebler's equation can also be considered. For example, Gruebler's equation is useful in determining mechanism link and joint combinations (which result in alternate mechanism solutions) for a given mobility [8]. *Number synthesis* involves the determination of alternate mechanism solutions for a given mobility. Expressing Gruebler's equations for planar and spatial mechanisms as functions of links and joints or as

$$f\left(L, J_1, J_2\right) = DOF_{PLANAR} \tag{3.4}$$

$$f\left(L, J_1, J_2, J_3, J_4, J_5\right) = DOF_{SPATIAL} \tag{3.5}$$

shows that, for a given mobility, an indefinite number of link and joint combinations—alternate mechanisms—exist. Number synthesis offers not only a means to assist in the creative design of mechanisms, but Equations 3.4 and 3.5 could also be implemented systematically [9,10]. By progressively increasing or decreasing the link and joint variables in Gruebler's equation for a given mobility, tables of concept mechanism solutions are produced.

While the number of mechanism links and joints required to achieve a specified mobility are determined through number synthesis, the specific mechanism design is not determined. With number synthesis, the user determines the specific mechanism design that incorporates the calculated links and joints.

Example 3.2

Problem Statement: Compile a table of single-DOF planar mechanisms having two, three, and four links.

Known Information: Equation 3.2, $DOF_{PLANAR} = 1$, $L = 2$, 3, and 4.

Solution Approach: For each value of variable L, variable J_1 is incrementally increased, and for each value of L and J_1, the corresponding value for the remaining unknown J_2 in Equation 3.2 is calculated. Using this systematic procedure, the resulting mechanism solutions in Table E.3.1 are calculated.

TABLE E.3.1

Two-, Three-, and Four-Link Single-DOF Planar Mechanisms

Mechanism Solution	L	J_1	J_2
1	2	0	2
2	2	1	0
3	3	0	5
4	3	1	3
5	3	2	1
6	4	0	8
7	4	1	6
8	4	2	4
9	4	3	2
10	4	4	0

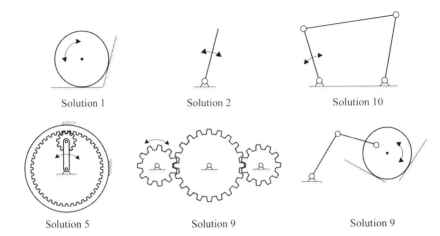

Solution 1   Solution 2   Solution 10

Solution 5   Solution 9   Solution 9

FIGURE E.3.3
Example two-, three-, and four-link single-DOF planar mechanism configurations.

Figure E.3.3 illustrates example mechanism configurations corresponding to mechanism solutions 1, 2, 5, 9, and 10 in Table E.3.1. Each of the mechanism solutions in this table does not necessarily represent a single mechanism configuration. For example, for mechanism solution 9 in Table E.3.1, two mechanism configurations are provided in Figure E.3.3.

3.4 Grashof's Criteria and Transmission Angle

Grashof's criteria are used to determine the link rotation behavior or link rotatability of four-bar mechanisms. These criteria are based on the lengths of the crank, coupler, follower, and ground links. Table 3.1 includes all of the Grashof and non-Grashof mechanism classifications. For the link length relationships in this table, the variables S and L are the shortest and longest mechanism link lengths, respectively, and the remaining two link lengths are the variables P and Q.

Figure 3.10 illustrates the link rotations for the *crank-rocker*, *double-crank* (also called *drag-link*), *double-rocker*, and *triple-rocker* mechanisms. As illustrated, only the crank link can undergo a complete rotation in a crank-rocker mechanism while both the crank and follower links can undergo complete rotations in a double-crank mechanism. In the double-rocker mechanism, the shortest link—the coupler link—is the driving link and undergoes a complete rotation, while no link rotates completely in the triple-rocker.* During the motion of a *change point* mechanism (not illustrated), specifically, twice per revolution, all of the links become simultaneously aligned. During this (theoretical) state, the output behavior of the change point mechanism is unpredictable [11].

Unlike the crank-rocker and drag-link mechanisms, the driving link in the double-rocker mechanism is not grounded. This characteristic can make the double-rocker mechanism less practical for design applications. When considering the double-rocker mechanism, the

---

* Figure 3.10c illustrates the path achieved by the shortest link in the double-rocker when a driver is applied to the labeled joint (of the shortest link).

TABLE 3.1

Grashof and Non-Grashof Mechanisms

Grashof Type	Link Length Relationship	Shortest Link
Crank-rocker	$S + L < P + Q$	Crank
Double-crank (drag-link)	$S + L < P + Q$	Ground
Double-rocker	$S + L < P + Q$	Coupler
Change point	$S + L = P + Q$	Any
Non-Grashof Type	**Link Length Relationship**	
Triple-rocker	$S + L > P + Q$	Any

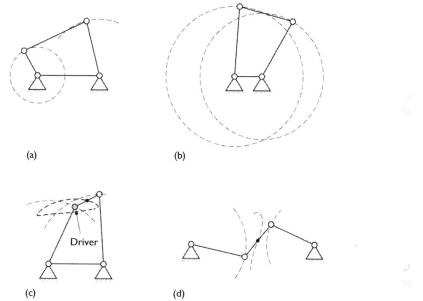

(a)   (b)

(c)   (d)

FIGURE 3.10 Link rotations of Grashof mechanisms: (a) crank-rocker, (b) drag-link (double-crank), (c) double-rocker, and (d) non-Grashof triple-rocker.

designer must overcome the technical challenges of affixing a drive system to a movable joint (and operating such a drive system).*

When force and torque transmission between the crank and follower links are of concern in four-bar mechanism design, knowing the *transmission angle* is critical. The transmission angle is the angle between the coupler and follower links. Figure 3.11a illustrates a planar four-bar mechanism and the transmission angle τ. When an input torque $\mathbf{T}_{in}$ is applied to the crank link, this link transmits force to the coupler which subsequently transmits force to the follower $\mathbf{F}_{follower}$. One component of this follower force ($\mathbf{F}_{follower} \sin (\tau)$ in Figure 3.11b) is normal to the follower link and results in the output torque $\mathbf{T}_{out}$. The other force component ($\mathbf{F}_{follower} \cos (\tau)$) is a columnar load acting along the length of the follower. Both follower force components are functions of the transmission angle. Attempts are often made to minimize the columnar component of the follower force in four-bar mechanism design, particularly

---

* Affixing a drive system (such as a motor or manual crank) to a grounded link joint (and operating such a system) can be more easily accomplished than to a nongrounded joint.

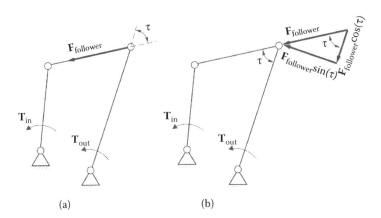

FIGURE 3.11
Planar four-bar mechanism with transmission angle and crank and follower loads.

in applications where the forces and torques transmitted are substantial.* A transmission angle of 90° is optimum because the resulting follower force has only a normal component ($\mathbf{F}_{follower}$ cos (τ) when $\tau = 90$°). A transmission angle range of $90° \pm 50°$ is generally preferred by designers [12]. It can be observed from the equations in Figure 3.11b that as the transmission angle decreases, the columnar load component increases and the normal load component decreases.

Follower-link forces are not governed by the transmission angle in dynamic force analyses (to be presented in Chapter 7). This is because body forces exist in the mechanism links due to gravity and link acceleration. Follower-link forces are also not governed by the transmission angle in static-force analyses (to be presented in Chapter 6) where link body forces (due to gravity) are comparable to link external loads. The ideal condition where transmission angles do govern follower-link forces is under static loading where the external link loads far exceed the link body forces. As the external static loads exceed the static body forces in a planar four-bar linkage, the transmission angle-based calculation of the follower forces becomes more accurate.

Example 3.3

Problem Statement: The link lengths for three planar four-bar mechanisms are given in Figure E.3.4 (all identical to the link dimensions given in Mechanism 1) and the driving links are labeled with rotation arrows. Determine the Grashof type for each mechanism.

Known Information: Figure E.3.4 and Table 3.1.

Solution Approach: Because the shortest link in Mechanism 1 is the ground link and the Grashof condition $S + L < P + Q$ is true, this mechanism is a *Grashof double-crank*. Because the shortest link in Mechanism 2 is the crank link and the Grashof condition $S + L < P + Q$ is true, this mechanism is a *Grashof crank-rocker*. Because the shortest link in Mechanism 3 is the coupler link and the Grashof condition $S + L < P + Q$ is true, this mechanism is a *Grashof double-rocker*.

---

* Unlike follower *normal* loads (which links having revolute joints are designed to accommodate), follower *columnar* loads ($\mathbf{F}_{follower}$ cos τ in Figure 3.11b) can result in follower buckling or excessive bearing forces in the follower revolute joints.

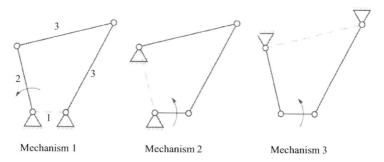

FIGURE E.3.4
Planar four-bar mechanism configurations (with dimensionless link lengths given for Mechanism 1).

3.5 Circuit Defect

While a Grashof mechanism is capable of producing full crank rotation, a non-Grashof mechanism is not. For the latter mechanism type, the crank can only rotate to the mechanism's *binding position*. When the binding position is reached, the mechanism "locks up" or is physically precluded from further movement (i.e., further movement in the given crank direction). Crank rotation beyond the binding position is only possible through mechanism disassembly.

Figure 3.12 illustrates a non-Grashof triple-rocker and one of its binding positions. The binding position is reached when link $\mathbf{a}_0$–$\mathbf{a}_1$ rotates by an angular displacement β. Because the mechanism locks up at β, further mechanism motion in the counterclockwise direction is only possible when the mechanism is disassembled and reassembled beyond the infeasible region shown in Figure 3.12.*

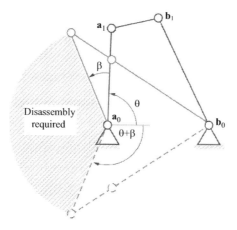

FIGURE 3.12
Planar four-bar mechanism (a non-Grashof triple-rocker) with circuit defect.

---

* For a given non-Grashof mechanism, the lock-up position can vary based on the designated crank link (e.g., link $\mathbf{b}_0$–$\mathbf{b}_1$ could be a crank link) as well as the crank rotation direction (clockwise instead of counterclockwise rotation).

If a mechanism is incapable of achieving a desired crank rotation (as, subsequently, a desired mechanism position) due to a lock-up condition, it has what is called a *circuit defect*.* Circuit defects are considered to be fatal to linkage operation (due to mechanism disassembly being required for full crank rotation).

3.6 Mechanism Inversion

The motion of a mechanism can vary, based on which joints are grounded. As illustrated in Figure E.3.4, the planar four-bar mechanisms given are all identical in terms of link length (and even identical in terms of initial link orientation), but vary only in terms of the joints that are grounded. The mechanisms in Figure E.3.4 are all *inversions* of each other because they are otherwise identical mechanisms that have different grounded joints.

Figure 3.13 illustrates all of the inversions of the planar four-bar, slider-crank, and 4R spherical mechanisms. Because the motion of a mechanism can vary based on which joints

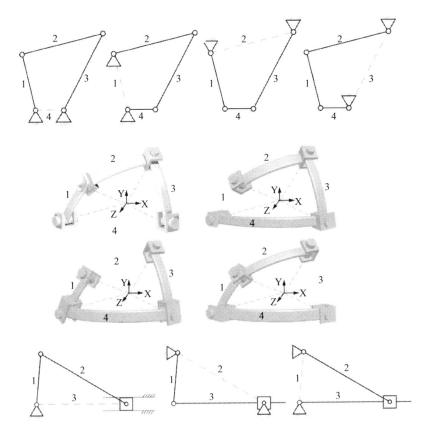

FIGURE 3.13
Planar four-bar, 4R spherical, and slider-crank mechanism inversions.

---

* A mechanism link loop (like loops in electrical and fluid flow systems) is also called a *circuit*.

are grounded, producing mechanism inversions is a systematic method to produce mechanisms that have unique motions, but include the same general linkage hardware.

3.7 Passive Degree of Freedom and Paradoxes

There are cases where the mobility values calculated from Equations 3.2 and 3.3 are misleading. For example, although Equation 3.3 produces a mobility value of 2 for the spatial RRSS and RSSR mechanisms, the RRSS mechanism is capable of controlled coupler motion and the RSSR is capable of controlled follower motion. The extra degree of freedom for these mechanisms is the free rotation of the S-S links about their length axes (Figure 3.14) and is called a *passive degree of freedom*. Because this degree of freedom is highly localized (limited to a single link) for these mechanisms, they have no effect on the kinematics of their links of interest (the coupler link for the RRSS and the follower link for the RSSR).

There are also cases where the mobility values calculated from Equations 3.2 and 3.3 are incorrect. For example, Equation 3.3 produces a mobility of –2 for the 4R spherical mechanism (which has a true mobility of 1) and Equation 3.2 produces a mobility of 0 for the pair of rolling cylinders in Figure 3.15 (which has a true mobility of 1). Mechanisms having true mobility values that violate Gruebler's equation are called *paradoxes* or *maverick mechanisms* [13,14]. Because Gruebler's equation cannot guarantee true mobility results for

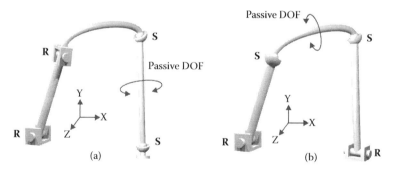

FIGURE 3.14
Passive DOF in spatial (a) RRSS and (b) RSSR mechanisms.

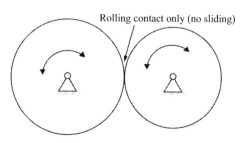

FIGURE 3.15
Rolling cylinder pair (with no sliding contact).

all mechanisms, it may be necessary for the user to validate the mobility of a mechanism (through kinematic analyses, for example).

3.8 Summary

The planar four-bar mechanism is one of the most widely utilized kinematic chains in everyday devices. The four interconnected links in this mechanism are the crank, coupler, follower, and ground links. The crank and follower links in the planar four-bar mechanism undergo pure rotation while the coupler link undergoes complex motion. The slider-crank mechanism is another commonly used kinematic chain, particularly in internal combustion engines. The four interconnected links in this mechanism are the crank, coupler, slider, and ground links. The crank link in the slider-crank mechanism undergoes pure rotation while the slider link undergoes pure translation. In the geared five-bar mechanism, the crank link and output link are coupled by a gear pair. Though not as commonly utilized in everyday devices as planar four-bar mechanisms, one advantage the planar five-bar mechanism has over the four-bar mechanism is that its intermediate links can trace paths of higher orders than the latter.

Attaching a dyad to the coupler of the planar four-bar mechanism produces the Stephenson III mechanism. Attaching the crank and follower of two planar four-bar mechanisms produces the Watt II mechanism. Both mechanisms are particular types of multiloop planar six-bar mechanisms. There are two classifications for planar multiloop six-bar mechanisms: Watt and Stephenson. Although these mechanisms are not as commonly utilized as planar four-bar mechanisms in everyday devices, they include at least two links that undergo complex motion—giving them greater capacities in kinematic synthesis than planar four-bar mechanisms.

Planar mechanisms are restricted to motion in two-dimensional or planar space. Spatial mechanisms can exhibit three-dimensional or spatial motion. Because spatial mechanisms have the capacity to exhibit spatial motion, they offer a greater variety of possible motions and are structurally more general than planar mechanisms. This textbook considers three types of four-bar spatial mechanisms: the revolute-revolute-spherical-spherical or RRSS, the revolute-revolute-revolute-revolute spherical or 4R spherical, and the revolute-spherical-spherical-revolute or RSSR mechanisms. The RRSS, 4R spherical, and RSSR are among the more basic four-bar spatial mechanisms in terms of the types of joints used and the required linkage assembly conditions for motion.

Mechanism links are interconnected by joints. Both the planar four-bar, geared five-bar, Watt, and Stephenson mechanisms include revolute joints: joints having a single rotational degree of freedom. The prismatic joint has a single translational degree of freedom and is used where 1-DOF sliding contact is needed. The cylindrical joint and spherical joint are other joint types utilized in spatial mechanisms. The cylindrical joint has a rotational and a translational degree of freedom, while the spherical joint has three rotational degrees of freedom. Joints having two degrees of freedom include the cam joint and gear joint, which both include rolling-sliding contact.

Gruebler's equation is used to calculate mechanism mobility or degrees of freedom (DOF). Determining the mobility of a mechanism enables the designer to determine if additional constraints are needed to reach the desired mobility, and if so, how many constraints are needed. In number synthesis, Gruebler's equations are implemented systematically to determine alternate mechanism solutions for a given mobility. Grashof's criteria are used to determine the link rotation behavior of four-bar mechanisms. Determining link rotation behavior is important in design, particularly when coupling a drive system to the crank link.

While a Grashof mechanism is capable of producing full crank rotation, a non-Grashof mechanism cannot. For the latter mechanism type, the crank can only rotate to the mechanism's binding position. Crank rotation beyond the binding position is only possible through mechanism disassembly. This defect associated with non-Grashof mechanisms is called a circuit defect. Circuit defects are considered to be fatal to linkage operation (due to mechanism disassembly being required for full crank rotation).

Knowing the transmission angle (the angle between the coupler and follower links) behavior is important when force and torque transmission between the crank and follower links are of concern in four-bar mechanism design. As the transmission angle decreases, the magnitude of the columnar load on the follower link increases and the magnitude of the load normal to the follower decreases (and vice versa). A transmission angle range of $90° \pm 50°$ is generally preferred by designers.

The motion of a mechanism can vary, based on which joints are grounded. Mechanisms of identical type and link dimensions, but different grounded links, are inversions of each other. Because the motion of a mechanism can vary based on which joints are grounded, producing mechanism inversions is a systematic method to produce mechanisms that have unique motions, but include the same general linkage hardware.

There are cases where the mobility values calculated by Gruebler's equation are misleading. The follower link of the spatial RRSS mechanism and the coupler link of the spatial RSSR mechanism have a passive degree of freedom: the free rotation of these links about their length axes. For example, although Gruebler's equation produces a mobility value of 2 for both mechanisms, the RRSS mechanism is capable of controlled coupler motion and the RSSR is capable of controlled follower motion.

There are also cases where the mobility values calculated by Gruebler's equation are incorrect. For example, Gruebler's equation produces a mobility value of –2 for the 4R spherical mechanism (which has a true mobility of 1). Mechanisms having true mobility values that violate Gruebler's equation are called paradoxes or maverick mechanisms. Because Gruebler's equation cannot guarantee true mobility results for all mechanisms, it may be necessary for the user to validate the mobility of a mechanism.

References

1. Sandor, G. N. and Erdman, A. G. 1984. *Advanced Mechanism Design: Analysis and Synthesis.* Volume 2, p. 4. Englewood Cliffs, NJ: Prentice-Hall.
2. Wunderlich, W. 1963. Höhere Koppelkurven. *Österreichisches Ingenieur Archiv*, 17: 162–165.
3. Sandor, G. N. and Erdman, A. G. 1984. *Advanced Mechanism Design: Analysis and Synthesis.* Volume 2, pp. 10–11. Englewood Cliffs, NJ: Prentice-Hall.
4. Capellen, W. M. 1966. Kinematics: A survey in retrospect and prospect. *Mechanism and Machine Theory*. 1: 211–228.
5. Russell, K., Shen, Q., and Sodhi, R. S. 2013. *Mechanism Design: Visual and Programmable Approaches.* Chapter 7. Boca Raton: CRC Press.
6. Suh, C. H. and Radcliffe, C. W. 1978. *Kinematics and Mechanisms Design.* Chapter 4. New York: John Wiley.

7. Norton, R. L. 2008. *Design of Machinery*. 4th edn, pp. 32–36. New York: McGraw-Hill.
8. Sandor, G. N. and Erdman, A. G. 1984. *Advanced Mechanism Design: Analysis and Synthesis*. Volume 2, p. 64. Englewood Cliffs, NJ: Prentice-Hall. 9. Ibid, pp. 64–75.
10. Norton, R. L. 2008. *Design of Machinery*. 4th edn, pp. 42–46. New York: McGraw-Hill.
11. Ibid, p. 58.
12. Wilson, C. E. and Sadler, J. P. 2003. *Kinematics and Dynamics of Machinery*. 3rd edn, p. 33. Saddle River, NJ: Prentice-Hall.
13. Norton, R. L. 2008. *Design of Machinery*. 4th edn, pp. 46–47. New York: McGraw-Hill.
14. Suh, C. H. and Radcliffe, C. W. 1978. *Kinematics and Mechanisms Design*. pp. 109–110. New York: John Wiley.

Problems

1. Planar four-bar linkages have many everyday applications (some are illustrated Figure 3.1). Identify and describe four additional everyday applications for the planar four-bar linkage.

2. a. Why is it important to know if a mechanism has a single degree of freedom?

   b. Why is a crank-rocker mechanism more useful than a double-rocker mechanism?

   c. Should the transmission angle for the planar four-bar linkage be close to 0°? Explain?.

3. For the two linkages illustrated in Figure P.3.1, which (if any) of the links can undergo a complete rotation relative to the other links? How do you know?

4. Repeat Problem 3 where one linkage has link lengths of 2.35 (ground), 3.25 (coupler), 2.25 and 2.15 cm and another linkage has link lengths of 2.75 (ground), 3.25 (coupler), 1.15 and 1.75 cm.

5. Determine the number of links and the mobility of each of the three planar mechanisms in Figure P.3.2.

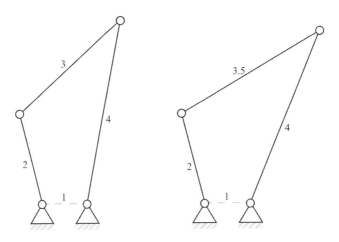

FIGURE P.3.1
Planar four-bar linkages with dimensionless link lengths.

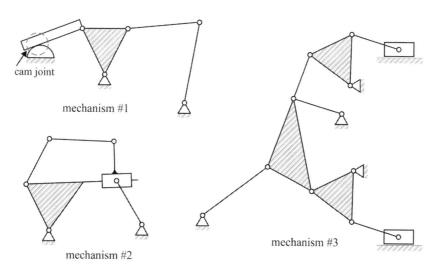

FIGURE P.3.2
Planar mechanisms.

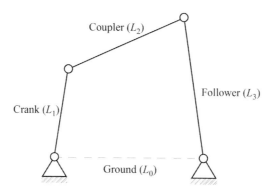

FIGURE P.3.3
Planar four-bar linkage.

6. For the planar four-bar linkage illustrated in Figure P.3.3, $L_2/L_1 = 1.5$ and $L_3/L_1 = 1.2$. Find the range of L_0/L_1 required for a *drag-link mechanism*.

7. Repeat Problem 6 for a *crank-rocker mechanism* where $L_2/L_1 = 2.75$ and $L_3/L_1 = 2.95$.

8. Compile a table of 1 DOF spatial mechanisms having two, three and four links (let $J_4 = J_5 = 0$ in Equation 3.3). Illustrate some of these mechanism solutions.

9. Compile a table of 1 DOF spatial mechanisms having five links (let $J_4 = J_5 = 0$ in Equation (3.3)). Illustrate some of these mechanism solutions.

10. Figure P.3.4 illustrates nine spatial mechanisms that include revolute (R), prismatic (P), cylindrical (C), and spherical (S) joints. Calculate the mobility of these mechanisms.

11. Figure P.3.5 illustrates five spatial robots that include revolute (R), prismatic (P), and cylindrical (C) joints. Calculate the mobility of these robots.

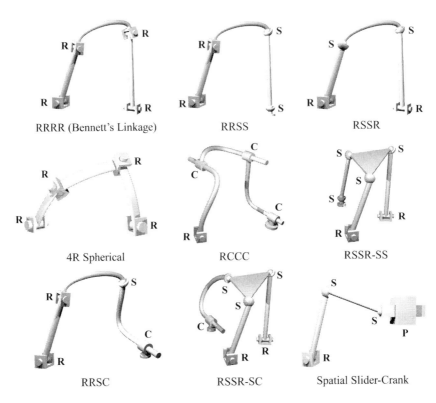

FIGURE P.3.4
Spatial mechanisms comprised of R, P, C, and S joints.

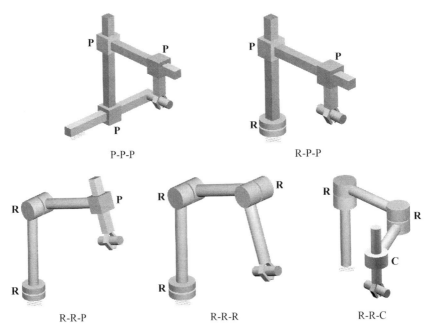

FIGURE P.3.5
Spatial robots comprised of R, P, and C joints.

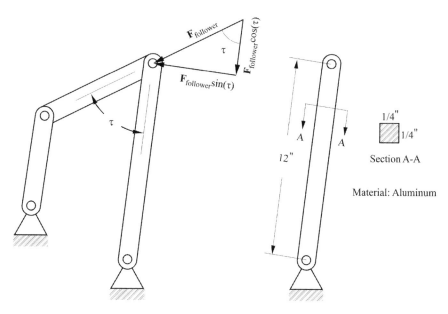

FIGURE P.3.6
Planar four-bar mechanism with transmission angle and crank and follower loads.

12. Euler's buckling load (F) for a columnar member with pinned ends is

$$F = \frac{\pi^2 EI}{L^2}$$

where E, I, and L are the modulus of elasticity, moment of inertia, and length, respectively, of the columnar member. Formulate an equation from Euler's equation and Figure P.3.6 to calculate the transmission angle corresponding to the follower-link buckling load. Calculate the transmission angle for a follower length of 12 in (0.3048 m), a ¼ in (0.635 cm) square follower cross section, a modulus of elasticity (for aluminum) of 10,000,000 psi (68.05 GPa), and a follower load ($\mathbf{F}_{follower}$) of 250 lbf (1112.05 N).

13. *Maverick Mechanisms* are mechanisms that defy Grubler's Equation (Grubler's Equation will produce misleading results for maverick mechanisms). *Passive Degrees of Freedom* are localized DOFs that have no effect on the overall mechanism kinematics. Determine which mechanisms in Figure P.3.4 are maverick mechanisms and locate the passive DOFs from among the mechanisms in Figure P.3.4.

14. Compile a table of 2 DOF planar mechanisms having four and five links (considering cam or gear joints as J_2). Illustrate one four-bar mechanism solution and one five-bar mechanism solution.

15. Compile a table of 2 DOF planar mechanisms having six and seven links (considering cam or gear joints as J_2). Illustrate one six-bar mechanism solution and one seven-bar mechanism solution.

4

Kinematic Analysis of Planar Mechanisms

CONCEPT OVERVIEW

In this chapter, the reader will gain a central understanding regarding

1. Link velocity and acceleration components in planar space
2. The *Newton–Raphson method* for a set of two simultaneous equations
3. Vector-loop-based displacement, velocity, and acceleration equation formulation and solution
4. Kinematics of mechanism link locations of interest
5. Instant centers in relative planar motion
6. Instant center generation and application in velocity analysis
7. Centrode generation and application in coupler motion replication
8. Configurations of closed-loop mechanisms
9. Relationships between general angular velocity and time
10. Cognate construction and application

4.1 Introduction

In a kinematic analysis, positions, displacements, velocities, and accelerations of mechanism links are determined either *qualitatively* or *quantitatively*. In a quantitative kinematic analysis, equations that fully describe the motion of the mechanism links are used. Qualitative methods include constructing and measuring mechanism schematics and polygons to determine the positions, velocities, and accelerations of mechanism links. As intended by the authors, the kinematic analysis methods presented in this textbook are all quantitative. Kinematic equations for the planar four-bar, slider-crank, geared five-bar, Watt II, and Stephenson III mechanisms are formulated in this chapter. Displacement equations are formulated by taking the sum of the closed vector loop(s) in each mechanism (as introduced in Section 2.2.2) [1]. Taking the first and second derivatives of the vector-loop displacement equations introduces mechanism link velocity and acceleration variables, respectively, and, ultimately, produces mechanism velocity and acceleration equations, respectively.

As noted in Section 3.1.4, the Watt II and Stephenson III mechanisms (Figure 3.6) are the planar multiloop mechanisms of choice for analysis in this textbook. Both mechanisms have two links that exhibit complex motion. As illustrated in Figure 3.6, Watt II and Stephenson III mechanisms are comprised of a planar four-bar mechanism and an additional dyad.*

Because the Watt II and Stephenson III mechanisms include the planar four-bar mechanism, the displacement, velocity, and acceleration equations for these planar multiloop mechanisms will include some of the variables and output from the planar four-bar mechanism equations. Displacement equations for the additional dyads in the Watt II and Stephenson III mechanisms are produced by formulating vector-loop equations that include these dyads. Taking the first and second derivatives of the resulting vector-loop displacement equations produces velocity and acceleration equations.

4.2 Numerical Solution Method for Two Simultaneous Equations

The displacement equations presented in this chapter form sets of two nonlinear simultaneous equations (where each set includes two unknown variables). Unlike linear simultaneous equations, which can be solved algebraically (see Section 2.4), nonlinear simultaneous equations cannot be solved in this way. Using a *root-finding method* (a method for calculating the unknown variables in a set of nonlinear equations), such equation sets can be solved numerically [2]. The *Newton–Raphson method* is one of the most common root-finding methods. A Newton–Raphson method flowchart for a set of two simultaneous equations is illustrated in Figure 4.1.[†]

Given a set of two simultaneous equations (f_1 and f_2) having two unknowns (V_1 and V_2), where $f_1(V_1, V_2) = 0$ and $f_2(V_1, V_2) = 0$, Figure 4.1 begins with initial values for the unknown variables being specified. The unknown variable residuals δV_1 and δV_2 are then calculated (by computing the negative product of the inverted Jacobian and the column matrix of $f_1(V_1, V_2)$ and $f_2(V_1, V_2)$). Updated values for the unknown variables are calculated by adding the variable residuals to the unknown variables, and f_1 and f_2 are calculated using the updated variables. As shown in Figure 4.1, the variables are updated repeatedly until the values calculated from the two equations (with the latest variable values) are smaller than a specified error term ε. The Newton–Raphson method can be codified (or currently exists) on a wide range of software platforms—including MATLAB—as a basic solver for simultaneous equations.

4.3 Link Velocity and Acceleration Components in Planar Space

Figure 4.2a illustrates an arbitrary grounded rotating link with rotation angle β. Figures 4.2b and c include the velocity and acceleration components at point $\mathbf{p}_1$ of this link, respectively. Given a link angular velocity (which we label $\dot{\beta}$), the velocity vector (which we label $\mathbf{V}_{p1}$) is produced. This velocity is tangent to the length $\mathbf{p}_0$–$\mathbf{p}_1$ and acts in the direction of $\dot{\beta}$.

---

* The additional dyad is connected to the follower link in the Watt II mechanism while the additional dyad is connected to the coupler link in the Stephenson III mechanism.

† While the *numerical* Newton–Raphson method can be used to calculate solutions for *linear* and *nonlinear* simultaneous equations, linear simultaneous equations can be also solved *analytically* (see Section 2.4).

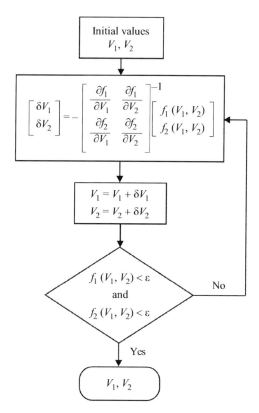

FIGURE 4.1
Newton–Raphson method flowchart (for two equations with two unknowns).

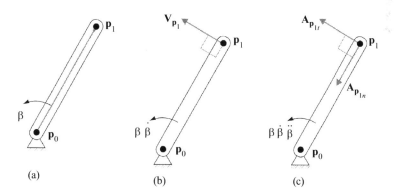

FIGURE 4.2
(a) Rotating link and its (a) velocity and (b) acceleration components.

Given a link angular velocity and an angular acceleration (which we label $\dot{\beta}$ and $\ddot{\beta}$, respectively), the acceleration vectors (which we label $\mathbf{A}_{p_{1t}}$ and $\mathbf{A}_{p_{1n}}$) are produced. The acceleration component $\mathbf{A}_{p_{1t}}$ is tangent to the length $\mathbf{p}_0$–$\mathbf{p}_1$ and acts in the direction of $\ddot{\beta}$. The acceleration component $\mathbf{A}_{p_{1n}}$ is along the length $\mathbf{p}_0$–$\mathbf{p}_1$ and acts in the direction toward the center of rotation (i.e., toward $\mathbf{p}_0$). The total acceleration at $\mathbf{p}_1$ is the sum of acceleration components $\mathbf{A}_{p_{1t}}$ and $\mathbf{A}_{p_{1n}}$.

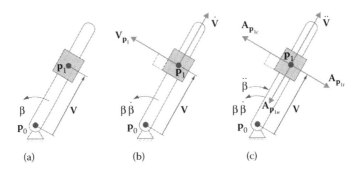

FIGURE 4.3
(a) Rotating-sliding link and its (a) velocity and (b) acceleration components.

Figure 4.3a illustrates an arbitrarily grounded rotating-sliding link with rotation angle β and sliding length $\mathbf{V}$.* Figures 4.3b and c include the velocity and acceleration components at point $\mathbf{p}_1$ of this link, respectively. Given a link angular velocity, the velocity vectors (which we label $\mathbf{V}_{\mathbf{p}_1}$ and $\dot{\mathbf{V}}$) are produced. The velocity component $\mathbf{V}_{\mathbf{p}_1}$ is tangent to the length $\mathbf{p}_0 - \mathbf{p}_1$ and acts in the direction of β. The velocity component $\dot{\mathbf{V}}$ (the sliding velocity) is along the length $\mathbf{p}_0 - \mathbf{p}_1$ and acts in the direction away from the center of rotation (i.e., away from $\mathbf{p}_0$). The total velocity at $\mathbf{p}_1$ is the sum of velocity components $\mathbf{V}_{\mathbf{p}_1}$ and $\dot{\mathbf{V}}$.

Given a link angular velocity and an angular acceleration, the acceleration vectors (which we Label $\mathbf{A}_{\mathbf{p}_{1t}}$, $\mathbf{A}_{\mathbf{p}_{1n}}$, $\mathbf{A}_{\mathbf{p}_{1c}}$, and $\ddot{\mathbf{V}}$) are produced. The acceleration component $\mathbf{A}_{\mathbf{p}_{1t}}$ is tangent to the length $\mathbf{p}_0 - \mathbf{p}_1$ and acts in the direction of β. The acceleration component $\mathbf{A}_{\mathbf{p}_{1t}}$ is along the length acts in the $\mathbf{p}_0 - \mathbf{p}_1$ and acts in the direction toward the center of rotation (i.e., toward $\mathbf{p}_0$).[†] The acceleration component $\mathbf{A}_{\mathbf{p}_{1n}}$ is also tangent to the length $\mathbf{p}_0 - \mathbf{p}_1$ but acts in the direction of β.[‡] The acceleration component $\ddot{\mathbf{V}}$ (the sliding acceleration) is along the length $\mathbf{p}_0 - \mathbf{p}_1$ and acts in the direction away from the center of rotation (i.e., away from $\mathbf{p}_0$). The total acceleration at $\mathbf{p}_1$ is the sum of acceleration components $\mathbf{A}_{\mathbf{p}_{1t}}$, $\mathbf{A}_{\mathbf{p}_{1n}}$, $\mathbf{A}_{\mathbf{p}_{1c}}$, and $\ddot{\mathbf{V}}$.

4.4 Four-Bar Mechanism Analysis

4.4.1 Displacement Equations

The planar four-bar mechanism consists of four links interconnected by revolute joints. As calculated from Gruebler's equation for planar mechanisms (with $L = 4$ and $J_1 = 4$), the planar four-bar mechanism has a single DOF. A single displacement equation is derived for this mechanism by taking the sum of the displaced mechanism vector loop in Figure 4.4. The equation produced from a clockwise sum of the displaced vector loop is

$$W_1 e^{i(\theta + \beta_j)} + V_1 e^{i(\rho + \alpha_j)} - U_1 e^{i(\sigma + \gamma_j)} - \mathbf{G}_1 = 0 \qquad (4.1)$$

---

* This type of link is used in slider-crank inversions (see Section 3.6 and Example 4.7).
† This acceleration is also known as *centripetal acceleration*.
‡ This acceleration is also known as *Coriolis acceleration*.

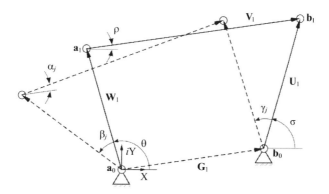

FIGURE 4.4
Planar four-bar mechanism displacement variables.

After expanding Equation 4.1 and grouping its real and imaginary terms as separate equations, the resulting planar four-bar mechanism displacement equations become

$$f_1(\alpha_j, \gamma_j) = W_1 \cos(\theta + \beta_j) + V_1 \cos(\rho + \alpha_j) - U_1 \cos(\sigma + \gamma_j) - G_{1x} = 0$$

$$f_2(\alpha_j, \gamma_j) = W_1 \sin(\theta + \beta_j) + V_1 \sin(\rho + \alpha_j) - U_1 \sin(\sigma + \gamma_j) - G_{1y} = 0$$

(4.2)

With the exception of the coupler displacement angle α_j and follower displacement angle γ_j, all other variables in the planar four-bar displacement equations are user prescribed. Because $f_1(\alpha_j, \gamma_j)$ and $f_2(\alpha_j, \gamma_j)$ in Equation 4.2 both include the unknown displacement angles α_j and γ_j, these equations form a set of nonlinear simultaneous equations from which the angles are calculated.

4.4.2 Velocity Equations

A single planar four-bar velocity equation is derived by differentiating the planar four-bar displacement equation. Differentiating Equation 4.1 with respect to time produces

$$i\dot{\beta}_j W_1 e^{i(\theta + \beta_j)} + i\dot{\alpha}_j V_1 e^{i(\rho + \alpha_j)} - i\dot{\gamma}_j U_1 e^{i(\sigma + \gamma_j)} = 0^*$$

(4.3)

Equation 4.4 includes the individual velocity variables from Equation 4.3. The velocity $\mathbf{V}_{a_1}$ is the *global* tangential velocity of the crank link vector—the tangential velocity of $\mathbf{a}_1$ with respect to $\mathbf{a}_0$ (see Figure 4.5). The velocity $\mathbf{V}_{b_1}$ is the global tangential velocity of the follower-link vector—the tangential velocity of $\mathbf{b}_1$ with respect to $\mathbf{b}_0$. Also, the velocity $\mathbf{V}_{b_1 - a_1}$ is the *relative* tangential velocity of the coupler link vector—the relative tangential velocity of $\mathbf{b}_1$ with respect to $\mathbf{a}_1$.

---

* Because the complex coefficient is fully distributed in Equation 4.3 (as well as in the velocity equations for the forthcoming planar mechanisms), it can be cancelled from the equation if preferred.

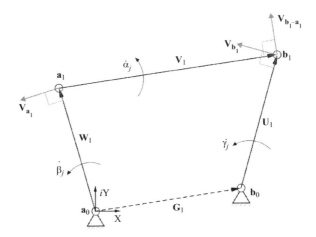

FIGURE 4.5
Planar four-bar mechanism velocity variables.

$$\mathbf{V}_{a_1} = i\dot{\beta}_j W_1 e^{i(\theta+\beta_j)}$$

$$\mathbf{V}_{b_1} = i\dot{\gamma}_j U_1 e^{i(\sigma+\gamma_j)} \tag{4.4}$$

$$\mathbf{V}_{b_1-a_1} = i\dot{\alpha}_j V_1 e^{i(\rho+\alpha_j)} = \mathbf{V}_{b_1} - \mathbf{V}_{a_1}$$

After moving the term $i\dot{\beta}_j W_1 e^{i(\theta+\beta_j)}$ to the right-hand side of Equation 4.3, expanding this equation, and grouping its real and imaginary terms as separate equations, the resulting planar four-bar mechanism velocity equation in matrix form is

$$\begin{bmatrix} -V_1\sin(\rho+\alpha_j) & U_1\sin(\sigma+\gamma_j) \\ V_1\cos(\rho+\alpha_j) & -U_1\cos(\sigma+\gamma_j) \end{bmatrix} \begin{Bmatrix} \dot{\alpha}_j \\ \dot{\gamma}_j \end{Bmatrix} = \dot{\beta}_j \begin{bmatrix} W_1\sin(\theta+\beta_j) \\ -W_1\cos(\theta+\beta_j) \end{bmatrix} \tag{4.5}$$

After including the prescribed mechanism variables and the unknown variables calculated from the planar four-bar displacement equations, Equation 4.5 can be solved (using Cramer's rule, for example) to calculate the coupler angular velocity $\dot{\alpha}_j$ and the follower angular velocity $\dot{\gamma}_j$.

4.4.3 Acceleration Equations

A single planar four-bar acceleration equation is derived by differentiating the planar four-bar velocity equation. Time differentiation of Equation 4.3 produces

$$-\dot{\beta}_j^2 W_1 e^{i(\theta+\beta_j)} + i\ddot{\beta}_j W_1 e^{i(\theta+\beta_j)} - \dot{\alpha}_j^2 V_1 e^{i(\rho+\alpha_j)} + i\ddot{\alpha}_j V_1 e^{i(\rho+\alpha_j)} - \left(-\dot{\gamma}_j^2 U_1 e^{i(\sigma+\gamma_j)} + i\ddot{\gamma}_j U_1 e^{i(\sigma+\gamma_j)}\right) = 0 \tag{4.6}$$

Equation 4.7 includes the individual acceleration variables from Equation 4.6. The accelerations $\mathbf{A}_{a_{1n}}$ and $\mathbf{A}_{a_{1t}}$ are the global *normal* and *tangential* accelerations, respectively, of the crank link vector—the normal and tangential accelerations of $\mathbf{a}_1$ with respect to $\mathbf{a}_0$ (see Figure 4.6).[*] The *total* acceleration $\mathbf{A}_{\mathbf{a}_1}$ is the sum of these normal and tangential accelerations. The accelerations $\mathbf{A}_{\mathbf{b}_{1n}}$ and $\mathbf{A}_{\mathbf{b}_{1t}}$ are the global normal and tangential accelerations, respectively, of the follower-link vector—the normal and tangential accelerations of $\mathbf{b}_1$ with respect to $\mathbf{b}_0$. The total acceleration $\mathbf{A}_{\mathbf{b}_1}$ is the sum of these normal and tangential accelerations. Also, the accelerations $\mathbf{A}_{(\mathbf{b}_1-\mathbf{a}_1)_n}$ and $\mathbf{A}_{(\mathbf{b}_1-\mathbf{a}_1)_t}$ are the relative normal and tangential accelerations, respectively, of $\mathbf{b}_1$ with respect to $\mathbf{a}_1$—the relative normal and tangential accelerations of the coupler link vector. The total relative acceleration $\mathbf{A}_{\mathbf{b}_1-\mathbf{a}_1}$ is the sum of these normal and tangential accelerations.

$$\mathbf{A}_{\mathbf{a}_1} = \mathbf{A}_{\mathbf{a}_{1t}} + \mathbf{A}_{\mathbf{a}_{1n}} = i\ddot{\beta}_j W_1 e^{i(\theta+\beta_j)} - \dot{\beta}_j^2 W_1 e^{i(\theta+\beta_j)}$$

$$\mathbf{A}_{\mathbf{b}_1} = \mathbf{A}_{\mathbf{b}_{1t}} + \mathbf{A}_{\mathbf{b}_{1n}} = i\ddot{\gamma}_j U_1 e^{i(\sigma+\gamma_j)} - \dot{\gamma}_j^2 U_1 e^{i(\sigma+\gamma_j)} \qquad (4.7)$$

$$\mathbf{A}_{\mathbf{b}_1-\mathbf{a}_1} = \mathbf{A}_{(\mathbf{b}_1-\mathbf{a}_1)_t} + \mathbf{A}_{(\mathbf{b}_1-\mathbf{a}_1)_n} = i\ddot{\alpha}_j V_1 e^{i(\rho+\alpha_j)} - \dot{\alpha}_j^2 V_1 e^{i(\rho+\alpha_j)} = \mathbf{A}_{\mathbf{b}_1} - \mathbf{A}_{\mathbf{a}_1}$$

After moving the terms $\dot{\beta}_j^2 W_1 e^{i(\theta+\beta_j)}$, $i\ddot{\beta}_j W_1 e^{i(\theta+\beta_j)}$, $\dot{\gamma}_j^2 U_1 e^{i(\sigma+\gamma_j)}$, and $\dot{\alpha}_j^2 V_1 e^{i(\rho+\alpha_j)}$ to the right-hand side of Equation 4.6, expanding this equation, and grouping its real and imaginary terms as separate equations, the resulting planar four-bar mechanism acceleration equation in matrix form is

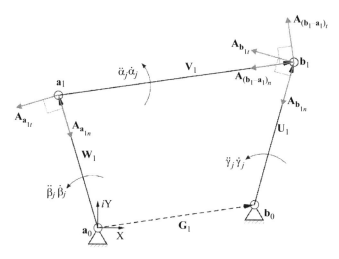

FIGURE 4.6
Planar four-bar mechanism acceleration variables.

---

[*] *Tangential acceleration* (which points along or in the opposite direction to the velocity vector) is the result of a change in the velocity vector magnitude, while *normal acceleration* or *centripetal acceleration* (which points toward the center of path curvature) is the result of a change in the velocity vector direction.

$$\begin{bmatrix} -V_1\sin(\rho+\alpha_j) & U_1\sin(\sigma+\gamma_j) \\ V_1\cos(\rho+\alpha) & -U_1\cos(\sigma+\gamma) \end{bmatrix}\begin{Bmatrix} \ddot{\alpha}_j \\ \ddot{\gamma}_j \end{Bmatrix} = \dot{\beta}_j^2\begin{bmatrix} W_1\cos(\theta+\beta_j) \\ W_1\sin(\theta+\beta_j) \end{bmatrix}$$

$$-\ddot{\beta}_j\begin{bmatrix} -W_1\sin(\theta+\beta_j) \\ W_1\cos(\theta+\beta_j) \end{bmatrix} - \dot{\gamma}_j^2\begin{bmatrix} U_1\cos(\sigma+\gamma_j) \\ U_1\sin(\sigma+\gamma_j) \end{bmatrix} + \dot{\alpha}_j^2\begin{bmatrix} V_1\cos(\rho+\alpha_j) \\ V_1\sin(\rho+\alpha_j) \end{bmatrix}$$

$$(4.8)$$

After including the prescribed mechanism variables and the unknowns calculated from the planar four-bar displacement and velocity equations, Equation 4.8 can be solved (using Cramer's rule, for example) to calculate the coupler angular acceleration $\ddot{\alpha}_j$ and the follower angular acceleration $\ddot{\gamma}_j$.

Appendix B.1 includes the MATLAB file user instructions for planar four-bar displacement, velocity, and acceleration analysis. In this MATLAB file (which is available for download at www. crcpress.com/product/isbn/9781498724937), solutions for Equations 4.2, 4.5, and 4.8 are calculated.

4.4.4 Kinematics of Coupler Locations of Interest

The solutions calculated from the planar four-bar displacement, velocity, and acceleration equations can be used in additional equations to calculate the displacement, velocity, and acceleration of any mechanism link location of interest. Like the planar four-bar kinematic equations, the additional equations are the result of formulating vector-loop equations and their derivatives.

For example, Figure 4.7 illustrates a planar four-bar mechanism with a coupler vector $\mathbf{L}_1$ that points to an arbitrary coupler location of interest $\mathbf{p}_1$. Equations 4.9 and 4.10 are produced from the loop $\mathbf{a}_0$–$\mathbf{a}_1$–$\mathbf{p}_1$ in Figure 4.7 (where Equation 4.10 is the result of expanding and separating Equation 4.9).

$$\mathbf{p}_{1j} = W_1 e^{i(\theta+\beta_j)} + L_1 e^{i(\delta+\alpha_j)} \tag{4.9}$$

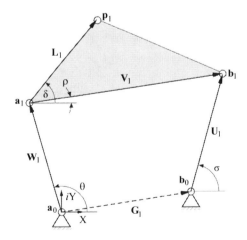

FIGURE 4.7
Planar four-bar mechanism with coupler location $\mathbf{p}_1$.

$$p_{1xj} = W_1 \cos(\theta + \beta_j) + L_1 \cos(\delta + \alpha_j)$$

$$p_{1yj} = W_1 \sin(\theta + \beta_j) + L_1 \sin(\delta + \alpha_j)$$ (4.10)

The velocities of the coupler location of interest can be calculated from Equation 4.12—the expanded form of Equation 4.11 (which is the first derivative of Equation 4.9).

$$\mathbf{V}_{p1j} = i\dot{\beta}_j W_1 e^{i(\theta + \beta_j)} + i\dot{\alpha}_j L_1 e^{i(\delta + \alpha_j)}$$ (4.11)

$$V_{p1xj} = -\dot{\beta}_j W_1 \sin(\theta + \beta_j) - \dot{\alpha}_j L_1 \sin(\delta + \alpha_j)$$

$$V_{p1yj} = \dot{\beta}_j W_1 \cos(\theta + \beta_j) + \dot{\alpha}_j L_1 \cos(\delta + \alpha_j)$$ (4.12)

The accelerations of the coupler location of interest can be calculated from Equation 4.14—the expanded form of Equation 4.13 (which is the second derivative of Equation 4.9).

$$\mathbf{A}_{p1j} = -\dot{\beta}_j^2 W_1 e^{i(\theta + \beta_j)} + i\ddot{\beta}_j W_1 e^{i(\theta + \beta_j)} - \dot{\alpha}_j^2 L_1 e^{i(\delta + \alpha_j)} + i\ddot{\alpha}_j L_1 e^{i(\delta + \alpha_j)}$$ (4.13)

$$A_{p1xj} = -\dot{\beta}_j^2 W_1 \cos(\theta + \beta_j) - \ddot{\beta}_j W_1 \sin(\theta + \beta_j) - \dot{\alpha}_j^2 L_1 \cos(\delta + \alpha_j) - \ddot{\alpha}_j L_1 \sin(\delta + \alpha_j)$$

$$A_{p1yj} = -\dot{\beta}_j^2 W_1 \sin(\theta + \beta_j) + \ddot{\beta}_j W_1 \cos(\theta + \beta_j) - \dot{\alpha}_j^2 L_1 \sin(\delta + \alpha_j) + \ddot{\alpha}_j L_1 \cos(\delta + \alpha_j)$$ (4.14)

The Appendix B.1 MATLAB file also includes Equations 4.10, 4.12, and 4.14, from which displacement, velocity, and acceleration values are calculated for a coupler-link location of interest.

Example 4.1

Problem Statement: Using the Appendix B.1 MATLAB file, calculate the displaced value of coupler point $\mathbf{p}_1$ for the planar four-bar mechanism configuration in Table E.4.1 for a 25° crank displacement angle.

Known Information: Table E.4.1 and Appendix B.1 MATLAB file.

Solution Approach: Figure E.4.1 includes the input specified (in bold text) in the Appendix B.1 MATLAB file. From the calculated output, it can be determined that the displaced coupler point $\mathbf{p}_1$ is $\mathbf{p}_j = 0.2127 + i1.8912$ at a 25° crank displacement angle.

Example 4.2

Problem Statement: Using the Appendix B.1 MATLAB file, plot the path traced by point $\mathbf{p}_1$ on the level-luffing crane (Figure E.4.2).* The mechanism assembly configuration is given in Table E.4.2. The crank displacement angle range is $\beta_j = 10°, 9°, ..., -32°$.

Known Information: Table E.4.2 and Appendix B.1 MATLAB file.

TABLE E.4.1

Planar Four-Bar Mechanism Configuration

W_1, θ	V_1, ρ	U_1, σ	G_{1x}, G_{1y}	L_1, δ
1.75, 90°	1.75, −22.4860°	1.75, 64.5895°	0.8660, −0.5	1, 6.4690°

---

* A level-luffing crane is a crane designed to trace a horizontal path for leveling applications.

```
W1 = 1.75*exp(i*90*pi/180);
V1 = 1.75*exp(-i*22.4860*pi/180);
G1 = 0.866 - i*0.5;
U1 = 1.75*exp(i*64.5895*pi/180);
L1 = exp(i*6.469*pi/180);

start_ang   = 0;
step_ang    = 1;
stop_ang    = 25;

angular_vel = 0 * ones(N+1,1);
angular_acc = 0 * ones(N+1,1);
```

FIGURE E.4.1
Specified input (in bold text) in the Appendix B.1 MATLAB file for Example 4.1.

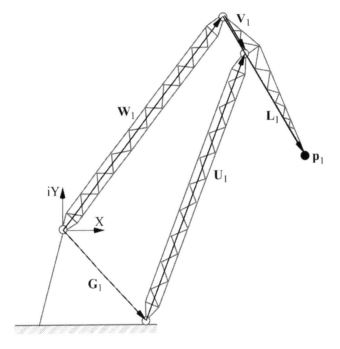

FIGURE E.4.2
Level-luffing crane mechanism.

TABLE E.4.2

Level-Luffing Crane Assembly Configuration

W_1, θ	V_1, ρ	U_1, σ	G_{1x}, G_{1y}	L_1, δ
1.96, 47.4041°	0.33, −51.6935°	2.09, 63.7721°	0.6075, −0.6909	1.19, −51.6935°

Solution Approach: Figure E.4.3 includes the input specified (in bold text) in the Appendix B.1 MATLAB file. In this particular mechanism, configuration the angles for V_1 and L_1 are identical because both vectors are parallel.* Figure E.4.4 illustrates the path traced by the coupler point p_1 using the given crank displacement angle range (with a crank rotation increment of −1°). Figure E.4.5 illustrates the level-luffing crane

---

* The angles for V_1 and L_1 (angles ρ and δ respectively) are identical in this example problem. Identical values for these angles however are not required for four-bar linkages in general.

```
W1 = 1.96*exp(i*47.4041*pi/180);
V1 = 0.33*exp(-i*51.6935*pi/180);
G1 = 0.6075 - i*0.6909;
U1 = 2.09*exp(i*63.7721*pi/180);
L1 = 1.19*exp(-i*51.6935*pi/180);

start_ang    = 10;
step_ang     = -1;
stop_ang     = -32;

angular_vel  = 0 * ones(N+1,1);
angular_acc  = 0 * ones(N+1,1);
```

FIGURE E.4.3
Specified input (in bold text) in the Appendix B.1 MATLAB file for Example 4.2.

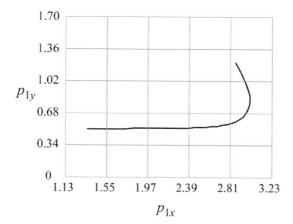

FIGURE E.4.4
Path traced by point $\mathbf{p}_1$ on the level-luffing crane mechanism.

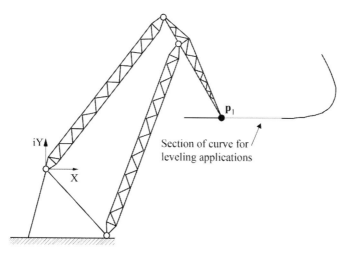

FIGURE E.4.5
Level-luffing crane mechanism with calculated coupler curve.

configuration with the calculated path. The lightly shaded section of curve in this figure is useful for leveling applications because this section maintains a near-constant level.

Example 4.3

Problem Statement: Using the Appendix B.1 MATLAB file, determine if the two door-linkage configurations in Table E.4.3 (Figure E.4.6a and b) will operate properly as the door closes.

Known Information: Appendix B.1 MATLAB file, Table E.4.3, and displacement angle range for crank link.

Solution Approach: One way to determine the operability of the door-linkage configurations is to calculate their coupler link (or follower link) displacement angles throughout the entire crank-link rotation range and check for circuit defects.* The data calculated from the Appendix B.1 MATLAB file includes the coupler and follower-link displacement angles α_j and γ_j, respectively. The crank rotation range corresponding to Figure E.4.6b is 0° (door fully open) to 90° (door fully closed).

Figure E.4.7 includes the input specified (in bold text) in the Appendix B.1 MATLAB file. Figure E.4.8 illustrates the calculated coupler link displacement angles with respect to the door (or crank) rotation angle (β) as the door closes. With configuration 1, a discontinuity appears after the door (the crank link of the linkage) exceeds an 84° displacement angle. This discontinuity corresponds to the mechanism exceeding its limiting position and subsequently experiencing a circuit defect (see Section 3.5). Using configuration 1, the door cannot close completely. In contrast, configuration 2 produces a coupler displacement angle curve that is continuous throughout the entire 90° door rotation range. Using configuration 2, the door closes completely.

TABLE E.4.3

Door-Linkage Assembly Configurations

	W_1, θ	V_1, ρ	U_1, σ	G_{1x}, G_{1y}
Config. 1	10, −90°	8, 8.9743°	9, −103.4799°	10, 0
Config. 2	8, −90°	8, 1.4850°	9, −120.0198°	12.5, 0

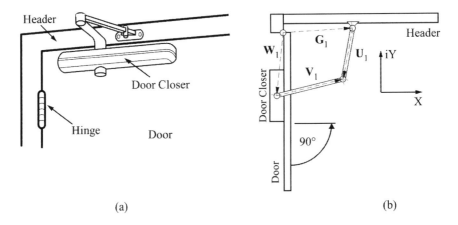

(a)   (b)

FIGURE E.4.6
(a) Door linkage and (b) kinematic model of four-bar door linkage.

---

* During circuit defects, no coupler and follower displacement angles appear in the Appendix B.1 MATLAB file. As a result, circuit defects will appear as discontinuities in coupler and follower displacement angle plots.

```
W1 = 10*exp(-i*90*pi/180);
V1 = 8*exp(i*8.9743*pi/180);
G1 = 10;
U1 = 9*exp(-i*103.4799*pi/180);
L1 = 0;

start_ang   = 0;
step_ang    = 1;
stop_ang    = 90;

angular_vel  = 0 * ones(N+1,1);
angular_acc  = 0 * ones(N+1,1);
```
(MATLAB File Input for Linkage Configuration 1)

```
W1 = 8*exp(-i*90*pi/180);
V1 = 8*exp(i*1.485*pi/180);
G1 = 12.5;
U1 = 9*exp(-i*120.0198*pi/180);
L1 = 0;

start_ang   = 0;
step_ang    = 1;
stop_ang    = 90;

angular_vel  = 0 * ones(N+1,1);
angular_acc  = 0 * ones(N+1,1);
```
(MATLAB File Input for Linkage Configuration 1)

FIGURE E.4.7
Specified input (in bold text) in the Appendix B.1 MATLAB file for Example 4.3.

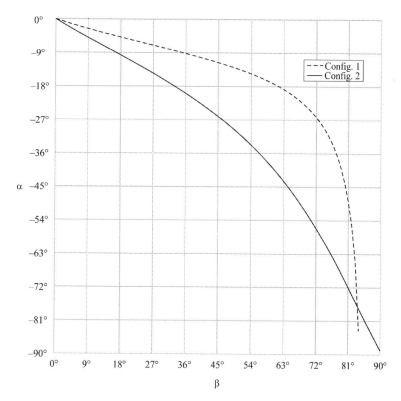

FIGURE E.4.8
Coupler displacement angles for Configurations 1 and 2.

4.4.5 Instant Center, Centrodes, and Centrode Generation

An *instant center* (IC) of velocity is a common point among two links (or two bodies in general) in planar motion which has the same instantaneous velocity in both bodies.[*] For a mechanical system of n links, the total number of ICs in the system, taking two links at a time, can be calculated from the equation

$$N_{IC} = \frac{n(n-1)}{2} \tag{4.15}$$

From Equation 4.15 it can be determined that both the planar four-bar and slider-crank mechanisms have a total number of six ICs each.

There is a systematic procedure to identify the locations of ICs for a given mechanism. Such a procedure is necessary due to the number of ICs that can exist, especially as the number of mechanism links increases. This procedure is based on the *Aronhold–Kennedy theorem*.[†] This theorem states the following:

> *For any three bodies in relative planar motion there will exist three instant centers and the three instant centers are collinear.*

Figure 4.8 illustrates a planar four-bar mechanism and each of its six ICs. Also illustrated in this figure is a diagram (called a *linear diagram* or *circle diagram*) used to keep track of the ICs that have been located. Numbers 1 through 4 marked around the circle in this diagram correspond to the four links in the planar four-bar mechanism, and the lines connecting the numbers (six lines in total) represent each IC.[‡] The lines are drawn one by one as each IC is established, and when every possible line connection has been drawn, the full number of ICs has been produced.

Using Figure 4.8 as an example, the procedure for finding ICs and completing the circle diagram is as follows:

1. Using Equation 4.15, calculate the total number of ICs for the given mechanism.
2. Number the mechanism links (e.g., 1, 2, 3, …) and mark these numbers on an arbitrary circle. The numbers should be equally spaced.
3. Determine as many ICs as possible by inspecting the mechanism. For example, since an IC is a common point among two links in planar motion, the four revolute joints in the planar four-bar mechanism are ICs. As each of these ICs is determined, a line should be drawn in the circle diagram to connect the link numbers corresponding to each IC. For example, the lines in the circle diagram that connect Numbers 1 and 2, 2 and 3, 3 and 4, and 4 and 1 correspond to ICs $\mathbf{I}_{1-2}$, $\mathbf{I}_{2-3}$, $\mathbf{I}_{3-4}$, and $\mathbf{I}_{1-4}$ in Figure 4.8.
4. Lines can be drawn along a link length between any two ICs. In Figure 4.8, such lines (the dashed lines) have been drawn along all four mechanism links. The point of intersection of any two lines, in accordance with the Aronhold–Kennedy theorem, is also an IC. These ICs correspond to the links that form a triangle with the drawn lines. For example, in Figure 4.8, because Links 3 and 1 form a triangle with the two lines associated with the top-most IC, this IC is labeled $\mathbf{I}_{1-3}$, and a line

---

[*] ICs are also called *poles*.
[†] This theorem (also called *Kennedy's rule*) was discovered in 1872 by Aronhold (from Germany) and independently in 1886 by Kennedy (from England).
[‡] For example, I_{1-2} in Figure 4.8 (the IC common between Links 1 and 2) is the straight line connecting Numbers 1 and 2 in the diagram in Figure 4.8 (etc.).

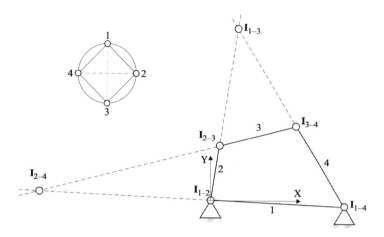

FIGURE 4.8
Planar four-bar mechanism, its ICs, and circle diagram.

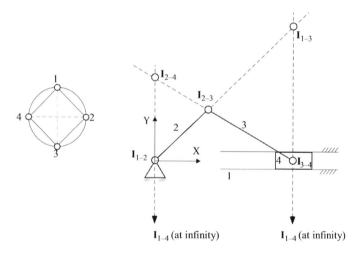

FIGURE 4.9
Slider-crank mechanism, its ICs, and circle diagram.

(a dashed line) is drawn between Numbers 1 and 3 in the circle diagram. Also, in Figure 4.8, because Links 2 and 4 form a triangle with the two lines associated with the left-most IC, this IC is labeled I_{2-4}, and a line (again, a dashed line) is drawn between Numbers 2 and 4 in the circle diagram.

Figure 4.9 illustrates the ICs and completed circle diagram for the slider-crank mechanism. If we recall, a planar four-bar mechanism having a follower link of infinite length becomes a slider-crank mechanism (see Figure 3.2).* A line drawn along the infinite-length follower link is perpendicular to the slider path and passes through the revolute joint

---

* One way to determine the ICs for cam follower systems and gear pairs is to produce an equivalent planar four-bar mechanism for these systems and determine the ICs for the equivalent four-bar mechanism.

attached to the slider (see Figure 4.9). The infinite-length follower link requires an infinite-length ground link, which produces a line parallel to the follower-link line that passes through the grounded revolute joint in Figure 4.9.

It is important to note that IC locations change with the position of a mechanism. Therefore, if ICs are required for a given mechanism over a crank rotation range, the IC location procedure must be repeated at each increment of the crank rotation range.

In addition to calculating velocities through vector-loop equations, velocities can also be calculated using ICs. Figure 4.10 includes a planar four-bar mechanism, its IC I_{1-3}, and its velocities V_{a_1}, V_{b_1}, and V_{p_1}. As illustrated in this figure, at an instant in time, the coupler link of the planar four-bar mechanism rotates about I_{1-3}. Given the angular velocity of the crank link $\dot{\beta}_j$, the magnitude of the velocity V_{a_1} (or $|V_{a_1}|$) can be calculated directly, since it is simply the product of the angular velocity and the link length. Once $|V_{a_1}|$ has been determined, the angular velocity $\dot{\alpha}_j$ in Figure 4.10 can be directly calculated, since $|V_{a_1}|$ is also the product of $\dot{\alpha}_j$ and the distance between the ICs I_{1-3} and I_{2-3}. Once $\dot{\alpha}_j$ has been determined, the magnitudes of velocities V_{p_1} and V_{b_1} can be directly calculated, since $|V_{p_1}|$ and $|V_{b_1}|$ are the products of $\dot{\alpha}_j$ and the distances between I_{1-3} and p_1, and I_{1-3} and b_1 (see Figure 4.10), respectively.

Another use for ICs for the planar four-bar mechanism (particularly I_{1-3}) is to replicate the motion of the coupler link. Because the locations of ICs vary with the position of the mechanism, a locus of ICs can be produced over a crank rotation range. A locus of ICs is called a *centrode*. Figure 4.11a illustrates the centrode produced for a given Grashof triple-rocker mechanism. The centrode produced for a given mechanism is called a *fixed centrode* because it is stationary. Figure 4.11b illustrates the centrode produced for the inverted triple-rocker mechanism. In this particular inversion, the coupler becomes the ground and the ground becomes the coupler (see Figure 3.13). The centrode produced for an inverted mechanism is called a *moving centrode* because this centrode can exhibit motion—specifically, rolling motion—over the fixed centrode.

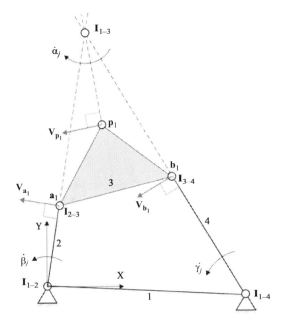

FIGURE 4.10
Planar four-bar velocity analysis using an IC.

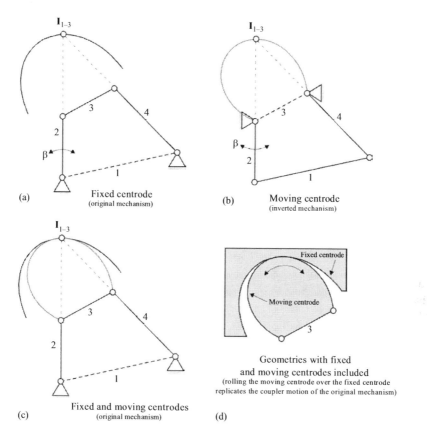

FIGURE 4.11
Fixed and moving centrode construction and use for a triple-rocker mechanism.

When illustrating the fixed and moving centrodes in the same illustration as shown in Figure 4.11c, it can be seen more clearly that the concave curvature of the fixed centrode, combined with the convex curvature of the moving centrode, enables the moving centrode to roll over the fixed centrode. This rolling motion replicates the coupler motion produced by the original mechanism itself. The fixed and moving centrodes can be incorporated into geometry (Figure 4.11d) to ultimately produce alternate mechanisms to replicate the coupler motion of their corresponding original planar four-bar mechanisms.

Appendix B.2 includes the MATLAB file user instructions for planar four-bar fixed and moving centrode generation. In this MATLAB file (which is available for download at www.crcpress.com/product/isbn/9781498724937), individual fixed and moving centrode points are calculated for the complete rotation range of a planar four-bar mechanism at $1°$ crank rotation increments. The Appendix B.2 MATLAB file can also be used to calculate the fixed and moving centrodes for slider-crank mechanisms (defined as a planar four-bar mechanism with a large follower length).* A slider-crank centrode generation example is included in Section 4.5.4.

---

* When using the Appendix B.2 MATLAB file for the slider-crank mechanism, the follower length should be long enough to produce an acceptable slider error. For example, using the follower fixed pivot coordinates $b_0 = (3, -100,000)$ for the four-bar mechanism coordinates $a_0 = (0, 0)$, $a_1 = (0.7071, 0.7071)$, $b_1 = (3, 0)$ produces a maximum sliding error (normal to the sliding direction) of 0.00001.

```
a0=[0, 0]';
a1=[0, 2]';
b0=[1, 0]';
b1=[1.8487, 1.2368]';
```

FIGURE E.4.9
Specified input (in bold text) in the Appendix B.2 MATLAB file for Example 4.4.

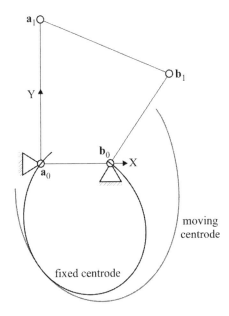

FIGURE E.4.10
Planar four-bar mechanism and sections of its fixed and moving centrodes.

Example 4.4

Problem Statement: Using the Appendix B.2 MATLAB file, plot the fixed and moving centrodes for a planar four-bar mechanism with the unitless dimensions $\mathbf{a}_0 = (0, 0)$, $\mathbf{a}_1 = (0, 2)$, $\mathbf{b}_0 = (1, 0)$, and $\mathbf{b}_1 = (1.8487, 1.2368)$.

Known Information: Known mechanism dimensions and Appendix B.2 MATLAB file.

Solution Approach: Figure E.4.9 includes the input specified (in bold text) in the Appendix B.2 MATLAB file. Figure E.4.10 illustrates the given planar four-bar mechanism and sections of the fixed and moving centrodes calculated from the Appendix B.2 MATLAB file.

4.5 Slider-Crank Mechanism Analysis

4.5.1 Displacement Equations

The planar slider-crank mechanism consists of four links interconnected by revolute joints, with the slider link connected to ground by a prismatic joint. As calculated from Gruebler's equation for planar mechanisms (with $L = 4$ and $J_1 = 4$), the planar slider-crank mechanism has a single DOF. A single displacement equation is derived for this mechanism by taking

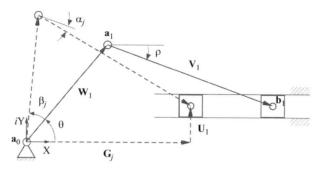

FIGURE 4.12
Slider-crank mechanism displacement variables.

the sum of the displaced mechanism vector loop in Figure 4.12. The equation produced from a clockwise vector-loop sum of the displaced vector loop is

$$e^{i(\theta+\beta_j)} + V_1 e^{i(\rho+\alpha_j)} - \mathbf{U}_1 - \mathbf{G}_j = 0 \tag{4.16}$$

As illustrated in Figure 4.12, vector $\mathbf{U}_1$ only has an imaginary component (the slider offset distance in the y-direction) and vector $\mathbf{G}_j$ only has a real component (the sliding distance in the x-direction). Having a zero $\mathbf{U}_1$ produces what is called an *in-line* slider-crank (a nonzero $\mathbf{U}_1$ produces an *offset* slider-crank).

After expanding Equation 4.16 and grouping its real and imaginary terms as separate equations, the resulting slider-crank mechanism displacement equations become

$$f_1(\alpha_j, G_j) = W_1 \cos(\theta+\beta_j) + V_1 \cos(\rho+\alpha_j) - G_j = 0$$

$$f_2(\alpha_j) = W_1 \sin(\theta+\beta_j) + V_1 \sin(\rho+\alpha_j) - U_{1y} = 0 \tag{4.17}$$

With the exception of the coupler displacement angle α_j and the slider displacement magnitude G_j, all other variables in the slider-crank displacement equations are user prescribed. Unlike Equation 4.2, which requires a numerical solution (at least in its given form), an analytical solution can be produced for Equation 4.17. The imaginary equation $f_2(\alpha_j)$ can be rearranged so that

$$\alpha_j = \sin^{-1}\left[\frac{U_{1y} - W_1 \sin(\theta+\beta_j)}{V_1}\right] - \rho \tag{4.18}$$

The coupler angle solutions from Equation 4.18 are used in $f_1(\alpha_j, G_j)$ and the corresponding sliding distances G_j are calculated.

4.5.2 Velocity Equations

A single slider-crank velocity equation is derived by differentiating the slider-crank displacement equation. Differentiating Equation 4.16 with respect to time produces

$$i\dot{\beta}_j W_1 e^{i(\theta+\beta_j)} + i\dot{\alpha}_j V_1 e^{i(\rho+\alpha_j)} - \dot{G}_j = 0 \tag{4.19}$$

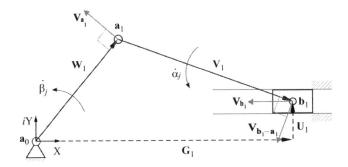

FIGURE 4.13
Slider-crank mechanism velocity variables.

Equation 4.20 includes the individual velocity variables from Equation 4.19. The velocity $\mathbf{V_{a_1}}$ is the *global* tangential velocity of the crank link vector—the velocity of $\mathbf{a_1}$ with respect to $\mathbf{a_0}$ (see Figure 4.13). The velocity $\mathbf{V_{b_1}}$ is the global sliding velocity of the slider link—the sliding velocity of $\mathbf{b_1}$ with respect to $\mathbf{a_0}$. Also, the velocity $\mathbf{V_{b_1-a_1}}$ is the *relative* tangential velocity of $\mathbf{b_1}$ with respect to $\mathbf{a_1}$—the relative velocity of the coupler link.

$$\mathbf{V_{a_1}} = i\dot{\beta}_j W_1 e^{i(\theta+\beta_j)}$$

$$\mathbf{V_{b_1}} = \dot{G}_j \tag{4.20}$$

$$\mathbf{V_{b_1-a_1}} = i\dot{\alpha}_j V_1 e^{i(\rho+\alpha_j)} = \mathbf{V_{b_1}} - \mathbf{V_{a_1}}$$

After moving the term $i\dot{\beta}_j W_1 e^{i(\theta+\beta_j)}$ to the right-hand side of Equation 4.19, expanding this equation and grouping its real and imaginary terms as separate equations, the resulting slider-crank mechanism velocity equation in matrix form is

$$\begin{bmatrix} -V_1\sin(\rho+\alpha_j) & -1 \\ V_1\cos(\rho+\alpha_j) & 0 \end{bmatrix} \begin{Bmatrix} \dot{\alpha}_j \\ \dot{G}_j \end{Bmatrix} = -\dot{\beta}_j \begin{bmatrix} -W_1\sin(\theta+\beta_j) \\ W_1\cos(\theta+\beta_j) \end{bmatrix} \tag{4.21}$$

After including the prescribed mechanism variables and the unknowns calculated from the slider-crank displacement equations, Equation 4.21 can be solved (using Cramer's rule, for example) to calculate the coupler angular velocity $\dot{\alpha}_j$ and the slider velocity $\dot{G}_j$.

Example 4.5

Problem Statement: Using the Appendix B.3 MATLAB file, calculate the slider position and velocity for the slider-bar mechanism configuration in Table E.4.4 for a 25° crank displacement angle and a constant rotation speed of 7 rad/s.

Known Information: Appendix B.3 MATLAB file, Table E.4.4, β and β̇.

Solution Approach: Figure E.4.11 includes the input specified (in bold text) in the Appendix B.3 MATLAB file. From the calculated output, it can be determined that the slider position and velocity are $\mathbf{G}_j = 77.28$ mm and $\dot{\mathbf{G}}_j = -193.10$ mm/s, respectively.

TABLE E.4.4

Slider-Crank Mechanism Configuration (with Link Lengths in mm)

W_1, θ	V_1, ρ	U_1
30, 90°	90, 0°	30

```
LW1 = 30;
theta = 90*pi/180;

LU1 = 30;

LV1 = 90;

start_ang  = 0;
step_ang   = 1;
stop_ang   = 25;

angular_vel  = 7 * ones(N+1,1);
angular_acc  = 0 * ones(N+1,1);
```

FIGURE E.4.11
Specified input (in bold text) in the Appendix B.3 MATLAB file for Example 4.5.

4.5.3 Acceleration Equations

A single slider-crank acceleration equation is derived by differentiating the slider-crank velocity equation. Time differentiation of the Equation 4.19 produces

$$-\dot{\beta}_j^2 W_1 e^{i(\theta+\beta_j)} + i\ddot{\beta}_j W_1 e^{i(\theta+\beta_j)} - \dot{\alpha}_j^2 V_1 e^{i(\rho+\alpha_j)} + i\ddot{\alpha}_j V_1 e^{i(\rho+\alpha_j)} - \ddot{G}_j = 0 \tag{4.22}$$

Equation 4.23 includes the individual acceleration terms from Equation 4.22. The accelerations $\mathbf{A}_{\mathbf{a}_{1n}}$ and $\mathbf{A}_{\mathbf{a}_{1t}}$ are the global normal and tangential accelerations, respectively, of the crank link vector—the normal and tangential accelerations of $\mathbf{a}_1$ with respect to $\mathbf{a}_0$ (see Figure 4.14). The total acceleration $\mathbf{A}_{\mathbf{a}_1}$ is the sum of these normal and tangential accelerations. The acceleration $\mathbf{A}_{\mathbf{b}_1}$ is the global sliding acceleration of the slider link—the sliding acceleration of $\mathbf{b}_1$ with respect to $\mathbf{a}_0$. Also, the accelerations $\mathbf{A}_{(\mathbf{b}_1-\mathbf{a}_1)_n}$ and $\mathbf{A}_{(\mathbf{b}_1-\mathbf{a}_1)_t}$ are the relative normal and tangential accelerations, respectively, of $\mathbf{b}_1$ with respect to $\mathbf{a}_1$—the

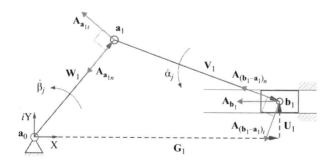

FIGURE 4.14
Slider-crank mechanism acceleration variables.

relative normal and tangential accelerations of the coupler link. The total relative acceleration $\mathbf{A}_{b_1-a_1}$ is the sum of these normal and tangential accelerations.

$$\mathbf{A}_{a_1} = \mathbf{A}_{a_{1t}} + \mathbf{A}_{a_{1n}} = i\ddot{\beta}_j^2 W_1 e^{i(\theta+\beta_j)} - \dot{\beta}_j^2 W_1 e^{i(\theta+\beta_j)}$$

$$\mathbf{A}_{b_1} = \ddot{G}_j \tag{4.23}$$

$$\mathbf{A}_{b_1-a_1} = \mathbf{A}_{(b_1-a_1)_t} + A_{(b_1-a_1)n} = i\ddot{\alpha}_j V_1 e^{i(\rho+\alpha_j)} - \dot{\alpha}_j^2 V_1 e^{i(\rho+\alpha_j)} = \mathbf{A}_{b_1} - \mathbf{A}_{a_1}$$

After moving the terms $\dot{\beta}_j^2 W_1 e^{i(\theta+\beta_j)}$, $i\ddot{\beta}_j W_1 e^{i(\theta+\beta_j)}$, and $\dot{\alpha}_j^2 V_1 e^{i(\rho+\alpha_j)}$ to the right-hand side of Equation 4.22, expanding this equation, and grouping its real and imaginary terms as separate equations, the resulting slider-crank mechanism acceleration equation in matrix form is

$$\begin{bmatrix} -V_1 \sin(\rho+\alpha_j) & -1 \\ V_1 \cos(\rho+\alpha_j) & 0 \end{bmatrix} \begin{Bmatrix} \ddot{\alpha}_j \\ \ddot{G}_j \end{Bmatrix} = \dot{\beta}_j^2 \begin{bmatrix} W_1 \cos(\theta+\beta_j) \\ W_1 \sin(\theta+\beta_j) \end{bmatrix} - \ddot{\beta}_j \begin{bmatrix} -W_1 \sin(\theta+\beta_j) \\ W_1 \cos(\theta+\beta_j) \end{bmatrix}$$

$$+ \dot{\alpha}_j^2 \begin{bmatrix} V_1 \cos(\rho+\alpha_j) \\ V_1 \sin(\rho+\alpha_j) \end{bmatrix} \tag{4.24}$$

After including the prescribed mechanism variables and the unknowns calculated from the slider-crank displacement and velocity equations, Equation 4.24 can be solved (using Cramer's rule, for example) to calculate the coupler angular acceleration $\ddot{\alpha}_j$ and the slider acceleration $\ddot{G}_j$.

Appendix B.3 includes the MATLAB file user instructions for slider-crank displacement, velocity, and acceleration analysis. In this MATLAB file (which is available for download at www.crcpress. com/product/isbn/9781498724937), solutions for Equations 4.17, 4.21, and 4.24 are calculated.

Example 4.6

Problem Statement: Using the Appendix B.3 MATLAB file, plot the displacement, velocity, and acceleration profiles for the piston in the crankshaft-connecting rod-piston linkage (Figure E.4.12). The mechanism assembly configuration and driving link parameters are given in Table E.4.5.

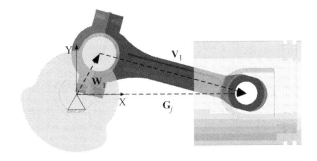

FIGURE E.4.12
Crankshaft-connecting rod-piston linkage.

TABLE E.4.5

Slider-Crank Mechanism Assembly Configuration (with Link Lengths in cm)

W_1, θ	V_1, ρ	$\dot{\beta}_0$	$\ddot{\beta}$
1, 45°	1.5, −28.1255°	100 rad/s	0 rad/s²

```
LW1 = 1;
theta = 45*pi/180;

LU1 = 0;

LV1 = 1.5;

start_ang  = 0;
step_ang   = 1;
stop_ang   = 720;

angular_vel  = 100 * ones(N+1,1);
angular_acc  = 0 * ones(N+1,1);
```

FIGURE E.4.13
Specified input (in bold text) in the Appendix B.3 MATLAB file for Example 4.6.

Known Information: Table E.4.5 and Appendix B.3 MATLAB file.

Solution Approach: As illustrated in Figure E.4.12, the crankshaft-connecting rod-piston linkage is an *in-line* slider crank mechanism (therefore $U_1 = 0$). Because variable G_x corresponds to piston displacement, the displacement, velocity, and acceleration profiles of the piston can be produced by calculating and plotting G_x, $\dot{G}_x$, and $\ddot{G}_x$, respectively.

Figure E.4.13 includes the input specified (in bold text) in the Appendix B.3 MATLAB file. Figure E.4.14 illustrates the piston displacement, velocity, and acceleration profiles with respect to the crank rotation angle (β) over the prescribed 720° crank rotation range.

Example 4.7

Problem Statement: Derive vector-loop displacement, velocity, and acceleration equations for the slider-crank inversion illustrated in Figure E.4.15(a). The starting and displaced mechanism variables are included in Figure E.4.15(b). Also verify the velocity and acceleration equations in MATLAB.

Known Information: Figure E.4.15 and vector-loop formulation procedure.

Solution Approach: The two unknowns in the inverted slider-crank mechanism are the sliding distance V_j and the follower displacement angle γ_j (which is the same displacement angle for the sliding distance vector V_j).* The angle formed by U_1 and V_j (angle ρ in Figure E.4.15(b)) remains constant throughout mechanism motion.

DISPLACEMENT EQUATION

Taking a clockwise vector-loop sum for the vector loop in Figure E.4.15(b) produces

$$W_1 e^{i(\theta+\beta_j)} - V_j e^{i(\sigma+\gamma_j+\rho)} - U_1 e^{i(\sigma+\gamma_j)} - \mathbf{G}_1 = 0 \qquad (4.25)$$

---

* The sliding distance is also the distance between points a_1 and b_1 in Figure E.4.11. Due to the sliding joint, the distance between these two points (the scalar length of V_j) can change throughout mechanism motion.

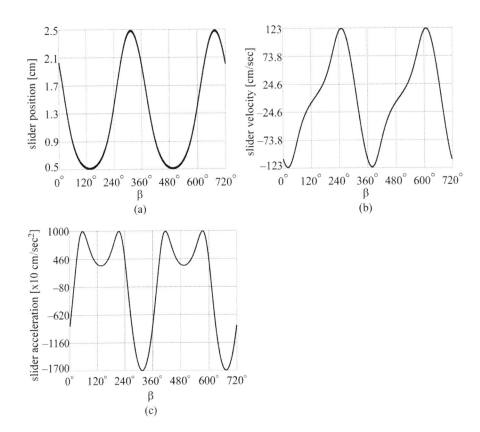

FIGURE E.4.14
Piston (a) displacement, (b) velocity, and (c) acceleration profiles.

VELOCITY EQUATION

Taking the derivative of Equation 4.25 produces

$$i\dot{\beta}_j W_1 e^{i(\theta+\beta_j)} - i\dot{\gamma}_j V_j e^{i(\sigma+\gamma_j+\rho)} - \dot{V}_j e^{i(\sigma+\gamma_j+\rho)} - i\dot{\gamma}_j U_1 e^{i(\sigma+\gamma_j)} = 0 \qquad (4.26)$$

Equation (4.27) includes the individual velocity variables from Equation (4.26). The velocity $\mathbf{V}_{a_1}$ is the *global* tangential velocity of the crank link vector—the tangential velocity of $\mathbf{a}_1$ with respect to $\mathbf{a}_0$ (see Figure E.4.16). The velocity $\mathbf{V}_{b_1}$ is the global tangential velocity of the follower-link vector—the tangential velocity of $\mathbf{b}_1$ with respect to $\mathbf{b}_0$. The velocity $\mathbf{V}_{b_1-a_1}$ is the *relative* tangential velocity of the coupler link vector—the relative tangential velocity of $\mathbf{b}_1$ with respect to $\mathbf{a}_1$. The velocity $\dot{\mathbf{V}}_j$ is the relative sliding velocity of $\mathbf{a}_1$ with respect to $\mathbf{b}_1$—the velocity of the sliding distance vector $\mathbf{V}_j$.

$$\mathbf{V}_{a_1} = i\dot{\beta}_j W_1 e^{i(\theta+\beta_j)}$$

$$\mathbf{V}_{b_1} = i\dot{\gamma}_j U_1 e^{i(\sigma+\gamma_j)}$$

$$\dot{\mathbf{V}}_j = \dot{V}_j e^{i(\sigma+\gamma_j+\rho)} \qquad (4.27)$$

$$\mathbf{V}_{b_1-a_1} = i\dot{\gamma}_j V_j e^{i(\sigma+\gamma_j+\rho)}$$

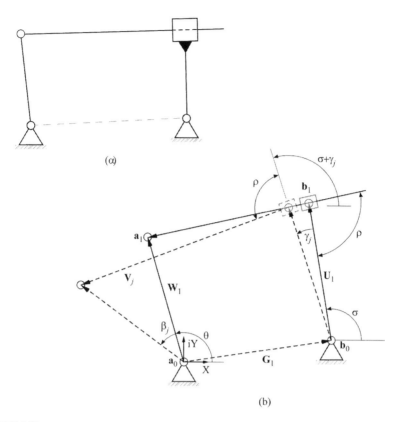

FIGURE E.4.15
(a) Slider-crank inversion and (b) mechanism displacement variables.

ACCELERATION EQUATION

Taking the derivative of Equation (4.26) produces

$$-\dot{\beta}_j^2 W_1 e^{i(\theta+\beta_j)} + i\ddot{\beta}_j W_1 e^{i(\theta+\beta_j)} + \dot{\gamma}_j^2 V_j e^{i(\sigma+\gamma_j+\rho)} - i2\dot{\gamma}_j \dot{V}_j e^{i(\sigma+\gamma_j+\rho)}$$

$$-i\ddot{\gamma}_j V_j e^{i(\sigma+\gamma_j+\rho)} - \ddot{V}_j e^{i(\sigma+\gamma_j+\rho)} + \dot{\gamma}_j^2 U_1 e^{i(\sigma+\gamma_j)} - i\ddot{\gamma}_j U_1 e^{i(\sigma+\gamma_j)} = 0 \tag{4.28}$$

Equation (4.29) includes the individual acceleration variables from Equation (4.28). The accelerations $\mathbf{A}_{\mathbf{a}_{1n}}$ and $\mathbf{A}_{\mathbf{a}_{1t}}$ are the global *normal* and *tangential* accelerations, respectively, of the crank link vector—the normal and tangential accelerations of $\mathbf{a}_1$ with respect to $\mathbf{a}_0$ (see Figure E.4.17). The *total* acceleration $\mathbf{A}_{\mathbf{a}_1}$ is the sum of these normal and tangential accelerations. The accelerations $\mathbf{A}_{\mathbf{b}_{1n}}$ and $\mathbf{A}_{\mathbf{b}_{1t}}$ are the global normal and tangential accelerations, respectively, of the follower link vector—the normal and tangential accelerations of $\mathbf{b}_1$ with respect to $\mathbf{b}_0$. The total acceleration $\mathbf{A}_{\mathbf{b}_1}$ is the sum of these normal and tangential accelerations. The accelerations $\mathbf{A}_{(\mathbf{b}_1-\mathbf{a}_1)_n}$ and $\mathbf{A}_{(\mathbf{b}_1-\mathbf{a}_1)_t}$ are the relative normal and tangential accelerations, respectively, of $\mathbf{b}_1$ with respect to $\mathbf{a}_1$—the relative normal and tangential accelerations of the coupler link vector. The acceleration $\dot{\mathbf{V}}_j$ is the relative sliding acceleration of $\mathbf{a}_1$ with respect to $\mathbf{b}_1$—the acceleration of the sliding distance vector $\mathbf{V}_j$. Lastly, the acceleration $\mathbf{A}_{(\mathbf{b}_1-\mathbf{a}_1)_c}$ is called the *Coriolis acceleration* [3]. This acceleration component is present whenever a link includes both sliding and angular velocities. While the Coriolis acceleration is tangent to the coupler link (like the relative tangential acceleration of $\mathbf{b}_1$ with respect to $\mathbf{a}_1$), its direction matches the direction of

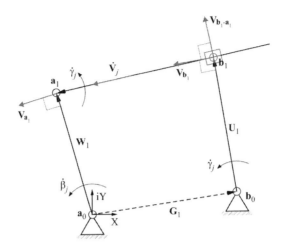

FIGURE E.4.16
Inverted slider-crank mechanism velocity variables.

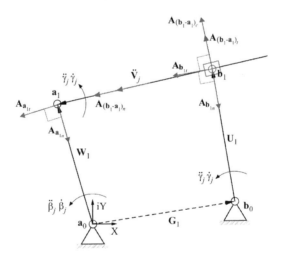

FIGURE E.4.17
Inverted slider-crank mechanism acceleration variables.

$\dot{\gamma}_j$ (while the direction of $\mathbf{A}_{(b_1-a_1)_t}$ matches the direction of $\ddot{\gamma}_j$). The total relative accelera-
tion of point $\mathbf{b}_1$ with respect to $\mathbf{a}_1$ is the sum of $\mathbf{A}_{(b_1-a_1)_n}$, $\mathbf{A}_{(b_1-a_1)_t}$, $\ddot{\mathbf{V}}_j$, and $\mathbf{A}_{(b_1-a_1)_c}$.

$$\mathbf{A}_{a_1} = \mathbf{A}_{a_{1t}} + \mathbf{A}_{a_{1n}} = i\ddot{\beta}_j W_1 e^{i(\theta+\beta_j)} - \dot{\beta}_j^2 W_1 e^{i(\theta+\beta_j)}$$

$$\mathbf{A}_{b_1} = \mathbf{A}_{b_{1t}} + \mathbf{A}_{b_{1n}} = \dot{\gamma}_j^2 U_1 e^{i(\sigma+\gamma_j)} - i\ddot{\gamma}_j U_1 e^{i(\sigma+\gamma_j)}$$

$$\mathbf{A}_{(b_1-a_1)_t} = i\ddot{\gamma}_j V_j e^{i(\sigma+\gamma_j+\rho)}$$

$$\mathbf{A}_{(b_1-a_1)_n} = \dot{\gamma}_j^2 V_j e^{i(\sigma+\gamma_j+\rho)}$$

$$\ddot{\mathbf{V}}_j = \ddot{V}_j e^{i(\sigma+\gamma_j+\rho)}$$

$$\mathbf{A}_{(b_1-a_1)_c} = i2\dot{\gamma}_j \dot{V}_j e^{i(\sigma+\gamma_j+\rho)}$$

```
>> syms W1 U1 G1 theta sigma rho t V(t) beta(t) gamma(t) real;
>> F = W1*exp(i*(theta + beta(t))) ...
- V(t)*exp(i*(sigma + gamma(t) + rho)) ...
- U1*exp(i*(sigma + gamma(t))) - G1;
>> dF = diff(F)

dF =

- exp(rho*i + sigma*i + gamma(t)*i)*diff(V(t), t)
+ W1*exp(theta*i + beta(t)*i)*diff(beta(t), t)*i
- U1*exp(sigma*i + gamma(t)*i)*diff(gamma(t), t)*i
- exp(rho*i + sigma*i + gamma(t)*i)*V(t)*diff(gamma(t), t)*i

>> ddF = diff(F, 2)

ddF =

- exp(rho*i + sigma*i + gamma(t)*i)*diff(V(t), t, t)
- exp(rho*i + sigma*i + gamma(t)*i)*diff(V(t), t)
*diff(gamma(t), t)*2*i
- W1*exp(theta*i + beta(t)*i)*diff(beta(t), t)^2
+ U1*exp(sigma*i + gamma(t)*i)*diff(gamma(t), t)^2
+ W1*exp(theta*i + beta(t)*i)*diff(beta(t), t, t)*i
- U1*exp(sigma*i + gamma(t)*i)*diff(gamma(t), t, t)*i
+ exp(rho*i + sigma*i + gamma(t)*i)*V(t)*diff(gamma(t), t)^2
- exp(rho*i + sigma*i + gamma(t)*i)*V(t)*diff(gamma(t), t, t)*i

>>
```

FIGURE E.4.18
Example 4.7 calculation procedure in MATLAB.

$$\mathbf{A}_{(b_1-a_1)} = \mathbf{A}_{(b_1-a_1)_t} + \mathbf{A}_{(b_1-a_1)_n} + \ddot{\mathbf{V}}_j + \mathbf{A}_{(b_1-a_1)_c} \tag{4.29}$$

EQUATION FORMULATION IN MATLAB

Figure E.4.18 includes the calculation procedure in MATLAB's command window. In this figure, the terms *diff(beta(t),t)*, *diff(V(t),t)* and *diff(gamma(t),t)* represent the terms $\dot{\beta}_j$, $\dot{V}_j$ and $\dot{\gamma}_j$ Equation (4.26), respectively. The terms *diff(beta(t),t,t)*, *diff(V(t),t,t)* and *diff(gamma(t),t,t)* in Figure E.4.18 represent the terms $\ddot{\beta}_j$, $\ddot{V}_j$ and $\ddot{\gamma}_j$ Equation (4.26), respectively.

4.5.4 Centrode Generation

The procedure to calculate ICs for the slider-crank mechanism are presented in Section 4.4.5. Like the planar four-bar mechanism, slider-crank mechanism ICs are useful for velocity analysis, and slider-crank mechanism fixed and moving centrodes are useful for replicating coupler motion.

By specifying the coordinates of a slider-crank mechanism in the Appendix B.2 MATLAB file, the fixed and moving centrodes for this mechanism are calculated.

Example 4.8

Problem Statement: Using the Appendix B.2 MATLAB file, plot the fixed and moving centrodes for a planar slider-crank mechanism with the unitless dimensions $\mathbf{a}_0 = (0, 0)$, $\mathbf{a}_1 = (0, 1)$ and $\mathbf{b}_1 = (2, 0)$.

Known Information: Known mechanism dimensions and Appendix B.2 MATLAB file.

Solution Approach: Because the dimensions for a planar four-bar mechanism are required for the Appendix B.2 MATLAB file, the coordinates of the fixed pivot $\mathbf{b}_0$ are required. The coordinates $\mathbf{b}_0 = (2, -100{,}000)$ were used because with this value, the moving pivot $\mathbf{b}_1$ (the slider) is accurate to five decimal places.*

Figure E.4.19 includes the input specified (in bold text) in the Appendix B.2 MATLAB file. Figure E.4.20 illustrates the given slider-crank mechanism and sections of the fixed and moving centrodes calculated from the Appendix B.2 MATLAB file.

```
a0=[0, 0]';
a1=[0, 1]';
b0=[2, -100000]';
b1=[2, 0]';
```

FIGURE E.4.19
Specified input (in bold text) in the Appendix B.2 MATLAB file for Example 4.8.

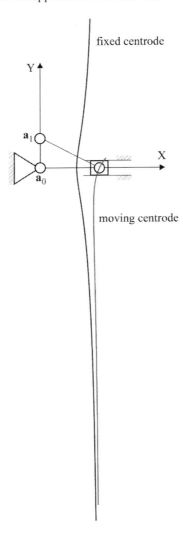

FIGURE E.4.20
Slider-crank mechanism and sections of its fixed and moving centrodes.

---

* Because the follower has an infinite length in a slider-crank mechanism, the accuracy of the slider increases as the value of b_{0y} increases.

4.6 Geared Five-Bar Mechanism Analysis

4.6.1 Displacement Equations

The geared five-bar mechanism consists of five links interconnected by revolute joints. Also included in this mechanism is a gear pair or gear train that couples the input and output links (see Figure 4.15). As calculated from Gruebler's equation for planar mechanisms (with $L = 5$, $J_1 = 5$, and $J_2 = 1$), the geared five-bar mechanism has a single DOF. A single displacement equation is derived for this mechanism by taking the sum of the displaced mechanism vector loop in Figure 4.15. The equation produced from a clockwise sum of the displaced vector loop is

$$W_1 e^{i(\theta+\beta_j)} + V_1 e^{i(\rho+\alpha_j)} - S_1 e^{i(\psi+v_j)} - U_1 e^{i(\sigma+\gamma_j)} - G_1 = 0 \tag{4.30}$$

The relationship between the displacement angles of the crank W_1 and the output link U_1 can be expressed as

$$\gamma_j = \pm \frac{\beta_j}{r} \tag{4.31}$$

where the variable r represents the *gear ratio* of the gear pair or train. The gear ratio can be defined as the ratio of the driven gear radius to the driving gear radius ($r = r_{driven}/r_{driving}$). If W_1 and U_1 rotate in the same direction (which occurs when an odd number of gears is used), the gear ratio is positive, and it is negative if they rotate in opposite directions (which occurs when an even number of gears is used).

After expanding Equation 4.30 and grouping its real and imaginary terms as separate equations, the resulting geared five-bar mechanism displacement equations become

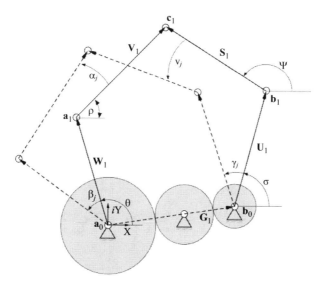

FIGURE 4.15
Geared five-bar mechanism displacement equation variables.

$$f_1(\alpha_j, v_j) = W_1 \cos(\theta + \beta_j) + V_1 \cos(\rho + \alpha_j) - S_1 \cos(\psi + v_j) - U_1 \cos(\sigma + \gamma_j) - G_{1x} = 0$$

$$f_2(\alpha_j, v_j) = W_1 \sin(\theta + \beta_j) + V_1 \sin(\rho + \alpha_j) - S_1 \sin(\psi + v_j) - U_1 \sin(\sigma + \gamma_j) - G_{1y} = 0$$

(4.32)

With the exception of the intermediate link displacement angles α_j and v_j, all other variables in the geared five-bar displacement equations are user prescribed. Because $f_1(\alpha_j, v_j)$ and $f_2(\alpha_j, v_j)$ in Equation 4.32 both include the unknown displacement angles α_j and v_j, these equations form a set of nonlinear simultaneous equations from which the angles are calculated.

4.6.2 Velocity Equations

A single-geared five-bar velocity equation is derived by differentiating the geared five-bar displacement equation. Differentiating Equation 4.30 with respect to time produces

$$i\dot{\beta}_j W_1 e^{i(\theta+\beta_j)} + i\dot{\alpha}_j V_1 e^{i(\rho+\alpha_j)} - i\dot{v}_j S_1 e^{i(\psi+v_j)} + i\dot{\gamma}_j U_1 e^{i(\sigma+\gamma_j)} = 0$$

(4.33)

The relationship between the angular velocities of the crank W_1 and the output link U_1 (the time derivative of Equation 4.31) becomes

$$\dot{\gamma}_j = \pm \frac{\dot{\beta}_j}{r}$$

(4.34)

Equation 4.35 includes the individual velocity terms from Equation 4.33. The velocity V_{a_1} is the *global* tangential velocity of the crank link vector—the tangential velocity of a_1 with respect to a_0 (see Figure 4.16). The velocity V_{b_1} is the global tangential velocity of the output link vector—the tangential velocity of b_1 with respect to b_0. Also, the velocity $V_{c_1-a_1}$ is the *relative* tangential velocity of c_1 with respect to a_1—the relative tangential velocity of this intermediate link. The global velocity of c_1 (V_{c_1}) can be calculated from the last equation in Equation 4.35.

$$V_{a_1} = i\dot{\beta}_j W_1 e^{i(\theta+\beta_j)}$$

$$V_{b_1} = i\dot{\gamma}_j U_1 e^{i(\sigma+\gamma_j)}$$

(4.35)

$$V_{c_1-a_1} = i\dot{\alpha}_j V_1 e^{i(\rho+\alpha_j)} = V_{c_1} - V_{a_1}$$

After moving the terms $i\dot{\beta}_j W_1 e^{i(\theta+\beta_j)}$ and $i\dot{\gamma}_j U_1 e^{i(\sigma+\gamma_j)}$ to the right-hand side of Equation 4.33, expanding this equation, and grouping its real and imaginary terms as separate equations, the resulting geared five-bar mechanism velocity equation in matrix form is

$$\begin{bmatrix} -V_1 \sin(\rho+\alpha_j) & S_1 \sin(\psi+v_j) \\ V_1 \cos(\rho+\alpha_j) & -S_1 \cos(\psi+v_j) \end{bmatrix} \begin{Bmatrix} \dot{\alpha}_j \\ \dot{v}_j \end{Bmatrix}$$

$$= \dot{\beta}_j \begin{bmatrix} W_1 \sin(\theta+\beta_j) \\ -W_1 \cos(\theta+\beta_j) \end{bmatrix} + \dot{\gamma}_j \begin{bmatrix} -U_1 \sin(\sigma+\gamma_j) \\ U_1 \cos(\sigma+\gamma_j) \end{bmatrix}$$

(4.36)

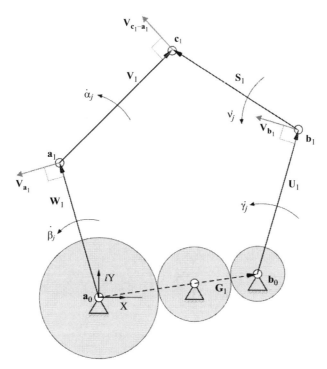

FIGURE 4.16
Geared five-bar mechanism velocity variables.

After including the prescribed mechanism variables and the unknowns calculated from the geared five-bar displacement equations, Equation 4.36 can be solved (using Cramer's rule, for example) to calculate the intermediate link angular velocities $\dot{\alpha}_j$ and $\dot{\upsilon}_j$.

4.6.3 Acceleration Equations

A single planar four-bar acceleration equation is derived by differentiating the planar four-bar velocity equation. Time differentiation of Equation 4.33 produces

$$-\dot{\beta}_j^2 W_1 e^{i(\theta+\beta_j)} + i\ddot{\beta}_j W_1 e^{i(\theta+\beta_j)} - \dot{\alpha}_j^2 V_1 e^{i(\rho+\alpha_j)} + i\ddot{\alpha}_j V_1 e^{i(\rho+\alpha_j)} - \left(-\dot{\upsilon}_j^2 S_1 e^{i(\psi+\upsilon_j)} + i\ddot{\upsilon}_j S_1 e^{i(\psi+\upsilon_j)}\right)$$
$$-\left(-\dot{\gamma}_j^2 U_1 e^{i(\sigma+\gamma_j)} + i\ddot{\gamma}_j U_1 e^{i(\sigma+\gamma_j)}\right) = 0 \tag{4.37}$$

The relationship between the angular accelerations of the crank W_1 and the output link U_1 (the time derivative of Equation 4.34) becomes

$$\ddot{\gamma}_j = \pm \frac{\ddot{\beta}_j}{r} \tag{4.38}$$

Equation 4.39 includes the individual acceleration terms from Equation 4.37. The accelerations $A_{a_{1n}}$ and $A_{a_{1t}}$ are the global normal and tangential accelerations, respectively, of

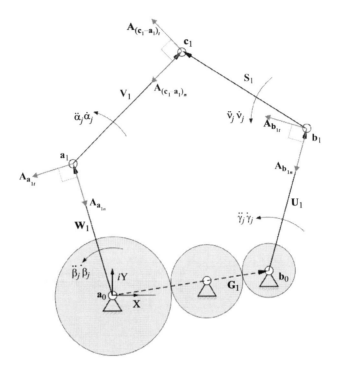

FIGURE 4.17
Geared five-bar mechanism acceleration variables.

the crank link vector—the normal and tangential accelerations of $\mathbf{a}_1$ with respect to $\mathbf{a}_0$ (see Figure 4.17). The *total* acceleration of $\mathbf{A}_{\mathbf{a}_1}$ is the sum of these normal and tangential accelerations. The accelerations $\mathbf{A}_{\mathbf{b}_{1n}}$ and $\mathbf{A}_{\mathbf{b}_{1t}}$ are the global normal and tangential accelerations, respectively, of the output link vector— the normal and tangential accelerations of $\mathbf{b}_1$ with respect to $\mathbf{b}_0$. The total acceleration $\mathbf{A}_{\mathbf{b}_1}$ is the sum of these normal and tangential accelerations. Also, the accelerations $\mathbf{A}_{(\mathbf{c}_1-\mathbf{a}_1)_n}$ and $\mathbf{A}_{(\mathbf{c}_1-\mathbf{a}_1)_t}$ are the relative normal and tangential accelerations, respectively, of $\mathbf{c}_1$ with respect to $\mathbf{a}_1$—the relative normal and tangential accelerations of this intermediate link. The total relative acceleration $\mathbf{A}_{\mathbf{c}_1-\mathbf{a}_1}$ is the sum of these normal and tangential accelerations. The global acceleration of $\mathbf{c}_1$ ($\mathbf{A}_{\mathbf{c}_1}$) can be calculated from the last equation in Equation 4.39.

$$\mathbf{A}_{\mathbf{a}_1} = \mathbf{A}_{\mathbf{a}_{1t}} + \mathbf{A}_{\mathbf{a}_{1n}} = i\ddot{\beta}_j W_1 e^{i(\theta+\beta_j)} - \dot{\beta}_j^2 W_1 e^{i(\theta+\beta_j)}$$

$$\mathbf{A}_{\mathbf{b}_1} = \mathbf{A}_{\mathbf{b}_{1t}} + \mathbf{A}_{\mathbf{b}_{1n}} = i\ddot{\gamma}_j U_1 e^{i(\sigma+\gamma_j)} - \dot{\gamma}_j^2 U_1 e^{i(\sigma+\gamma_j)} \tag{4.39}$$

$$\mathbf{A}_{\mathbf{c}_1-\mathbf{a}_1} = \mathbf{A}_{(\mathbf{c}_1-\mathbf{a}_1)_t} + \mathbf{A}_{(\mathbf{c}_1-\mathbf{a}_1)_n} = i\alpha_j V_1 e^{i(\rho+\alpha_j)} - \alpha_j^2 V_1 e^{i(\rho+\alpha_j)} = \mathbf{A}_{\mathbf{c}_1} - \mathbf{A}_{\mathbf{a}_1}$$

After moving the terms $\dot{\beta}_j^2 W_1 e^{i(\theta+\beta_j)}, i\ddot{\beta}_j W_1 e^{i(\theta+\beta_j)}, \dot{\gamma}_j^2 U_1 e^{i(\sigma+\gamma_j)}, i\ddot{\gamma}_j U_1 e^{i(\sigma+\gamma_j)}, \dot{\alpha}_j^2 V_1 e^{i(\rho+\alpha_j)}$, and $\dot{v}_j^2 S_1 e^{i(\psi+v_j)}$ to the right-hand side of Equation 4.37, expanding this equation, and grouping its real and imaginary terms as separate equations, the resulting geared five-bar mechanism acceleration equation in matrix form becomes

$$
\begin{bmatrix} -V_1 \sin(\rho + \alpha_j) & S_1 \sin(\psi + v_j) \\ V_1 \cos(\rho + \alpha_j) & -S_1 \cos(\psi + v_j) \end{bmatrix} \begin{Bmatrix} \ddot{\alpha}_j \\ \ddot{v}_j \end{Bmatrix} = \dot{\beta}_j^2 \begin{bmatrix} W_1 \cos(\theta + \beta_j) \\ W_1 \sin(\theta + \beta_j) \end{bmatrix} - \ddot{\beta}_j \begin{bmatrix} -W_1 \sin(\theta + \beta_j) \\ W_1 \cos(\theta + \beta_j) \end{bmatrix}
$$

$$
-\dot{\gamma}_j^2 \begin{bmatrix} U_1 \cos(\sigma + \gamma_j) \\ U_1 \sin(\sigma + \gamma_j) \end{bmatrix} + \ddot{\gamma}_j \begin{bmatrix} -U_1 \sin(\sigma + \gamma_j) \\ U_1 \cos(\sigma + \gamma_j) \end{bmatrix} + \dot{\alpha}_j^2 \begin{bmatrix} V_1 \cos(\rho + \alpha_j) \\ V_1 \sin(\rho + \alpha_j) \end{bmatrix} + \dot{v}_j^2 \begin{bmatrix} S_1 \cos(\psi + v_j) \\ S_1 \sin(\psi + v_j) \end{bmatrix}
$$

$$
(4.40)
$$

After including the prescribed mechanism variables and the unknowns calculated from the geared five-bar displacement and velocity equations, Equation 4.40 can be solved (using Cramer's rule, for example) to calculate the intermediate link angular accelerations $\ddot{\alpha}_j$ and $\ddot{v}_j$.

Appendices B.4 and B.5 include the MATLAB file user instructions for geared five-bar displacement, velocity, and acceleration analysis. The Appendix B.4 file considers a mechanism having two gears (or negative gear ratios) and the Appendix B.5 file considers a mechanism having three gears (or positive gear ratios). In these MATLAB files (which are available for download at www.crcpress. com/product/isbn/9781498724937), solutions for Equations 4.32, 4.36, and 4.40 are calculated.

4.6.4 Kinematics of Intermediate Link Locations of Interest

Equations 4.10, 4.12, and 4.14 can be directly applied to calculate the displacements, velocities, and accelerations of an arbitrary intermediate link location of interest $\mathbf{p}_1$ of a geared five-bar mechanism (see Figure 4.18). The Appendix B.4 and B.5 MATLAB files include

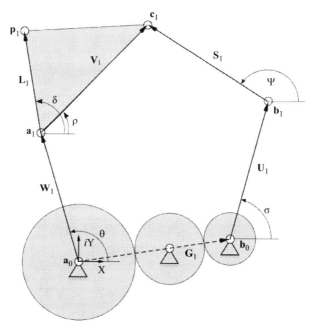

FIGURE 4.18
Geared five-bar mechanism with intermediate link location $\mathbf{p}_1$.

TABLE E.4.6

Geared Five-Bar Mechanism Configuration (with Link Lengths in m)

W_1, θ	V_1, ρ	U_1, σ	S_1, ψ	G_{1x}, G_{1y}	L_1, δ
0.35, 90°	0.525, 54.7643°	0.35, 60°	0.525, 115.0279°	0.35, 0	0.35, −15.7645°

```
W1 = 0.35*exp(i*90*pi/180);
V1 = 0.525*exp(i*54.7643*pi/180);
L1 = 0.35*exp(-i*15.7645*pi/180);
G1 = 0.35;
U1 = 0.35*exp(i*60*pi/180);
S1 = 0.525*exp(i*115.0279*pi/180);
ratio = 2;

start_ang   = 0;
step_ang    = 1;
stop_ang    = 45;

angular_vel  = 1.5 * ones(N+1,1);
angular_acc  = 0 * ones(N+1,1);
```

FIGURE E.4.21
Specified input (in bold text) in the Appendix B.5 MATLAB file for Example 4.9.

Equations 4.10, 4.12, and 4.14, from which displacement, velocity, and acceleration values are calculated for $\mathbf{p}_1$.

Example 4.9

Problem Statement: Using the Appendix B.5 MATLAB file, calculate the displaced value of link point $\mathbf{p}_1$ and its velocity for the geared five-bar mechanism configuration in Table E.4.6 for a 45° crank displacement angle and a constant rotation speed of 1.5 rad/s. The gear ratio is $r = +2$.

   Known Information: Appendix B.5 MATLAB file, Table E.4.6, β, and $\dot{\beta}$.

   Solution Approach: Figure E.4.21 includes the input specified (in bold text) in the Appendix B.5 MATLAB file. From the calculated output, it can be determined that the displaced coupler point $\mathbf{p}_1$ is $\mathbf{p}_j = (0.0972 + i0.1865)\,\text{m}$ and its velocity is $\mathbf{V}_{\mathbf{p}_j} = (-0.3480 - i0.2398)\,\text{m/s}$ at a 45° crank displacement angle.

Example 4.10

Problem Statement: Using the Appendix B.5 MATLAB file, calculate the location, velocity and acceleration values of $\mathbf{p}_1$ at β = 0°, 15°, 30°, ..., 90° given the geared five-bar mechanism dimensions in Table E.4.7. The initial angular velocity and angular acceleration of the driving link are $\dot{\beta}_0 = 1\,\text{rad/s}$ and $\ddot{\beta} = 0.1\,\text{rad/s}^2$, respectively. The gear ratio is $r = +2$.

   Known Information: Table E.4.7 and Appendix B.5 MATLAB file.

TABLE E.4.7

Geared Five-Bar Mechanism Dimensions (with Lengths in m)

W_1, θ	V_1, ρ	U_1, σ	S_1, ψ	G_{1x}, G_{1y}	L_1, δ
1, 90°	1.5, 32.7304°	1.5, 45°	1.5, 149.9837°	1.5, 0	1, 74.1400°

```
W1 = 1.0*exp(i*90*pi/180);
V1 = 1.5*exp(i*32.7304*pi/180);
L1 = 1.0*exp(i*74.1400*pi/180);
G1 = 1.5;
U1 = 1.5*exp(i*45*pi/180);
S1 = 1.5*exp(i*149.9847*pi/180);
ratio = 2;

start_ang   = 0;
step_ang    = 1;
stop_ang    = 90;

angular_vel = 1 * ones(N+1,1);
angular_acc = 0.1 * ones(N+1,1);
```

FIGURE E.4.22

Specified input (in bold text) in the Appendix B.5 MATLAB file for Example 4.10.

TABLE E.4.8

Calculated Geared Five-Bar Point Positions, Velocities, and Accelerations

β	p_{1j} (m)	$V_{p_{1j}}$ (m/s)	$A_{p_{1j}}$ (m/s²)
0°	0.2733, 1.9619	−0.8980, −0.0290	−0.2352, −0.9733
15°	0.0360, 1.9215	−0.9273, −0.2851	0.0102, −0.9959
30°	−0.1949, 1.8184	−0.8956, −0.5302	0.2336, −0.9326
45°	−0.4068, 1.6609	−0.8202, −0.7419	0.3516, −0.7683
60°	−0.5944, 1.4624	−0.7440, −0.8976	0.2326, −0.5178
75°	−0.7657, 1.2386	−0.7199, −0.9973	0.0079, −0.3769
90°	−0.9304, 0.9976	−0.6957, −1.0978	0.2980, −0.5166

Solution Approach: Figure E.4.22 includes the input specified (in bold text) in the Appendix B.5 MATLAB file. Table E.4.8 includes the values of p_{1j}, $V_{p_{1j}}$ and $A_{p_{1j}}$ calculated from the Appendix B.5 MATLAB file over the given crank rotation range.

4.7 Watt II Mechanism Analysis

The Watt II mechanism consists of six links interconnected by revolute joints. As calculated from Gruebler's equation for planar mechanisms (with $L = 6$ and $J_1 = 7$), the Watt II mechanism has a single DOF.

From Figure 4.19a, it can be seen that the Watt II mechanism is essentially two planar four-bar mechanisms that share a common link—the follower link of one planar four-bar mechanism being the crank link of the other mechanism.* Because of this particular construction, the displacement, velocity, and acceleration equations presented in Section 4.4 for the planar four-bar mechanism can be used to analyze the Watt II mechanism.

Figure 4.19b includes the displacement variables for the Watt II mechanism. Mechanism loop $\mathbf{W}_1$–$\mathbf{V}_1$–$\mathbf{U}_1$–$\mathbf{G}_1$ is the planar four-bar mechanism loop presented in Section 4.4. For this mechanism loop, the equations in Section 4.4 can be used directly. Mechanism loop

---

* The Watt II mechanism can also be described as essentially a planar four-bar mechanism with a dyad attached to its follower link.

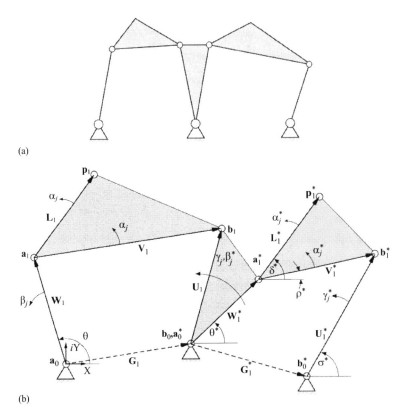

FIGURE 4.19
(a) Watt II mechanism and (b) mechanism displacement variables.

$\mathbf{W}_1^* - \mathbf{V}_1^* - \mathbf{U}_1^* - \mathbf{G}_1^*$ is the additional planar four-bar mechanism loop. Because vectors $\mathbf{W}_1^*$ (the crank of mechanism $\mathbf{W}_1^* - \mathbf{V}_1^* - \mathbf{U}_1^* - \mathbf{G}_1^*$) and $\mathbf{U}_1$ (the follower of mechanism $\mathbf{W}_1 - \mathbf{V}_1 - \mathbf{U}_1 - \mathbf{G}_1$) share the same link, the follower angular displacement, velocity, and acceleration values calculated for the first planar four-bar mechanism are the crank angular displacement, velocity, and acceleration values for the second planar four-bar mechanism (therefore $\gamma_j = \beta_j^*$, $\dot{\gamma}_j = \dot{\beta}_j^*$, and $\ddot{\gamma}_j = \ddot{\beta}_j^*$).

The equations in Section 4.4 can also be used to analyze mechanism loop $\mathbf{W}_1^* - \mathbf{V}_1^* - \mathbf{U}_1^* - \mathbf{G}_1^*$. When using the planar four-bar displacement equations for this mechanism loop, the results must be offset by a value of $\mathbf{G}_1^*$, since the grounded pivot of $\mathbf{W}_1^*$ (pivot $\mathbf{a}_0^*$ in Figure 4.19b) is offset from the coordinate system origin by this value. As a result, Equations 4.9 and 4.10, when utilized to calculate the coordinates of $\mathbf{p}_1^*$ would take on the form

$$\mathbf{p}_1^* = \mathbf{G}_1 + W_1^* e^{i\left(\theta^* + \beta_j^*\right)} + L_1^* e^{i\left(\delta^* + \alpha_j^*\right)} \tag{4.41}$$

$$p_{1xj}^* = G_{1x} + W_1^* \cos\left(\theta^* + \beta_j^*\right) + L_1^* \cos\left(\delta^* + \alpha_j^*\right)$$
$$p_{1yj}^* = G_{1y} + W_1^* \sin\left(\theta^* + \beta_j^*\right) + L_1^* \sin\left(\delta^* + \alpha_j^*\right) \tag{4.42}$$

The velocity and acceleration equations in Section 4.4, on the other hand, can be used directly for mechanism loop $\mathbf{W}_1^* - \mathbf{V}_1^* - \mathbf{U}_1^* - \mathbf{G}_1^*$.

Appendix B.6 includes the MATLAB file user instructions for Watt II displacement, velocity, and acceleration analysis. In this MATLAB file (which is available for download at www.crcpress.com/product/isbn/9781498724937), solutions for both Watt II mechanism loops are calculated using the planar four-bar displacement, velocity, and acceleration equations in Section 4.4. This MATLAB file also includes Equation 4.42 and its derivatives for the analysis coupler point $\mathbf{p}_1^*$.

Example 4.11

Problem Statement: Using the Appendix B.6 MATLAB file, calculate the displaced value of coupler point $\mathbf{p}_1^*$ for the Watt II mechanism configuration in Tables E.4.1 and E.4.9 for a –55° crank displacement angle.

Known Information: Tables E.4.1, E.4.9, and Appendix B.6 MATLAB file.

Solution Approach: Figure E.4.23 includes the input specified (in bold text) in the Appendix B.6 MATLAB file. From the calculated output, it can be determined that the displaced coupler point $\mathbf{p}_1$ is $\mathbf{p}_j = 0.7063 + i0.2577$ at a –55° crank displacement angle.

Example 4.12

Problem Statement: Using the Appendix B.6 MATLAB file, calculate the location and acceleration values of $\mathbf{p}_1^*$ at $\beta = 0°, -15°, -30, \dots, -90°$ given the Watt II mechanism dimensions in Table E.4.10. The initial angular velocity and angular acceleration of $\mathbf{W}_1$ are $\dot{\beta}_0 = -1.5 \, \text{rad/s}$ and $\ddot{\beta} = -0.25 \, \text{rad/s}^2$, respectively.

Known Information: Table E.4.10 and Appendix B.6 MATLAB file.

Solution Approach: Figure E.4.24 includes the input specified (in bold text) in the Appendix B.6 MATLAB file. Table E.4.11 includes the values of $\mathbf{p}_{1j}^*$, $\mathbf{V}_{\mathbf{p}_{1j}}^*$ and $\mathbf{A}_{\mathbf{p}_{1j}}^*$ calculated from the Appendix B.6 MATLAB file over the given crank rotation range.

TABLE E.4.9

Watt II Mechanism Configuration

W_1^*, θ^*	V_1^*, ρ^*	U_1^*, σ^*	G_{1x}^*, G_{1y}^*	L_1^*, δ^*
1, 45°	1.25, 16.6249°	1.25, 58.4069°	1.25, 0	1.5, 34.8197°

```
W1 = 1.75*exp(i*90*pi/180);
V1 = 1.75*exp(-i*22.4860*pi/180);
G1 = 0.866 - i*0.5;
U1 = 1.75*exp(i*64.5895*pi/180);
L1 = exp(i*6.469*pi/180);

W1s = exp(i*45*pi/180);
V1s = 1.25*exp(i*16.6249*pi/180);
G1s = 1.25;
U1s = 1.25*exp(i*58.4069*pi/180);
L1s = 1.5*exp(i*34.8197*pi/180);

start_ang  = 0;
step_ang   = -1;
stop_ang   = -55;

angular_vel  = 0 * ones(N+1,1);
angular_acc  = 0 * ones(N+1,1);
```

FIGURE E.4.23

Specified input (in bold text) in the Appendix B.6 MATLAB file for Example 4.11.

TABLE E.4.10

Watt II Mechanism Dimensions (with Link Lengths in m)

W_1, θ	V_1, ρ	U_1, σ	G_{1x}, G_{1y}	L_1, δ
1, 90°	1.5, 19.3737°	1.5, 93.2461°	1.5, 0	1, 60.7834°
W_1^*, θ^*	V_1^*, ρ^*	U_1^*, σ^*	G_{1x}^*, G_{1y}^*	L_1^*, δ^*
1, 45°	1.5, 7.9416°	1.5, 60.2717°	1.4489, −0.3882	1, 49.3512°

```
W1 = exp(i*90*pi/180);
V1 = 1.5*exp(i*19.3737*pi/180);
G1 = 1.5;
U1 = 1.5*exp(i*93.2461*pi/180);
L1 = exp(i*60.7834*pi/180);

W1s = exp(i*45*pi/180);
V1s = 1.5*exp(i*7.9416*pi/180);
G1s = 1.4489 - i*0.3882;
U1s = 1.5*exp(i*60.2717*pi/180);
L1s = exp(i*49.3512*pi/180);

start_ang   = 0;
step_ang    = -1;
stop_ang    = -90;

angular_vel  = -1.5 * ones(N+1,1);
angular_acc  = -0.25 * ones(N+1,1);
```

FIGURE E.4.24
Specified input (in bold text) in the Appendix B.6 MATLAB file for Example 4.12.

TABLE E.4.11

Calculated Watt II Point Positions, Velocities, and Accelerations

β	p_{1j}^* (m)	$V_{p_{1j}}$ (m/s)	$A_{p_{1j}}$ (m/s²)
0°	2.8585, 1.4658	0.5291, −0.5525	−1.1396, −0.3436
−15°	2.9309, 1.3677	0.3029, −0.5747	−1.4555, 0.1176
−30°	2.9610, 1.2759	0.0591, −0.5039	−1.3376, 0.7658
−45°	2.9566, 1.2078	−0.0796, −0.3041	−0.1181, 1.7982
−60°	2.9509, 1.1899	0.0535, 0.1407	1.4468, 4.1631
−75°	2.9606, 1.2790	−0.1487, 1.1232	−7.2492, 8.3587
−90°	2.7922, 1.5254	−2.2811, 1.7817	−13.9847, −4.4822

4.8 Stephenson III Mechanism Analysis

4.8.1 Displacement Equations

The Stephenson III mechanism consists of six links interconnected by revolute joints. As calculated from Gruebler's equation for planar mechanisms (with $L = 6$ and $J_1 = 7$), the Stephenson III mechanism has a single DOF.

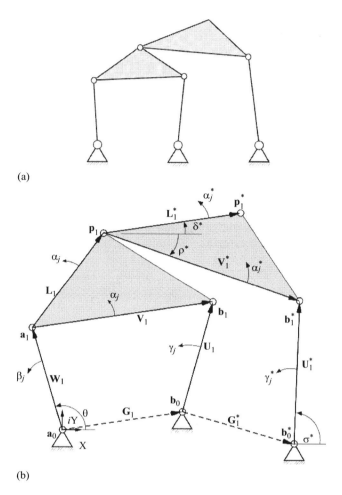

(a)

(b)

FIGURE 4.20
(a) Stephenson III mechanism and (b) mechanism displacement variables.

From Figure 4.20a, it can be seen that the Stephenson III mechanism is essentially a planar four-bar mechanism with a dyad connected to the coupler link. Because of this particular construction, the displacement, velocity, and acceleration equations presented in Section 4.4 for the planar four-bar mechanism can be used to analyze the planar four-bar mechanism loop of the Stephenson III mechanism.

To analyze the $\mathbf{U}_1^* - \mathbf{V}_1^*$ dyad (Figure 4.20b), a vector loop must be formed that includes this dyad. The equation produced from the clockwise vector-loop sum $\mathbf{W}_1 - \mathbf{L}_1 - \mathbf{V}_1^* - \mathbf{U}_1^* - \mathbf{G}_1^* - \mathbf{G}_1$ is

$$W_1 e^{i(\theta + \beta_j)} + L_1 e^{i(\delta + \alpha_j)} + V_1^* e^{i(\rho^* + \alpha_j^*)} - U_1^* e^{i(\sigma^* + \gamma_j^*)} - \mathbf{G}_1^* - \mathbf{G}_1 = 0 \tag{4.43}$$

After expanding Equation 4.43 and grouping its real and imaginary terms as separate equations, the resulting Stephenson III mechanism displacement equations become

$$f_1\left(\alpha_j^*, \gamma_j^*\right) = W_1 \cos\left(\theta + \beta_j\right) + L_1 \cos\left(\delta + \alpha_j\right) + V_1^* \cos\left(\rho^* + \alpha_j^*\right)$$

$$- U_1^* \cos\left(\sigma^* + \gamma_j^*\right) - G_{1x}^* - G_{1x} = 0$$

$$f_2\left(\alpha_j^*, \gamma_j^*\right) = W_1 \sin\left(\theta + \beta_j\right) + L_{11} \sin\left(\delta + \alpha_j\right) + V_1^* \sin\left(\rho^* + \alpha_j^*\right)$$

$$- U_1^* \sin\left(\sigma^* + \gamma_j^*\right) - G_{1y}^* - G_{1y} = 0$$

(4.44)

With the exception of the $\mathbf{U}_1^* - \mathbf{V}_1^*$ dyad displacement angles α_j^* and γ_j^*, all other variables in the Stephenson III mechanism displacement equations are either user prescribed or calculated from Equation 4.2. Because $f_1(\alpha_j^*, \gamma_j^*)$ and $f_2(\alpha_j^*, \gamma_j^*)$ in Equation 4.44 both include the unknown $\mathbf{U}_1^* - \mathbf{V}_1^*$ dyad displacement angles α_j^* and γ_j^* these equations form a set of nonlinear simultaneous equations from which the angles are calculated.

Example 4.13

Problem Statement: Using the Appendix B.7 MATLAB file, calculate the displacement angle of vector $\mathbf{V}_1^*$ for the Stephenson III mechanism configuration in Tables E.4.1 and E.4.12 for a 50° crank displacement angle.

Known Information: Tables E.4.1, E.4.12, and Appendix B.7 MATLAB file.

Solution Approach: Figure E.4.25 includes the input specified (in bold text) in the Appendix B.7 MATLAB file. From the calculated output, it can be determined that the displacement angle of vector $\mathbf{V}_1^*$ is $\alpha_j^* = 7.1215°$ at a 50° crank displacement angle.

TABLE E.4.12

Stephenson III Mechanism Configuration

V_1^*, ρ^*	U_1^*, σ^*	G_{1x}^*, G_{1y}^*
1.75, −8.8397°	1.75, 65.6031°	1.1340, 0.5

```
W1 = 1.75*exp(i*90*pi/180);
V1 = 1.75*exp(-i*22.4860*pi/180);
G1 = 0.866 - i*0.5;
U1 = 1.75*exp(i*64.5895*pi/180);
L1 = exp(i*6.469*pi/180);

V1s = 1.75*exp(-i*8.8397*pi/180);
G1s = 1.1340 + i*0.5;
U1s = 1.75*exp(i*65.6031*pi/180);
L1s = 0;

start_ang    = 0;
step_ang     = 1;
stop_ang     = 50;

angular_vel  = 0 * ones(N+1,1);
angular_acc  = 0 * ones(N+1,1);
```

FIGURE E.4.25
Specified input (in bold text) in the Appendix B.7 MATLAB file for Example 4.13.

4.8.2 Velocity Equations

Because the Stephenson III mechanism loop $\mathbf{W}_1 - \mathbf{V}_1 - \mathbf{U}_1 - \mathbf{G}_1$ is a planar four-bar mechanism, a velocity analysis for this mechanism loop can be conducted using the planar four-bar velocity equations in Section 4.4. In this section, velocity equations for the Stephenson III mechanism dyad $\mathbf{U}_1^* - \mathbf{V}_1^*$ (Figure 4.21) are presented.

A single Stephenson III velocity equation is derived by differentiating the Stephenson III displacement equation. Differentiating Equation 4.43 with respect to time produces

$$i\dot{\beta}_j W_1 e^{i(\theta+\beta_j)} + i\dot{\alpha}_j L_1 e^{i(\delta+\alpha_j)} + i\dot{\alpha}_j^* V_1^* e^{i(\rho^*+\alpha_j^*)} - i\dot{\gamma}_j^* U_1^* e^{i(\sigma^*+\gamma_j^*)} = 0 \qquad (4.45)$$

Equation 4.46 includes the individual velocity terms from Equation 4.45. The velocity $\mathbf{V}_{\mathbf{b}_1^*}$ is the *global* tangential velocity of vector $\mathbf{U}_1^*$—the tangential velocity of $\mathbf{b}_1^*$ with respect to $\mathbf{b}_0^*$ (see Figure 4.21). The velocity $\mathbf{V}_{\mathbf{p}_1-\mathbf{b}_1^*}$ is the *relative* tangential velocity of $\mathbf{p}_1$ with respect to $\mathbf{b}_1^*$—the relative tangential velocity of vector $\mathbf{V}_1^*$. The global velocity of $\mathbf{p}_1(\mathbf{V}_{\mathbf{p}_1})$ can be calculated from the last equation in Equation 4.46.

$$\mathbf{V}_{\mathbf{b}_1^*} = i\dot{\gamma}_j^* U_1^* e^{i(\sigma^*+\gamma_j^*)}$$

$$\mathbf{V}_{\mathbf{p}_1-\mathbf{b}_1^*} = i\dot{\alpha}_j^* V_1^* e^{i(\rho^*+\alpha_j^*)} = V_{\mathbf{p}_1} - V_{\mathbf{b}_1^*} \qquad (4.46)$$

After moving the terms $i\dot{\beta}_j W_1 e^{i(\theta+\beta_j)}$ and $i\dot{\alpha}_j L_1 e^{i(\delta+\alpha_j)}$ to the right-hand side of Equation 4.45, expanding this equation, and grouping its real and imaginary terms as separate equations, the resulting Stephenson III mechanism velocity equation in matrix form becomes

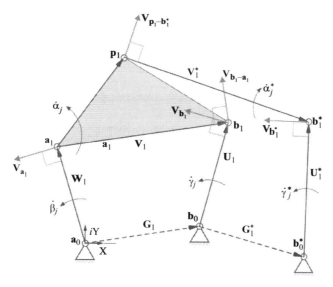

FIGURE 4.21
Stephenson III mechanism velocity variables.

$$
\begin{bmatrix}
-V_1^* \sin\left(\rho^* + \alpha_j^*\right) & U_1^* \sin\left(\sigma^* + \gamma_j^*\right) \\
V_1^* \cos\left(\rho^* + \alpha_j^*\right) & -U_1^* \cos\left(\sigma^* + \gamma_j^*\right)
\end{bmatrix}
\begin{Bmatrix}
\dot{\alpha}_j^* \\
\dot{\gamma}_j^*
\end{Bmatrix}
$$

$$
= \dot{\beta}_j
\begin{bmatrix}
W_1 \sin\left(\theta + \beta_j\right) \\
-W_1 \cos\left(\theta + \beta_j\right)
\end{bmatrix}
+ \dot{\alpha}_j
\begin{bmatrix}
L_1 \sin\left(\delta + \alpha_j\right) \\
-L_1 \cos\left(\delta + \alpha_j\right)
\end{bmatrix}
\tag{4.47}
$$

After including the prescribed mechanism variables and the unknowns calculated from the Stephenson III displacement equations, Equation 4.47 can be solved (using Cramer's rule, for example) to calculate the $U_1^* - V_1^*$ dyad angular velocities $\dot{\alpha}_j^*$ and $\dot{\gamma}_j^*$.

4.8.3 Acceleration Equations

A single planar four-bar acceleration equation is derived by differentiating the planar four-bar velocity equation. Time differentiation of Equation 4.45 produces

$$
-\dot{\beta}_j^2 W_1 e^{i(\theta + \beta_j)} + i\ddot{\beta}_j W_1 e^{i(\theta + \beta_j)} - \dot{\alpha}_j^2 L_1 e^{i(\delta + \alpha_j)} + i\ddot{\alpha}_j L_1 e^{i(\delta + \alpha_j)} - \left(\dot{\alpha}_j^*\right)^2 V_1^* e^{i\left(\rho^* + \alpha_j^*\right)}
$$

$$
+ i\ddot{\alpha}_j^* V_1^* e^{i\left(\rho^* + \alpha_j^*\right)} - \left(-\left(\dot{\gamma}_j^*\right)^2 U_1^* e^{i\left(\sigma^* + \gamma_j^*\right)} + i\ddot{\gamma}_j^* U_1^* e^{i\left(\sigma^* + \gamma_j^*\right)}\right) = 0
\tag{4.48}
$$

Equation 4.49 includes the individual acceleration terms from Equation 4.48. The accelerations $A_{b_{1n}^*}$ and $A_{b_{1t}^*}$ are the global normal and tangential accelerations, respectively, of vector U_1^*—the normal and tangential accelerations of b_1^* with respect to b_0^* (see Figure 4.22). The *total* acceleration of $A_{b_1^*}$ is the sum of these normal and tangential accelerations. The accelerations $A_{\left(p_1 - b_1^*\right)_n}$ and $A_{\left(p_1 - b_1^*\right)_t}$ are the relative normal and tangential

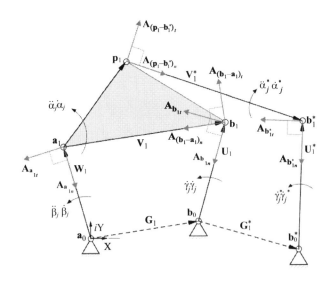

FIGURE 4.22
Stephenson III mechanism acceleration variables.

accelerations, respectively, of $\mathbf{p}_1$ with respect to $\mathbf{b}_1^*$—the relative normal and tangential accelerations of vector $\mathbf{V}_1^*$. The total relative acceleration $\mathbf{A}_{p_1-b_1^*}$ is the sum of these normal and tangential accelerations. The global acceleration of $\mathbf{p}_1(\mathbf{A}_{p_1})$ can be calculated from the last equation in Equation 4.49.

$$\mathbf{A}_{b_1^*} = \mathbf{A}_{b_{1t}^*} + \mathbf{A}_{b_{1n}^*} = i\ddot{\gamma}_j^* U_1^* e^{i\left(\sigma^*+\gamma_j^*\right)} - \left(\dot{\gamma}_j^*\right)^2 U_1^* e^{i\left(\sigma^*+\gamma_j^*\right)}$$

$$\mathbf{A}_{p_1-b_1^*} = \mathbf{A}_{\left(p_1-b_1^*\right)_t} + \mathbf{A}_{\left(p_1-b_1^*\right)_n} = i\ddot{\alpha}_j^* V_1^* e^{i\left(\rho^*+\alpha_j^*\right)} - \left(\dot{\alpha}_j^*\right)^2 V_1^* e^{i\left(\rho^*+\alpha_j^*\right)} = \mathbf{A}_{p_1} - \mathbf{A}_{b_1^*}$$

(4.49)

After moving all the terms except $i\ddot{\alpha}_j^* V_1^* e^{i\left(\rho^*+\alpha_j^*\right)}$ and $i\ddot{\gamma}_j^* U_1^* e^{i\left(\sigma^*+\gamma_j^*\right)}$ to the right-hand side of Equation 4.48, expanding this equation, and grouping its real and imaginary terms as separate equations, the resulting Stephenson III mechanism acceleration equation in matrix form becomes

$$\begin{bmatrix} -V_1^* \sin\left(\rho^*+\alpha_j^*\right) & U_1^* \sin\left(\sigma^*+\gamma_j^*\right) \\ V_1^* \cos\left(\rho^*+\alpha_j^*\right) & -U_1^* \cos\left(\sigma^*+\gamma_j^*\right) \end{bmatrix} \begin{Bmatrix} \ddot{\alpha}_j^* \\ \ddot{\gamma}_j^* \end{Bmatrix} = \dot{\beta}_j^2 \begin{bmatrix} W_1 \cos\left(\theta+\beta_j\right) \\ W_1 \sin\left(\theta+\beta_j\right) \end{bmatrix}$$

$$-\ddot{\beta}_j \begin{bmatrix} -W_1 \sin\left(\theta+\beta_j\right) \\ W_1 \cos\left(\theta+\beta_j\right) \end{bmatrix} + \dot{\alpha}_j^2 \begin{bmatrix} L_1 \cos\left(\delta+\alpha_j\right) \\ L_1 \sin\left(\delta+\alpha_j\right) \end{bmatrix} - \ddot{\alpha}_j \begin{bmatrix} -L_1 \sin\left(\delta+\alpha_j\right) \\ L_1 \cos\left(\delta+\alpha_j\right) \end{bmatrix}$$

$$+\left(\dot{\alpha}_j^*\right)^2 \begin{bmatrix} V_1^* \cos\left(\rho^*+\alpha_j^*\right) \\ V_1^* \sin\left(\rho^*+\alpha_j^*\right) \end{bmatrix} - \left(\dot{\gamma}_j^*\right)^2 \begin{bmatrix} U_1^* \cos\left(\sigma^*+\gamma_j^*\right) \\ U_1^* \sin\left(\sigma^*+\gamma_j^*\right) \end{bmatrix}$$

(4.50)

After including the prescribed mechanism variables and the unknowns calculated from the Stephenson III displacement and velocity equations, Equation 4.50 can be solved (using Cramer's rule, for example) to calculate the $\mathbf{U}_1^* - \mathbf{V}_1^*$ dyad angular accelerations $\ddot{\alpha}_j^*$ and $\ddot{\gamma}_j^*$.

Appendix B.7 includes the MATLAB file user instructions for Stephenson III displacement, velocity, and acceleration analysis. In this MATLAB file (which is available for download at www.crcpress. com/product/isbn/9781498724937), solutions for Equations 4.44, 4.47, and 4.50 are calculated.

4.8.4 Kinematics of Intermediate Link Locations of Interest

The results calculated from the Stephenson III displacement, velocity, and acceleration equations can be used in additional equations to calculate the displacement, velocity, and acceleration of particular intermediate link locations of interest. These equations are the result of formulating vector-loop equations and their derivatives.

Figure 4.18b illustrates a Stephenson III mechanism with an intermediate link vector $\mathbf{L}_1^*$ that points to an arbitrary intermediate link location of interest $\mathbf{p}_1^*$. Equations 4.51 and 4.52 are produced from the loop $\mathbf{a}_0 - \mathbf{a}_1 - \mathbf{p}_1 - \mathbf{p}_1^*$ in Figure 4.18b (where Equation 4.52 is the result of expanding and separating Equation 4.51).

$$\mathbf{p}_{1j}^* = W_1 e^{i(\theta+\beta_j)} + L_1 e^{i(\delta+\alpha_j)} + L_1^* e^{i(\delta^*+\alpha_j^*)} \tag{4.51}$$

$$p_{1xj}^* = W_1 \cos(\theta+\beta_j) + L_1 \cos(\delta+\alpha_j) + L_1^* \cos(\delta^*+\alpha_j^*)$$
$$p_{1yj}^* = W_1 \sin(\theta+\beta_j) + L_1 \sin(\delta+\alpha_j) + L_1^* \sin(\delta^*+\alpha_j^*) \tag{4.52}$$

The velocities of the intermediate link location of interest can be calculated from Equation 4.54—the expanded form of Equation 4.53 (which is the first derivative of Equation 4.51).

$$\mathbf{V}_{p_{1j}^*} = i\dot{\beta}_j W_1 e^{i(\theta+\beta_j)} + i\dot{\alpha}_j L_1 e^{i(\delta+\alpha_j)} + i\dot{\alpha}_j^* L_1^* e^{i(\delta^*+\alpha_j^*)} \tag{4.53}$$

$$V_{p_{1xj}^*} = -\dot{\beta}_j W_1 \sin(\theta+\beta_j) - \dot{\alpha}_j L_1 \sin(\delta+\alpha_j) - \dot{\alpha}_j^* L_1^* \sin(\delta^*+\alpha_j^*)$$
$$V_{p_{1yj}^*} = \dot{\beta}_j W_1 \cos(\theta+\beta_j) + \dot{\alpha}_j L_1 \cos(\delta+\alpha_j) + \dot{\alpha}_j^* L_1^* \cos(\delta^*+\alpha_j^*) \tag{4.54}$$

The accelerations of the intermediate link location of interest can be calculated from Equation 4.56—the expanded form of Equation 4.55 (which is the second derivative of Equation 4.51).

$$\mathbf{A}_{p_{1j}^*} = -\dot{\beta}_j^2 W_1 e^{i(\theta+\beta_j)} + i\ddot{\beta}_j W_1 e^{i(\theta+\beta_j)} - \dot{\alpha}_j^2 L_1 e^{i(\delta+\alpha_j)} + i\ddot{\alpha}_j L_1 e^{i(\delta+\alpha_j)} - \left(\dot{\alpha}_j^*\right)^2 L_1^* e^{i(\delta^*+\alpha_j^*)}$$
$$+ i\ddot{\alpha}_j^* L_1^* e^{i(\delta^*+\alpha_j^*)} \tag{4.55}$$

$$A_{p_{1xj}^*} = -\dot{\beta}_j^2 W_1 \cos(\theta+\beta_j) - \ddot{\beta}_j W_1 \sin(\theta+\beta_j) - \dot{\alpha}_j^2 L_1 \cos(\delta+\alpha_j) - \ddot{\alpha}_j L_1 \sin(\delta+\alpha_j)$$
$$- \left(\dot{\alpha}_j^*\right)^2 L_1^* \cos(\delta^*+\alpha_j^*) - \ddot{\alpha}_j^* L_1^* \sin(\delta^*+\alpha_j^*) \tag{4.56}$$

$$A_{p_{1yj}^*} = -\dot{\beta}_j^2 W_1 \sin(\theta+\beta_j) + \ddot{\beta}_j W_1 \cos(\theta+\beta_j) - \dot{\alpha}_j^2 L_1 \sin(\delta+\alpha_j) + \ddot{\alpha}_j L_1 \cos(\delta+\alpha_j)$$
$$- \left(\dot{\alpha}_j^*\right)^2 L_1^* \sin(\delta^*+\alpha_j^*) + \ddot{\alpha}_j^* L_1^* \cos(\delta^*+\alpha_j^*)$$

The Appendix B.7 MATLAB file also includes Equations 4.52, 4.54, and 4.56, from which displacement, velocity, and acceleration values are calculated for the intermediate link locations of interest.

Example 4.14

Problem Statement: Using the Appendix B.7 MATLAB file, calculate the path traced by point $\mathbf{p}_1^*$ of the Stephenson III mechanism given in Table E.4.13 for a complete crank rotation.

    Known Information: Table E.4.13 and Appendix B.7 MATLAB file.

    Solution Approach: Figure E.4.26 includes the input specified (in bold text) in the Appendix B.7 MATLAB file. Figure E.4.27 includes the Stephenson III mechanism and the $\mathbf{p}_1^*$ path calculated from the Appendix B.7 MATLAB file.

TABLE E.4.13

Stephenson III Mechanism Dimensions

W_1, θ	V_1, ρ	U_1, σ	G_{1x}, G_{1y}	L_1, δ
1, 90°	1.5, 19.3737°	1.5, 93.2461°	1.5, 0	1, 60.7834°
L_1^*, δ^*	V_1^*, ρ^*	U_1^*, σ^*	G_{1x}^*, G_{1y}^*	
1, 63.7091°	2, 17.1417°	2, 76.4844°	0.4318, 0.5176	

```
W1  = exp(i*90*pi/180);
V1  = 1.5*exp(i*19.3737*pi/180);
G1  = 1.5;
U1  = 1.5*exp(i*93.2461*pi/180);
L1  = exp(i*60.7834*pi/180);

V1s = 2*exp(i*17.1417*pi/180);
G1s = 0.4318 + i*0.5176;
U1s = 2*exp(i*76.4844*pi/180);
L1s = exp(i*63.7091*pi/180);

start_ang    = 0;
step_ang     = 1;
stop_ang     = 360;

angular_vel  = 0 * ones(N+1,1);
angular_acc  = 0 * ones(N+1,1);
```

FIGURE E.4.26
Specified input (in bold text) in the Appendix B.7 MATLAB file for Example 4.14.

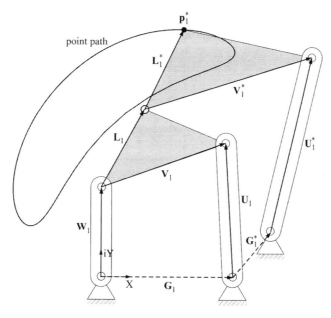

FIGURE E.4.27
Stephenson III mechanism and $\mathbf{p}_1^*$ path.

TABLE E.4.14

Geared Five-Bar Driver Velocities and Point Positions, Velocities, and Accelerations

β	$\dot{\beta}$ (rad/s)	p_{1j} (m)	$V_{p_{1j}}$ (m/s)	$A_{p_{1j}}$ (m/s^2)
0°	1	0.2733, 1.9619	−0.8980, −0.0290	−0.2352, −0.9733
15°	1.0258	0.0360, 1.9215	−0.9273, −0.2851	0.0102, −0.9959
30°	1.0511	−0.1949, 1.8184	−0.8956, −0.5302	0.2336, −0.9326
45°	1.0757	−0.4068, 1.6609	−0.8202, −0.7419	0.3516, −0.7683
60°	1.0997	−0.5944, 1.4624	−0.7440, −0.8976	0.2326, −0.5178
75°	1.1233	−0.7657, 1.2386	−0.7199, −0.9973	0.0079, −0.3769
90°	1.1464	−0.9304, 0.9976	−0.6957, −1.0978	0.2980, −0.5166

4.9 Time and Driver Angular Velocity

In the Chapter 4, 7, and 10 MATLAB files, the driving link angular velocity $\dot{\beta}$ is calculated according to $\dot{\beta} = \dot{\beta}_0 + \ddot{\beta}t$, where $\dot{\beta}_0, \ddot{\beta}$, and t are the initial angular velocity, angular acceleration, and time, respectively.* Under a constant velocity condition (where $\ddot{\beta} = 0$ and subsequently $\dot{\beta} = \dot{\beta}_0$), the time increment becomes $t = \beta / \dot{\beta}$. Under a condition where both an initial angular velocity and an angular acceleration are present, the time increment becomes

$$t = \frac{\left(-\dot{\beta}_0 \pm \sqrt{\dot{\beta}_0^2 + 2\ddot{\beta}\beta}\right)}{\ddot{\beta}},$$

where only the smallest positive value of the two solutions is valid.

Example 4.15

Problem Statement: For the problem in Example 4.10, include $\dot{\beta}$ in Table E.4.8.
  Known Information: Example 4.10 and Appendix B.4 MATLAB file.
  Solution Approach: Table E.4.14 includes the values of $\dot{\beta}$ calculated using $\dot{\beta}_0$ and $\ddot{\beta}$ from Example 4.10, the t values from the Appendix B.5 MATLAB file and the angular velocity equation $\dot{\beta} = \dot{\beta}_0 + \ddot{\beta}t$ presented in Section 4.9.

4.10 Mechanism Configurations

For a given crank link orientation of single-loop four- or five-bar mechanisms, there are two ways to assemble the remaining links. These distinct assemblies are called *assembly configurations* (also called *open* and *crossed* configurations) and can be easily determined graphically for planar mechanisms. Figure 4.23 illustrates the assembly configurations of the planar four-bar, geared five-bar, and slider-crank mechanisms. Reflecting $\mathbf{b}_1$ in the planar four-bar mechanism about an axis that passes through $\mathbf{a}_1$ and $\mathbf{b}_0$ produces the alternate

---

* In Chapter 10, the crank rotation variables, $\dot{\beta}_0 \dot{\beta}$, and $\ddot{\beta}$ become $\dot{\theta}_0, \dot{\theta}$, and $\ddot{\theta}$, respectively.

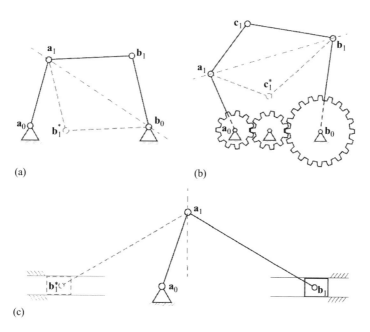

FIGURE 4.23
(a) Planar four-bar, (b) geared five-bar, and (c) slider-crank mechanism assembly configurations.

assembly configuration $\mathbf{a}_0 - \mathbf{a}_1 - \mathbf{b}_1^* - \mathbf{b}_0$. Reflecting $\mathbf{c}_1$ in the geared five-bar mechanism about an axis that passes through $\mathbf{a}_1$ and $\mathbf{b}_1$ produces the alternate assembly configuration $\mathbf{a}_0 - \mathbf{a}_1 - \mathbf{c}_1^* - \mathbf{b}_1 - \mathbf{b}_0$. Reflecting $\mathbf{b}_1$ in the slider-crank mechanism about an axis that passes through $\mathbf{a}_1$ and is perpendicular to the slider path produces the alternate assembly configuration $\mathbf{a}_0 - \mathbf{a}_1 - \mathbf{b}_1^*$.

In the type of planar four-bar, geared five-bar, and slider-crank mechanism displacement equations where mechanism motion is calculated *algebraically*, two sets of unknown link displacement angles are calculated for each given crank link displacement angle [3].* These two sets of displacement angle solutions correspond to the two mechanism assembly configurations. This can be observed in the algebraic spatial mechanism equations that appear in Chapter 10.

4.11 Constructing Cognates

For a given planar four-bar path mechanism, there are alternate four-bar mechanisms of different dimensions that will trace coupler point curves identical to the given four-bar mechanism. These alternate mechanisms are called *cognates* [4, 5]. A well-known schematic to construct two cognates for a given four-bar mechanism is the *Cayley diagram*. The construction of this schematic begins by repositioning the crank and follower links of the given four-bar mechanism so that the crank, coupler, and follower links are collinear

---

* The kinematic equations presented in this chapter are solved *numerically* given the initial mechanism assembly configuration. Therefore, kinematic analyses are performed on the particular mechanism assembly configuration specified by the user.

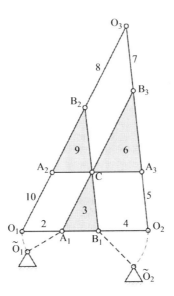

FIGURE 4.24
Cayley diagram construction.

(mechanism O_1–A_1–B_1–O_1 in Figure 4.24). Next, lines are constructed that are parallel to all sides of the links in the original linkage. In Figure 4.24, the two newly constructed cognates are visible (mechanisms O_1–A_2– B_2–O_3 and O_2–A_3–B_3–O_3).

After constructing the Cayley diagram, the fixed pivots of the original four-bar linkage are returned to their original placements, making this linkage and the cognates movable (as positioned in the Cayley diagram, mechanism O_1–A_1–B_1–O_2 and cognates are immovable). When returning the fixed pivots of the original planar four-bar mechanism, the crank, coupler, and follower lengths of the cognates should be maintained. The repositioned Cayley diagram is called a *Roberts diagram* Figure 4.25a). Figure 4.25b illustrates the three separate cognates. All three cognates will trace the same curve at coupler point C.

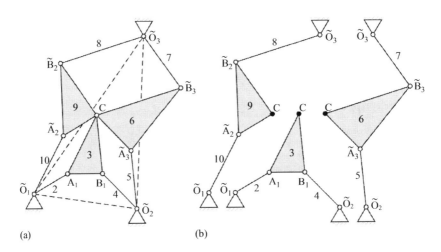

(a)   (b)

FIGURE 4.25
(a) Roberts diagram and (b) separate cognates.

Example 4.16

Problem Statement: Construct the cognates for the Grashof drag-link mechanism with the unitless dimensions $\mathbf{a}_0 = (0, 0)$, $\mathbf{a}_1 = (0, 1.75)$, $\mathbf{b}_0 = (1, 0)$, $\mathbf{b}_1 = (1.4646, 1.4262)$, and $\mathbf{p}_1 = (0.8751, 2.2340)$.

   Known Information: Section 4.11 and the given mechanism dimensions.

   Solution Approach: Figure E.4.28 includes the given planar four-bar mechanism and its Cayley diagram. Figure E.4.29 includes the Roberts diagram and separate cognates.

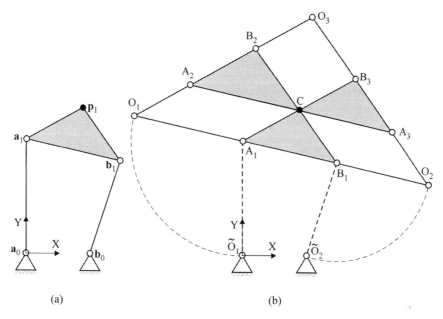

(a)   (b)

FIGURE E.4.28
(a) Planar four-bar mechanism and (b) Cayley diagram.

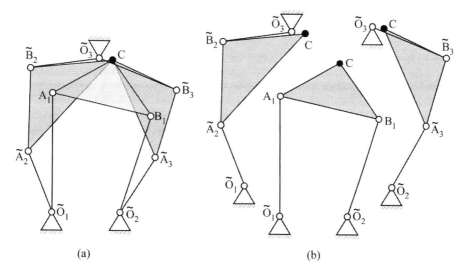

(a)   (b)

FIGURE E.4.29
(a) Roberts diagram and (b) separate cognates.

4.12 Planar Mechanism Kinematic Analysis and Modeling in Simmechanics®

It has been noted throughout this chapter that Appendices B.1 and B.2 through B.6 include user instructions for the planar four-bar, slider-crank, geared five-bar, Watt II, and Stephenson III mechanisms in MATLAB files, respectively. In these files, the displacement, velocity, and acceleration equations formulated in this chapter are solved. These MATLAB files provide a means for the user to efficiently conduct planar four-bar, slider-crank, geared five-bar, Watt II, and Stephenson III kinematic analyses by solving their displacement, velocity, and acceleration equations.

This textbook also utilizes SimMechanics as an alternate approach for kinematic analysis. SimMechanics is a MATLAB toolbox that provides a physical modeling environment for the mechanical modeling and simulation of rigid, multibody systems. The MATLAB and SimMechanics files that accompany this textbook use distinct approaches for kinematic analysis. In the Appendix B MATLAB files, for example, solutions are predominantly calculated analytically from closed-form kinematic equations. In SimMechanics, however, *Newtonian equations* (produced by building mechanism links and joints in a physical modeling environment) are solved using *ordinary differential equation* (ODE) solvers [6].

A library of SimMechanics files is available for download at www.crcpress.com/product/isbn/9781498724937 to conduct displacement, velocity, and acceleration analyses on the planar four-bar, slider-crank, geared five-bar, Watt II, and Stephenson III mechanisms. With these files, the user specifies the mechanism link dimensions and driving link parameters (e.g., crank displacements, velocities, or accelerations) and the displacements, velocities or accelerations of the mechanism locations of interest are measured. Additionally, the motion of the mechanism itself is simulated. The SimMechanics file user instructions for the planar four-bar, slider-crank, geared five-bar, Watt II, and Stephenson III mechanisms are given in Appendices H.1 through H.6, respectively.

Example 4.17

Problem Statement: Using the Appendix H.1 SimMechanics files, plot the transmission angle versus crank displacement angle plot for the planar four-bar mechanism in Table E.4.15 for a complete crank rotation.

  Known Information: Table E.4.15 and Appendix H.1 SimMechanics files.

  Solution Approach: Figure E.4.30 includes the input specified (in bold text) in the Appendix H.1 SimMechanics file. Figure E.4.31 illustrates a plot of the transmission angle data measured in the Appendix H.1 SimMechanics files.

TABLE E.4.15

Planar Four-Bar Mechanism Dimensions

W_1, θ	V_1, ρ	U_1, σ	G_{1x}, G_{1y}
1, 90°	1.5, 4.2451°	1.5, 88.2046°	1.4489, −0.3882

```
W1 = 1*exp(i*90*pi/180);
V1 = 1.5*exp(i*4.2451*pi/180);
G1 = 1.4489 - i*0.3882;
U1 = 1.5*exp(i*88.2046*pi/180);
L1 = 0;

start_ang    = 0;
step_ang     = 1;
stop_ang     = 360;

angular_vel  = 0;
angular_acc  = 0;
```

FIGURE E.4.30

Specified input (in bold text) in the Appendix H.1 SimMechanics file for Example 4.17.

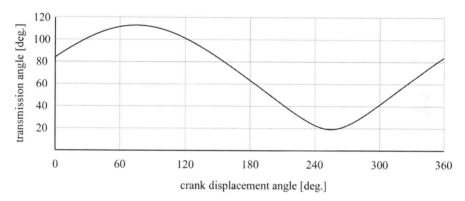

FIGURE E.4.31

Transmission angle vs. crank displacement angle plot.

TABLE E.4.16

Slider-Crank Mechanism Dimensions (with Lengths in cm)

W_1, θ	V_1	U_1
1, 90°	3	−0.6583

Example 4.18

Problem Statement: Using the Appendix H.2 SimMechanics files, plot slider position, velocity, and acceleration versus crank angular displacement plots for the slider-crank mechanism in Table E.4.16 for a complete crank rotation. The initial angular velocity and angular acceleration of the crank link are 1 rad/s and 0.1 rad/s^2, respectively.

Known Information: Table E.4.15, driving link parameters and Appendix H.2 SimMechanics files.

Solution Approach: Figure E.4.32 includes the input specified (in bold text) in the Appendix H.2 SimMechanics file. Figure E.4.33 illustrates plots of the slider position, velocity, and acceleration data measured in the Appendix H.2 SimMechanics files.

Example 4.19

Problem Statement: Repeat Example 4.10 using the Appendix H.3 SimMechanics files. In this example, $\beta = 0°, -15°, -30°, \ldots, -90°, \dot{\beta}_0 = -1$ rad/s, and $\ddot{\beta} = -1.0$ rad/s^2, respectively.

Known Information: Table E.4.7 and Appendix H.3 SimMechanics files.

```
LW1 = 1;
theta = 90*pi/180;

LU1 = -0.6583;

LV1 = 3;

start_ang   = 0;
step_ang    = 1;
stop_ang    = 360;

angular_vel  = 1;
angular_acc  = 0.1;
```

FIGURE E.4.32
Specified input (in bold text) in the Appendix H.2 SimMechanics file for Example 4.18.

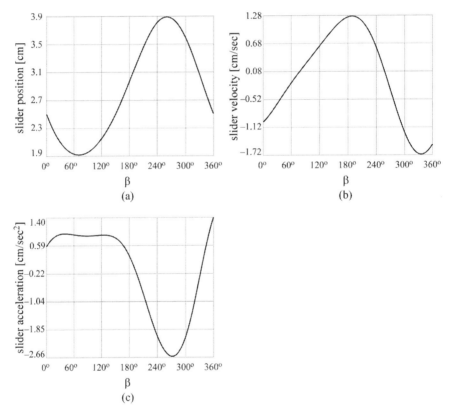

FIGURE E.4.33
Slider (a) position, (b) velocity, and (c) acceleration plots.

Solution Approach: Because the Appendix H.3 SimMechanics files consider geared five-bar mechanisms where the crank and output links rotate in opposite directions, the gear ratio for this problem is $r = -2$.* Figure E.4.34 includes the input specified (in bold text) in the Appendix H.3 SimMechanics file. Table E.4.17 includes the values of $\mathbf{p}_{1j}$, $\mathbf{V}_{\mathbf{p}_{1j}}$, and $\mathbf{A}_{\mathbf{p}_{1j}}$ measured from Appendix H.3 SimMechanics files.

---

* The Appendix H.4 SimMechanics file considers geared five-bar mechanism where the crank and output links rotate in the same direction, while the Appendix H.3 file considers links that rotate in the opposite direction.

```
W1 = 1.0*exp(i*90*pi/180);
V1 = 1.5*exp(i*32.7304*pi/180);
L1 = 1.0*exp(i*74.1400*pi/180);
G1 = 1.5;
U1 = 1.5*exp(i*45*pi/180);
S1 = 1.5*exp(i*149.9847*pi/180);
ratio = -2;

start_ang   = 0;
step_ang    = -15;
stop_ang    = -90;

angular_vel  = -1;
angular_acc  = -0.1;
```

FIGURE E.4.34
Specified input (in bold text) in the Appendix H.3 SimMechanics file for Example 4.19.

TABLE E.4.17

Measured Geared Five-Bar Point Positions, Velocities, and Accelerations

β	p_{1j} (m)	V_{p1j} (m/s)	A_{p1j} (m/s²)
0°	0.2733, 1.9619	−0.1473, 0.326	0.4799, −2.5867
−14.999°	0.2391, 1.9657	−0.1531, −0.288	−0.376, −2.3558
−29.996°	0.1848, 1.8151	−0.2857, −0.9225	−0.5636, −2.7124
−44.995°	0.102, 1.5034	−0.3576, −1.6103	0.1527, −2.6826
−59.992°	0.0314, 1.0511	−0.1897, −2.0727	1.0793, −0.8176
−74.999°	0.0135, 0.5637	0.0062, −1.9739	0.3128, 1.4234
−90.001°	0.0119, 0.1535	−0.0653, −1.5665	−0.8276, 1.8546

Example 4.20

Problem Statement: Using the Appendix H.5 SimMechanics files, plot the path traced by point p_1^* of the Watt II mechanism given in Example 4.12 for a complete crank rotation.

Known Information: Table E.4.10 and Appendix H.5 SimMechanics files.

Solution Approach: Figure E.4.35 includes the input specified (in bold text) in the Appendix H.5 SimMechanics file. Figure E.4.36 includes the Watt II mechanism and the p_1^* path calculated from the Appendix H.5 SimMechanics files.

```
W1 = exp(i*90*pi/180);
V1 = 1.5*exp(i*19.3737*pi/180);
G1 = 1.5;
U1 = 1.5*exp(i*93.2461*pi/180);
L1 = exp(i*60.7834*pi/180);

W1s = exp(i*45*pi/180);
V1s = 1.5*exp(i*7.9416*pi/180);
G1s = 1.4489 - i*0.3882;
U1s = 1.5*exp(i*60.2717*pi/180);
L1s = exp(i*49.3512*pi/180);

start_ang   = 0;
step_ang    = 1;
stop_ang    = 360;

angular_vel  = 0;
angular_acc  = 0;
```

FIGURE E.4.35
Specified input (in bold text) in the Appendix H.5 SimMechanics file for Example 4.20.

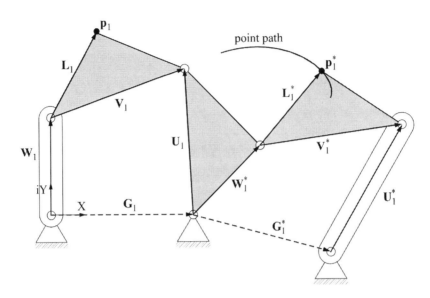

FIGURE E.4.36
Watt II mechanism and $\mathbf{p}_1^*$ path.

Example 4.21

Problem Statement: Using the Appendix H.6 SimMechanics files for the mechanism in Example 4.14, measure the position of vector $\mathbf{V}_1^*$ over a complete crank rotation (at 45° rotation increments).

Known Information: Table E.4.13 and Appendix H.6 SimMechanics files.

Solution Approach: One way to define the position of vector $\mathbf{V}_1^*$ of the Stephenson III mechanism is by point $\mathbf{p}_1$ and angle α^*. Figure E.4.37 includes the input specified (in bold text) in the Appendix H.6 SimMechanics file. Table E.4.18 includes the $\mathbf{V}_1^*$ position data measured from the Appendix H.6 SimMechanics files.

```
W1 = exp(i*90*pi/180);
V1 = 1.5*exp(i*19.3737*pi/180);
G1 = 1.5;
U1 = 1.5*exp(i*93.2461*pi/180);
L1 = exp(i*60.7834*pi/180);

V1s = 2*exp(i*17.1417*pi/180);
G1s = 0.4318 + i*0.5176;
U1s = 2*exp(i*76.4844*pi/180);
L1s = exp(i*63.7091*pi/180);

start_ang   = 0;
step_ang    = 45;
stop_ang    = 360;

angular_vel  = 0;
angular_acc  = 0;
```

FIGURE E.4.37
Specified input (in bold text) in the Appendix H.6 SimMechanics file for Example 4.21.

TABLE E.4.18

Measured Position Data for Stephenson III Mechanism Dyad Vector V_1^*

β	P_1	α*
0°	0.4881, 1.8728	0°
45.014°	−0.2543, 1.5985	8.9846°
89.996°	−0.7406, 0.9658	20.694°
134.98°	−0.8565, 0.282	33.297°
180.02°	−0.6177, −0.2139	47.332°
225.01°	−0.179, −0.2437	58.569°
269.99°	0.4725, 0.8493	38.084°
315°	1.0664, 1.6403	−0.2703°

4.13 Summary

The method of vector-loop closure is employed in this chapter to formulate kinematic equations for the planar four-bar mechanism, slider-crank mechanism, geared five-bar mechanism, Watt II, and Stephenson III mechanisms. With this method, a sum of the closed loop of mechanism link vectors is taken, expanded, grouped into real and imaginary components, and expressed as a system of two equations to calculate two unknown mechanism variables. With this approach, equations that fully describe the position, displacement, velocity, and acceleration of each mechanism link are formulated. Taking the first and second derivatives of the mechanism displacement equations produces velocity and acceleration equations, respectively. The unknown variables calculated from these kinematic equations can be used in additional vector-loop equations to calculate the positions, displacements, velocities, and accelerations of additional mechanism locations of interest.

The displacement equations presented in this chapter form sets of two nonlinear simultaneous equations. Unlike linear simultaneous equations, nonlinear simultaneous equations cannot be solved algebraically. Using a root-finding method, such equation sets can be solved numerically. The Newton–Raphson method is one of the most common root-finding methods. In the Appendix B.1 and B.3 through B.6 MATLAB files, the displacement, velocity, and acceleration equations for the planar four-bar, slider-crank, geared five-bar, Watt II, and Stephenson III mechanisms, respectively, are solved numerically.

An instant center (or IC) of velocity is a common point among two bodies in planar motion which has the same instantaneous velocity in both bodies. A procedure (based on the Aronhold–Kennedy theorem) has been developed to help locate the ICs for a given mechanism. ICs can conduct velocity analyses as well as replicate coupler motion. A locus of ICs is called a *centrode*. The centrode produced for a given mechanism is called a fixed centrode because it is stationary. The centrode produced for the inverted mechanism is called a moving centrode because it can exhibit motion—rolling motion over the fixed centrode. The fixed and moving centrodes can be incorporated into geometry, to ultimately produce alternate mechanisms to replicate the coupler motion of their corresponding four-bar mechanisms. In the Appendix B.2 MATLAB file, individual fixed and moving centrode points are calculated for the complete rotation range of a planar four-bar mechanism, as well as slider-crank mechanisms defined as planar four-bar mechanisms.

For a given crank link orientation of most single-loop four- or five-bar mechanisms, there are two distinct assembly configurations (the open and crossed configurations). These configurations can be easily determined graphically for planar mechanisms by reflecting particular links about particular axes.

For a given planar four-bar mechanism, there are alternate four-bar mechanisms (called cognates) that will trace coupler curves identical to the original mechanism. The Cayley diagram is a well-known schematic to construct two cognates for a given four-bar mechanism.

This textbook also utilizes SimMechanics as an alternate approach for simulation-based kinematic analyses. Using the Appendix H.1 through H.6 SimMechanics files, the user can conduct displacement, velocity, and acceleration analyses on the planar four-bar, slider-crank, geared five- bar, Watt II, and Stephenson III mechanisms, respectively, as well as simulate mechanism motion.

References

1. Kimbrell, J. T. 1991. *Kinematics Analysis and Synthesis*. Chapter 2. New York: McGraw-Hill.
2. Chapra, S. C. and Canale, R. P. 1998. *Numerical Methods for Engineers*. Chapters 5–8. New York: McGraw-Hill.
3. Norton, R. L. 2008. *Design of Machinery*. 4th edn, pp. 355–360. New York: McGraw-Hill.
4. Ibid, pp. 132–137.
5. Waldron, K. J. and Kinzel, G. L. 2004. *Kinematics, Dynamics and Design of Machinery*. 2nd edn, pp. 318–320. Saddle River, NJ: Prentice-Hall.
6. Mathworks. What you can do with SimMechanics software. http://www.mathworks.com / help/physmod/sm/mech/gs/what-you-can-do-with-simmechanics-software.html. Accessed July 16, 2015.

Additional Reading

Sandor, G. N. and Erdman, A. G. 1984. *Advanced Mechanism Design: Analysis and Synthesis*. Volume 2, p. 64. Englewood Cliffs, NJ: Prentice-Hall.
Waldron, K. J. and Kinzel, G. L. 2004. *Kinematics, Dynamics and Design of Machinery*. 2nd edn, Chapter 5. Saddle River, NJ: Prentice-Hall.
Wilson, C. E. and Sadler, J. P. 2003. *Kinematics and Dynamics of Machinery*. 3rd edn, Chapter 2. Saddle River, NJ: Prentice-Hall.

Problems

1. Figure P.4.1 illustrates a planar four-bar mechanism used to guide a hatch from the closed-hatch position to the opened-hatch position. The dimensions for the illustrated mechanism are included in Table P.4.1. Using a crank rotation increment

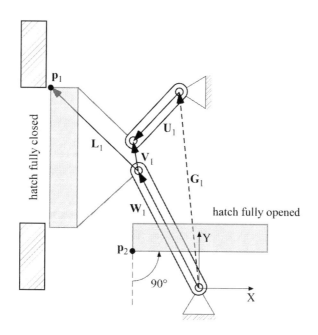

FIGURE P.4.1
Hatch mechanism.

TABLE P.4.1

Hatch Mechanism Dimensions

W_1, θ	V_1, ρ	U_1, σ	G_{1x}, G_{1y}	L_1, δ
4.4127, 118.7982°	1.0214, 101.2268°	2.2807, 225.1319°	−0.7156, 6.4851	4.1345, 139.4559°

of −1°, determine the crank rotation range required to reach the opened-hatch position as well as the value of point $\mathbf{p}_1$ at the opened-hatch position (using the Appendix B.1 or H.1 files).

2. Figure P.4.2 illustrates a planar four-bar mechanism used to guide a bucket from the loading position to the unloading position. The dimensions for the illustrated mechanism are included in Table P.4.2. Considering a crank rotation increment of −0.1°, determine the crank rotation range required to reach the unloading position as well as the value of point $\mathbf{p}_1$ at the unloading position (using the Appendix B.1 or H.1 files).

3. Using the Appendix B.1 or H.1 files, produce a velocity magnitude versus crank angular displacement plot and an acceleration magnitude versus crank angular displacement plot for point $\mathbf{p}_1$ of the leveling crane illustrated in Figure P.4.3 over a −35° crank rotation range. The initial crank angular velocity and angular acceleration are −1 rad/s and −0.1 rad/s², respectively.

4. Using the Appendix B.2 file, plot the fixed and moving centrodes for the mechanism in Problem 1. Limit your minimum and maximum *x*- and *y*-axis plot ranges to ±10.

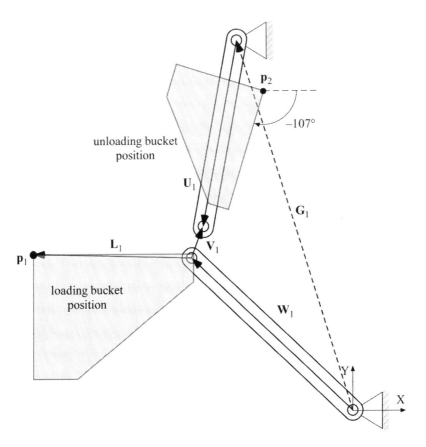

FIGURE P.4.2
Loading–unloading mechanism.

TABLE P.4.2

Loading–Unloading Mechanism Dimensions

W_1, θ	V_1, ρ	U_1, σ	G_{1x}, G_{1y}	L_1, δ
2.9777, 136.588°	0.4032, 70.2408°	2.513, 259.5885°	−1.5726, 4.8975	2.1995, 179.5331°

5. Using the Appendix B.2 file, plot the fixed and moving centrodes for the mechanism in Problem 2. Limit your minimum and maximum x- and y-axis plot ranges to ±5.

6. Figure P.4.4 illustrates a planar four-bar mechanism used to guide a component from the initial position to the assembled position. The dimensions for the illustrated mechanism are included in Table P.4.3. Considering a crank rotation increment of +0.025°, determine the crank rotation range required to reach the assembled position as well as the value of point $\mathbf{p}_1$ at the assembled position (using the Appendix B.1 or H.1 files).

7. Figure P.4.5 illustrates a planar four-bar mechanism used to guide a digging bucket from the initial position to the final position. The dimensions for the illustrated mechanism are included in Table P.4.4. Considering a crank rotation increment

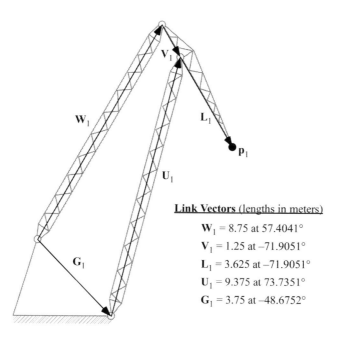

Link Vectors (lengths in meters)
W_1 = 8.75 at 57.4041°
V_1 = 1.25 at −71.9051°
L_1 = 3.625 at −71.9051°
U_1 = 9.375 at 73.7351°
G_1 = 3.75 at −48.6752°

FIGURE P.4.3
Leveling crane.

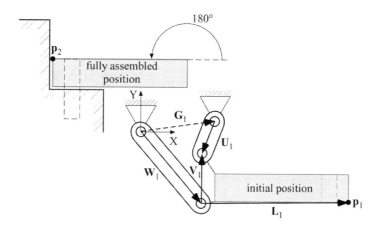

FIGURE P.4.4
Component-assembly mechanism.

TABLE P.4.3

Component-Assembly Mechanism Dimensions

W_1, θ	V_1, ρ	U_1, σ	G_{1x}, G_{1y}	L_1, δ
3.0645, 310.6493°	1.6039, 89.216°	1.1359, −112.147°	2.4465, 0.3309	4.8525, 0.7042°

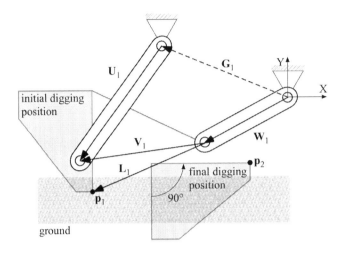

FIGURE P.4.5
Digging mechanism.

TABLE P.4.4

Digging Mechanism Dimensions

W_1, θ	V_1, ρ	U_1, σ	G_{1x}, G_{1y}	L_1, δ
4.1332, 207.829°	5.3857, 187.8282°	6.0475, 233.487°	−5.3924, 2.1975	5.1839, 202.3921°

TABLE P.4.5

Planar Four-Bar Mechanism Configurations

W_1	V_1, ρ	U_1, σ	G_1	L_1, δ
2.1, 90°	2.4, −16.0138°	2.4, 58.1173°	1.0392, −0.6	2.4, 12.9412°
0.7361, −0.425	1.7, 58.1173°	1.7, −16.0138°	1.4875, 90°	1.7, 12.9412°

of −0.2°, determine the crank rotation range required to reach the final digging position as well as the value of point $\mathbf{p}_1$ at the final digging position (using the Appendix B.1 or H.1 files).

8. For the two planar four-bar mechanism configurations given in Table P.4.5, plot the paths traced by point $\mathbf{p}$ over a complete crank rotation range (using the Appendix B.1 or H.1 files).

9. Using the Appendix B.2 file, plot the fixed and moving centrodes for the mechanism in Problem 6. Limit your minimum and maximum x- and y-axis plot ranges to ±10.

10. Using the Appendix B.2 file, plot the fixed and moving centrodes for the mechanism in Problem 7. Limit your minimum and maximum x- and y-axis plot ranges to ±20.

11. Using the Appendix B.1 or H.1 files, calculate the rotation range of the designated coupler link in the planar four-bar folding chair linkage from the fully opened position to the fully closed position illustrated in Figure P.4.6.

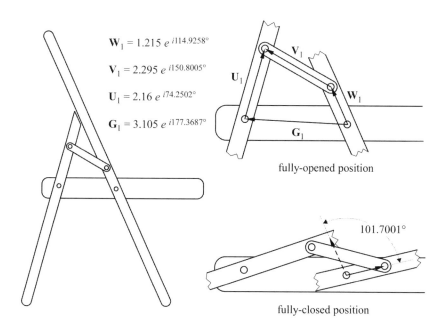

$$W_1 = 1.215\ e^{\ i114.9258°}$$

$$V_1 = 2.295\ e^{\ i150.8005°}$$

$$U_1 = 2.16\ e^{\ i74.2502°}$$

$$G_1 = 3.105\ e^{\ i177.3687°}$$

fully-opened position

101.7001°

fully-closed position

FIGURE P.4.6
Folding-chair linkage in (left) fully open and (right) fully closed position.

12. Using the Appendix B.1 or H.1 files, determine if the two folding chair linkage designs illustrated in Figure P.4.7 will produce properly-folded chairs (a chair that permits a crank rotation of −101.7°) and if not, why?

13. For the two planar four-bar mechanism configurations given in Problem 8, produce transmission angle versus crank displacement angle plots over a complete crank rotation range (using the Appendix B.1 or H.1 files).

14. For the two planar four-bar mechanism configurations given in Problem 8 (with link lengths in meters), produce tables of the velocity and acceleration vectors of P_1 over a complete crank rotation range at 30° crank rotation increments (using the Appendix B.1 or H.1 files). The initial crank angular velocity and angular acceleration are 1 rad/s and 0.25 rad/s², respectively.

15. Figure P.4.8 illustrates a planar four-bar mechanism used to guide a wiping blade. For this mechanism, produce a table of the angular displacements, velocities, and accelerations of the follower link over a 45° crank rotation range at 5° crank rotation increments (using the Appendix B.1 or H.1 files). The crank angular velocity and acceleration are 1.25 rad/s and 0.15 rad/s², respectively.

16. For the slider-crank mechanism configuration given in Table P.4.6, produce slider displacement, velocity, and acceleration (versus crank displacement angle) plots over a complete crank rotation range (using the Appendix B.3 or H.2 files).

17. For the slider-crank mechanism configuration given in Table P.4.6, produce a table of the angular displacement, velocity, and acceleration of V_1 over a complete crank rotation range at 30° crank rotation increments (using the Appendix B.3 or H.2 files).

18. Using the Appendix B.2 file, plot the fixed and moving centrodes for the mechanism in Problem 16 (let $b_0 = (b_{1x}, -1{,}000{,}000)$). Limit your minimum and maximum *x*- and *y*-axis plot ranges to ±20.

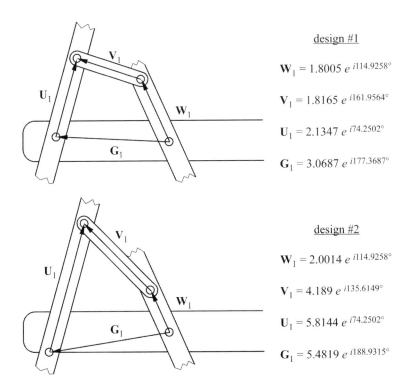

$\underline{\text{design \#1}}$

$\mathbf{W}_1 = 1.8005 \; e^{\; i114.9258°}$

$\mathbf{V}_1 = 1.8165 \; e^{\; i161.9564°}$

$\mathbf{U}_1 = 2.1347 \; e^{\; i74.2502°}$

$\mathbf{G}_1 = 3.0687 \; e^{\; i177.3687°}$

$\underline{\text{design \#2}}$

$\mathbf{W}_1 = 2.0014 \; e^{\; i114.9258°}$

$\mathbf{V}_1 = 4.189 \; e^{\; i135.6149°}$

$\mathbf{U}_1 = 5.8144 \; e^{\; i74.2502°}$

$\mathbf{G}_1 = 5.4819 \; e^{\; i188.9315°}$

FIGURE P.4.7
Folding-chair linkage designs.

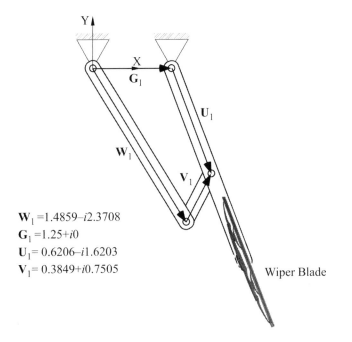

$\mathbf{W}_1 = 1.4859 - i2.3708$
$\mathbf{G}_1 = 1.25 + i0$
$\mathbf{U}_1 = 0.6206 - i1.6203$
$\mathbf{V}_1 = 0.3849 + i0.7505$

Wiper Blade

FIGURE P.4.8
Wiper-blade mechanism.

TABLE P.4.6

Slider-Crank Mechanism Configuration

W_1, θ	V_1	U_1	$\dot{\beta}$ (rad/s)	$\ddot{\beta}$ (rad/s²)
3.175 cm, 90°	9.525 cm	0.7938 cm	1.5	0.15

TABLE P.4.7

Slider-Crank Mechanism Configuration (with Unitless Link Lengths)

W_1, θ	V_1	U_1
1.35, 90°	1.6875	0.3375

19. Using the Appendix B.2 file, plot the fixed and moving centrodes for the mechanism in Table P.4.7 (let $\mathbf{b}_0 = (b_{1x}, -1,000,000)$). Limit your minimum and maximum *x*- and *y*-axis plot ranges to ±10.

20. Using the Appendix B.3 or H.2 files, produce piston velocity versus crankshaft rotation plots for the engine linkage illustrated in Figure P.4.9 at crankshaft speeds of 1250, 2187.5, and 3750 rpm (130.9, 229.07, and 392.7 rad/s). The ratio of the coupler length to the crank length is 3:1.

21. For the slider-crank mechanism configuration given in Table P.4.8, produce slider displacement, velocity, and acceleration (versus crank displacement angle) plots over a complete crank rotation range (using the Appendix B.3 or H.2 files).

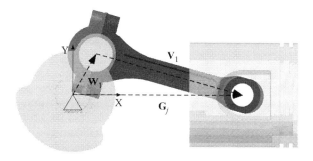

FIGURE P.4.9
Slider-crank mechanism used in a crankshaft-connecting rod-piston mechanism.

TABLE P.4.8

Slider-Crank Mechanism Configuration

W_1, θ	V_1	U_1	$\dot{\beta}$ (rad/s)	$\ddot{\beta}$ (rad/s²)
4.76 cm, 60°	6.35 cm	0 cm	1.5	0

22. Using the Appendix B.2 file, plot the fixed and moving centrodes for the mechanism in Problem 20 (let $\mathbf{b}_0 = (b_{1x}, -1,000,000)$). Limit your minimum and maximum x- and y-axis plot ranges to ±10.

23. Using the Appendix B.2 file, plot the fixed and moving centrodes for the mechanism in Problem 21 (let $\mathbf{b}_0 = (b_{1x}, -1,000,000)$). Limit your minimum and maximum x- and y-axis plot ranges to ±20.

24. Using the Appendix B.5 or H.4 files, plot the path traced by point $\mathbf{p}_1$ over a complete crank rotation range for the geared five-bar mechanism configuration given in Table P.4.9. The gear ratio is +2.

25. For the geared five-bar mechanism configuration given in Table P.4.9, plot the angular displacements of $\mathbf{V}_1$ and $\mathbf{S}_1$ (versus crank displacement) over a complete crank rotation range (using the Appendix B.5 or H.4 files). The gear ratio is +2.

26. For the geared five-bar mechanism configuration given in Table P.4.10, produce tables of the velocity and acceleration of point $\mathbf{p}_1$ over a complete crank rotation range at 30° crank rotation increments (using the Appendix B.4 or H.3 files).

27. For the geared five-bar mechanism configuration given in Table P.4.10, produce tables of the angular velocity and acceleration of $\mathbf{V}_1$ and $\mathbf{S}_1$ over a complete crank rotation range at 30° crank rotation increments (using the Appendix B.4 or H.3 files).

28. Figure P.4.10 illustrates a Watt II mechanism used in a concept adjustable chair. Using the Appendix B.6 or H.5 files, calculate the corresponding angular displacements of the head-rest ($\mathbf{V}_1$) and leg-rest ($\mathbf{V}_1^*$) components for a given total 20° angular displacement range of the base-rest component (at 1° crank rotation increments). The base rest includes vectors ($\mathbf{U}_1$) and ($\mathbf{W}_1^*$).

29. For the Watt II mechanism configuration given in Table P.4.11, plot the path traced by points $\mathbf{p}_1$ and $\mathbf{p}_1^*$ over a complete crank rotation range (using the Appendix B.6 or H.5 files).

TABLE P.4.9

Geared Five-Bar Mechanism Configuration

W_1, θ	V_1, ρ	U_1, σ	S_1, ψ	G_{1x}, G_{1y}	L_1, δ
1.3, 60°	9.1, 173.6421°	5.2, 150°	11.7, 182.2848°	7.8, 0	7.8, 203.6421°

TABLE P.4.10

Geared Five-Bar Mechanism Configuration

W_1, θ	V_1, ρ	U_1, σ	S_1, ψ	G_1
3.302 cm, 90°	13.208 cm, 140.2031°	4.953 cm, 75°	19.812 cm, 154.0128°	6.604 cm, −15°

L_1, δ	$\dot{\beta}$ (rad/s)	$\ddot{\beta}$ (rad/s²)	Gear Ratio
9.906 cm, 180°	1	0.5	−2

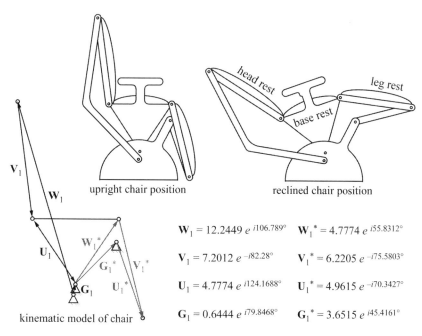

$\mathbf{W}_1 = 12.2449\ e^{\ i106.789°}$ $\mathbf{W}_1{}^* = 4.7774\ e^{\ i55.8312°}$

$\mathbf{V}_1 = 7.2012\ e^{\ -i82.28°}$ $\mathbf{V}_1{}^* = 6.2205\ e^{\ -i75.5803°}$

$\mathbf{U}_1 = 4.7774\ e^{\ i124.1688°}$ $\mathbf{U}_1{}^* = 4.9615\ e^{\ -i70.3427°}$

$\mathbf{G}_1 = 0.6444\ e^{\ i79.8468°}$ $\mathbf{G}_1{}^* = 3.6515\ e^{\ i45.4161°}$

FIGURE P.4.10
Watt II mechanism used in an adjustable chair.

TABLE P.4.11

Watt II Mechanism Dimensions

W_1, θ	V_1, ρ	U_1, σ	G_{1x}, G_{1y}	L_1, δ
0.85, 90°	1.275, 19.3737°	1.275, 93.2461°	1.275, 0	0.85, 60.7834°
W_1^*, θ^*	V_1^*, ρ^*	U_1^*, σ^*	G_{1x}^*, G_{1y}^*	L_1^*, δ^*
1.7, 25°	1.7, −59.4144°	1.7, −10.8507°	0.7361, −0.425	1.275, −11.2247°

30. For the Watt II mechanism configuration given in Table P.4.11, produce tables of the angular velocity and acceleration of $\mathbf{V}_1$ and $\mathbf{V}_1^*$ over a complete crank rotation range at −30° crank rotation increments (using the Appendix B.6 or H.5 files). The initial crank angular velocity and angular acceleration are −0.75 rad/s and −0.25 rad/s², respectively.

31. For the Watt II mechanism configuration given in Table P.4.11 (with link lengths in meters), produce tables of the velocity and acceleration of points $\mathbf{p}_1$ and $\mathbf{p}_1^*$ over a complete crank rotation range at 30° crank rotation increments (using the Appendix B.6 or H.5 files). The initial crank angular velocity and angular acceleration are 1 rad/s and 0.5 rad/s², respectively.

32. Figure P.4.11 illustrates a Stephenson III mechanism used to guide a gripping tool. The dimensions for the illustrated mechanism are included in Table P.4.12. Considering a crank rotation increment of 1°, determine the crank rotation range required to simultaneously achieve a −75° upper jaw rotation and a −15° lower jaw rotation (using the Appendix B.7 or H.6 files).

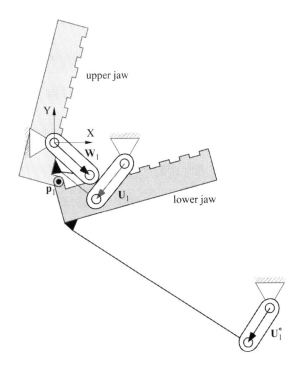

FIGURE P.4.11
Stephenson III gripper mechanism.

TABLE P.4.12

Stephenson III Gripper Mechanism Dimensions

W_1, θ	V_1, ρ	U_1, σ	G_{1x}, G_{1y}	L_1, δ
1.7995, 318.4178°	0.9091, 286.4362°	1.6482, 230.6143°	2.6491, −0.7924	1.2304, 189.1904°
V_1^*, ρ^*	U_1^*, σ^*	L_1^*, δ^*	G_{1x}^*, G_{1y}^*	
9.2542, 319.9647°	1.5166, 236.6216°	5.7298, 11.8113°	5.4021, −5.2847	

33. Figure P.4.12 illustrates a Stephenson III mechanism used to guide a digging tool. The dimensions for the illustrated mechanism are included in Table P.4.13. Considering a crank rotation increment of 1°, determine the crank rotation range required to simultaneously achieve a 60° digging arm rotation and a 75° digging bucket rotation (using the Appendix B.7 or H.6 files).

34. For the Stephenson III mechanism configuration given in Table P.4.14, produce a table of the location, velocity, and acceleration of $\mathbf{p}_1^*$ over a complete crank rotation range at 30° crank rotation increments (using the Appendix B.7 or H.6 files). The initial crank angular velocity and angular acceleration are 1.35 rad/s and 0.6 rad/s², respectively.

35. For the Stephenson III mechanism configuration given in Table P.4.14, produce a table of the angular position, velocity, and acceleration of $\mathbf{V}_1^*$ over a complete crank rotation range at −30° crank rotation increments (using the Appendix B.7 or H.6 files). The initial crank angular velocity and angular acceleration are −0.35 rad/s and −0.05 rad/s², respectively.

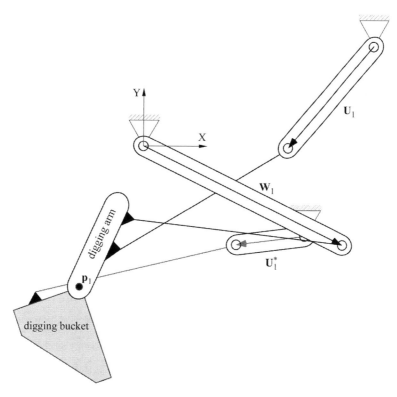

FIGURE P.4.12
Stephenson III gripper mechanism.

TABLE P.4.13

Stephenson III Digging Mechanism Dimensions

W_1, θ	V_1, ρ	U_1, σ	G_{1x}, G_{1y}	L_1, δ
4.2366, −25.6775°	2.0805, 120.3504°	2.5252, 229.6156°	4.4031, 1.8831	5.1069, 189.0097°
V_1^*, ρ^*	U_1^*, σ^*	L_1^*, δ^*	G_{1x}^*, G_{1y}^*	
3.0994, 14.6953°	1.3340, 187.0724°	1.9944, 65°	−1.3069, −3.5681	

TABLE P.4.14

Stephenson III Mechanism Dimensions

W_1, θ	V_1, ρ	U_1, σ	G_{1x}, G_{1y}	L_1, δ
1.45 m, 90°	2.175 m, 19.3737°	2.175 m, 93.2461°	2.175 m, 0 m	4.35 m, −5.1593°
V_1^*, ρ^*	U_1^*, σ^*	G_{1x}^*, G_{1y}^*	L_1^*, δ^*	
2.9 m, −46.9725°	2.9 m, −6.6563°	1.2557 m, −0.725 m	2.175 m, 1.2172°	

5

Dimensional Synthesis

CONCEPT OVERVIEW

In this chapter, the reader will gain a central understanding regarding

1. The study of *dimensional synthesis* and the distinctions between kinematic analysis and dimensional synthesis

2. Categories of dimensional synthesis

3. Types of mechanism defects and defect-elimination methods in dimensional synthesis

4. The formulation of linear simultaneous equation sets for planar four-bar motion generation

5. Distinctions between motion generation and path generation

6. The formulation of linear simultaneous equation sets for Stephenson III motion generation

7. The formulation of linear simultaneous equation sets for planar four-bar function generation

8. Distinctions between *finitely separated positions* (FSPs) and *multiply separated positions* (MSPs)

9. The formulation of linear simultaneous equation sets for planar four-bar function generation with FSPs and MSPs

10. Preparation of results from planar four-bar and Stephenson III dimensional synthesis for planar four-bar and Stephenson III kinematic analysis

5.1 Introduction

As noted in Section 1.5, *dimensional synthesis* is a category in *kinematic synthesis* where the objective is to calculate the mechanism dimensions required to achieve a prescribed mechanism motion sequence [1].* The calculated mechanism dimensions include link lengths, link positions (also called "rigid-body" positions), and joint coordinates.† The

---

* Another basic description of dimensional synthesis is *the design of a mechanism to produce a desired output motion for a given input motion*.

† Mechanism links are also called "rigid bodies" because mechanism links are generally assumed to be rigid (nondeforming) in kinematic synthesis.

parameters pertaining to the prescribed mechanism motion sequence include link positions, path points, and displacement angles. In contrast to *kinematic analysis*—where mechanism dimensions are known and the resulting mechanism motion sequence is calculated—in dimensional synthesis, the mechanism motion sequence is known and the mechanism dimensions are calculated (Figure 5.1) [2].

Dimensional synthesis includes three distinct subcategories: *motion generation, path generation,* and *function generation* (Figure 5.2). In motion generation, mechanism dimensions are calculated to achieve prescribed rigid-body positions, while prescribed rigid-body path points are achieved in path generation.* In function generation, mechanism dimensions are calculated to achieve prescribed crank and follower-link displacement angles.[†]

An overview of published research in dimensional synthesis will reveal an assortment of *qualitative* and *quantitative* methods for motion, path, and function generation [3]. Qualitative methods include graphical techniques, which can provide a wealth of information with virtually no computational effort [4]. Quantitative methods include mathematical models that can be solved analytically or numerically by way of solution algorithms and root-finding methods [5]. As intended by the authors, the dimensional synthesis methods presented in this chapter are all quantitative.

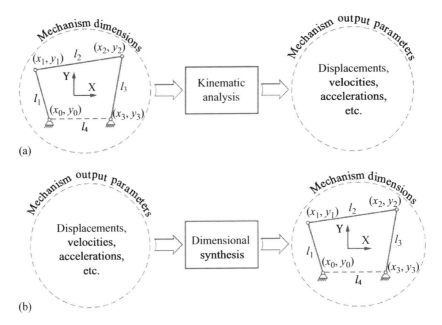

FIGURE 5.1
(a) Kinematic analysis vs. (b) dimensional synthesis.

---

* In four-bar motion and path generation, the coupler link is the rigid body for which positions and path points are prescribed.
† The crank and follower displacement angles are often prescribed in accordance to a mathematical function—hence the name *function generation*.

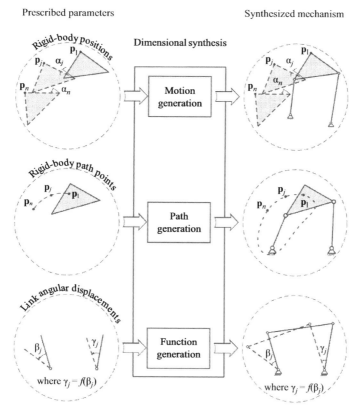

FIGURE 5.2
Subcategories of dimensional synthesis.

5.2 Branch and Order Defects

Although motion and path generation ensure that the synthesized mechanisms will achieve prescribed rigid-body positions and path points, respectively, they do not guarantee

- The synthesized mechanism will achieve the prescribed rigid-body parameters without a change in its original assembly configuration (thus requiring mechanism disassembly).
- The synthesized mechanism will achieve the prescribed rigid-body parameters in the intended order.

These two uncertainties are defects inherent in motion and path generation, mechanism synthesis requiring rigid-body positions and rigid-body path points, respectively.

The first noted defect is commonly called a *branch defect* [6]. Figure 5.3a illustrates the two assembly configurations of a planar four-bar mechanism: configurations a-b-c-d and a-b-c*-d.* In kinematic synthesis, a *branch* represents the rigid-body positions or path

---

* The assembly configurations a-b-c-d and a-b-c*-d of the planar four-bar mechanism are called the *open* and *crossed* configurations, respectively.

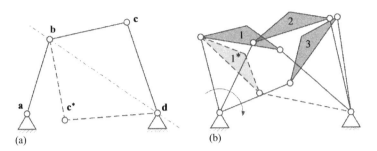

FIGURE 5.3
(a) Four-bar assembly configurations and (b) branch defect.

points that are achieved by a single mechanism assembly configuration. When both mechanism assembly configurations are required to achieve all of the rigid-body positions or path points, this can introduce a potential design problem, because the mechanism would require disassembly and reassembly from one assembly configuration to the other during operation to achieve all of the prescribed rigid-body output.* Figure 5.3b illustrates a planar four-bar branch defect. In this figure, rigid-body positions 1-2-3 are achieved by planar four-bar configuration a-b-c-d, while position 1* is achieved by configuration a-b-c*-d. So, if the design application requires that this mechanism achieves the positions in the order 1*-1-2-3 continuously, this would not be possible with a single mechanism assembly configuration due to the given branch defect. Branch defects are inherent in analytical motion generation [7].

The second noted defect is commonly called an *order defect* [8]. When rigid-body positions or path points are prescribed, it is typically desired that the synthesized mechanism achieve the prescribed rigid-body output in the given order in which they were prescribed. If, for example, the rigid-body path points are prescribed in the order 1-2-3-4-1 (Figure 5.4a) but the synthesized mechanism achieves the points in the order 1-2-4-3-1 (Figure 5.4b) this can present a potential problem—especially if the prescribed point order is required to trace a particular coupler curve. As shown in Figure 5.4, the order in which the rigid-body path points are achieved determines the profile of the path achieved.

Branch- and order-defect elimination by way of constraint equations, graphical methods, or particular prescribed values are often employed in motion and path generation

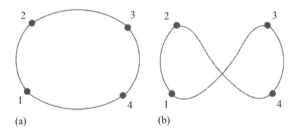

FIGURE 5.4
Order difference with four rigid-body path points (orders (a) 1-2-3-4 and (b) 1-2-4-3).

---

* The *branch defect* is distinct from the *circuit defect* (presented in Section 3.5) although the branch and circuit defect both require mechanism disassembly. With the circuit defect, the mechanism is reassembled in another position of the same assembly configuration (and not from one assembly configuration to another as with the branch defect).

equations to produce motion and path generator solutions that are branch-defect free and order-defect free [9]. Several common branch- and order-defect elimination methods are presented in Section 5.4.

5.3 Planar Four-Bar Motion Generation: Three Precision Positions

Figure 5.5(a) illustrates a planar four-bar mechanism and three coupler positions achieved by the mechanism. The location and orientation or the *position* of the coupler is defined by the coordinates of rigid-body point $\mathbf{p}_j$ and rigid-body displacement angle α_j. In motion generation (or *quantitative* motion generation, to be more exact), the mechanism dimensions required to achieve precision positions are calculated.* To further convey motion generation, Figure 5.5b illustrates the operation sequence of an aircraft landing gear. For effective operation, the aircraft landing gear should be guided from within the airframe to a position where the wheels can properly contact the ground. The planar four-bar motion generation equations presented in this section are useful for calculating the four-bar mechanism dimensions required to achieve a set of three precision positions.

Figure 5.6 illustrates both dyads of a planar four-bar mechanism in a starting position (Position 1) and a displaced position (Position j). The left- and right-side dyads include the vector chains $\mathbf{W}$–$\mathbf{Z}$ and $\mathbf{U}$–$\mathbf{S}$, respectively. By taking the vector sum between the starting and displaced positions for each dyad, vector-loop equations are derived [10, 11]. After taking the counterclockwise vector sum for each dyad (starting with $\mathbf{W}_1$ for the left-side dyad and $\mathbf{U}_1$ for the right-side dyad), the vector-loop equations for the four-bar mechanism dyads become

$$W_1 e^{i\theta} + Z_1 e^{i\phi} + P_{j1} e^{i\delta_j} - Z_1 e^{i(\phi+\alpha_j)} - W_1\, e^{j(\theta+\beta_j)} = 0 \tag{5.1}$$

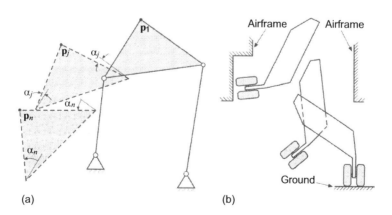

(a)   (b)

FIGURE 5.5

(a) Four-bar mechanism and coupler positions and (b) aircraft landing-gear example.

---

* In motion generation, the prescribed rigid-body positions are also called *precision positions*.

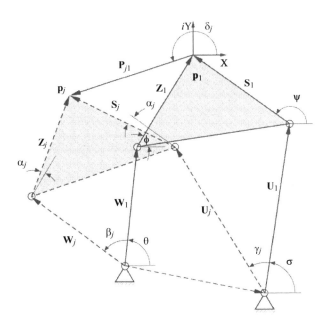

FIGURE 5.6
Four-bar mechanism in starting and (dashed) displaced positions.

$$U_1 e^{i\sigma} + S_1 e^{i\psi} + P_{j1} e^{i\delta_j} - S_1 e^{i(\psi + \alpha_j)} - U_1 e^{i(\sigma + \gamma_j)} = 0 \tag{5.2}$$

where $P_{j1} e^{i\delta_j} = \mathbf{p}_j - \mathbf{p}_1$

After factoring the terms for the starting dyad position (which are unknown terms), the resulting *standard-form* vector-loop equations become

$$W_1 e^{i\theta}\left(e^{i\beta_j} - 1\right) + Z_1 e^{i\phi}\left(e^{i\alpha_j} - 1\right) = P_{j1} e^{i\delta_j} \tag{5.3}$$

$$U_1 e^{i\sigma}\left(e^{i\gamma_j} - 1\right) + S_1 e^{i\psi}\left(e^{i\alpha_j} - 1\right) = P_{j1} e^{i\delta_j} \tag{5.4}$$

Expanding Equation 5.3 and grouping the real and imaginary terms as separate equations produces the equation set

$$W_1 \cos\theta\left(\cos\beta_j - 1\right) - W_1 \sin\theta \sin\beta_j + Z_1 \cos\phi\left(\cos\alpha_j - 1\right) - Z_1 \sin\phi \sin\alpha_j = P_{j1} \cos\delta_j$$
$$W_1 \sin\theta\left(\cos\beta_j - 1\right) + W_1 \cos\theta \sin\beta_j + Z_1 \sin\phi\left(\cos\alpha_j - 1\right) + Z_1 \cos\phi \sin\alpha_j = P_{j1} \sin\delta_j \tag{5.5}$$

After specifying $W_1 \cos\theta = W_{1x}$, $W_1 \sin\theta = W_{1y}$, $Z_1 \cos\phi = Z_{1x}$, and $Z_1 \sin\phi = Z_{1y}$, Equation 5.5 becomes

$$W_{1x}\left(\cos\beta_j - 1\right) - W_{1y}\sin\beta_j + Z_{1x}\left(\cos\alpha_j - 1\right) - Z_{1y}\sin\alpha_j = P_{j1}\cos\delta_j$$
$$W_{1y}\left(\cos\beta_j - 1\right) + W_{1x}\sin\beta_j + Z_{1y}\left(\cos\alpha_j - 1\right) + Z_{1x}\sin\alpha_j = P_{j1}\sin\delta_j \tag{5.6}$$

Likewise, after separating the real and imaginary terms in the right-side dyad in Equation 5.4 and specifying $U_1 \cos \sigma = U_{1x}$, $U_1 \sin \sigma = U_{1y}$, $S_1 \cos \psi = S_{1x}$, and $S_1 \sin \psi = S_{1y}$, the equation becomes

$$U_{1x}\left(\cos\gamma_j - 1\right) - U_{1y}\sin\gamma_j + S_{1x}\left(\cos\alpha_j - 1\right) - S_{1y}\sin\alpha_j = P_{j1}\cos\delta_j$$

$$U_{1y}\left(\cos\gamma_j - 1\right) + U_{1x}\sin\gamma_j + S_{1y}\left(\cos\alpha_j - 1\right) + S_{1x}\sin\alpha_j = P_{j1}\sin\delta_j$$

$$(5.7)$$

When expressed in matrix form for three precision positions (therefore, $j = 2, 3$), Equations 5.6 and 5.7 become Equations 5.8 and 5.9, respectively, when expressed in matrix form.

$$\begin{bmatrix} \cos\beta_2 - 1 & -\sin\beta_2 & \cos\alpha_2 - 1 & -\sin\alpha_2 \\ \sin\beta_2 & \cos\beta_2 - 1 & \sin\alpha_2 & \cos\alpha_2 - 1 \\ \cos\beta_3 - 1 & -\sin\beta_3 & \cos\alpha_3 - 1 & -\sin\alpha_3 \\ \sin\beta_3 & \cos\beta_3 - 1 & \sin\alpha_3 & \cos\alpha_3 - 1 \end{bmatrix} \begin{Bmatrix} W_{1x} \\ W_{1y} \\ Z_{1x} \\ Z_{1y} \end{Bmatrix} = \begin{bmatrix} P_{21}\cos\delta_2 \\ P_{21}\sin\delta_2 \\ P_{31}\cos\delta_3 \\ P_{31}\sin\delta_3 \end{bmatrix} \quad (5.8)$$

$$\begin{bmatrix} \cos\gamma_2 - 1 & -\sin\gamma_2 & \cos\alpha_2 - 1 & -\sin\alpha_2 \\ \sin\gamma_2 & \cos\gamma_2 - 1 & \sin\alpha_2 & \cos\alpha_2 - 1 \\ \cos\gamma_3 - 1 & -\sin\gamma_3 & \cos\alpha_3 - 1 & -\sin\alpha_3 \\ \sin\gamma_3 & \cos\gamma_3 - 1 & \sin\alpha_3 & \cos\alpha_3 - 1 \end{bmatrix} \begin{Bmatrix} U_{1x} \\ U_{1y} \\ S_{1x} \\ S_{1y} \end{Bmatrix} = \begin{bmatrix} P_{21}\cos\delta_2 \\ P_{21}\sin\delta_2 \\ P_{31}\cos\delta_3 \\ P_{31}\sin\delta_3 \end{bmatrix} \quad (5.9)$$

Equations 5.8 and 5.9 can be solved using Cramer's rule to calculate the scalar components of dyads $\mathbf{W}_1$–$\mathbf{Z}_1$ and $\mathbf{U}_1$–$\mathbf{S}_1$, respectively. In Equation 5.8, angles β_2 and β_3 are the two "free choices" that are prescribed along with the precision positions. By the term "free choice," we mean that the variable value specified is entirely according to the user's own preferences. In Equation 5.9, angles γ_2 and γ_3 are the free choices that are prescribed along with the precision positions. Because an infinite variety of unique combinations of β_2, β_3, γ_2, and γ_3 can be specified, the number of possible dyad solutions from Equations 5.8 and 5.9 is also infinite.

The resulting mechanism calculated from Equations 5.8 and 5.9 not only achieves the precision positions precisely, but does so according to the prescribed dyad displacement angles β and γ. By including the precision positions and prescribing the dyad displacement angles, it is ensured that the calculated mechanism solutions are free of order defects because both the precision positions and corresponding dyad displacements are specified.*

Example 5.1

Problem Statement: Synthesize a planar four-bar mechanism to guide the landing gear through the three precision positions in Figure E.5.1.

---

* Motion generation with prescribed dyad displacement angles is called motion generation with *prescribed timing*.

Known Information: Equations 5.8 and 5.9, and Table E.5.1.

Solution Approach: From the prescribed coupler points $\mathbf{p}_j$, vectors $\mathbf{p}_{j1}$ and vector angles δ_j are calculated.* These variables, along with the prescribed dyad displacement angles in Table E.5.1, are used in Equations 5.8 and 5.9 to calculate (through Cramer's rule) the dyad vectors for the planar four-bar mechanism ($\mathbf{W}_1$, $\mathbf{Z}_1$, $\mathbf{U}_1$, and $\mathbf{S}_1$).

Figure E.5.2 includes the calculation procedure in the MATLAB command window and Figure E.5.3 illustrates the resulting planar four-bar mechanism. Being an analytically calculated result, this mechanism achieves the precision positions precisely.

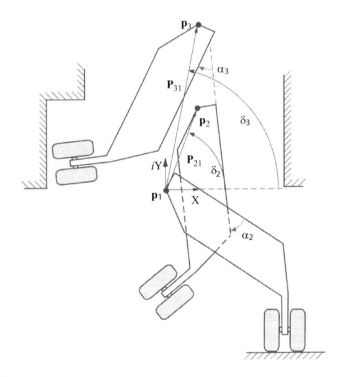

FIGURE E.5.1

Three landing-gear precision positions.

TABLE E.5.1

Landing-Gear Precision Position Parameters and Dyad Displacement Angles

Precision Position	$\mathbf{p}_j$	α_j (°)	β_j (°)	γ_j (°)
1	0, 0			
2	0.292, 0.734	−51.7124	18	−40
3	0.299, 1.461	−84.9734	38	−87

---

* $\mathbf{P}_{j1} = \mathbf{p}_j − \mathbf{p}_1$, and δ_j is the angle vector $\mathbf{P}_{j1}$ makes with the positive *x*-axis.

```
>> p1 = [0, 0]';
>> p2 = [0.292, 0.734]';
>> p3 = [0.299, 1.461]';
>> P21 = p2 - p1;
>> P31 = p3 - p1;
>> A2 = -51.7124*pi/180;
>> A3 = -84.9734*pi/180;
>> CA2 = cos(A2); SA2 = sin(A2);
>> CA3 = cos(A3); SA3 = sin(A3);
>> B2 = 18*pi/180;
>> B3 = 38*pi/180;
>> CB2 = cos(B2); SB2 = sin(B2);
>> CB3 = cos(B3); SB3 = sin(B3);
>> WZ = inv([CB2 - 1, -SB2, CA2 - 1, -SA2
SB2, CB2 - 1, SA2, CA2 - 1
CB3 - 1, -SB3, CA3 - 1, -SA3
SB3, CB3 - 1, SA3, CA3 - 1])*[P21(1), P21(2), P31(1), P31(2)]'

WZ =

    2.0580
   -0.8054
   -0.1324
    0.1191


>> G2 = -40*pi/180;
>> G3 = -87*pi/180;
>> CG2 = cos(G2); SG2 = sin(G2);
>> CG3 = cos(G3); SG3 = sin(G3);
>> US = inv([CG2 - 1, -SG2, CA2 - 1, -SA2
SG2, CG2 - 1, SA2, CA2 - 1
CG3 - 1, -SG3, CA3 - 1, -SA3
SG3, CG3 - 1, SA3, CA3 - 1])*[P21(1), P21(2), P31(1), P31(2)]'

US =

    0.5808

   -1.8615

   -1.5053

    1.3400

>>
```

FIGURE E.5.2
Example 5.1 W_1–Z_1 and U_1–S_1 calculation procedure in MATLAB.

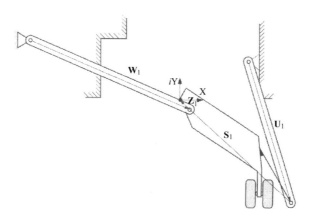

FIGURE E.5.3
Synthesized planar four-bar motion generator.

5.4 Order- and Branch-Defect Elimination

Although branch and order defects are inherent in analytical motion generation, several conventional practices and construction methods are used to produce motion and path generators that are branch- and order-defect free. These practices and construction methods include *prescribed timing*, *Filemon's construction*, and *Waldron's construction*.

Motion generation with prescribed timing prevents order defects because, with this practice, each coupler position and its corresponding dyad displacement angles are prescribed. The planar four-bar motion generation equation sets in Section 5.3 (Equations 5.8 and 5.9) include displacement angle variables for both mechanism dyads (angles β and α for dyad $\mathbf{W}$–$\mathbf{Z}$ and angles γ and α for dyad $\mathbf{U}$–$\mathbf{S}$). By specifying both the precision positions and the corresponding dyad displacement angles, the order in which the precision positions are achieved is maintained. Motion generation with prescribed timing was demonstrated in Example 5.1.

In the early 1970s, Filemon introduced a construction method to ensure that planar four-bar crank motion generator solutions (specifically, Grashof crank-rocker and drag-link solutions) were branch-defect free [12, 13]. Filemon's work shows that as long as the *moving pivot* (the nongrounded revolute joint) of the mechanism driving link is selected outside the wedge-shaped region produced using her construction method, the resulting planar four-bar motion generator will pass through all of the precision positions without disassembly (therefore making it a nonbranching solution).*

In Filemon's construction, the output link is synthesized first. The follower link moving pivot and fixed pivot will be defined as $\mathbf{b}_1$ and $\mathbf{b}_0$, respectively, where $\mathbf{b}_1 = -\mathbf{S}_1$ and $\mathbf{b}_0 = -(\mathbf{S}_1 + \mathbf{U}_1)$ ... Relative inverse displacements of the follower link are taken to sweep a planar wedge-shaped region (with angles $\angle \mathbf{b}_{0j}\mathbf{b}_1\mathbf{b}_0$).† The positions that the follower fixed pivot can take relative to the follower moving pivot are computed as

$$\mathbf{b}_{0j} = \left[M_j \right] \mathbf{b}_0 \tag{5.10}$$

where

$$[M_j] = \begin{bmatrix} 1 & 0 & 0 \\ 0 & 1 & 0 \\ 0 & 0 & 1 \end{bmatrix} \begin{bmatrix} \cos\alpha_j & -\sin\alpha_j & P_{j1x} \\ \sin\alpha_j & \cos\alpha_j & P_{j1y} \\ 0 & 0 & 1 \end{bmatrix}^{-1} \quad j = 2,3,\ldots \tag{5.11}$$

and $\mathbf{b}_0 = (b_{0x}, b_{0y}, 1)T$.

Given the precision position variables α_j and $\mathbf{P}_{j1}$, Equation 5.10 calculates the ground link displacements of the inverted four-bar mechanism. When synthesizing the input link (having already constructed the wedge-shaped region), an input link having a moving pivot that lies outside the region should be chosen. Filemon's construction can be applied in motion generation with three precision positions and beyond.

---

* In linkages, the grounded revolute joints are also known as *fixed pivots* and the nongrounded revolute joints are also known as *moving pivots*.
† If the angle of the wedge-shaped region is 180° or greater, the region will fill the entire 2D space (and no solution will be available under Filemon's construction).

```
>> p1 = [0, 0]';
>> p2 = [0.292, 0.734]';
>> p3 = [0.299, 1.461]';
>> P21 = p2 - p1;
>> P31 = p3 - p1;
>> A2 = -51.7124*pi/180;
>> A3 = -84.9734*pi/180;
>> CA2 = cos(A2); SA2 = sin(A2);
>> CA3 = cos(A3); SA3 = sin(A3);
>> U1 =[0.5808, -1.8615]';
>> S1 = [-1.5053, 1.3400]';
>> S1U1 = -(S1 + U1);
>> b0 = [S1U1(1,1), S1U1(2,1), 1]';
>> b02 =([1 0 0;0 1 0;0 0 1]*...
inv([CA2 -SA2 P21(1,1);SA2 CA2 P21(2,1);0 0 1]))*b0

b02 =

    0.5587
    0.3648
    1.0000

>> b03 =([1 0 0;0 1 0;0 0 1]*...
inv([CA3 -SA3 P31(1,1);SA3 CA3 P31(2,1);0 0 1]))*b0

b03 =

    0.9907
    0.5408
    1.0000

>>
```

FIGURE E.5.4
Example 5.2 b_{02} and b_{03} calculation procedure in MATLAB.

Example 5.2

Problem Statement: Calculate a nonbranching landing-gear mechanism for Example 5.1 using Filemon's construction.

Known Information: Example 5.1 and Equation 5.10.

Solution Approach: The synthesized follower (U_1–S_1) in Example 5.1 is the selected output link in this example. From Table E.5.1 in Example 5.1, the coupler displacement angles α_j and the x and y-components of the coupler point vectors P_{j1} are known.* Figure E.5.4 includes the calculation procedure in MATLAB's command window. The lines passing through b_1 and b_{02} and through b1 and b03 form the borders for the planar wedge-shaped region (Figure E.5.5).

The moving pivot for the driving link dyad (W_1–Z_1) must lie outside the wedge-shaped region. Figure E.5.6 illustrates the (W_1–Z_1) solution from Equation 5.8 with $\beta_2 = 10°$ and $\beta_3 = 40°$. Because the four-bar motion generator in this figure is a Grashof crank-rocker and its driving link dyad lies outside the wedge-shaped region, it is branch-defect free.

In the mid-1970s, Waldron introduced a feasible region construction method for selecting non-branching planar four-bar motion generators for three precision positions [14, 15]. Waldron's work shows that follower link moving pivots selected outside the three circles produced using his construction method, as well as follower link moving pivots selected in regions where any two circles overlap, result in nonbranching motion generator solutions. Once a suitable follower-link dyad is calculated according to Waldron's construction

---

* Because $P_{j1} = p_j - p_1$, if p_1 is specified as $p_1 = (0, 0)$, then the x- and y-components of the precision point vectors P_{j1} are identical to the x- and y-components of the precision points p_j.

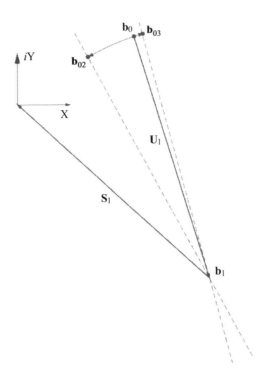

FIGURE E.5.5
Wedge-shaped region borders formed by $\mathbf{b}_{02}$ and $\mathbf{b}_{03}$ (form $\mathbf{U}_1$).

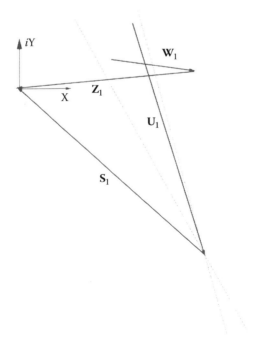

FIGURE E.5.6
Wedge-shaped region with calculated dyads $\mathbf{W}_1$–$\mathbf{Z}_1$ and $\mathbf{U}_1$–$\mathbf{S}_1$.

method, a corresponding suitable crank link dyad is calculated according to Filemon's construction.

Given three precision positions, three rotation centers or *poles* exist to rotate from Position 1 to 2 (pole $\mathbf{p}_{12}$), Position 1 to 3 (pole $\mathbf{p}_{13}$) and Position 2 to 3 (pole $\mathbf{p}_{23}$) [16].[*] These poles form a triangle (called a *pole triangle*) $\Delta\mathbf{p}_{12}\mathbf{p}_{13}\mathbf{p}_{23}$ [17]. The reflection of pole $\mathbf{p}_{23}$ about the triangle side $\mathbf{p}_{12}\mathbf{p}_{13}$ produces a new pole (called an *image pole*) $\mathbf{p}'_{23}$. In Waldron's construction, each side of the triangle $\Delta\mathbf{p}_{12}\mathbf{p}_{13}\mathbf{p}'_{23}$ is equal to a diameter of each circle.

The rotation pole equation corresponding to the two rigid-body positions in Figure 4.6 is defined as

$$\text{pole } \mathbf{p}_{jk} = P_{j1}e^{i\delta_j} - \left[\frac{P_{k1}e^{i\delta_k} - P_{j1}e^{i\delta_j}}{e^{i(\alpha_k - \alpha_j)} - 1} \right] \tag{5.12}$$

where $P_{11} = \delta_1 = \alpha_1 = 0$ [18]. Equation 5.12 is used to calculate pole $\mathbf{p}_{12}$, pole $\mathbf{p}_{13}$, and pole $\mathbf{p}_{23}$.

Example 5.3

Problem Statement: Calculate a nonbranching landing-gear mechanism for Example 5.1 using Waldron's construction and Filemon's construction.

Known Information: Example 5.1 and Equation 5.12.

Solution Approach: Figure E.5.7 includes the calculation procedure for $\mathbf{p}_{12}$, $\mathbf{p}_{13}$, and $\mathbf{p}_{23}$ in MATLAB's command window. Reflecting pole $\mathbf{p}_{23}$ about the triangle edge containing poles $\mathbf{p}_{12}$ and $\mathbf{p}_{13}$ produces the image pole (pole $\mathbf{p}'_{23}$ in Figure E.5.8a). Figure E.5.8b includes the three-circle diagram resulting from Waldron's construction. Suitable

```
>> p1 = [0, 0]';
>> p2 = [0.292, 0.734]';
>> p3 = [0.299, 1.461]';
>> P21 = norm(p2 - p1);
>> P31 = norm(p3 - p1);
>> A2 = -51.7124*pi/180;
>> A3 = -84.9734*pi/180;
>> D2 = atan2(p2(2,1), p2(1,1));
>> D3 = atan2(p3(2,1), p3(1,1));
>> pole_p12 = -(P21*exp(i*D2))/(exp(i*A2) - 1)

pole_p12 =

   0.9033 + 0.0657i

>> pole_p13 = -(P31*exp(i*D3))/(exp(i*A3) - 1)

pole_p13 =

   0.9471 + 0.5673i

>> pole_p23 = P21*exp(i*D2) - (P31*exp(i*D3) - P21*exp(i*D2))/...
(exp(i*(A3-A2)) - 1)

pole_p23 =

   1.5125 + 1.0858i

>>
```

FIGURE E.5.7
Example 5.3 $\mathbf{p}_{12}$, $\mathbf{p}_{13}$, and $\mathbf{p}_{23}$ calculation procedure in MATLAB.

---

[*] The center of rotation between two positions of a body in planar motion is called a *pole*.

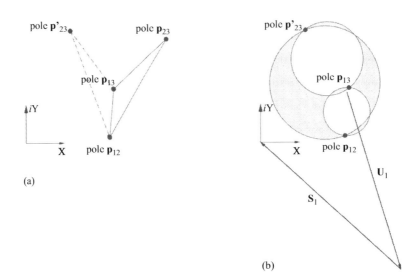

FIGURE E.5.8
(a) Calculated poles with image pole and (b) three-circle diagram with calculated dyad U_1–S_1.

follower link moving pivots lie either outside the three circles or in regions where any two the three circles overlap.

It can be determined from Figure E.5.8 that the follower link dyad (the U_1–S_1 dyad) solution from Example 5.1 is satisfactory, since its moving pivot lies outside the three circles. The crank link dyad (the W_1–Z_1 dyad) solution from Example 5.2 is also satisfactory since its moving pivot lies outside the wedge-shaped region. The resulting four-bar motion generator, a Grashof crank-rocker, is branch-defect free.

5.5 Path Generation versus Motion Generation

Figure 5.7a illustrates a planar four-bar mechanism and the curve traced by coupler point p_1. In path generation, the mechanism dimensions required to achieve prescribed coupler

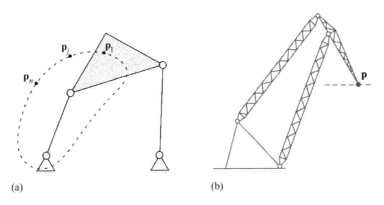

FIGURE 5.7
Path generation concepts. (a) Four-bar mechanism and coupler path points. (b) Level-luffing crane mechanism.

path points are calculated.* To further convey the concept of path generation, Figure 5.7b illustrates a *level-luffing crane*, a mechanism that remains at a constant level during motion. For effective operation, the extremity of the level-luffing crane should follow a horizontal path.

There is a distinct difference between path generation and motion generation. In motion generation, *precision positions* are achieved, while *precision points* are achieved in path generation. Because coupler path points rather than coupler positions are prescribed in path generation, coupler displacement angles are not of particular concern. Equations 5.8 and 5.9 can be directly applied for path generation for three precision points, since the mechanisms calculated from these equations achieve prescribed coupler points in addition to prescribed coupler displacement angles.[†] Because coupler displacement angles are required in Equations 5.8 and 5.9, the user is free to specify the coupler displacement angles arbitrarily.

5.6 Stephenson III Motion Generation: Three Precision Positions

It can be observed from Figure 5.8 (as well as the illustrations in Section 4.7) that the Stephenson III mechanism is a planar four-bar mechanism with a dyad ($U_1^* - V_1^*$ in Figure 5.8) attached to its coupler point.

Because the $U_1^* - V_1^*$ dyad and the coupler it is attached to share a common point p_1, this point can be used when prescribing precision positions for the dyads W_1-Z_1, U_1-S_1, and

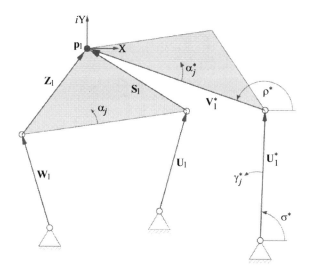

FIGURE 5.8
Stephenson III mechanism in starting and (dashed) displaced positions.

---

* In path generation, the prescribed rigid-body path points are also called *precision points*.
† Although Equations 5.8 and 5.9 can be used for path generation, the user is limited to three precision points (making them too limited and subsequently impractical for detailed paths).

$\mathbf{U}_1^* - \mathbf{V}_1^*$. However, while the displacement angle α_j is associated with dyads $\mathbf{W}_1$–$\mathbf{Z}_1$ and $\mathbf{U}_1$–$\mathbf{S}_1$, the dyad $\mathbf{U}_1^* - \mathbf{V}_1^*$ has its own rigid-body displacement angle (see Figure 5.8).

After taking the counterclockwise vector sum for dyad $\mathbf{U}_1^* - \mathbf{V}_1^*$ (starting with $\mathbf{U}_1^*$) in the same manner as illustrated in Figure 5.6, the vector-loop equation for this dyad becomes

$$U_1^* e^{i\sigma^*} + V_1^* e^{i\rho^*} + P_{j1} e^{i\delta_j} - V_1^* e^{i\left(\rho^* + \alpha_j^*\right)} - U_1^* e^{i\left(\sigma^* + \gamma_j^*\right)} = 0 \qquad (5.13)$$

where $P_{j1} e^{i\delta_j} = \mathbf{p}_j - \mathbf{p}_1$

After factoring the terms for the starting dyad position, the resulting standard-form vector-loop equation becomes

$$U_1^* e^{i\sigma^*} \left(e^{i\gamma_j^*} - 1 \right) + V_1^* e^{i\rho^*} \left(e^{i\alpha_j^*} - 1 \right) = P_{j1} e^{i\delta_j} \qquad (5.14)$$

Expanding Equation 5.14 and grouping the real and imaginary terms as separate equations produces the equation set

$$U_1^* \cos\sigma^* \left(\cos\gamma_j^* - 1 \right) - U_1^* \sin\sigma^* \sin\gamma_j^* + V_1^* \cos\rho^* \left(\cos\alpha_j^* - 1 \right) - V_1^* \sin\rho^* \sin\alpha_j^* = P_{j1} \cos\delta_j$$

$$U_1^* \sin\sigma^* \left(\cos\gamma_j^* - 1 \right) + U_1^* \cos\sigma^* \sin\gamma_j^* + V_1^* \sin\rho^* \left(\cos\alpha_j^* - 1 \right) - V_1^* \cos\rho^* \sin\alpha_i^* = P_{j1} \sin\delta_j$$

$$(5.15)$$

After specifying $U_1^* \cos\sigma^* = U_{1x}^*$, $U_1^* \sin\sigma^* = U_{1y}^*$, $V_1^* \cos\rho^* = V_{1x}^*$, and $V_1^* \sin\rho^* = V_{1y}^*$,

Equation 5.15 becomes

$$U_{1x}^* \left(\cos\gamma_j^* - 1 \right) - U_{1y}^* \sin\gamma_j^* + V_{1x}^* \left(\cos\alpha_j^* - 1 \right) - V_{1y}^* \sin\alpha_j^* = P_{j1} \cos\delta_j$$

$$(5.16)$$

$$U_{1y}^* \left(\cos\gamma_j^* - 1 \right) - U_{1x}^* \sin\gamma_j^* + V_{1y}^* \left(\cos\alpha_j^* - 1 \right) - V_{1x}^* \sin\alpha_i^* = P_{j1} \sin\delta_j$$

When expressed in matrix form for three precision positions (therefore, $j = 2, 3$), Equation 5.16 becomes Equation 5.17 when expressed in matrix form.

$$\begin{bmatrix} \cos\gamma_2^* - 1 & -\sin\gamma_2^* & \cos\alpha_2^* - 1 & -\sin\alpha_2^* \\ \sin\gamma_2^* & \cos\gamma_2^* - 1 & \sin\alpha_2^* & \cos\alpha_2^* - 1 \\ \cos\gamma_3^* - 1 & -\sin\gamma_3^* & \cos\alpha_3^* - 1 & \sin\alpha_3^* \\ \sin\gamma_3^* & \cos\gamma_3^* - 1 & \sin\alpha_3^* & \cos\alpha_3^* - 1 \end{bmatrix} \begin{Bmatrix} U_{1x}^* \\ U_{1y}^* \\ V_{1x}^* \\ V_{1y}^* \end{Bmatrix} = \begin{bmatrix} P_{21}\cos\delta_2 \\ P_{21}\sin\delta_2 \\ P_{31}\cos\delta_3 \\ P_{31}\sin\delta_3 \end{bmatrix} \qquad (5.17)$$

Like Equations 5.8 and 5.9, Equation 5.17 can be solved using Cramer's rule to calculate the scalar components of dyad $\mathbf{U}_1^* - \mathbf{V}_1^*$. In Equation 5.17, angles γ_2^* and γ_3^* are the two "free choices" that are prescribed along with the precision positions.

Example 5.4

Problem Statement: Synthesize a Stephenson III mechanism to guide the gripper through the three precision positions in Figure E.5.9.

Known Information: Equations 5.8, 5.9, and 5.17, and Table E.5.2.

Solution Approach: Table E.5.2 includes the precision positions and dyad displacement angles.

Figure E.5.10 includes the calculation procedure in MATLAB's command window for dyads $\mathbf{W}_1-\mathbf{Z}_1$ and $\mathbf{U}_1-\mathbf{S}_1$. Figure E.5.11 includes the calculation procedure in MATLAB's command window for the $\mathbf{U}_1^* - \mathbf{V}_1^*$ dyad. Figure E.5.12 illustrates the resulting Stephenson III mechanism.

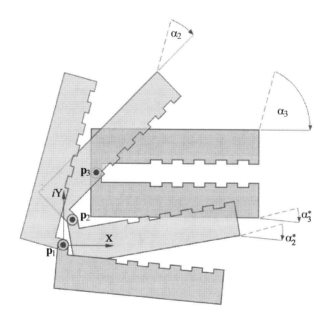

FIGURE E.5.9
Three gripper precision positions.

TABLE E.5.2

Landing-Gear Precision Position Parameters and Dyad Displacement Angles

Precision Position	p_j	α_j (°)	α_j^* (°)	β_j (°)	γ_j (°)	γ_j^* (°)
1	0, 0					
2	0.1815, 0.4882	−30	15	15	−20	10
3	0.6647, 1.4078	−75	5	40	−35	30

```
>> p1 = [0, 0]';
>> p2 = [0.1815, 0.4882]';
>> p3 = [0.6647, 1.4078]';
>> P21 = p2 - p1;
>> P31 = p3 - p1;
>> A2 = -30*pi/180;
>> A3 = -75*pi/180;
>> CA2 = cos(A2); SA2 = sin(A2);
>> CA3 = cos(A3); SA3 = sin(A3);
>> B2 = 15*pi/180;
>> B3 = 40*pi/180;
>> CB2 = cos(B2); SB2 = sin(B2);
>> CB3 = cos(B3); SB3 = sin(B3);
>> WZ = inv([CB2 - 1, -SB2, CA2 - 1, -SA2
SB2, CB2 - 1, SA2, CA2 -1
CB3 - 1, -SB3, CA3 - 1, -SA3
SB3, CB3 - 1, SA3, CA3 - 1])*[P21(1), P21(2), P31(1), P31(2)]'

WZ =

    0.5854
   -1.2248
   -0.4926
   -0.3631

>> G2 = -20*pi/180;
>> G3 = -35*pi/180;
>> CG2 = cos(G2); SG2 = sin(G2);
>> CG3 = cos(G3); SG3 = sin(G3);
>> US = inv([CG2 - 1, -SG2, CA2 - 1, -SA2
SG2, CG2 - 1, SA2, CA2 - 1
CG3 - 1, -SG3, CA3 - 1, -SA3
SG3, CG3 - 1, SA3, CA3 - 1])*[P21(1), P21(2), P31(1), P31(2)]'

US =

    0.1905
    1.8923
   -1.0184
   -1.1813

>>
```

FIGURE E.5.10
Example 5.4 $\mathbf{W}_1$–$\mathbf{Z}_1$ and $\mathbf{U}_1$–$\mathbf{S}_1$ calculation procedure in MATLAB.

```
>> p1 = [0, 0]';
>> p2 = [0.1815, 0.4882]';
>> p3 = [0.6647, 1.4078]';
>> P21 = p2 - p1;
>> P31 = p3 - p1;
>> A2s = 15*pi/180;
>> A3s = 5*pi/180;
>> CA2s = cos(A2s); SA2s = sin(A2s);
>> CA3s = cos(A3s); SA3s = sin(A3s);
>> G2s = 10*pi/180;
>> G3s = 30*pi/180;
>> CG2s = cos(G2s); SG2s = sin(G2s);
>> CG3s = cos(G3s); SG3s = sin(G3s);
>> UsVs = inv([CG2s - 1, -SG2s, CA2s - 1, -SA2s
SG2s, CG2s- 1, SA2s, CA2s - 1
CG3s - 1, -SG3s, CA3s - 1, -SA3s
SG3s, CG3s - 1, SA3s, CA3s - 1])*[P21(1), P21(2), P31(1), P31(2)]'

UsVs =

    2.2215
   -2.0095
    0.3400
    0.4718

>>
```

FIGURE E.5.11
Example 5.4 $\mathbf{U}_1^*$ – $\mathbf{V}_1^*$ calculation procedure in MATLAB.

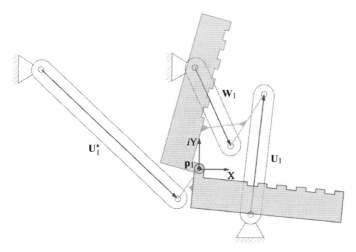

FIGURE E.5.12
Synthesized Stephenson III motion generator.

5.7 Planar Four-Bar Function Generation: Three Precision Points

In function generation, the mechanism dimensions required to achieve prescribed link displacement angles are calculated.* The displacement angles are commonly defined according to a mathematical function. In four-bar function generation, the angular displacements of both the crank and follower links are typically where the angular displacement of the follower link is a function of the crank link angular displacement (Figure 5.6a).† To further convey function generation, Figure 5.9b illustrates a four-bar lawn sprinkling mechanism (where the sprinkler is affixed to the follower link). For effective operation, the sprinkler

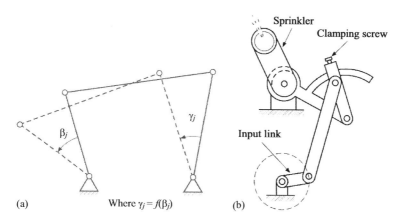

FIGURE 5.9
(a) Four-bar mechanism in starting and displaced positions and (b) lawn sprinkling mechanism.

---

* In addition to link angular displacements, derivative quantities such as angular velocities and angular accelerations are also among the prescribed parameters that are considered in function generation.
† Prescribed displacement angles are also called *precision points* in function generation.

should oscillate within a particular range to avoid over-sprinkling or under-sprinkling the lawn area. The planar four-bar function generation models presented in this section are useful for calculating the planar four-bar mechanism dimensions required to achieve prescribed link angular displacements.

As previously noted, the precision points in four-bar function generation are commonly crank and follower displacement angles where the follower angles are defined as a function of the crank angles. In Figure 5.9a, the crank displacement angles β_j correspond to the independent function variable x and the follower displacement angles γ_j correspond to a user-defined function $f(x)$ [19, 20].

The curve in Figure 5.10 represents an example function to be achieved through function generation. From this function, precision points (p_j) are selected. Because only a finite number of precision points can be prescribed through analytical function generation, the resulting function generator will achieve $f(x)$ at p_j, or simply $f(x_j)$, as opposed to achieving the function continuously.

The abscissa and ordinate ranges for the function are Δx and Δy, respectively, and the corresponding crank and follower displacement angle ranges are $\Delta\beta$ and $\Delta\gamma$, respectively (Figure 5.10). Linear relationships between the crank and follower precision points and the specific function precision points can be expressed as

$$\frac{\beta_j - \beta_1}{x_j - x_1} = \frac{\Delta\beta}{\Delta x}$$

$$\frac{\gamma_i - \gamma_1}{y_j - y_1} = \frac{\Delta\gamma}{\Delta y}$$

(5.18)

Angles β_1 and γ_1 are zero because they are the displacement angles in the initial mechanism position. Solving for β_j and γ_j, Equation 5.18 can be expressed as Equation 5.19 where β_j and γ_j are the crank and follower precision points corresponding to the coordinates x_j and y_j of the function precision points p_j. Crank and follower-link displacement angles are among the precision points for the planar four-bar function generation equations in this chapter.

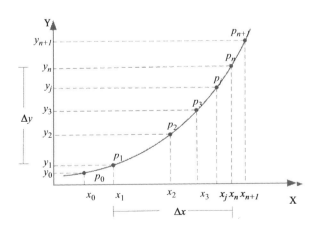

FIGURE 5.10
Function curve with precision points.

$$\beta_j = \frac{\Delta\beta}{\Delta x}\left(x_j - x_1\right)$$

$$\gamma_j = \frac{\Delta\gamma}{\Delta y}\left(y_j - y_1\right)$$

(5.19)

The general vector-loop closure equation for the four-bar mechanism in Figure 5.11 in the *j*th position is

$$W_1 e^{i(\theta+\beta_j)} + V_1 e^{i(\rho+\alpha_j)} - U_1 e^{i(\sigma+\gamma_j)} - G_1 = 0$$

(5.20)

Being always in-line with the *x*-axis, vector $\mathbf{G}_1$ only has the real component G_{1x}. Because the mechanism link *proportions* affect its link rotation angles (which are of interest in function generation) and not the *scale* of the mechanism, a single mechanism link length variable can be specified.* After specifying $G_{1x} = 1$ and moving it to the right-hand side of the equation, Equation 5.20 becomes

$$W_1 e^{i(\theta+\beta_j)} + V_1 e^{i(\rho+\alpha_j)} - U_1 e^{i(\sigma+\gamma_j)} = 1$$

(5.21)

Expanding Equation 5.21 and grouping the real and imaginary terms as separate equations produces

$$W_1 \cos\theta\cos\beta_j - W_1 \sin\theta\sin\beta_j + V_1 \cos\rho\cos\alpha_j - V_1 \sin\rho\sin\alpha_j - U_1 \cos\sigma\cos\gamma_j$$

$$+ U_1 \sin\sigma\,\sin\gamma_j = 1$$

$$W_1 \sin\theta\cos\beta_j - W_1 \cos\theta\sin\beta_j + V_1 \sin\rho\cos\alpha_j + V_1 \cos\rho\sin\alpha_j$$

$$- U_1 \sin\sigma\cos\gamma_j - U_1 \cos\sigma\,\sin\gamma_j = 0$$

(5.22)

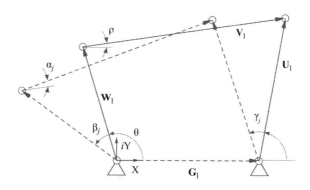

FIGURE 5.11
Four-bar mechanism in starting and (dashed) displaced positions.

---

* This also means that scaled versions of a given function generator will produce identical results.

After setting $W_1 \cos\theta = W_{1x}$, $W_1 \sin\theta = W_{1y}$, $V_1 \cos\rho = V_{1x}$, $V_1 \sin\rho = V_{1y}$, $U_1 \cos\sigma = U_{1x}$, and $U_1 \sin\sigma = U_{1y}$, Equation 5.22 becomes

$$W_{1x}\cos\beta_j - W_{1y}\sin\beta_j + V_{1x}\cos\alpha_j - V_{1y}\sin\alpha_j - U_{1x}\cos\gamma_j + U_{1y}\sin\gamma_j = 1$$

$$W_{1y}\cos\beta_j + W_{1x}\sin\beta_j + V_{1y}\cos\alpha_j + V_{1x}\sin\alpha_j - U_{1y}\cos\gamma_j - U_1\sin\gamma_j = 0$$

(5.23)

When expressed in matrix form for three precision points, Equation 5.23 becomes

$$\begin{bmatrix} 1 & 0 & 1 & 0 & -1 & 0 \\ 0 & 1 & 0 & 1 & 0 & -1 \\ \cos\beta_2 & -\sin\beta_2 & \cos\alpha_2 & -\sin\alpha_2 & -\cos\gamma_2 & \sin\gamma_2 \\ \sin\beta_2 & \cos\beta_2 & \sin\alpha_2 & \cos\alpha_2 & -\sin\gamma_2 & -\cos\gamma_2 \\ \cos\beta_3 & -\sin\beta_3 & \cos\alpha_3 & -\sin\alpha_3 & -\cos\gamma_3 & \sin\gamma_3 \\ \sin\beta_3 & \cos\beta_3 & \sin\alpha_3 & \cos\alpha_3 & -\sin\gamma_3 & -\cos\gamma_3 \end{bmatrix} \begin{Bmatrix} W_{1x} \\ W_{1y} \\ V_{1x} \\ V_{1y} \\ U_{1x} \\ U_{1y} \end{Bmatrix} = \begin{bmatrix} 1 \\ 0 \\ 1 \\ 0 \\ 1 \\ 0 \end{bmatrix}$$

(5.24)

The first two rows in Equation 5.24 are the result of displacement angles β_1, α_1, and γ_1 (corresponding to the initial mechanism position) all being zero in Equation 5.23.

Therefore, with three precision points, Equation 5.23 forms Equation 5.24—a set of six linear equations (that can be solved using Cramer's rule) to calculate the scalar components of link vectors $\mathbf{W}_1$, $\mathbf{V}_1$, and $\mathbf{U}_1$. In addition to prescribing the crank and follower displacement angles, the coupler displacement angles α_2 and α_3 are also prescribed in Equation 5.24. The coupler displacement angles are typically incidental in function generation, since only the crank and follower displacement angles are typically of concern. Because an infinite variety of unique α_2 and α_3 combinations can be specified, the number of possible mechanism solutions from Equation 5.24 is also infinite.

Example 5.5

Problem Statement: Synthesize a planar four-bar sprinkler mechanism to achieve a follower rotation range of 60° for a corresponding crank rotation range of 180°.

Known Information: Crank and follower displacement angle ranges and Equation 5.24.

Solution Approach: The crank and follower rotation ranges have been equally divided into the displacement angles given in Table E.5.3. This table also includes arbitrarily specified coupler-link displacement angles.

Figure E.5.13 includes the calculation procedure in MATLAB's command window for vectors $\mathbf{W}_1$, $\mathbf{V}_1$, and $\mathbf{U}_1$. The synthesized function generator is illustrated in Figure E.5.14. Because the function generator was calculated analytically, it achieves the precision points precisely.

TABLE E.5.3

Four-Bar Sprinkler Mechanism Precision Points

Precision Point	β_j (°)	γ_j (°)	α_j (°)
1	0	0	0
2	90	30	−5
3	180	60	10

```
>> B2 = 90*pi/180;
>> B3 = 180*pi/180;
>> CB2 = cos(B2); SB2 = sin(B2);
>> CB3 = cos(B3); SB3 = sin(B3);
>> G2 = 30*pi/180;
>> G3 = 60*pi/180;
>> CG2 = cos(G2); SG2 = sin(G2);
>> CG3 = cos(G3); SG3 = sin(G3);
>> A2 = -5*pi/180;
>> A3 = 10*pi/180;
>> CA2 = cos(A2); SA2 = sin(A2);
>> CA3 = cos(A3); SA3 = sin(A3);
>> WVU = inv([1, 0, 1, 0, -1, 0
0, 1, 0, 1, 0, -1
CB2, -SB2, CA2, -SA2, -CG2, SG2
SB2, CB2, SA2, CA2, -SG2, -CG2
CB3, -SB3, CA3, -SA3, -CG3, SG3
SB3, CB3, SA3, CA3, -SG3, -CG3])*[1, 0, 1, 0, 1, 0]'

WVU =

    0.2670
    0.0816
    1.0598
    0.4564
    0.3268
    0.5380

>>
```

FIGURE E.5.13
Example 5.5 W_1–Z_1–U_1 calculation procedure in MATLAB.

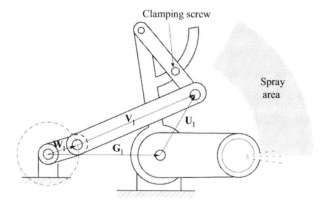

FIGURE E.5.14
Synthesized planar four-bar function generator.

5.8 Planar Four-Bar Function Generation: FSPs and MSPs

In addition to link displacement angles, it is also possible to synthesize a function generator to achieve prescribed link angular velocities and accelerations.* Figure 5.12 includes
the function generator angular velocity and angular acceleration variables—the first and
second derivatives of angular displacements β, α, and γ.

---

* Angular displacements (being discrete, finitely separated quantities) are called *finitely separated positions* (FSPs)
  and derivative quantities such as angular velocities and accelerations are *multiply separated positions* (MSPs).

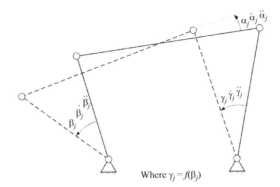

FIGURE 5.12
Planar four-bar function generator with angular velocity and acceleration variables.

By differentiating, expanding, and separating vector-loop Equation 5.21, mechanism velocity and acceleration constraint equations are derived for the planar four-bar function generator. Differentiating Equation 5.21 and cancelling the complex coefficient produces

$$\dot{\beta}_j W_1 e^{i(\theta+\beta_j)} + \dot{\alpha}_j V_1 e^{i(\rho+\alpha_j)} - \dot{\gamma}_j U_1 e^{i(\sigma+\gamma_j)} = 0* \tag{5.25}$$

Expanding Equation 5.25 and grouping the real and imaginary terms into separate equations produces

$$\dot{\beta}_j \left(W_{1x} \cos\beta_j - W_{1y} \sin\beta_j \right) + \dot{\alpha}_j \left(V_{1x} \cos\alpha_j - V_{1y} \sin\alpha_j \right) - \dot{\gamma}_j \left(U_{1x} \cos\gamma_j - U_{1y} \sin\gamma_j \right) = 0$$
$$\dot{\beta}_j \left(W_{1x} \sin\beta_j + W_{1y} \cos\beta_j \right) + \dot{\alpha}_j \left(V_{1x} \sin\alpha_j + V_{1y} \cos\alpha_j \right) + \dot{\gamma}_j \left(U_{1x} \sin\gamma_j - U_{1y} \cos\gamma_j \right) = 0 \tag{5.26}$$

Equation 5.26 constitutes a planar four-bar function generator velocity constraint (introducing link velocity terms $\dot{\alpha}$, $\dot{\beta}$, and $\dot{\gamma}$). Differentiating Equation 5.25 produces

$$i\ddot{\beta}_j W_1 e^{i(\theta+\beta_j)} - \dot{\beta}_j^2 W_1 e^{i(\theta+\beta_j)} + i\ddot{\alpha}_j V_1 e^{i(\rho+\alpha_j)} - \dot{\alpha}_j^2 V_1 e^{i(\rho+\alpha_j)} - i\ddot{\gamma}_j U_1 e^{i(\sigma+\gamma_j)} + \dot{\gamma}_j^2 U_1 e^{i(\sigma+\gamma_j)} = 0 \tag{5.27}$$

Expanding Equation 5.27 and grouping the real and imaginary terms into separate equations produces

$$-\ddot{\beta}_j \left(W_{1x} \sin\beta_j + W_{1y} \cos\beta_j \right) - \dot{\beta}_j^2 \left(W_{1x} \cos\beta_j - W_{1y} \sin\beta_j \right) - \ddot{\alpha}_j \left(V_{1x} \sin\alpha_j + V_{1y} \cos\alpha_j \right)$$
$$- \dot{\alpha}_j^2 \left(V_{1x} \cos\alpha_j - V_{1y} \sin\alpha_j \right) + \ddot{\gamma}_j \left(U_{1x} \sin\gamma_j + U_{1y} \cos\gamma_j \right) + \dot{\gamma}_j^2 \left(U_{1x} \cos\gamma_j - U_{1y} \sin\gamma_j \right) = 0$$
$$\ddot{\beta}_j \left(W_{1x} \cos\beta_j - W_{1y} \sin\beta_j \right) - \dot{\beta}_j^2 \left(W_{1x} \sin\beta_j + W_{1y} \cos\beta_j \right) + \ddot{\alpha}_j \left(V_{1x} \cos\alpha_j - V_{1y} \sin\alpha_j \right)$$
$$- \dot{\alpha}_j^2 \left(V_{1x} \sin\alpha_j + V_{1y} \cos\alpha_j \right) - \ddot{\gamma}_j \left(U_{1x} \cos\gamma_j - U_{1y} \sin\gamma_j \right) + \dot{\gamma}_j^2 \left(U_{1x} \sin\gamma_j + U_{1y} \cos\gamma_j \right) = 0 \tag{5.28}$$

---

* Whenever the complex coefficient (or any other term) is distributed throughout an equation, it can be cancelled if preferred.

Equation 5.28 constitutes a planar four-bar function generator acceleration constraint—introducing link angular acceleration terms $\ddot{\alpha}, \ddot{\beta}$, and $\ddot{\gamma}$ in addition to the link angular velocity terms introduced in Equation 5.26.

As observed in Equation 5.24, three precision points for planar four-bar function generation produce a set of six scalar equations to calculate mechanism variables $\mathbf{W}_1$, $\mathbf{V}_1$, and $\mathbf{U}_1$. A set of six equations to calculate mechanism variables $\mathbf{W}_1$, $\mathbf{V}_1$, and $\mathbf{U}_1$ can also be formed from Equation 5.23 (with $\beta = \alpha = \gamma = 0$) and any combination of Equations 5.23, 5.26, or 5.28. With such an equation set, the user can specify not only link angular displacements (FSPs), but also corresponding link angular velocities, or accelerations (MSPs). This equation set can be solved using Cramer's rule. Being an analytically calculated solution, the resulting mechanism will precisely achieve the prescribed angular displacements, velocities, or accelerations.

Example 5.6

Problem Statement: Given a constant driving link angular velocity of 1 rad/s and angular acceleration of zero, synthesize a planar four-bar mechanism to achieve a follower angular velocity and acceleration of −0.75 rad/s and −0.1 rad/s², respectively.*

Known Information: Equations 5.23, 5.26, 5.28, $\dot{\beta} = 1$ rad/s, $\ddot{\beta} = 0$ rad/s², $\dot{\gamma} = -0.75$ rad/s, and $\ddot{\gamma} = -0.1$ rad/s².

Solution Approach: A set of six equations to calculate mechanism variables $\mathbf{W}_1$, $\mathbf{V}_1$, and $\mathbf{U}_1$ is formed from Equation 5.23 (with $\beta = \alpha = \gamma = 0$), Equation 5.23 (with β, α, and γ) and Equation 5.28.[†] When expressed in matrix form, these equations become

$$
\begin{bmatrix}
1 & 0 & 1 & 0 & -1 & 0 \\
0 & 1 & 0 & 1 & 0 & -1 \\
\dot{\beta}c_\beta & -\dot{\beta}s_\beta & \dot{\alpha}c_\alpha & -\dot{\alpha}s_\alpha & -\dot{\gamma}c_\gamma & \dot{\gamma}s_\gamma \\
\dot{\beta}s_\beta & \dot{\beta}c_\beta & \dot{\alpha}s_\alpha & \dot{\alpha}c_\alpha & -\dot{\gamma}s_\gamma & -\dot{\gamma}c_\gamma \\
-\ddot{\beta}s_\beta - \dot{\beta}^2 c_\beta & -\ddot{\beta}c_\beta + \dot{\beta}^2 s_\beta & -\ddot{\alpha}s_\alpha - \dot{\alpha}^2 c_\alpha & -\ddot{\alpha}c_\alpha + \dot{\alpha}^2 s_\alpha & \ddot{\gamma}s_\gamma + \dot{\gamma}^2 c_\gamma & \ddot{\gamma}c_\gamma - \dot{\gamma}^2 s_\gamma \\
\ddot{\beta}c_\beta - \dot{\beta}^2 s_\beta & -\ddot{\beta}s_\beta - \dot{\beta}^2 c_\beta & \ddot{\alpha}c_\alpha - \dot{\alpha}^2 s_\alpha & -\ddot{\alpha}s_\alpha - \dot{\alpha}^2 c_\alpha & -\ddot{\gamma}c_\gamma + \dot{\gamma}^2 s_\gamma & \ddot{\gamma}s_\gamma + \dot{\gamma}^2 c_\gamma
\end{bmatrix}
\begin{Bmatrix}
W_{1x} \\
W_{1y} \\
V_{1x} \\
V_{1y} \\
U_{1x} \\
U_{1y}
\end{Bmatrix}
=
\begin{bmatrix}
1 \\
0 \\
0 \\
0 \\
0 \\
0
\end{bmatrix}
$$

(5.29)

In Equation 5.29, $c_{\text{ang}} = \cos(\text{ang})$ and $s_{\text{ang}} = \sin(\text{ang})$. The angular velocity and acceleration of the coupler link were arbitrarily specified as $\dot{\alpha} = -1.5$ rad/s and $\ddot{\alpha} = -1$ rad/s². Figure E.5.15 includes the calculation procedure in MATLAB's command window for vectors $\mathbf{W}_1$, $\mathbf{V}_1$, and $\mathbf{U}_1$. Plotting the follower angular velocity and acceleration profiles (Figure E.5.16) of the planar four-bar function generator solution (Figure E.5.17) confirms that the prescribed MSPs are achieved precisely at $\beta = 0$.

Example 5.7

Problem Statement: Given a driving link angular displacement of 35° and a constant driving link angular velocity of 1 rad/s, synthesize a planar four-bar mechanism to achieve a follower angular displacement and velocity of −20° and −0.75 rad/s, respectively.[‡]

Known Information: Equations 5.23, 5.26, $\beta = 35°$, $\dot{\beta} = 1$ rad/s, $\gamma = -20°$ and $\dot{\gamma} = -0.75$ rad/s.

---

* The velocity and acceleration of $\mathbf{V}_1$ are arbitrarily specified as $\dot{\alpha} = -1.5$ rad/s and $\ddot{\alpha} = -1$ rad/s².
† To ensure the function generator solution forms a closed loop, Equation 5.20 (with $\beta = \alpha = \gamma = 0$) must be included in the solution equation set.
‡ The displacement and velocity of $\mathbf{V}_1$ are arbitrarily specified as $\alpha = -10°$ and $\dot{\alpha} = -1.5$ rad/s.

```
>> B = 0;
>> dB = 1;
>> ddB = 0;
>> CB = cos(B); SB = sin(B);
>> A = 0;
>> dA = -1.5;
>> ddA = -1;
>> CA = cos(A); SA = sin(A);
>> G = 0;
>> dG = -0.75;
>> ddG = -0.1;
>> CG = cos(G); SG = sin(G);
>> WVU = inv([1, 0, 1, 0, -1, 0
0, 1, 0, 1, 0, -1
dB*CB, -dB*SB, dA*CA, -dA*SA, -dG*CG, dG*SG
dB*SB, dB*CB, dA*SA, dA*CA, -dG*SG, -dG*CG
-ddB*SB - dB^2*CB, -ddB*CB + dB^2*SB, -ddA*SA - dA^2*CA,...
-ddA*CA + dA^2*SA, ddG*SG + dG^2*CG, ddG*CG - dG^2*SG
ddB*CB - dB^2*SB, -ddB*SB - dB^2*CB, ddA*CA - dA^2*SA,...
-ddA*SA - dA^2*CA, -ddG*CG + dG^2*SG, ddG*SG + dG^2*CG
])*[1, 0, 0, 0, 0, 0]'

WVU =

    0.2818
    0.0540
   -0.3424
    0.1260
   -1.0605
    0.1801

>>
```

FIGURE E.5.15
Example 5.6 W_1–Z_1–U_1 calculation procedure in MATLAB.

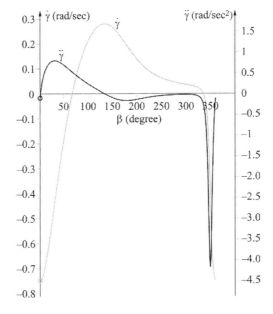

FIGURE E.5.16
Follower angular velocity and acceleration profiles for synthesized function generator.

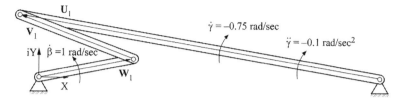

FIGURE E.5.17
Synthesized planar four-bar function generator.

Solution Approach: A set of six equations to calculate mechanism variables $\mathbf{W}_1$, $\mathbf{V}_1$, and $\mathbf{U}_1$ is formed from Equation 5.23 (with $\beta = \alpha = \gamma = 0$), Equation 5.23 (with β, α, and γ) and Equation 5.26. When expressed in matrix form, these equations become

$$
\begin{bmatrix}
1 & 0 & 1 & 0 & -1 & 0 \\
0 & 1 & 0 & 1 & 0 & -1 \\
\cos\beta & -\sin\beta & \cos\alpha & -\sin\alpha & -\cos\gamma & \sin\gamma \\
\sin\beta & \cos\beta & \sin\alpha & \cos\alpha & -\sin\gamma & -\cos\gamma \\
\dot{\beta}\cos\beta & -\dot{\beta}\sin\beta & \dot{\alpha}\cos\alpha & -\dot{\alpha}\sin\alpha & -\dot{\gamma}\cos\gamma & \dot{\gamma}\sin\gamma \\
\dot{\beta}\sin\beta & \dot{\beta}\cos\beta & \dot{\alpha}\sin\alpha & \dot{\alpha}\cos\alpha & -\dot{\gamma}\sin\gamma & -\dot{\gamma}\cos\gamma
\end{bmatrix}
\begin{Bmatrix}
W_{1x} \\ W_{1y} \\ V_{1x} \\ V_{1y} \\ U_{1x} \\ U_{1y}
\end{Bmatrix}
=
\begin{bmatrix}
1 \\ 0 \\ 1 \\ 0 \\ 0 \\ 0
\end{bmatrix}
$$

(5.30)

Figure E.5.18 includes the calculation procedure in MATLAB's command window for vectors $\mathbf{W}_1$, $\mathbf{V}_1$, and $\mathbf{U}_1$. The synthesized planar four-bar function generator solution is illustrated in Figure E.5.19.

```
>> B = 35*pi/180;
>> dB = 1;
>> CB = cos(B); SB = sin(B);
>> A = -10*pi/180;
>> dA = -1.5;
>> CA = cos(A); SA = sin(A);
>> G = -20*pi/180;
>> dG = -0.75;
>> CG = cos(G); SG = sin(G);
>> WVU = inv([1, 0, 1, 0, -1, 0
0, 1, 0, 1, 0, -1
CB, -SB, CA, -SA, -CG, SG
SB, CB, SA, CA, -SG, -CG
dB*CB, -dB*SB, dA*CA, -dA*SA, -dG*CG, dG*SG
dB*SB, dB*CB, dA*SA, dA*CA, -dG*SG, -dG*CG
])*[1, 0, 1, 0, 0, 0]'

WVU =

    0.3675
   -0.1534
   -0.1252
    0.1796
   -0.7577
    0.0261

>>
```

FIGURE E.5.18
Example 5.7 $\mathbf{W}_1 - \mathbf{V}_1 - \mathbf{U}_1$ calculation procedure in MATLAB.

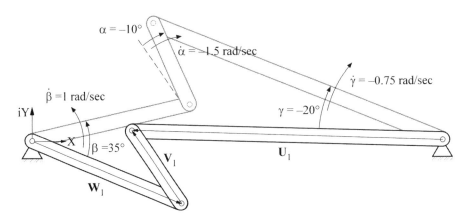

FIGURE E.5.19
Synthesized planar four-bar function generator.

5.9 Mechanism Dimensions: From Dimensional Synthesis to Kinematic Analysis

After synthesizing a mechanism, it is commonly analyzed kinematically. This is usually done to verify that the mechanism is free of defects. The vector solutions from the planar four-bar and Stephenson III dimensional synthesis equations can be incorporated in their respective kinematic analysis equations. Figure 5.13 includes the dimensional synthesis vectors and the kinematic analysis vectors for the planar four-bar and Stephenson III mechanisms. Figures 5.13a and 5.13b are for planar four-bar and Stephenson III motion generation and kinematic analysis respectively and Figure 5.13c is for planar four-bar function generation and kinematic analysis. In four-bar and Stephenson III motion generation, it is not certain that the $W_1 - Z_1$ will be the true driving dyad. When a kinematic analysis of a synthesized four-bar or Stephenson II mechanism yields incorrect results (with respect to the prescribed values used for synthesis), one should also drive these mechanisms using the $U_1 - S_1$ dyad along with its displacement angles to fully determine if the synthesized mechanism has a defect.

Table 5.1 includes the vectors and vector expressions for planar four-bar and Stephenson III motion generation and kinematic analysis.* The coupler displacement angles α_j and α_j^* specified in planar four-bar and Stephenson III motion generation require no changes for the kinematic analysis of these mechanisms. Adding the vector sum $W_1 + Z_1$ to the precision points p_1, p_2, and p_3 as shown in Table 5.1 expresses these precision points in the reference frame used for kinematic analysis. One can also subtract the kinematic analysis-calculated p_1 value from any remaining kinematic analysis-calculated p_j values to express them in the reference frame used for kinematic synthesis.

In planar four-bar function generation (Figure 5.13c), the synthesis vectors W_1, V_1, U_1, and G_1 can be directly incorporated in the analysis equations without any additional calculations. From this, we can see that $W_1 = W_{1x} + iW_{1y}$, $V_1 = V_{1x} + iV_{1y}$, $U_1 = U_{1x} + iU_{1y}$, and $G_1 = G_{1x} + iG_{1y}$ in the planar four-bar kinematic analysis equations.

---

* Although link dimensions mainly appear in *polar exponential* form in the Appendix B, C, D MATLAB files and user instructions, they can be specified in any of the rectangular and complex forms given in Equation 2.1.

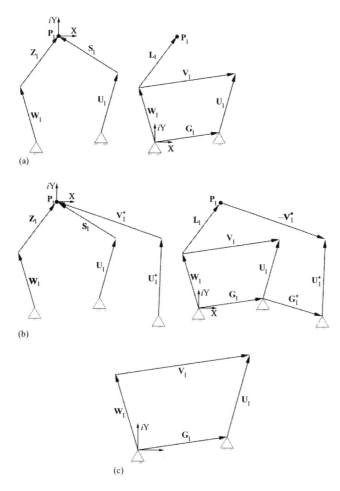

FIGURE 5.13
Synthesis and analysis vectors for (a, c) four-bar and (b) Stephenson III mechanisms.

TABLE 5.1

Vectors and Vector Expressions in Four-Bar and Stephenson III Synthesis and Analysis

Motion Generation		Kinematic Analysis	
Vector	**Vector Expression**	**Vector**	**Vector Expression**
$\mathbf{W}_1$	$\mathbf{W}_1 = W_{1x} + iW_{1y}$	$\mathbf{W}_1$	$\mathbf{W}_1 = W_{1x} + iW_{1y}$
$\mathbf{Z}_1$	$\mathbf{Z}_1 = Z_{1x} + iZ_{1y}$	$\mathbf{L}_1$	$\mathbf{L}_1 = Z_{1x} + iZ_{1y}$
$\mathbf{U}_1$	$\mathbf{U}_1 = U_{1x} + iU_{1y}$	$\mathbf{U}_1$	$\mathbf{U}_1 = U_{1x} + iU_{1y}$
$\mathbf{S}_1$	$\mathbf{S}_1 = S_{1x} + iS_{1y}$	$\mathbf{V}_1$	$\mathbf{V}_1 = \mathbf{Z}_1 - \mathbf{S}_1$
		$\mathbf{G}_1$	$\mathbf{G}_1 = \mathbf{W}_1 + \mathbf{Z}_1 - \mathbf{S}_1 - \mathbf{U}_1$
$\mathbf{U}_1^*$	$\mathbf{U}_1^* = U_{1x}^* + iU_{1y}^*$	$\mathbf{U}_1^*$	$\mathbf{U}_1^* = U_{1x}^* + iU_{1y}^*$
$\mathbf{V}_1^*$	$\mathbf{V}_1^* = V_{1x}^* + iV_{1y}^*$	$\mathbf{V}_1^*$	$\mathbf{V}_1^* = -\mathbf{V}_1^*$
		$\mathbf{G}_1^*$	$\mathbf{G}_1^* = \mathbf{W}_1 + \mathbf{Z}_1 + \left(-\mathbf{V}_1^*\right) - \mathbf{U}_1^* = -\mathbf{G}_1$
$\mathbf{p}_1$	$\mathbf{p}_1 = p_{1x} + ip_{1y}$	$\mathbf{p}_1$	$\mathbf{p}_1 = \mathbf{p}_1 + \mathbf{W}_1 + \mathbf{Z}_1$
$\mathbf{p}_2$	$\mathbf{p}_2 = p_{2x} + ip_{2y}$	$\mathbf{p}_2$	$\mathbf{p}_2 = \mathbf{p}_2 + \mathbf{W}_1 + \mathbf{Z}_1$ or $\mathbf{p}_2 = \mathbf{p}_2 - \mathbf{p}_1$
$\mathbf{p}_3$	$\mathbf{p}_3 = p_{3x} + ip_{3y}$	$\mathbf{p}_3$	$\mathbf{p}_3 = \mathbf{p}_3 + \mathbf{W}_1 + \mathbf{Z}_1$ or $\mathbf{p}_3 = \mathbf{p}_3 - \mathbf{p}_1$

Example 5.8

Problem Statement: Using the Appendix B.1 MATLAB file, determine if the mechanism solution in Example 5.1 achieves its precision positions.

  Known Information: Example 5.1 and Appendix B.1 MATLAB file.

  Solution Approach: Figure E.5.20 includes the input specified (in bold text) in the Appendix B.1 MATLAB file. Table E.5.4 includes the coupler positions and follower displacement angles achieved by the planar four-bar mechanism synthesized in Example 5.1. The values of $\mathbf{p}_1$, $\mathbf{p}_2$, and $\mathbf{p}_3$ (in the $\mathbf{p}_j$ column in Table E.5.4) calculated from the Appendix B.1 MATLAB file are actually offset by $\mathbf{W}_1 + \mathbf{Z}_1$, since the coordinate system origin for kinematic analysis is as shown in Figure 5.13a (right image). By subtracting $\mathbf{p}_1$ from $\mathbf{p}_1$, $\mathbf{p}_2$, and $\mathbf{p}_3$ (as shown in the $\mathbf{p}_j - \mathbf{p}_1$ column in Table E.5.4), the coordinate system origin matches the origin shown in Figure 5.13a (left image) and the resulting values in this column can be compared directly to the $\mathbf{p}_j$ column data in Table E.5.1.

Example 5.9

Problem Statement: Using the Appendix B.7 MATLAB file, determine if the mechanism solution in Example 5.4 achieves its precision positions.

  Known Information: Example 5.4 and Appendix B.7 MATLAB file.

  Solution Approach: Figure E.5.21 includes the input specified (in bold text) in the Appendix B.7 MATLAB file. Table E.5.5 includes the coupler positions and follower displacement angles achieved by the planar four-bar mechanism synthesized in Example 5.4. The values of $\mathbf{p}_1$, $\mathbf{p}_2$, and $\mathbf{p}_3$ (in the $\mathbf{p}_j$ column in Table E.5.5) calculated from the Appendix B.7 MATLAB file are actually offset by $\mathbf{W}_1 + \mathbf{Z}_1$ since the coordinate system origin for kinematic analysis is as shown in Figure 5.13b (right image). By subtracting $\mathbf{p}_1$ from $\mathbf{p}_1$, $\mathbf{p}_2$, and $\mathbf{p}_3$ (as shown in the $\mathbf{p}_j - \mathbf{p}_1$ column in Table E.5.5), the coordinate system origin matches the origin shown in Figure 5.13b (left image) and the resulting values in this column can be compared directly to the $\mathbf{p}_j$ column data in Table E.5.2.

```
W1 = 2.058 - i*0.8054;
Z1 = -0.1324 + i*0.1191;
U1 = 0.5808 - i*1.8615;
S1 = -1.5053 + i*1.34;
V1 = Z1 - S1;
G1 = W1 + Z1 - S1 - U1;
L1 = Z1;

start_ang  = 0;
step_ang   = 1;
stop_ang   = 38;

angular_vel  = 0 * ones(N+1,1);
angular_acc  = 0 * ones(N+1,1);
```

FIGURE E.5.20
Specified input (in bold text) in the Appendix B.1 MATLAB file for Example 5.1.

TABLE E.5.4

Coupler Positions and Follower Angles Achieved by Example 5.1 Solution

β_j (°)	$\mathbf{p}_j$	α_j (°)	γ_j (°)	$\mathbf{p}_j - \mathbf{p}_1$
0	1.9256, −0.6863	0	0	0, 0
18	2.2176, 0.0477	−51.716	−40.002	0.292, 0.734
38	2.2246, 0.7747	−84.98	−87.004	0.299, 1.461

```
W1 = 0.5854 - i*1.2248;
Z1 = -0.4926 - i*0.3631;
U1 = 0.1905 + i*1.8923;
S1 = -1.0184 - i*1.1813;
V1 = Z1 - S1;
G1 = W1 + Z1 - S1 - U1;
L1 = Z1;

U1s = 2.2215 - i*2.0095;
V1s = -(0.3400 + i*0.4718);
G1s = W1 + Z1 + V1s - U1s - G1;
L1s = 0;

start_ang    = 0;
step_ang     = 1;
stop_ang     = 40;

angular_vel  = 0 * ones(N+1,1);
angular_acc  = 0 * ones(N+1,1);
```

FIGURE E.5.21
Specified input (in bold text) in the Appendix B.7 MATLAB file for Example 5.4.

TABLE E.5.5

Coupler Positions and Follower Angles Achieved by Example 5.4 Solution

β_j (°)	p_j	α_j (°)	α_j^* (°)	γ_j (°)	γ_j^* (°)	$p_j - p_1$
0	0.0928, −1.5879	0	0	0	0	0, 0
15	0.2743, −1.0997	−30	14.999	−20	10	0.1815, 0.4882
40	0.7575, −0.1801	−75.001	5	−35	29.999	0.6647, 1.4078

Example 5.10

Problem Statement: Using the Appendix B.1 MATLAB file, determine if the mechanism solution in Example 5.5 achieves its precision points.

Known Information: Example 5.5 and Appendix B.1 MATLAB file.

Solution Approach: Figure E.5.22 includes the input specified (in bold text) in the Appendix B.1 MATLAB file. Table E.5.6 includes the link displacement angles achieved by the planar four-bar mechanism synthesized in Example 5.5. The values in this table match the precision points in Table E.5.3.

```
W1 = 0.2670 + i*0.0816;
V1 = 1.0598 + i*0.4564;
U1 = 0.3268 + i*0.5380;
G1 = 1;
L1 = 0;

start_ang    = 0;
step_ang     = 1;
stop_ang     = 180;

angular_vel  = 0 * ones(N+1,1);
angular_acc  = 0 * ones(N+1,1);
```

FIGURE E.5.22
Specified input (in bold text) in the Appendix B.1 MATLAB file for Example 5.5.

TABLE E.5.6

Precision Points Achieved by Example 5.5
Solution

β_j (°)	γ_j (°)	α_j (°)
0	0	0
90	30.007	−4.999
180	60.008	10.004

Example 5.11

Problem Statement: Using the Appendix B.1 MATLAB file, determine if the mechanism solution in Example 5.6 achieves its precision points.

  Known Information: Example 5.6 and Appendix B.1 MATLAB file.

  Solution Approach: Figure E.5.23 includes the input specified (in bold text) in the Appendix B.1 MATLAB file. Table E.5.7 includes the link displacement angles achieved by the planar four-bar mechanism synthesized in Example 5.6. The values in this table match the precision points in Figure E.5.15.

Example 5.12

Problem Statement: Using the Appendix B.1 MATLAB file, determine if the mechanism solution in Example 5.7 achieves its precision points.

  Known Information: Example 5.7 and Appendix B.1 MATLAB file.

  Solution Approach: Figure E.5.24 includes the input specified (in bold text) in the Appendix B.1 MATLAB file. Table E.5.8 includes the link displacement angles achieved by the planar four-bar mechanism synthesized in Example 5.7. The values in this table match the precision points in Figure E.5.18.

```
W1 = 0.2818 + i*0.0540;
V1 = -0.3424 + i*0.1260;
U1 = -1.0605 + i*0.1801;
G1 = 1;
L1 = 0;

start_ang   = 0;
step_ang    = 1;
stop_ang    = 1;

angular_vel  = 1 * ones(N+1,1);
angular_acc  = 0 * ones(N+1,1);
```

FIGURE E.5.23
Specified input (in bold text) in the Appendix B.1 MATLAB file for Example 5.6.

TABLE E.5.7

Precision Points Achieved by Example 5.6 Solution

β_j (°)	α_j (°)	$\dot{\alpha}_j$ (rad/s)	$\ddot{\alpha}_j$ (rad/s²)	γ_j (°)	$\dot{\gamma}_j$ (rad/s)	$\ddot{\gamma}_j$ (rad/s²)
0	−0.0701	−1.495	−1.0388	−0.0172	−0.7482	−0.1116

```
W1 = 0.3675 - i*0.1534;
V1 = -0.1252 + i*0.1796;
U1 = -0.7577 + i*0.0261;
G1 = 1;
L1 = 0;

start_ang    = 0;
step_ang     = 1;
stop_ang     = 35;

angular_vel  = 1 * ones(N+1,1);
angular_acc  = 0 * ones(N+1,1);
```

FIGURE E.5.24

Specified input (in bold text) in the Appendix B.1 MATLAB file for Example 5.7.

TABLE E.5.8

Precision Points Achieved by Example 5.7 Solution

β_j (°)	α_j (°)	$\dot{\alpha}_j$ (rad/s)	γ_j (°)	$\dot{\gamma}_j$ (rad/s)
35	−10.012	−1.4996	−20.007	−0.7497

5.10 Summary

In kinematic analysis, mechanism dimensions are known and the resulting mechanism motion sequence is calculated. In dimensional synthesis (a category of kinematic synthesis), a mechanism motion sequence is prescribed and the mechanism dimensions required to achieve them are calculated. Dimensional synthesis includes three subcategories: motion generation, path generation, and function generation. In motion generation, mechanism link positions are prescribed (specifically, coupler-link positions for the planar four-bar mechanism). In planar four-bar path generation, coupler-link path points are prescribed while crank and follower displacement angles are prescribed in planar four-bar function generation. Prescribed coupler positions are called precision positions, while precision points are prescribed coupler points (this is the chief distinction between motion and path generation).

Order defects and branch defects are two uncertainties inherent in motion and path generation. With the order defect, the synthesized mechanism will not achieve the precision positions/points in the intended order. With the branch defect, the synthesized mechanism will not achieve its precision positions/points in a single mechanism branch. A variety of methods have been developed to ensure order- and branch-defect-free solutions. For example, motion generation with prescribed timing (where dyad rotation angles are prescribed) ensures order-defect-free mechanism solutions.

Linear simultaneous equation sets for planar four-bar motion generation are formulated using vector-loop closure for the starting and displaced position of a mechanism dyad. These equations are applicable for three precision positions. By including an equation set corresponding to an additional dyad connected to the coupler point of the planar four-bar mechanism, Stephenson III motion generators are produced.

In planar four-bar function generation, displacement angles are typically specified for the crank and follower links. Often the displacement angles correspond to a mathematical function (where the crank and follower displacement angles are particular input and output function values, respectively). Simultaneous equation sets for planar four-bar function generation are formulated by taking the vector-loop sum of the entire mechanism. The resulting equations can be used to calculate the dimensions of a planar four-bar mechanism to precisely achieve three prescribed crank and follower displacement angle pairs (also called precision points).

In addition to link displacement angles (or FSPs), it is also possible to synthesize function generators to achieve or approximate prescribed link angular velocities or accelerations (or *MSPs*). Differentiating, expanding, and separating the planar four-bar vector-loop equation produces velocity and acceleration equations (after the first and second derivatives, respectively). Any combination of the velocity equation or the acceleration equation and the initial-position vector-loop equation produces an equation set to analytically calculate the mechanism dimensions required to precisely achieve prescribed FSPs or MSPs.

References

1. Sandor, G. N. and Erdman, A. G. 1984. *Advanced Mechanism Design: Analysis and Synthesis.* Volume 2, p. 51. Englewood Cliffs, NJ: Prentice-Hall.
2. Martin, G. H. 1969. *Kinematics and Dynamics of Machines.* P. 319. New York: McGraw-Hill.
3. Erdman, A. G. 1993. *Modern Kinematics-Developments in the Last Forty Years.* Chapters 4–5. New York: John Wiley.
4. Kimbrell, J. T. 1991. *Kinematics Analysis and Synthesis.* Chapter 7. New York: McGraw-Hill.
5. Ibid, Chapter 8.
6. Mallik, A. K., Amitabha, G., and Dittrich, G. 1994. *Kinematic Analysis and Synthesis of Mechanisms.* pp. 306–308. Boca Raton: CRC Press.
7. Russell, K., Shen, Q., and Sodhi, R. 2014. *Mechanism Design: Visual and Programmable Approaches.* Chapter 4. Boca Raton: CRC Press.
8. Mallik, A. K., Amitabha, G., and Dittrich, G. 1994. *Kinematic Analysis and Synthesis of Mechanisms.* pp. 306–308. Boca Raton: CRC Press.
9. Russell, K., Shen, Q., and Sodhi, R. 2014. *Mechanism Design: Visual and Programmable Approaches.* Chapters 4–5. Boca Raton: CRC Press.
10. Norton, R. L. 2008. *Design of Machinery.* 4th edn, Chapter 5. New York: McGraw-Hill.
11. Sandor, G. N. and Erdman, A. G. 1984. *Advanced Mechanism Design: Analysis and Synthesis.* Volume 2, pp. 179–180. Englewood Cliffs, NJ: Prentice-Hall.
12. McCarthy, M. J. 2000. *Geometric Design of Linkages.* pp. 110–111. New York: Springer.
13. Filemon, E. 1972. Useful ranges of centerpoint curves for design of crank-and-rocker linkages. *Mechanism and Machine Theory* 7: 47–53.
14. McCarthy, M. J. 2000. *Geometric Design of Linkages.* pp. 111–112. New York: Springer.
15. Waldron, K. L. 1976. Elimination of the branch problem in graphical Burmester mechanism synthesis for four finitely separated positions. *Transactions of the ASME, Journal of Engineering for Industry* 98: 176–182.
16. Sandor, G. N. and Erdman, A. G. 1984. *Advanced Mechanism Design: Analysis and Synthesis.* Volume 2, p. 76. Englewood Cliffs, NJ: Prentice-Hall.
17. McCarthy, M. J. 2000. *Geometric Design of Linkages.* pp. 51–52. New York: Springer.
18. Sandor, G. N. and Erdman, A. G. 1984. *Advanced Mechanism Design: Analysis and Synthesis.* Volume 2, pp. 198–199. Englewood Cliffs, NJ: Prentice-Hall.
19. Ibid, pp. 52–53.
20. Waldron, K. J. and Kinzel, G. L. 2004. *Kinematics, Dynamics and Design of Machinery.* 2nd edn, pp. 287–288. Saddle River, NJ: Prentice-Hall.

Additional Reading

Norton, R. L. 2008. *Design of Machinery*. 4th edn, pp. 256–259. New York: McGraw-Hill.
Sandor, G. N. and Erdman, A. G. 1984. *Advanced Mechanism Design: Analysis and Synthesis*. Volume 2, pp. 127–133. Englewood Cliffs, NJ: Prentice-Hall.
Waldron, K. J. and Kinzel, G. L. 2004. *Kinematics, Dynamics and Design of Machinery*. 2nd edn, pp. 283–294. Saddle River, NJ: Prentice-Hall.

Problems

1. Synthesize a branch defect-free planar four-bar mechanism for the three hatch positions in Figure P.5.1. Branch defect elimination can be verified through *Filemon's Construction, Waldron's Construction* or a displacement analysis (using the Appendix B.1 or H.1 files).

2. Synthesize a branch defect-free planar four-bar mechanism for the three load–unload bucket positions in Figure P.5.2. Branch defect elimination can be verified through *Filemon's Construction, Waldron's Construction,* or a displacement analysis (using the Appendix B.1 or H.1 files).

3. Synthesize a branch defect-free planar four-bar mechanism for the three stamping tool positions in Figure P.5.3. Branch defect elimination can be verified through *Filemon's Construction, Waldron's Construction,* or a displacement analysis (using the Appendix B.1 or H.1 files).

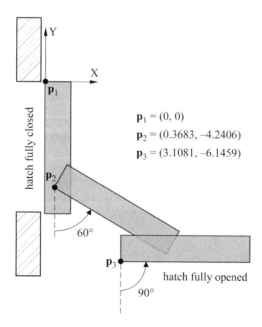

$\mathbf{p}_1 = (0, 0)$
$\mathbf{p}_2 = (0.3683, -4.2406)$
$\mathbf{p}_3 = (3.1081, -6.1459)$

FIGURE P.5.1
Tree hatch positions.

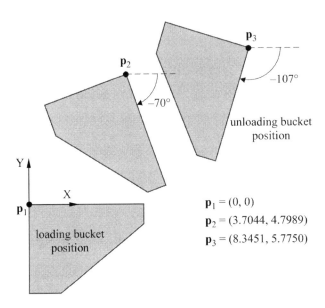

$\mathbf{p}_1 = (0, 0)$
$\mathbf{p}_2 = (3.7044, 4.7989)$
$\mathbf{p}_3 = (8.3451, 5.7750)$

FIGURE P.5.2
Three load–unload bucket positions.

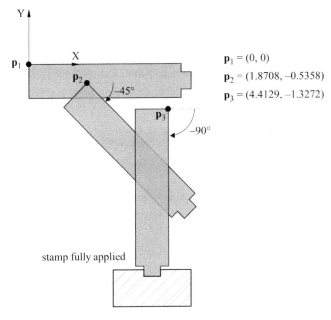

$\mathbf{p}_1 = (0, 0)$
$\mathbf{p}_2 = (1.8708, -0.5358)$
$\mathbf{p}_3 = (4.4129, -1.3272)$

FIGURE P.5.3
Three stamping tool positions.

4. Synthesize a branch defect-free planar four-bar mechanism for the three folding wing positions in Figure P.5.4. Branch defect elimination can be verified through *Filemon's Construction, Waldron's Construction,* or a displacement analysis (using the Appendix B.1 or H.1 files).

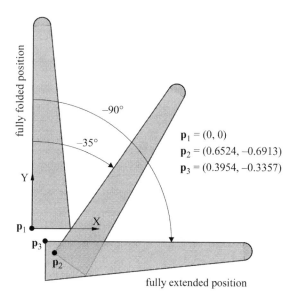

FIGURE P.5.4
Three folding wing positions.

5. Synthesize a branch defect-free planar four-bar mechanism for the three lower blade positions in Figure P.5.5. Branch defect elimination can be verified through *Filemon's Construction, Waldron's Construction,* or a displacement analysis (using the Appendix B.1 or H.1 files).

6. Synthesize a branch defect-free planar four-bar mechanism for the three brake pad positions in Figure P.5.6. Branch defect elimination can be verified through *Filemon's Construction, Waldron's Construction,* or a displacement analysis (using the Appendix B.1 or H.1 files).

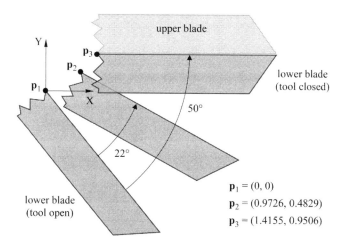

FIGURE P.5.5
Three lower blade positions.

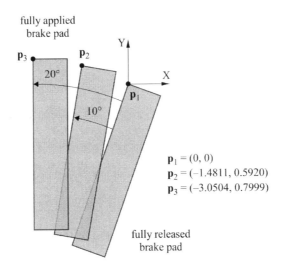

FIGURE P.5.6
Three brake pad positions.

7. Synthesize a branch defect-free planar four-bar mechanism for the three digger bucket positions in Figure P.5.7. Branch defect elimination can be verified through *Filemon's Construction, Waldron's Construction,* or a displacement analysis (using the Appendix B.1 or H.1 files).

8. Synthesize a branch defect-free planar four-bar mechanism for the three latch positions in Figure P.5.8. Branch defect elimination can be verified through *Filemon's Construction, Waldron's Construction,* or a displacement analysis (using the Appendix B.1 or H.1 files).

9. Synthesize a branch defect-free planar four-bar mechanism for the three top gripper positions in Figure P.5.9. Branch defect elimination can be verified through *Filemon's Construction, Waldron's Construction,* or a displacement analysis (using the Appendix B.1 or H.1 files).

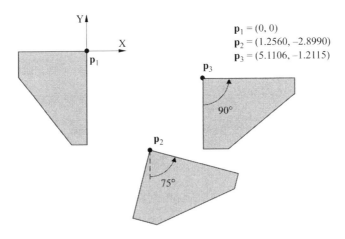

FIGURE P.5.7
Three digger bucket positions.

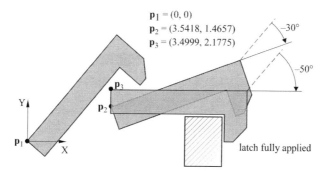

FIGURE P.5.8
Three latch positions.

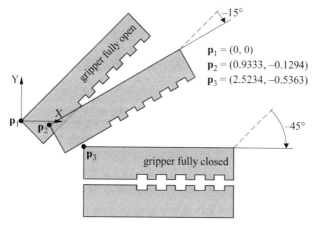

FIGURE P.5.9
Three top gripper positions.

10. Synthesize a branch defect-free planar four-bar mechanism for the three assembly component positions in Figure P.5.10. Branch defect elimination can be verified through *Filemon's Construction*, *Waldron's Construction*, or a displacement analysis (using the Appendix B.1 or H.1 files).

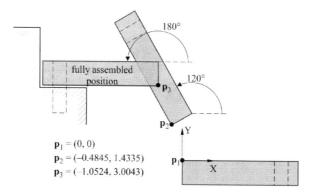

FIGURE P.5.10
Three assembly component positions.

11. Repeat Problem 1 with $\mathbf{p}_2 = (0.785, -4.71)$ and $\mathbf{p}_3 = (1.57, -6.28)$.

12. Repeat Problem 2 with $\mathbf{p}_2 = (1.5, 4.5)$ and $\mathbf{p}_3 = (3, 6)$.

13. Repeat Problem 4 with $\mathbf{p}_2 = (1.45, -0.725)$ and $\mathbf{p}_3 = (0.725, -0.725)$.

14. Repeat Problem 6 with $\mathbf{p}_2 = (-1.47, 1.47)$ and $\mathbf{p}_3 = (-2.205, 2.94)$.

15. Repeat Problem 9 with $\mathbf{p}_2 = (0.35, -1.4)$ and $\mathbf{p}_3 = (1.4, -1.75)$.

16. Synthesize a Stephenson III mechanism for the three gripper positions in Figure P.5.11. Verify that the gripper positions are achieved (and that the mechanism is free of branch defects) through a displacement analysis using the Appendix B.7 or H.6 files.

17. Synthesize a Stephenson III mechanism for the three seat positions in Figure P.5.12. Verify that the seat positions are achieved (and that the mechanism is free of branch defects) through a displacement analysis using the Appendix B.7 or H.6 files.

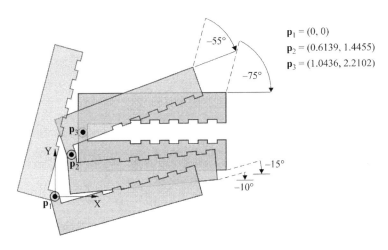

$\mathbf{p}_1 = (0, 0)$
$\mathbf{p}_2 = (0.6139, 1.4455)$
$\mathbf{p}_3 = (1.0436, 2.2102)$

FIGURE P.5.11
Three gripper positions.

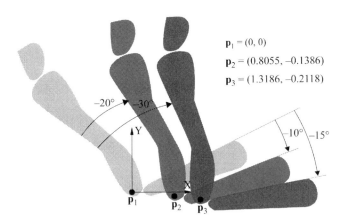

$\mathbf{p}_1 = (0, 0)$
$\mathbf{p}_2 = (0.8055, -0.1386)$
$\mathbf{p}_3 = (1.3186, -0.2118)$

FIGURE P.5.12
Three seat positions.

18. Synthesize a Stephenson III mechanism for the three digger arm positions in Figure P.5.13. Verify that the digger arm positions are achieved (and that the mechanism is free of branch defects) through a displacement analysis using the Appendix B.7 or H.6 files.

19. Synthesize a Stephenson III mechanism for the three cutting tool positions in Figure P.5.14. Verify that the cutting tool positions are achieved (and that the mechanism is free of branch defects) through a displacement analysis using the Appendix B.7 or H.6 files.

20. Repeat Problem 16 with $\mathbf{p}_2 = (1.4, 1.4)$ and $\mathbf{p}_3 = (2.1, 1.75)$.

21. Repeat Problem 17 with $\mathbf{p}_2 = (0.735, -0.735)$ and $\mathbf{p}_3 = (2.94, -1.1025)$.

22. Repeat Problem 18 with $\mathbf{p}_2 = (2.9, -1.45)$ and $\mathbf{p}_3 = (4.35, -0.3625)$.

23. Repeat Problem 19 with $\mathbf{p}_2 = (1.5, 1.5)$ and $\mathbf{p}_3 = (3, 1.875)$.

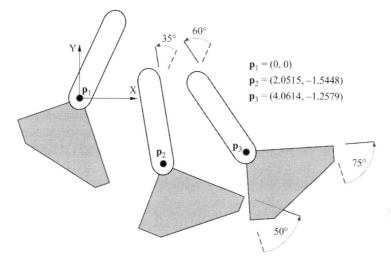

FIGURE P.5.13
Three digger arm positions.

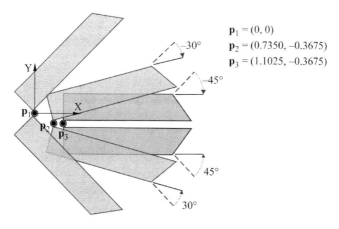

FIGURE P.5.14
Three cutting tool positions.

24. Using the MATLAB calculation procedure in Example 5.5 (Figure E.5.13), synthesize a planar four-bar mechanism for a 75° total wiper blade rotation range (Figure P.5.15) and a 45° total crank rotation range (where the wiper blade is affixed to the follower link).

25. Using the MATLAB calculation procedure in Example 5.5 (Figure E.5.13), synthesize a planar four-bar mechanism for a 75° wheel rotation range (Figure P.5.16) and a 50° crank rotation range (where the wheel is affixed to the follower link).

26. Using the MATLAB calculation procedure in Example 5.5 (Figure E.5.13), synthesize a planar four-bar mechanism for a 90° total valve rotation range (Figure P.5.17) and a 40° total crank rotation range (where the valve is affixed to the follower link).

27. Using the MATLAB calculation procedure in Example 5.5 (Figure E.5.13), synthesize a planar four-bar mechanism for a 130° total hatch rotation range (Figure P.5.18) and a 65° total crank rotation range (where the hatch is affixed to the follower link).

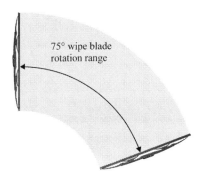

FIGURE P.5.15
Wiper blade rotation range.

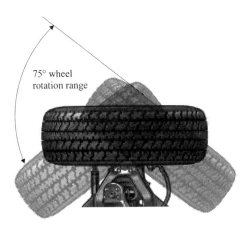

FIGURE P.5.16
Wheel rotation range.

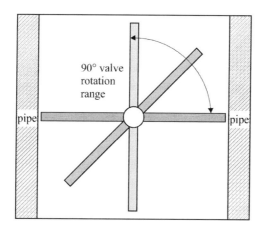

FIGURE P.5.17
Pipe valve rotation range.

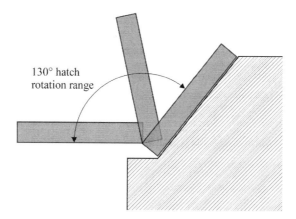

FIGURE P.5.18
Hatch rotation range.

28. Using Equation (5.19) and the MATLAB calculation procedure in Example 5.5 (Figure E.5.13), synthesize a planar four-bar mechanism for the function $f(x) = \cos x$ at precision points $x_2 = 30°$ and $x_3 = 90°$ (where $x_1 = 0°$, $\Delta\beta = 60°$ and $\Delta\gamma = 135°$).

29. Using Equation (5.19) and the MATLAB calculation procedure in Example 5.5 (Figure E.5.13), synthesize a planar four-bar mechanism for the function $f(x) = \log x$ at precision points $x_2 = 5$ and $x_3 = 10$ (where $x_1 = 1$, $\Delta\beta = 45°$ and $\Delta\gamma = 100°$).

30. Using Equation (5.19) and the MATLAB calculation procedure in Example 5.5 (Figure E.5.13), synthesize a planar four-bar mechanism for the function $f(x) = \tan x$ at precision points $x_2 = 35°$ and $x_3 = 45°$ (where $x_1 = 0°$, $\Delta\beta = 75°$ and $\Delta\gamma = 125°$).

31. Using the MATLAB calculation procedure in Example 5.6 (Figure E.5.15), synthesize a planar four-bar mechanism to achieve a follower angular velocity and acceleration of $\dot{\gamma} = 1.25\,\text{rad/s}$ and $\ddot{\gamma} = -0.25\,\text{rad/s}^2$, respectively, for a crank angular velocity and acceleration of $\dot{\beta} = 2\,\text{rad/s}$ and $\ddot{\beta} = 0.5\,\text{rad/s}^2$, respectively, and a coupler angular velocity and acceleration of $\dot{\alpha} = 0.65\,\text{rad/s}$ and $\ddot{\alpha} = 0.10\,\text{rad/s}^2$, respectively.

32. Using the MATLAB calculation procedure in Example 5.6 (Figure E.5.15), synthesize a planar four-bar mechanism to achieve a follower angular velocity and acceleration of $\dot{\gamma} = -1.5\,\text{rad/s}$ and $\ddot{\gamma} = 0.25\,\text{rad/s}^2$, respectively, for a crank angular velocity and acceleration of $\dot{\beta} = 2\,\text{rad/s}$ and $\ddot{\beta} = -0.5\,\text{rad/s}^2$, respectively, and a coupler angular velocity and acceleration of $\dot{\alpha} = -0.65\,\text{rad/s}$ and $\ddot{\alpha} = -0.35\,\text{rad/s}^2$, respectively.

33. Using the MATLAB calculation procedure in Example 5.7 (Figure E.5.18), synthesize a planar four-bar mechanism to achieve a follower angular displacement and velocity of $\gamma = -35°$ and $\dot{\gamma} = -0.25\,\text{rad/s}$, respectively, for a crank angular displacement and velocity of $\beta = 55°$ and $\dot{\beta} = 0.5\,\text{rad/s}$, respectively, and a coupler angular displacement and velocity of $\alpha = -15°$ and $\dot{\alpha} = -0.35\,\text{rad/s}$, respectively.

34. Using the MATLAB calculation procedure in Example 5.7 (Figure E.5.18), synthesize a planar four-bar mechanism to achieve a follower angular displacement and velocity of $\gamma_2 = \pi/4\,\text{rad}$ and $\dot{\gamma}_2 = 1.15\,\text{rad/s}$ and a coupler angular displacement and velocity of $\alpha_2 = \pi/18\,\text{rad}$ and $\dot{\alpha}_2 = 0.25\,\text{rad/s}$. The angular displacement and constant angular velocity for the crank are $\beta_2 = \pi/6\,\text{rad}$ and $\dot{\beta} = 1\,\text{rad/s}$.

35. Using the MATLAB calculation procedure in Example 5.7 (Figure E.5.18), synthesize a planar four-bar mechanism to achieve a follower angular displacement and velocity of $\gamma = -75°$ and $\dot{\gamma} = -3.25\,\text{rad/s}$, respectively, for a crank angular displacement and velocity of $\beta = 65°$ and $\dot{\beta} = 2.5\,\text{rad/s}$, respectively, and a coupler angular displacement and velocity of $\alpha = 35°$ and $\dot{\alpha} = 1.95\,\text{rad/s}$, respectively.

6

Static Force Analysis of Planar Mechanisms

CONCEPT OVERVIEW

In this chapter, the reader will gain a central understanding regarding

1. Criteria for static force analysis and its applications
2. Link static loads in 2D space
3. Formulation and solution of linear simultaneous equation sets for static force analysis
4. The effects of gear train inclusion in the static force analysis of five-bar mechanisms

6.1 Introduction

As explained in Chapter 1, additional analyses often follow a kinematic analysis. In terms of structural force analyses for mechanical systems, a *static force analysis* (Figure 1.1) is the most basic type of analysis to consider beyond kinematics. In this type of analysis, loads such as forces and torques are considered for each mechanism link according to Newton's first law ($\sum \mathbf{F} = \sum \mathbf{M} = 0$) [1–3].

A static force analysis is the only type of force analysis required if the mechanism operates in a static state.* In Figure 6.1a, the lock pliers holding the solid object are an example of a static state because, when holding the solid object, the pliers are not in motion. A static force analysis may also be applicable when the mechanism operates in a *quasi-static* state. An example of this state is given in the cutting tool illustrated in Figure 6.1b. Although the mechanism is in motion while the blades of the cutting tool shear the material, the dynamic forces produced during this motion can be so small (because the motion can be so slow) that they can be neglected. Under a condition where dynamic forces are small enough to be negligible in a mechanism, a quasi-static state exists and static force assumptions are suitable.

Even if the mechanism motion is truly dynamic, static force assumptions can be useful as a preliminary force analysis. For example, static forces or stresses can be multiplied by certain scale factors to account for dynamic events such as impact and fatigue [4].

Although it is possible for out-of-plane forces and moments to exist in a planar mechanism (due to mechanism mass and force imbalances in the z-direction), the equation

---

* When a mechanism is in a static state, it is said to be *statically determinate*.

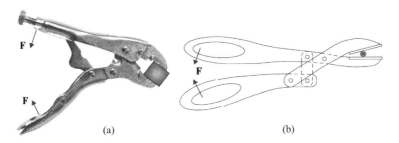

(a)   (b)

FIGURE 6.1
Applied forces in (a) lock pliers and (b) cutting tool.

systems presented in this chapter consider forces in the x-y plane only.* The equation systems presented in this chapter consider the reaction forces due to externally applied loads as well as the additional forces produced by gravity on the link masses (link weights). Although link weights are not neglected in the forthcoming equations, as the externally applied loads exceed the link weights, the reaction force-producing effect of these weights becomes increasingly negligible. The contribution of link weights generally becomes more important as the scale of the mechanism increases. For example, it would be essential to consider link weights in the static force analysis of a multistory level-luffing crane mechanism (see Figure E.4.1), but not essential in the analysis of the hand tools in Figure 6.1. Lastly, in this chapter, the mechanism links are considered to be *rigid* or nondeforming in the mechanism static force equations.[†]

6.2 Static Loading in Planar Space

Figure 6.2 illustrates arbitrarily grounded rotating and translating planar links under static loads. Loads, even distributed loads, can be represented as force vectors applied to link points. For example, force vectors $\mathbf{F}_{p_0}$, $\mathbf{F}_{p_1}$, and $m\mathbf{g}$ are applied to link points $\mathbf{p}_0$, $\mathbf{p}_1$ and the link's CG (center of gravity), respectively, in Figure 6.2a. Also, a torque $\mathbf{T}_{a_0}$ is applied about point $\mathbf{p}_0$ in Figure 6.2a.[‡] Static forces and torques are either applied externally (like force vector $\mathbf{F}$ in Figure 6.2b) or they are reactions to externally applied loads (like force vectors $\mathbf{F}_{p_1}$ and $\mathbf{F}_f$ in Figure 6.2b).

Link static equilibrium is achieved in accordance to Newton's first law. With this law, when the sum of all link forces is zero (or $\Sigma\mathbf{F} = 0$) and the sum of all link moments is zero (or $\Sigma\mathbf{M} = 0$), link static equilibrium is achieved.[§§] Because the force vectors include both x- and y-direction components, the forces' sums are taken in both directions ($\Sigma F_x = 0$ and $\Sigma F_y = 0$).

The conditions from Newton's first law must be satisfied for each mechanism link in order to achieve static equilibrium in the entire mechanism.

---

* Whether out-of-plane mechanism forces and moments should be considered in an analysis depends in part on the amount of mass and force imbalance preset in the mechanism as well as the overall mechanism scale.
† If springs are included in the mechanism design, their deflection should be considered since spring force is proportional to spring deflection.
‡ The torque is actually applied about an axis, but in planar space, the axis can be represented by a point (since the axis is normal to the plane).
§ The words *torque and moment* are used in this chapter since they are synonymous.

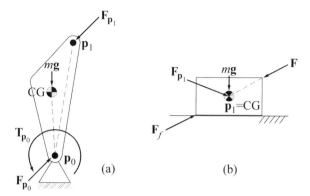

FIGURE 6.2
Static loads on (a) grounded rotating and (b) translating planar links.

6.3 Four-Bar Mechanism Analysis

Figure 6.3a illustrates a planar four-bar mechanism where a force $\mathbf{F_{p_1}}$ is applied to the coupler-link point $\mathbf{p_1}$. To maintain static equilibrium, a torque $\mathbf{T_{a_0}}$ is applied about the crank link revolute joint $\mathbf{a_0}$. Vectors $\mathbf{R_1}$, $\mathbf{R_2}$, and $\mathbf{R_3}$ point from $\mathbf{a_0}$, $\mathbf{a_1}$, and $\mathbf{b_0}$, respectively, to the center of gravity of each link. The loads on the individual planar four-bar mechanism links are illustrated in Figure 6.3b. Because the joints at $\mathbf{a_1}$ and $\mathbf{b_1}$ are shared among two links, the forces at $\mathbf{a_1}$ and $\mathbf{b_1}$ must be equal but opposite (resulting in $\pm\mathbf{F_{a_1}}$ and $\pm\mathbf{F_{b_1}}$ in Figure 6.3b). The remaining force and torque variables, however, remain positive for simplicity.[*] This approach is repeated for the mechanisms in Sections 6.4 through 6.7.

Taking the sum of the forces and moments for each link, according to the static equilibrium conditions $\Sigma\mathbf{F} = 0$ and $\Sigma\mathbf{M} = 0$, produces two static equilibrium equations for each link. Expanding these two equations and separating the force equation into two equations, where $\Sigma F_x = 0$ and $\Sigma F_y = 0$, ultimately produces three static equilibrium equations for each link.

Using the first two static equilibrium conditions for the crank link (where the moment sum is taken about $\mathbf{a_0}$) produces

$$\mathbf{F_{a_0}} + \mathbf{F_{a_1}} + m_1\mathbf{g} = 0$$

$$\mathbf{T_{a_0}} + \mathbf{W_1} \times \mathbf{F_{a_1}} + \mathbf{R_1} \times m_1\mathbf{g} = 0^\dagger \qquad (6.1)$$

Expanding and separating Equation 6.1 produces

$$F_{a_{0x}} + F_{a_{1x}} = 0$$

$$F_{a_{0y}} + F_{a_{1y}} + m_1 g = 0 \qquad (6.2)$$

$$T_{a_0} - F_{a_{1x}} W_1 \sin(\theta + \beta_j) + F_{a_{1y}} W_1 \cos(\theta + \beta_j) + m_1 g \left(R_{1x} \cos\beta_j - R_{1y} \sin\beta_j\right) = 0$$

---

[*] The signs of the calculated static force and torque variables are not determined by the signs prescribed to them during equation formulation. They are determined by the mechanism position and the applied load values.

[†] The expression $\mathbf{A} \times \mathbf{B}$ is the *cross product* of planar vectors $\mathbf{A}$ and $\mathbf{B}$. When expanded, $\mathbf{A} \times \mathbf{B} = A_x B_y - A_y B_x$.

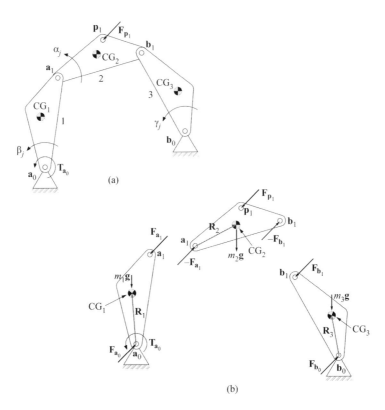

FIGURE 6.3
(a) Planar four-bar mechanism and (b) link static forces and torque.

Since gravity is directed along the y-axis (which is identical to the imaginary axis), the expansion of the cross product $\mathbf{R}_1 \times m_1\mathbf{g}$ in Equation 6.2 only includes the expansion of the product of m_1g and the real component of $\mathbf{R}_1$ where $\mathbf{R}_1 = (R_{1x} + iR_{1y})e^{i\beta_j}$.

   Using the first two static equilibrium conditions for the coupler link (where the moment sum is taken about $\mathbf{a}_1$) produces

$$-\mathbf{F}_{a_1} - \mathbf{F}_{b_1} + \mathbf{F}_{p_1} + m_2\mathbf{g} = 0$$

$$-\mathbf{V}_1 \times \mathbf{F}_{b_1} + \mathbf{L}_1 \times \mathbf{F}_{p_1} + \mathbf{R}_2 \times m_2\mathbf{g} = 0. \tag{6.3}$$

Expanding and separating Equation 6.3 produces

$$-F_{a_{1x}} - F_{b_{1x}} + F_{p_{1x}} = 0$$

$$-F_{a_{1y}} - F_{b_{1y}} + F_{p_{1y}} + m_2g = 0$$

$$F_{b_{1x}} V_1 \sin(\rho + \alpha_j) - F_{b_{1y}} V_1 \cos(\rho + \alpha_j) - F_{p_{1x}} L_1 \sin(\delta + \alpha_j) + F_{p_{1y}} L_1 \cos(\delta + \alpha_j)$$

$$+ m_2g\left(R_{2x}\cos\alpha_j - R_{2y}\sin\alpha_j\right) = 0. \tag{6.4}$$

Since gravity is directed along the y-axis, the expansion of the cross product $\mathbf{R}_2 \times m_2\mathbf{g}$ in Equation 6.4 only includes the expansion of the product of m_2g and the real component of $\mathbf{R}_2$ where $\mathbf{R}_2 = (R_{2x} + iR_{2y})e^{i\alpha_j}$.

Using the first two static equilibrium conditions for the follower link (where the moment sum is taken about $\mathbf{b}_0$) produces

$$\mathbf{F}_{b_0} + \mathbf{F}_{b_1} + m_3\mathbf{g} = 0$$

$$\mathbf{U}_1 \times \mathbf{F}_{b_1} + \mathbf{R}_3 \times m_3\mathbf{g} = 0. \tag{6.5}$$

Expanding and separating Equation 6.5 produces

$$F_{b_0x} + F_{b_1x} = 0$$

$$F_{b_0y} + F_{b_1y} + m_3 g = 0 \tag{6.6}$$

$$-F_{b_1x}U_1 \sin(\sigma + \gamma_j) + F_{b_1y}U_1 \cos(\sigma + \gamma_j) + m_3 g\left(R_{3x}\cos\gamma_j - R_{3y}\sin\gamma_j\right) = 0$$

Again, due to the y-axis direction of gravity, the expansion of the cross product $\mathbf{R}_3 \times m_3\mathbf{g}$ in Equation 6.6 only includes the expansion of the product of $m_3 g$ and the real component of $\mathbf{R}_3$ where $\mathbf{R}_3 = (R_{3x} + iR_{3y})e^{i\gamma_j}$.

Expressing Equations 6.2, 6.4, and 6.6 in a combined matrix form produces

$$
\begin{bmatrix}
0 & 1 & 0 & 1 & 0 & 0 & 0 & 0 & 0 \\
0 & 0 & 1 & 0 & 1 & 0 & 0 & 0 & 0 \\
1 & 0 & 0 & -W_y & W_x & 0 & 0 & 0 & 0 \\
0 & 0 & 0 & -1 & 0 & 0 & 0 & -1 & 0 \\
0 & 0 & 0 & 0 & -1 & 0 & 0 & 0 & -1 \\
0 & 0 & 0 & 0 & 0 & 0 & 0 & V_y & -V_x \\
0 & 0 & 0 & 0 & 0 & 1 & 0 & 1 & 0 \\
0 & 0 & 0 & 0 & 0 & 0 & 1 & 0 & 1 \\
0 & 0 & 0 & 0 & 0 & 0 & 0 & -U_y & U_x
\end{bmatrix}
\begin{Bmatrix}
T_{a_0} \\
F_{a_0x} \\
F_{a_0y} \\
F_{a_1x} \\
F_{a_1y} \\
F_{b_0x} \\
F_{b_0y} \\
F_{b_1x} \\
F_{b_1y}
\end{Bmatrix}
$$

$$
=
\begin{bmatrix}
0 \\
-m_1 g \\
-m_1 g\left(R_{1x}\cos\beta_j - R_{1y}\sin\beta_j\right) \\
-F_{p_1x} \\
-F_{p_1y} - m_2 g \\
\begin{bmatrix} F_{p_1x}L_y - F_{p_1y}L_x \\ -m_2 g\left(R_{2x}\cos\alpha_j - R_{2y}\sin\alpha_j\right) \end{bmatrix} \\
0 \\
-m_3 g \\
-m_3 g\left(R_{3x}\cos\gamma_j - R_{3y}\sin\gamma_j\right)
\end{bmatrix}
\tag{6.7}
$$

where:

$W_x = W_1 \cos(\theta + \beta_j)$
$W_y = W_1 \sin(\theta + \beta_j)$
$L_x = L_1 \cos(\delta + \alpha_j)$
$L_y = L_1 \sin(\delta + \alpha_j)$
$V_x = V_1 \cos(\rho + \alpha_j)$
$V_y = V_1 \sin(\rho + \alpha_j)$
$U_x = U_1 \cos(\sigma + \gamma_j)$
$U_y = U_1 \sin(\sigma + \gamma_j).^*$

Equation 6.7 can be solved using Cramer's rule to determine the unknown forces and torque. The unknown planar four-bar displacement angles α_j and γ_j are the same angles calculated from the planar four-bar displacement equations in Section 4.4.1. Given β_j, α_j, and γ_j solutions, the corresponding static forces and torques can be calculated from Equation 6.7.

Appendix C.1 includes the MATLAB® file user instructions for planar four-bar static force analysis. In this MATLAB file (which is available for download at www.crcpress.com/product/ isbn/9781498724937), solutions for Equation 6.7 are calculated.

Example 6.1

Problem Statement: When vector $\mathbf{W}_1$ of the planar four-bar stamping mechanism (Figure E.6.1a) is rotated $\beta = 60°$, a reaction force of $\mathbf{F}_{p_2} = (0, 4500)\,\text{N}$ is applied (Figure E.6.1b) due to the stamping event. Tables E.6.1 and E.6.2 include the dimensions and mass properties of the planar four-bar stamping mechanism in the initial position

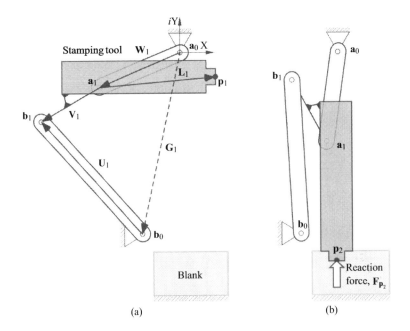

(a) (b)

FIGURE E.6.1
Stamping mechanism in (a) initial and (b) stamping positions.

---

* This method of calculating mechanism forces via matrix manipulation is called the *matrix method*. With this method, link force equations are quickly derived.

(Figure E.6.1a). Using the Appendix C.1 MATLAB file, calculate the static torque and forces generated in the stamping mechanism in the stamping position (Figure E.6.1b). Gravity is −9.81 m/s².

Known Information: Tables E.6.1, E.6.2, and Appendix C.1 MATLAB file.

Solution Approach: Figure E.6.2 includes the input specified (in bold text) in the Appendix C.1 MATLAB file. Table E.6.3 includes the static torque and forces calculated for the planar four-bar stamping mechanism using the Appendix C.1 MATLAB file.

TABLE E.6.1

Stamping Mechanism Dimensions in Initial Position (with Link Lengths in m)

W_1, θ	V_1, ρ	U_1, σ
0.2013, −157.8291°	0.1583, −150.1267°	0.3455, 133.0953°
G_{1x}, G_{1y}	L_1, δ	
−0.0876, −0.4071	0.27, 4.7572°	

TABLE E.6.2

Planar Four-Bar Mechanism Dynamic Parameters (with Length in m and mass in kg)

R_1	−0.0932 − i0.038	m_1	8
R_2	0.0955 + i0.0159	m_2	40
R_3	−0.118 + i0.1261	m_3	12

```
W1 = 0.2013*exp(-i*157.8291*pi/180);
V1 = 0.1583*exp(-i*150.1267*pi/180);
G1 = -0.0876 -i*0.4071;
U1 = 0.3455*exp(i*133.0953*pi/180);
L1 = 0.27*exp(i*4.7572*pi/180);

Fp1 = [0, 4500];
g = -9.81;

R1 = -0.0932 -i*0.038;
R2 = 0.0955 + i*0.0159;
R3 = -0.118+ i*0.1261;

m1 = 8;
m2 = 40;
m3 = 12;

start_ang  = 0;
step_ang   = 1;
stop_ang   = 60;
```

FIGURE E.6.2

Specified input (in bold text) in the Appendix C.1 MATLAB file for Example 6.1.

TABLE E.6.3

Calculated Stamping Mechanism Static Forces (in N) and Torque (in N-m)

T_{a_0}	F_{a_0}	F_{a_1}	F_{b_0}	F_{b_1}
162.51	74.14, −5347.8	−74.14, 5426.3	−74.14, 1436.4	74.14, −1318.7

6.4 Slider-Crank Mechanism Analysis

Figure 6.4a illustrates a slider-crank mechanism where a force $\mathbf{F}$ is applied to the slider link revolute joint $\mathbf{b}_1$. To maintain static equilibrium, a torque $\mathbf{T}_{a_0}$ is applied about the crank link revolute joint $\mathbf{a}_0$. Vectors $\mathbf{R}_1$ and $\mathbf{R}_2$ point from $\mathbf{a}_0$ and $\mathbf{a}_1$, respectively, to the center of gravity of each link. The loads on the individual slider-crank mechanism links are illustrated in Figure 6.4b. Taking the sum of the forces and moments for each link according to the static equilibrium conditions $\Sigma\mathbf{F} = 0$ and $\Sigma\mathbf{M} = 0$ produces two static equilibrium equations for each link. Expanding these two equations and separating the force equation into two equations, where $\Sigma F_x = 0$ and $\Sigma F_y = 0$, ultimately produces three static equilibrium equations for each link. Using these conditions for the crank link (where the moment sum is taken about $\mathbf{a}_0$) produces Equation 6.2.

Using the first two static equilibrium conditions for the coupler link (where the moment sum is taken about $\mathbf{a}_1$) produces

$$-\mathbf{F}_{a_1} - \mathbf{F}_{b_1} + m_2\mathbf{g} = 0$$
$$-\mathbf{V}_1 \times \mathbf{F}_{b_1} + \mathbf{R}_2 \times m_2\mathbf{g} = 0$$

(6.8)

Expanding and separating Equation 6.8 produces

$$-F_{a_{1x}} - F_{b_{1x}} = 0$$
$$-F_{a_{1y}} - F_{b_{1y}} + m_2 g = 0$$

(6.9)

$$F_{b_{1x}} V_1 \sin(\rho + \alpha_j) - F_{b_{1y}} V_1 \cos(\rho + \alpha_j) + m_2 g\left(R_{2x}\cos\alpha_j - R_{2y}\sin\alpha_j\right) = 0$$

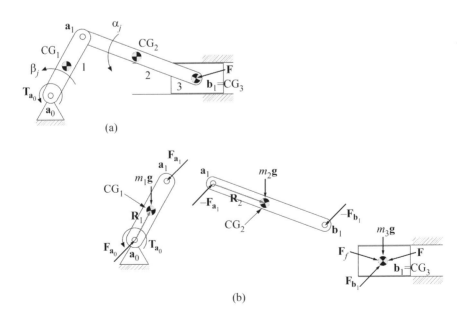

(a)

(b)

FIGURE 6.4
(a) Slider-crank mechanism and (b) link static forces and torque.

Since gravity is directed along the y-axis, the expansion of the cross product $\mathbf{R}_2 \times m_2\mathbf{g}$ in Equation 6.9 only includes the expansion of the product of m_2g and the real component of $\mathbf{R}_2$ where $\mathbf{R}_2 = (R_{2x} + iR_{2y})e^{i\alpha_j}$.

Using only the static equilibrium condition $\Sigma\mathbf{F} = 0$ for the slider link produces

$$\mathbf{F} + \mathbf{F}_{b_1} + \mathbf{F}_f + m_3\mathbf{g} = 0^* \tag{6.10}$$

In Equation 6.10, the x- and y-components of vector $\mathbf{F}_f$ are the *friction force* $\pm\mu F_{normal}$ and the *normal force* F_{normal}, respectively. Expanding and separating Equation 6.10 produces

$$F_x + F_{b_{1x}} \pm \mu F_{normal} = 0$$

$$F_y + F_{b_{1y}} + F_{normal} + m_3g = 0 \tag{6.11}$$

Expressing Equations 6.2, 6.9, and 6.11 in a combined matrix form produces

$$
\begin{bmatrix}
0 & 1 & 0 & 1 & 0 & 0 & 0 & 0 \\
0 & 0 & 1 & 0 & 1 & 0 & 0 & 0 \\
1 & 0 & 0 & -W_y & W_x & 0 & 0 & 0 \\
0 & 0 & 0 & -1 & 0 & -1 & 0 & 0 \\
0 & 0 & 0 & 0 & -1 & 0 & -1 & 0 \\
0 & 0 & 0 & 0 & 0 & V_y & -V_x & 0 \\
0 & 0 & 0 & 0 & 0 & 1 & 0 & \pm\mu \\
0 & 0 & 0 & 0 & 0 & 0 & 1 & 1
\end{bmatrix}
\begin{Bmatrix}
T_{a_0} \\
F_{a_{0x}} \\
F_{a_{0y}} \\
F_{a_{1x}} \\
F_{a_{1y}} \\
F_{b_{1x}} \\
F_{b_{1y}} \\
F_{normal}
\end{Bmatrix}
$$

$$
=
\begin{bmatrix}
0 \\
-m_1g \\
-m_1g\left(R_{1x}\cos\beta_j - R_{1y}\sin\beta_j\right) \\
0 \\
-m_2g \\
-m_2g\left(R_{2x}\cos\alpha_j - R_{2y}\sin\alpha_j\right) \\
-F_x \\
-F_y - m_3g
\end{bmatrix} \tag{6.12}
$$

where:
$W_x = W_1\cos(\theta + \beta_j)$
$W_y = W_1\sin(\theta + \beta_j)$
$V_x = V_1\cos(\rho + \alpha_j)$
$V_y = V_1\sin(\rho + \alpha_j)^\dagger$

---

* Because the slider does not rotate, the static equilibrium condition $\Sigma\mathbf{M} = 0$ is not included in Equation 6.10.
† Equation 6.12 could be solved using both signs for $\pm\mu$ to determine the maximum static loads.

Equation 6.12 can be solved using Cramer's rule to determine the unknown forces and torque. The unknown slider-crank displacement angles α_j are the same angles calculated from the slider-crank displacement equations in Section 4.5.1. Given β_j and α_j solutions, the corresponding static forces and torques can be calculated from Equation 6.12.

Appendix C.2 includes the MATLAB® file user instructions for slider-crank static force analysis. In this MATLAB file (which is available for download at www.crcpress.com/product/ isbn/9781498724937), solutions for Equation 6.12 are calculated.

Example 6.2

Problem Statement: Using the Appendix C.2 MATLAB file, calculate the static torque and forces generated in the in-line slider-crank mechanism in Tables E.6.4 and E.6.5 where a force of $\mathbf{F} = (-50, 0)$ N is applied and gravity is -9.81 m/s².

　　Known Information: Tables E.6.4, E.6.5, and Appendix C.2 MATLAB file.

　　Solution Approach: Figure E.6.3 includes the input specified (in bold text) in the Appendix C.2 MATLAB file. Table E.6.6 includes the static torque and forces calculated for the slider-crank mechanism using the Appendix C.2 MATLAB file.

TABLE E.6.4

Slider-Crank Mechanism Dimensions (with Link Lengths in m)

W_1, θ	V_1	U_1	μ
0.04, 45°	0.06	0	0.1

TABLE E.6.5

Slider-Crank Mechanism Dynamic Parameters (with Length in m and Mass in kg)

$\mathbf{R}_1$	0	m_1	0.05
$\mathbf{R}_2$	$0.0265 - i0.0141$	m_2	0.025
		m_3	0.075

```
LW1 = 0.04;
theta = 45*pi/180;

LU1 = 0;

LV1 = 0.06;

F = [-50, 0];
mu = 0.1;
g = -9.81;

R1 = 0;
R2 = 0.0265 - i*0.0141;

m1 = 0.05;
m2 = 0.025;
m3 = 0.075;

start_ang   = 0;
step_ang    = 1;
stop_ang    = 0;
```

FIGURE E.6.3
Specified input (in bold text) in the Appendix C.2 MATLAB file for Example 6.2.

TABLE E.6.6

Calculated Slider-Crank Mechanism Static Forces (in N) and Torque (in N-m)

T_{a_0}	F_{a_0}	F_{a_1}	F_{b_1}	F_{normal}	$F_{friction}$
−2.053	47.381, −24.714	−47.381, 25.204	47.381, −25.449	26.815	2.6185

6.5 Geared Five-Bar Mechanism Analysis

We will begin the formulation of a static force equation system for the geared five-bar mechanism by first formulating a static force equation system for a five-bar mechanism *without* gears. Because such a system has two degrees of freedom, static equilibrium is achieved by independently constraining the rotations of the links containing vectors $\mathbf{W}_1$ and $\mathbf{U}_1$.

Figure 6.5a illustrates a five-bar mechanism where a force $\mathbf{F}_{p_1}$ is applied to the intermediate link point $\mathbf{p}_1$. To maintain static equilibrium, torques $\mathbf{T}_{a_0}$ and $\mathbf{T}_{b_0}$ are applied about both grounded revolute joints $\mathbf{a}_0$ and $\mathbf{b}_0$, respectively. Vectors $\mathbf{R}_1$ through $\mathbf{R}_4$ point from $\mathbf{a}_0$, $\mathbf{a}_1$, $\mathbf{b}_1$, and $\mathbf{b}_0$, respectively, to the center of gravity of each link. The loads on the individual five-bar mechanism links are illustrated in Figure 6.5b. Taking the sum of the forces and moments for each link according to the static equilibrium conditions $\Sigma \mathbf{F} = 0$ and $\Sigma \mathbf{M} = 0$ produces two static equilibrium equations for each link. Expanding these two equations and separating the force equation into two equations, where $\Sigma F_x = 0$, $\Sigma F_y = 0$, ultimately produces three static equilibrium equations for each link. Using these conditions for the crank link $\mathbf{a}_0 - \mathbf{a}_1$ (where the moment sum is taken about $\mathbf{a}_0$) produces Equation 6.2.

Using the first two static equilibrium conditions for the intermediate link $\mathbf{a}_1 - \mathbf{c}_1$ (where the moment sum is taken about $\mathbf{a}_1$) produces

$$-\mathbf{F}_{a_1} - \mathbf{F}_{c_1} + \mathbf{F}_{p_1} + m_2 \mathbf{g} = 0$$

$$-\mathbf{V}_1 \times \mathbf{F}_{c_1} + \mathbf{L}_1 \times \mathbf{F}_{p_1} + \mathbf{R}_2 \times m_2 \mathbf{g} = 0 \tag{6.13}$$

Expanding and separating Equation 6.13 produces

$$-F_{a_{1x}} - F_{c_{1x}} + F_{p_{1x}} = 0$$

$$-F_{a_{1y}} - F_{c_{1y}} + F_{p_{1y}} + m_2 g = 0$$

$$F_{c_{1x}} V_1 \sin(\rho + \alpha_j) - F_{c_{1y}} V_1 \cos(\rho + \alpha_j) - F_{p_{1x}} L_1 \sin(\delta + \alpha_j) + F_{p_{1y}} L_1 \cos(\delta + \alpha_j) \tag{6.14}$$

$$+ m_2 g \left(R_{2x} \cos \alpha_j - R_{2y} \sin \alpha_j \right) = 0$$

Since gravity is directed along the y-axis, the expansion of the cross product $\mathbf{R}_2 \times m_2 \mathbf{g}$ in Equation 6.14 only includes the expansion of the product of $m_2 g$ and the real component of $\mathbf{R}_2$ where $\mathbf{R}_2 = (R_{2x} + iR_{2y})e^{i\alpha_j}$.

Using the first two static equilibrium conditions for the intermediate link $\mathbf{b}_1 - \mathbf{c}_1$ (where the moment sum is taken about $\mathbf{b}_1$) produces

$$\mathbf{F}_{c_1} - \mathbf{F}_{b_1} + m_3 \mathbf{g} = 0$$

$$\mathbf{S}_1 \times \mathbf{F}_{c_1} + \mathbf{R}_3 \times m_3 \mathbf{g} = 0 \tag{6.15}$$

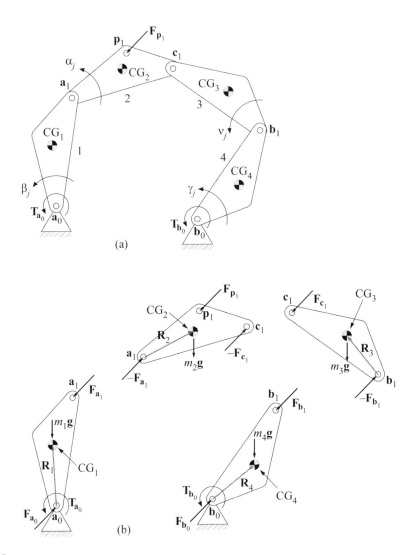

FIGURE 6.5
(a) Five-bar mechanism and (b) link static forces and torques.

Expanding and separating Equation 6.15 produces

$$F_{c_{1x}} - F_{b_{1x}} = 0$$

$$F_{c_{1y}} - F_{b_{1y}} + m_3 g = 0 \tag{6.16}$$

$$-F_{c_{1x}} S_1 \sin(\psi + v_j) + F_{c_{1y}} S_1 \cos(\psi + v_j) + m_3 g \left(R_{3x} \cos v_j - R_{3y} \sin v_j \right) = 0$$

Since gravity is directed along the y-axis, the expansion of the cross product $\mathbf{R}_3 \times m_3\mathbf{g}$ in Equation 6.16 only includes the expansion of the product of $m_3 g$ and the real component of $\mathbf{R}_3$ where $\mathbf{R}_3 = (R_{3x} + i R_{3y})e^{iv_j}$.

Using the first two static equilibrium conditions for the crank link $\mathbf{b}_0 - \mathbf{b}_1$ (where the moment sum is taken about $\mathbf{b}_0$) produces

$$\mathbf{F}_{b_0} + \mathbf{F}_{b_1} + m_4\mathbf{g} = 0$$

$$\mathbf{T}_{b_0} + \mathbf{U}_1 \times \mathbf{F}_{b_1} + \mathbf{R}_4 \times m_4\mathbf{g} = 0$$

(6.17)

Expanding and separating Equation 6.17 produces

$$F_{b_0x} + F_{b_1x} = 0$$

$$F_{b_0y} + F_{b_1y} + m_4g = 0$$

(6.18)

$$T_{b_0} - F_{b_1x}U_1\sin(\sigma + \gamma_j) + F_{b_1y}U_1\cos(\sigma + \gamma_j) + m_4g\left(R_{4x}\cos\gamma_j - R_{4y}\sin\gamma_j\right) = 0$$

Since gravity is directed along the y-axis, the expansion of the cross product $\mathbf{R}_4 \times m_4\mathbf{g}$ in Equation 6.18 only includes the expansion of the product of m_4g and the real component of $\mathbf{R}_4$ where $\mathbf{R}_4 = (R_{4x} + iR_{4y})e^{i\gamma_j}$.

Expressing Equations 6.2, 6.14, 6.16, and 6.18 in a combined matrix form produces

$$\begin{bmatrix}
0 & 1 & 0 & 1 & 0 & 0 & 0 & 0 & 0 & 0 & 0 & 0 \\
0 & 0 & 1 & 0 & 1 & 0 & 0 & 0 & 0 & 0 & 0 & 0 \\
1 & 0 & 0 & -W_y & W_x & 0 & 0 & 0 & 0 & 0 & 0 & 0 \\
0 & 0 & 0 & -1 & 0 & -1 & 0 & 0 & 0 & 0 & 0 & 0 \\
0 & 0 & 0 & 0 & -1 & 0 & -1 & 0 & 0 & 0 & 0 & 0 \\
0 & 0 & 0 & 0 & 0 & V_y & -V_x & 0 & 0 & 0 & 0 & 0 \\
0 & 0 & 0 & 0 & 0 & 1 & 0 & 0 & 0 & 0 & -1 & 0 \\
0 & 0 & 0 & 0 & 0 & 0 & 1 & 0 & 0 & 0 & 0 & -1 \\
0 & 0 & 0 & 0 & 0 & -S_y & S_x & 0 & 0 & 0 & 0 & 0 \\
0 & 0 & 0 & 0 & 0 & 0 & 0 & 0 & 1 & 0 & 1 & 0 \\
0 & 0 & 0 & 0 & 0 & 0 & 0 & 0 & 0 & 1 & 0 & 1 \\
0 & 0 & 0 & 0 & 0 & 0 & 0 & 1 & 0 & 0 & -U_y & U_x
\end{bmatrix}$$

$$\times \begin{Bmatrix}
T_{a_0} \\
F_{a_0x} \\
F_{a_0y} \\
F_{a_1x} \\
F_{a_1y} \\
F_{c_1x} \\
F_{c_1y} \\
T_{b_0} \\
F_{b_0x} \\
F_{b_0y} \\
F_{b_1x} \\
F_{b_1y}
\end{Bmatrix} = \begin{bmatrix}
0 \\
-m_1g \\
-m_1g\left(R_{1x}\cos\beta_j - R_{1y}\sin\beta_j\right) \\
-F_{p_1x} \\
-F_{p_1y} - m_2g \\
F_{p_1x}L_y - F_{p_1y}L_x - m_2g\left(R_{2x}\cos\alpha_j - R_{2y}\sin\alpha_j\right) \\
0 \\
-m_3g \\
-m_3g\left(R_{3x}\cos v_j - R_{3y}\sin v_j\right) \\
0 \\
-m_4g \\
-m_4g\left(R_{4x}\cos\gamma_j - R_{4y}\sin\gamma_j\right)
\end{bmatrix}$$

(6.19)

where:

$$W_x = W_1 \cos(\theta + \beta_j)$$
$$W_y = W_1 \sin(\theta + \beta_j)$$
$$L_x = L_1 \cos(\delta + \alpha_j)$$
$$L_y = L_1 \sin(\delta + \alpha_j)$$
$$V_x = V_1 \cos(\rho + \alpha_j)$$
$$V_y = V_1 \sin(\rho + \alpha_j)$$
$$S_x = S_1 \cos(\psi + v_j)$$
$$S_y = S_1 \sin(\psi + v_j)$$
$$U_x = U_1 \cos(\sigma + \gamma_j)$$
$$U_y = U_1 \sin(\sigma + \gamma_j)$$

Equation 6.19 can be solved using Cramer's rule to determine the unknown forces and torques. The unknown five-bar displacement angles α_j and v_j are the same angles calculated from the geared five-bar displacement equations in Section 4.6.1. Given β_j, α_j, and v_j solutions, the corresponding static forces and torques can be calculated from Equation 6.19.

As noted at the start of this section, Equation 6.19 calculates the static forces and torques for a five-bar mechanism without gears. The inclusion of a gear pair or gear train, however, will affect the calculated values of T_{a_0}, $\mathbf{F}_{a_0}$, and $\mathbf{F}_{b_0}$.

Including a gear pair or train in the five-bar mechanism reduces it to a single degree of freedom. To achieve static equilibrium, a new torque is applied about $\mathbf{a}_0$ (while the calculated torque T_{b_0} is still applied about $\mathbf{b}_0$). This new, gear-based torque (which we call T'_{a_0}) includes T_{a_0} and T_{b_0} from Equation 6.19 and can be expressed as

$$T'_{a_0} = T_{a_0} + \frac{r_1}{r_2} T_{b_0} = T_{a_0} + \frac{1}{r} T_{b_0} \tag{6.20}$$

where the gear ratio r is the ratio of the radius of the driven gear to the driving gear.[*]

Therefore, with the inclusion of a gear pair or a gear train in the five-bar mechanism, the static torque T'_{a_0} from Equation 6.20 replaces T_{a_0} from Equation 6.19, while the static torque T_{b_0} calculated from Equation 6.19 remains unchanged.

Figure 6.6a illustrates a gear pair used in the geared five-bar mechanism. By using a gear pair, the links containing vectors $\mathbf{W}_1$ and $\mathbf{U}_1$ rotate in opposite directions. The force transmitted by Gear 2 (force $\mathbf{F}$ in Figure 6.6a) must be included among the components of $\mathbf{F}_{a_0}$ and $\mathbf{F}_{b_0}$ calculated in Equation 6.19.

In Equations 6.21 and 6.22, the components of the transmitted gear pair force $\mathbf{F}$ (which is calculated through the torque T_{b_0} from Equation 6.19) are included in $\mathbf{F}_{a_0}$ and $\mathbf{F}_{b_0}$, respectively.

$$\mathbf{F}'_{a_0} = \mathbf{F}_{a_0} + \frac{T_{b_0}}{r_2} e^{i\left(\frac{\pi}{2} + ang\right)} = \mathbf{F}_{a_0} + T_{b_0} \left(\frac{1 + |r|}{r |\mathbf{G}_1|}\right) e^{i\left(\frac{\pi}{2} + ang\right)} \tag{6.21}$$

$$\mathbf{F}'_{b_0} = \mathbf{F}_{b_0} - \frac{T_{b_0}}{r_2} e^{i\left(\frac{\pi}{2} + ang\right)} = \mathbf{F}_{b_0} - T_{b_0} \left(\frac{1 + |r|}{r |\mathbf{G}_1|}\right) e^{i\left(\frac{\pi}{2} + ang\right)} \tag{6.22}$$

Figure 6.6b illustrates a three-gear train used in the geared five-bar mechanism.[†] By using a three-gear train, the links containing vectors $\mathbf{W}_1$ and $\mathbf{U}_1$ rotate in the same

---

[*] The radius of a gear is commonly referred to as the *pitch radius* (see Chapter 8).
[†] The gear train (Figure 6.6b) considered in this text for the geared five-bar mechanism is comprised of three gears, where the middle gear and the gear affixed to $\mathbf{a}_0$ have identical radii.

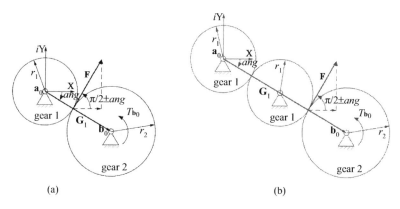

FIGURE 6.6
(a) Gear pair and (b) three-gear train used in geared five-bar mechanism.

direction. The force transmitted by Gear 2 (force **F** in Figure 6.6b) must be included among the components of $\mathbf{F_{a_0}}$ and $\mathbf{F_{b_0}}$ calculated in Equation 6.19.

In Equations 6.23 and 6.24, the components of the transmitted three-gear-train force **F** (which is calculated through the torque T_{b_0} from Equation 6.19) are included in $\mathbf{F_{a_0}}$ and $\mathbf{F_{b_0}}$, respectively.

$$\mathbf{F'_{a_0}} = \mathbf{F_{a_0}} + \frac{T_{b_0}}{r_2} e^{i\left(\frac{\pi}{2} + ang\right)} = \mathbf{F_{a_0}} + T_{b_0} \left(\frac{3 + |r|}{r\,|\mathbf{G_1}|}\right) e^{i\left(\frac{\pi}{2} + ang\right)} \tag{6.23}$$

$$\mathbf{F'_{b_0}} = \mathbf{F_{b_0}} + \frac{T_{b_0}}{r_2} e^{i\left(\frac{\pi}{2} + ang\right)} = \mathbf{F_{b_0}} + T_{b_0} \left(\frac{3 + |r|}{r\,|\mathbf{G_1}|}\right) e^{i\left(\frac{\pi}{2} + ang\right)} \tag{6.24}$$

Therefore, with the inclusion of gears in the five-bar mechanism, the static forces $\mathbf{F'_{a_0}}$ and $\mathbf{F'_{b_0}}$ from Equations 6.21 and 6.22 replace $\mathbf{F_{a_0}}$ and $\mathbf{F_{b_0}}$ calculated from Equation 6.19, when a gear pair is included, while $\mathbf{F'_{a_0}}$ and $\mathbf{F'_{b_0}}$ from Equation 6.23 and 6.24 replace $\mathbf{F_{a_0}}$ and $\mathbf{F_{b_0}}$, when a three-gear train is included.

Appendices C.3 and C.4 include the MATLAB file user instructions for geared five-bar static force analysis (for two and three gears, respectively). In this MATLAB file (which is available for download at www.crcpress.com/product/isbn/9781498724937), solutions for Equation 6.19 and Equations 6.20 through 6.24 are calculated.

Example 6.3

Problem Statement: Using the Appendix C.4 MATLAB file, calculate the static forces and driver torque generated in the geared five-bar mechanism in Tables E.6.7 and E.6.8 where a force of $\mathbf{F_{p_1}} = (-2500, -3000)\,\text{N}$ is applied at $\beta = 60°$. The gear ratio is $r = +2$, and the gravity is $-9.81\,\text{m/s}^2$.

Known Information: Tables E.6.7, E.6.8, and Appendix C.4 MATLAB file.

Solution Approach: Figure E.6.4 includes the input specified (in bold text) in the Appendix C.4 MATLAB file. Table E.6.9 includes the static forces and torques calculated for the geared five-bar mechanism using the Appendix C.4 MATLAB file.

TABLE E.6.7

Geared Five-Bar Mechanism Dimensions (with Link Lengths in m)

W_1, θ	V_1, ρ	U_1, σ	S_1, σ	G_{1x}, G_{1y}	L_1, δ
0.5, 90°	0.75, 32.7304°	0.75, 45°	0.75, 149.9837°	0.75, 0	0.5, 74.1400°

TABLE E.6.8

Geared Five-Bar Mechanism Dynamic Parameters (with Length in m and mass in kg)

R_1	$i0.0831$	m_1	22.54
R_2	$0.2558 + i0.2955$	m_2	29.785
R_3	$-0.3247 + i0.1876$	m_3	12.075
R_4	$0.0356 + i0.0356$	m_4	75.67

```
w1 = 0.5*exp(i*90*pi/180);
v1 = 0.75*exp(i*32.7304*pi/180);
G1 = 0.75;
U1 = 0.75*exp(i*45*pi/180);
L1 = 0.5*exp(i*74.1400*pi/180);
S1 = 0.75*exp(i*149.9847*pi/180);

ratio = 2;
Fp1 = [-2500,-3000];
g = -9.81;

R1 = i*0.0831;
R2 = 0.2558 + i*0.2955;
R3 = -0.3247 + i*0.1876;
R4 = 0.0356 + i*0.0356;

m1 = 22.54;
m2 = 29.785;
m3 = 12.075;
m4 = 75.67;

start_ang   = 0;
step_ang    = 1;
stop_ang    = 60;
```

FIGURE E.6.4
Specified input (in bold text) in the Appendix C.4 MATLAB file for Example 6.3.

TABLE E.6.9

Calculated Geared Five-Bar Mechanism Static Forces (in N) and Torques (in N-m)

T'_{a_0}	F'_{a_0}	F_{a_1}	F_{c_1}	F_{b_1}
−2207	543.23, −1144.8	−543.23, −3173.8	−1956.8, −118.43	−1956.8, −236.88

T_{b_0}	F'_{b_0}
−1361.9	1956.8, −3560.5

6.6 Watt II Mechanism Analysis

Figure 6.7a illustrates a Watt II mechanism where forces F_{p_1} and $F_{p_1^*}$ are applied to the intermediate link points p_1 and p_1^*. Vectors R_1 through R_5 point from a_0, a_1, b_0, a_1^*, and b_0^*,

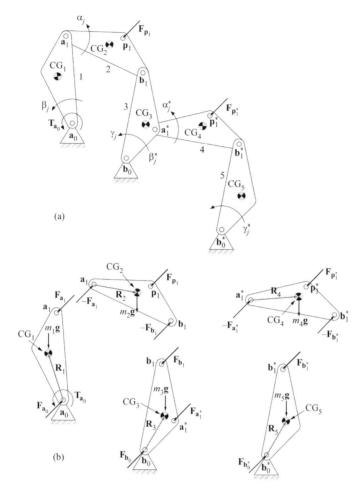

FIGURE 6.7
(a) Watt II mechanism and (b) link static forces and torque.

respectively, to the center of gravity of each link. To maintain static equilibrium, a torque $\mathbf{T}_{a_0}$ is applied about the crank link revolute joint $\mathbf{a}_0$. The loads on the individual Watt II mechanism link are illustrated in Figure 6.7b. Taking the sum of the forces and moments for each link according to the static equilibrium conditions $\sum \mathbf{F} = 0$ and $\sum \mathbf{M} = 0$ produces two static equilibrium equations for each link. Expanding these two equations and separating the force equation into two equations, where $\sum F_x = 0$ and $\sum F_y = 0$, ultimately produces three static equilibrium equations for each link. Because the Watt II mechanism includes the planar four-bar mechanism, the static equilibrium equations given in Equations 6.2 and 6.4 are used for the crank and coupler links, respectively, of the planar four-bar mechanism loop $\mathbf{a}_0$–$\mathbf{a}_1$–$\mathbf{b}_1$–$\mathbf{b}_0$ in the Watt II mechanism.

Using the first two static equilibrium conditions for the follower link of the planar four-bar mechanism loop $\mathbf{a}_0$– $\mathbf{a}_1$– $\mathbf{b}_1$– $\mathbf{b}_0$ (where the moment sum is taken about $\mathbf{b}_0$) produces

$$\mathbf{F}_{b_0} + \mathbf{F}_{b_1} + \mathbf{F}_{a_1^*} + m_3\mathbf{g} = 0$$

$$\mathbf{U}_1 \times \mathbf{F}_{b_1} + \mathbf{W}_1^* \times \mathbf{F}_{a_1^*} + \mathbf{R}_3 \times m_3\mathbf{g} = 0$$

(6.25)

Expanding and separating Equation 6.25 produces

$$F_{b_0x} + F_{b_1x} + F_{a_1^*x} = 0$$

$$F_{b_0y} + F_{b_1y} + F_{a_1^*y} + m_3 g = 0$$

$$-F_{b_1x} U_1 \sin(\sigma + \gamma_j) + F_{b_1y} U_1 \cos(\sigma + \gamma_j) - F_{a_1^*x} W_1^* \sin(\theta^* + \beta_j^*) + F_{a_1^*y} W_1^* \cos(\theta^* + \beta_j^*)$$

$$+m_3 g \left(R_{3x} \cos \gamma_j - R_{3y} \sin \gamma_j \right) = 0$$

(6.26)

Since gravity is directed along the *y*-axis, the expansion of the cross product $\mathbf{R}_3 \times m_3\mathbf{g}$ in Equation 6.26 only includes the expansion of the product of $m_3 g$ and the real component of $\mathbf{R}_3$ where $\mathbf{R}_3 = (R_{3x} + iR_{3y})e^{i\gamma_j}$.

Using the first two static equilibrium conditions for the coupler link of the planar four-bar mechanism loop $\mathbf{b}_0 - \mathbf{a}_1^* - \mathbf{b}_1^* - \mathbf{b}_0^*$ (where the moment sum is taken about $\mathbf{a}_1^*$) produces

$$-\mathbf{F}_{a_1^*} - \mathbf{F}_{b_1^*} + \mathbf{F}_{p_1^*} + m_4\mathbf{g} = 0$$

$$-\mathbf{V}_1^* \times \mathbf{F}_{b_1^*} + \mathbf{L}_1^* \times \mathbf{F}_{p_1^*} + \mathbf{R}_4 \times m_4\mathbf{g} = 0$$

(6.27)

Expanding and separating Equation 6.27 produces

$$-F_{a_1^*x} - F_{b_1^*x} + F_{p_1^*x} = 0$$

$$-F_{a_1^*y} - F_{b_1^*y} + F_{p_1^*y} + m_4 g = 0$$

$$F_{b_1^*x} V_1^* \sin(\rho^* + \alpha_j^*) - F_{b_1^*y} V_1^* \cos(\rho^* + \alpha_j^*) - F_{p_1^*x} L_1^* \sin(\delta^* + \alpha_j^*) + F_{p_1^*y} L_1^* \cos(\delta^* + \alpha_j^*)$$

$$+m_4 g \left(R_{4x} \cos \alpha_j^* - R_{4y} \sin \alpha_j^* \right) = 0$$

(6.28)

Since gravity is directed along the *y*-axis, the expansion of the cross product $\mathbf{R}_4 \times m_4\mathbf{g}$ in Equation 6.28 only includes the expansion of the product of $m_4 g$ and the real component of $\mathbf{R}_4$ where $\mathbf{R}_4 = (R_{4x} + iR_{4y})e^{i\alpha_j^*}$.

Using the first two static equilibrium conditions for the follower of the planar four-bar mechanism loop $\mathbf{b}_0 - \mathbf{a}_1^* - \mathbf{b}_1^* - \mathbf{b}_0^*$ (where the moment sum is taken about $\mathbf{b}_0^*$) produces

$$\mathbf{F}_{b_0^*} + \mathbf{F}_{b_1^*} + m_5\mathbf{g} = 0$$

$$\mathbf{U}_1^* \times \mathbf{F}_{b_1^*} + \mathbf{R}_5 \times m_5\mathbf{g} = 0$$

(6.29)

Expanding and separating Equation 6.29 produces

$$F_{b_0^*x} + F_{b_1^*x} = 0$$

$$F_{b_0^*y} + F_{b_1^*y} + m_5 g = 0$$

$$-F_{b_1^*x} U_1^* \sin(\sigma^* + \gamma_j^*) + F_{b_1^*y} U_1^* \cos(\sigma^* + \gamma_j^*) + m_5 g \left(R_{5x} \cos \gamma_j^* - R_{5y} \sin \gamma_j^* \right) = 0$$

(6.30)

Since gravity is directed along the y-axis, the expansion of the cross product $\mathbf{R}_5 \times m_5\mathbf{g}$ in Equation 6.30 only includes the expansion of the product of m_5g and the real component of $\mathbf{R}_5$ where $\mathbf{R}_5 = (R_{5x} + iR_{5y})e^{i\gamma_j^*}$.

Expressing Equations 6.2, 6.4, 6.26, 6.28, and 6.30 in a combined matrix form produces

$$
\begin{bmatrix}
0 & 1 & 0 & 1 & 0 & 0 & 0 & 0 & 0 & 0 & 0 & 0 & 0 & 0 & 0 \\
0 & 0 & 1 & 0 & 1 & 0 & 0 & 0 & 0 & 0 & 0 & 0 & 0 & 0 & 0 \\
1 & 0 & 0 & -W_y & W_x & 0 & 0 & 0 & 0 & 0 & 0 & 0 & 0 & 0 & 0 \\
0 & 0 & 0 & -1 & 0 & 0 & 0 & -1 & 0 & 0 & 0 & 0 & 0 & 0 & 0 \\
0 & 0 & 0 & 0 & -1 & 0 & 0 & 0 & -1 & 0 & 0 & 0 & 0 & 0 & 0 \\
0 & 0 & 0 & 0 & 0 & 0 & 0 & V_y & -V_x & 0 & 0 & 0 & 0 & 0 & 0 \\
0 & 0 & 0 & 0 & 0 & 1 & 0 & 1 & 0 & 1 & 0 & 0 & 0 & 0 & 0 \\
0 & 0 & 0 & 0 & 0 & 0 & 1 & 0 & 1 & 0 & 1 & 0 & 0 & 0 & 0 \\
0 & 0 & 0 & 0 & 0 & 0 & 0 & -U_y & U_x & -W_y^* & W_x^* & 0 & 0 & 0 & 0 \\
0 & 0 & 0 & 0 & 0 & 0 & 0 & 0 & 0 & -1 & 0 & 0 & 0 & -1 & 0 \\
0 & 0 & 0 & 0 & 0 & 0 & 0 & 0 & 0 & 0 & -1 & 0 & 0 & 0 & -1 \\
0 & 0 & 0 & 0 & 0 & 0 & 0 & 0 & 0 & 0 & 0 & 0 & 0 & V_y^* & -V_x^* \\
0 & 0 & 0 & 0 & 0 & 0 & 0 & 0 & 0 & 0 & 0 & 1 & 0 & 1 & 0 \\
0 & 0 & 0 & 0 & 0 & 0 & 0 & 0 & 0 & 0 & 0 & 0 & 1 & 0 & 1 \\
0 & 0 & 0 & 0 & 0 & 0 & 0 & 0 & 0 & 0 & 0 & 0 & 0 & -U_y^* & U_x^*
\end{bmatrix}
$$

$$
\times
\begin{Bmatrix}
T_{a_0} \\
F_{a_0x} \\
F_{a_0y} \\
F_{a_1x} \\
F_{a_1y} \\
F_{b_0x} \\
F_{b_0y} \\
F_{b_1x} \\
F_{b_1y} \\
F_{a_1x}^* \\
F_{a_1y}^* \\
F_{b_0x}^* \\
F_{b_0y}^* \\
F_{b_1x}^* \\
F_{b_1y}^*
\end{Bmatrix}
=
\begin{bmatrix}
0 \\
-m_1 g \\
-m_1 g\left(R_{1x}\cos\beta_j - R_{1y}\sin\beta_j\right) \\
-F_{p_1x} \\
-F_{p_1y} - m_2 g \\
F_{p_1x}L_y - F_{p_1y}L_x - m_2 g\left(R_{2x}\cos\alpha_j - R_{2y}\sin\alpha_j\right) \\
0 \\
-m_3 g \\
-m_3 g\left(R_{3x}\cos\gamma_j - R_{3y}\sin\gamma_j\right) \\
-F_{p_1x}^* \\
-F_{p_1y}^* - m_4 g \\
F_{p_1x}^* L_y^* - F_{p_1y}^* L_x^* - m_4 g\left(R_{4x}\cos\alpha_j^* - R_{4y}\sin\alpha_j^*\right) \\
0 \\
-m_5 g \\
-m_5 g\left(R_{5x}\cos\gamma_j^* - R_{5y}\sin\gamma_j^*\right)
\end{bmatrix}
$$

$$(6.31)$$

where:

variables W_x, W_y, L_x, L_y, V_x, V_y, U_x, and U_y are identical to those used in Equation 6.7 and

$$W_x^* = W_1^* \cos\left(\theta^* + \beta_j^*\right)$$

$$W_y^* = W_1^* \sin\left(\theta^* + \beta_j^*\right)$$

$$L_x^* = L_1^* \cos\left(\delta^* + \alpha_j^*\right)$$

$$L_y^* = L_1^* \sin\left(\delta^* + \alpha_j^*\right)$$

$$V_x^* = V_1^* \cos\left(\rho^* + \alpha_j^*\right)$$

$$V_y^* = V_1^* \sin\left(\rho^* + \alpha_j^*\right)$$

$$U_x^* = U_1^* \cos\left(\sigma^* + \gamma_j^*\right)$$

$$U_y^* = U_1^* \sin\left(\sigma^* + \gamma_j^*\right)$$

Equation 6.31 can be solved using Cramer's rule to determine the unknown forces and torque. The unknown Watt II displacement angles α_j, γ_j, α_j^*, and γ_j^* are the same angles calculated from the planar four-bar displacement equations in Section 4.4.1. Given β_j, α_j, γ, α_j^*, and γ_j^* solutions, the corresponding static forces and torques can be calculated from Equation 6.31.

Appendix C.5 includes the MATLAB® file user instructions for Watt II static force analysis. In this MATLAB file (which is available for download at www.crcpress.com/product/isbn/9781498724937), solutions for Equation 6.31 are calculated.

Example 6.4

Problem Statement: Using the Appendix C.5 MATLAB file, calculate the static forces and torque generated in the Watt II mechanism in Tables E.6.10 and E.6.11 where forces of $\mathbf{F}_{P_1} = (2500, 3000)\,\text{N}$ and $\mathbf{F}_{P_1^*} = (-1500, 2000)\,\text{N}$ are applied at $\beta = 100°$ and the gravity is $-9.81\,\text{m/s}^2$.

Known Information: Tables E.6.10, E.6.11, and Appendix C.5 MATLAB file.

Solution Approach: Figure E.6.5 includes the input specified (in bold text) in the Appendix C.5 MATLAB file. Table E.6.12 includes the static forces and torque calculated for the Watt II mechanism using the Appendix C.5 MATLAB file.

TABLE E.6.10

Watt II Mechanism Dimensions (with Link Lengths in m)

W_1, θ	V_1, ρ	U_1, σ	G_{1x}, G_{1y}	L_1, δ
0.5, 90°	0.75, 19.3737°	0.75, 93.2461°	0.75, 0	0.5, 60.7834°
$W_1^*, \theta*$	$V_1^*, \rho*$	$U_1^*, \sigma*$	G_{1x}^*, G_{1y}^*	$L_1^*, \delta*$
0.5, 45°	0.75, 7.9416°	0.75, 60.2717°	0.7244, −0.1941	0.5, 49.3512°

TABLE E.6.11

Watt II Mechanism Dynamic Parameters (with Length in m and Mass in kg)

R_1	$i0.25$	m_1	8.05
R_2	$0.3172 + i0.2284$	m_2	29.785
R_3	$0.1037 + i0.3675$	m_3	33.81
R_4	$0.3562 + i0.161$	m_4	29.785
R_5	$0.186 + i0.3257$	m_5	12.075

```
W1 = 0.5*exp(i*90*pi/180);
V1 = 0.75*exp(i*19.3737*pi/180);
G1 = 0.75;
U1 = 0.75*exp(i*93.2461*pi/180);
L1 = 0.5*exp(i*60.7834*pi/180);

W1s = 0.5*exp(i*45*pi/180);
V1s = 0.75*exp(i*7.9416*pi/180);
G1s = 0.7244 - i*0.1941;
U1s = 0.75*exp(i*60.2717*pi/180);
L1s = 0.5*exp(i*49.3512*pi/180);

Fp1 = [2500, 3000];
Fp1s = [-1500, 2000];
g = -9.81;

R1 = i*0.25;
R2 = 0.3172 + i*0.2284;
R3 = 0.1037 + i*0.3675;
R4 = 0.3562 + i*0.161;
R5 = 0.1860 + i*0.3257;

m1 = 8.05;
m2 = 29.785;
m3 = 33.81;
m4 = 29.785;
m5 = 12.075;

start_ang  = 0;
step_ang   = 1;
stop_ang   = 100;
```

FIGURE E.6.5
Specified input (in bold text) in the Appendix C.5 MATLAB file for Example 6.4.

TABLE E.6.12

Calculated Watt II Mechanism Static Forces (in N) and Torque (in N-m)

T_{a_0}	F_{a_0}	F_{a_1}	F_{b_0}
1323.5	−412.06, −2721	412.06, 2800	−741.83, 211.14

F_{b_1}	$F_{a_1^*}$	$F_{b_0^*}$	$F_{b_1^*}$
2087.9, −92.193	−1346.1, 212.73	153.89, −1376.6	−153.89, 1495.1

6.7 Stephenson III Mechanism Analysis

Figure 6.8a illustrates a Stephenson III mechanism where a force $\mathbf{F}_{\mathbf{p}_1^*}$ is applied to the intermediate link point. $\mathbf{p}_1^*$ To maintain static equilibrium, a torque $\mathbf{T}_{\mathbf{a}_0}$ is applied about the crank link revolute joint $\mathbf{a}_0$. Vectors $\mathbf{R}_1$ through $\mathbf{R}_5$ point from $\mathbf{a}_0$, $\mathbf{a}_1$, $\mathbf{b}_0$, $\mathbf{p}_1$, and $\mathbf{b}_0^*$, respectively, to the center of gravity of each link. The loads on the individual Stephenson III mechanism link are illustrated in Figure 6.8b. Taking the sum of the forces and moments for each link according to the static equilibrium conditions $\Sigma \mathbf{F} = 0$ and $\Sigma \mathbf{M} = 0$ produces two static equilibrium equations for each link. Expanding these two equations and separating the force equation into two equations, where $\Sigma F_x = 0$, $\Sigma F_y = 0$, ultimately produces three

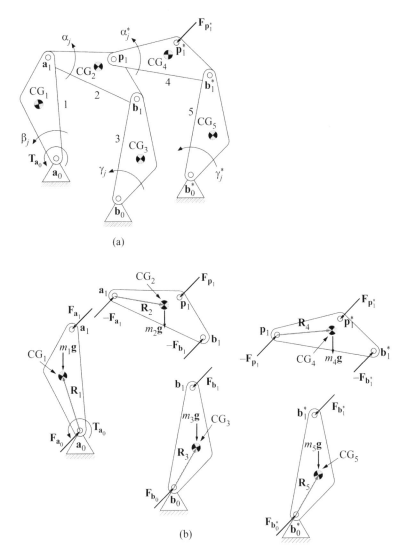

(a)

(b)

FIGURE 6.8

(a) Stephenson III mechanism and (b) link static forces and torque.

static equilibrium equations for each link. Because the Stephenson III mechanism includes the planar four-bar mechanism, the static equilibrium equations given in Equations 6.2, 6.4, and 6.6 are used for the crank and coupler links, respectively, of the planar four-bar mechanism loop $\mathbf{a}_0 - \mathbf{a}_1 - \mathbf{b}_1 - \mathbf{b}_0$ in the Stephenson III mechanism.

Using the first two static equilibrium conditions for the dyad link containing vector $\mathbf{R}_4$ (where the moment sum is taken about $\mathbf{p}_1$) produces

$$-\mathbf{F}_{p_1} - \mathbf{F}_{b_1^*} + \mathbf{F}_{p_1^*} + m_4\mathbf{g} = 0$$

$$-\mathbf{V}_1^* \times \mathbf{F}_{b_1^*} + \mathbf{L}_1^* \times \mathbf{F}_{p_1^*} + \mathbf{R}_4 \times m_4\mathbf{g} = 0$$

(6.32)

Expanding and separating Equation 6.32 produces

$$-F_{p_{1x}} - F_{b_{1x}^*} + F_{p_{1x}^*} = 0$$

$$-F_{p_{1y}} - F_{b_{1y}^*} + F_{p_{1y}^*} + m_4g = 0$$

$$F_{b_{1x}^*} V_1^* \sin\left(\rho^* + \alpha_j^*\right) - F_{b_{1y}^*} V_1^* \cos\left(\rho^* + \alpha_j^*\right) - F_{p_{1x}^*} L_1^* \sin\left(\delta^* + \alpha_j^*\right) + F_{p_{1y}^*} L_1^* \cos\left(\delta^* + \alpha_j^*\right)$$

$$+ m_4g\left(R_{4x} \cos\alpha_j^* - R_{4y} \sin\alpha_j^*\right) = 0$$

(6.33)

Since gravity is directed along the *y*-axis, the expansion of the cross product $\mathbf{R}_4 \times m_4\mathbf{g}$ in Equation 6.33 only includes the expansion of the product of m_4g and the real component of $\mathbf{R}_4$ where $\mathbf{R}_4 = \left(R_{4x} + iR_{4y}\right)e^{i\alpha_j^*}$.

Using the first two static equilibrium conditions for the dyad link containing vector $\mathbf{R}_5$ (where the moment sum is taken about $\mathbf{b}_0^*$) produces

$$\mathbf{F}_{b_0^*} + \mathbf{F}_{b_1^*} + m_5\mathbf{g} = 0$$

$$\mathbf{U}_1^* \times \mathbf{F}_{b_1^*} + \mathbf{R}_5 \times m_5\mathbf{g} = 0.$$

(6.34)

Expanding and separating Equation 6.34 produces

$$F_{b_{0x}^*} + F_{b_{1x}^*} = 0$$

$$F_{b_{0y}^*} + F_{b_{1y}^*} + m_5g = 0$$

$$-F_{b_{1x}^*} U_1^* \sin\left(\sigma^* + \gamma_j^*\right) + F_{b_{1y}^*} U_1^* \cos\left(\sigma^* + \gamma_j^*\right) + m_5g\left(R_{5x} \cos\gamma_j^* - R_{5y} \sin\gamma_j^*\right) = 0.$$

(6.35)

Since gravity is directed along the *y*-axis, the expansion of the cross product $\mathbf{R}_5 \times m_5\mathbf{g}$ in Equation 6.35 only includes the expansion of the product of m_5g and the real component of $\mathbf{R}_5$ where $\mathbf{R}_5 = \left(R_{5x} + iR_{5y}\right)e^{i\gamma_j^*}$.

Expressing Equations 6.2, 6.4, 6.6, 6.33, and 6.35 in a combined matrix form produces

$$
\begin{bmatrix}
0 & 1 & 0 & 1 & 0 & 0 & 0 & 0 & 0 & 0 & 0 & 0 & 0 & 0 & 0 \\
0 & 0 & 1 & 0 & 1 & 0 & 0 & 0 & 0 & 0 & 0 & 0 & 0 & 0 & 0 \\
1 & 0 & 0 & -W_y & W_x & 0 & 0 & 0 & 0 & 0 & 0 & 0 & 0 & 0 & 0 \\
0 & 0 & 0 & -1 & 0 & 0 & 0 & -1 & 0 & 1 & 0 & 0 & 0 & 0 & 0 \\
0 & 0 & 0 & 0 & -1 & 0 & 0 & 0 & -1 & 0 & 1 & 0 & 0 & 0 & 0 \\
0 & 0 & 0 & 0 & 0 & 0 & 0 & V_y & -V_x & -L_y & L_x & 0 & 0 & 0 & 0 \\
0 & 0 & 0 & 0 & 0 & 1 & 0 & 1 & 0 & 0 & 0 & 0 & 0 & 0 & 0 \\
0 & 0 & 0 & 0 & 0 & 0 & 1 & 0 & 1 & 0 & 0 & 0 & 0 & 0 & 0 \\
0 & 0 & 0 & 0 & 0 & 0 & 0 & -U_y & U_x & 0 & 0 & 0 & 0 & 0 & 0 \\
0 & 0 & 0 & 0 & 0 & 0 & 0 & 0 & 0 & -1 & 0 & 0 & 0 & -1 & 0 \\
0 & 0 & 0 & 0 & 0 & 0 & 0 & 0 & 0 & 0 & -1 & 0 & 0 & 0 & -1 \\
0 & 0 & 0 & 0 & 0 & 0 & 0 & 0 & 0 & 0 & 0 & 0 & 0 & V_y^* & -V_x^* \\
0 & 0 & 0 & 0 & 0 & 0 & 0 & 0 & 0 & 0 & 0 & 1 & 0 & 1 & 0 \\
0 & 0 & 0 & 0 & 0 & 0 & 0 & 0 & 0 & 0 & 0 & 0 & 1 & 0 & 1 \\
0 & 0 & 0 & 0 & 0 & 0 & 0 & 0 & 0 & 0 & 0 & 0 & 0 & -U_y^* & U_x^*
\end{bmatrix}
$$

$$
\times
\begin{Bmatrix}
T_{a0} \\
F_{a0x} \\
F_{a0y} \\
F_{a1x} \\
F_{a1y} \\
F_{b0x} \\
F_{b0y} \\
F_{b1x} \\
F_{b1y} \\
F_{p1x} \\
F_{p1y} \\
F_{b0x}^* \\
F_{b0y}^* \\
F_{b1x}^* \\
F_{b1y}^*
\end{Bmatrix}
=
\begin{bmatrix}
0 \\
-m_1 g \\
-m_1 g\left(R_{1x}\cos\beta_j - R_{1y}\sin\beta_j\right) \\
0 \\
-m_2 g \\
-m_2 g\left(R_{2x}\cos\alpha_j - R_{2y}\sin\alpha_j\right) \\
0 \\
-m_3 g \\
-m_3 g\left(R_{3x}\cos\gamma_j - R_{3y}\sin\gamma_j\right) \\
-F_{p1x}^* \\
-F_{p1y}^* - m_4 g \\
F_{p1x}^* L_y^* - F_{p1y}^* L_x^* - m_4 g\left(R_{4x}\cos\alpha_j^* - R_{4y}\sin\alpha_j^*\right) \\
0 \\
-m_5 g \\
-m_5 g\left(R_{5x}\cos\gamma_j^* - R_{5y}\sin\gamma_j^*\right)
\end{bmatrix}
$$

$$(6.36)$$

where:
variables W_x, W_y, L_x, L_y, V_x, V_y, U_x, and U_y are identical to those used in Equation 6.7

$$L_x^* = L_1^* \cos\left(\delta^* + \alpha_j^*\right)$$

$$L_y^* = L_1^* \sin\left(\delta^* + \alpha_j^*\right)$$

$$V_x^* = V_1^* \cos\left(\rho^* + \alpha_j^*\right)$$

$$V_y^* = V_1^* \sin\left(\rho^* + \alpha_j^*\right)$$

$$U_x^* = U_1^* \cos\left(\sigma^* + \gamma_j^*\right)$$

$$U_y^* = U_1^* \sin\left(\sigma^* + \gamma_j^*\right)$$

Equation 6.36 can be solved using Cramer's rule to determine the unknown forces and torque. The unknown Stephenson III displacement angles α, γ, α_j^*, and γ_j^* are the same angles calculated from the Stephenson III displacement equations in Sections 4.4.1 and 4.8.1. Given β, α, γ, α_j^*, and γ_j^* solutions, the corresponding static forces and torques can be calculated from Equation 6.36.

Appendix C.6 includes the MATLAB file user instructions for Stephenson III static force analysis. In this MATLAB file (which is available for download at www.crcpress.com/product/ isbn/9781498724937), solutions for Equation 6.36 are calculated.

Example 6.5

Problem Statement: When vector $\mathbf{W}_1$ of the Stephenson III gripper mechanism is rotated by $\beta = 40°$, a reaction force of $\mathbf{F}_{\mathbf{p}_i}^* = \left(0, -40\right)\text{N}$ is applied. Tables E.6.13 and E.6.14 include the dimensions and mass properties of Stephenson III gripper mechanism in the initial position (Figure E.6.6). Gravity is $-9.81\,\text{m/s}^2$. Using the Appendix C.6 MATLAB file, calculate the static torque and forces generated in the gripper mechanism.

Known Information: Tables E.6.13, E.6.14, and Appendix C.6 MATLAB file.

Solution Approach: Figure E.6.7 includes the input specified (in bold text) in the Appendix C.5 MATLAB file. Table E.6.15 includes the static torque and forces calculated for the Stephenson III gripper mechanism using the Appendix C.6 MATLAB file.

TABLE E.6.13

Gripping Mechanism Dimensions in Initial Position (with Link Lengths in m)

W_1, θ	V_1, ρ	U_1, σ	G_{1x}, G_{1y}	L_1, δ
0.0136, −64.4543°	0.0097, 57.2740°	0.0190, 84.2513°	0.0092, −0.023	0.0061, −143.6057°

L_1^*, δ^*	V_1^*, ρ^*	U_1^*, σ^*	G_{1x}^*, G_{1y}^*	
0.0222, −5°	0.0058, −125.7782°	0.03, −42.1315°	−0.0339, 0.0225	

TABLE E.6.14

Stephenson III Mechanism Dynamic Parameters (with Length in m and Mass in kg)

R_1	$0.0029 - i0.0061$	m_1	8
R_2	$-0.0058 + i0.0103$	m_2	45
R_3	$0.001 + i0.0095$	m_3	8
R_4	$0.0128 - i0.0055$	m_4	45
R_5	$0.0111 - i0.01$	m_5	18

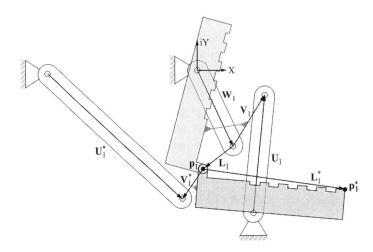

FIGURE E.6.6
Specified input (in bold text) in the Appendix C.5 MATLAB file for Example 6.5.

TABLE E.6.15

Calculated Gripping Mechanism Static Forces (in N) and Torque (in N-m)

T_{a_0}	F_{a_0}	F_{a_1}	F_{b_0}
3.2998	−1489.3, 983.37	1489.3, −904.89	338.3, 430.88

F_{b_1}	F_{p_1}	$F_{b_0}^*$	$F_{b_1}^*$
−338.3, −352.4	1151, −815.83	1151, −157.8	−1151, 334.38

```
W1 = 0.0136*exp(i*-64.4543*pi/180);
V1 = 0.0097*exp(i*57.2740*pi/180);
G1 = 0.0092 - i*0.023;
U1 = 0.019*exp(i*84.2513*pi/180);
L1 = 0.0061*exp(-i*143.6057*pi/180);

V1s = 0.0058*exp(-i*125.7782*pi/180);
G1s = -0.0339 + i*0.0225;
U1s = 0.03*exp(-i*42.1315*pi/180);
L1s = 0.0222*exp(-i*5*pi/180);

Fp1s = [0, -40];
g = -9.81;

R1 = 0.0029 - i*0.0061;
R2 = -0.0058 + i*0.0103;
R3 = 0.001 + i*0.0095;
R4 = 0.0128 - i*0.0055;
R5 = 0.0111 - i*0.01;

m1 = 8;
m2 = 45;
m3 = 8;
m4 = 45;
m5 = 18;

start_ang    = 0;
step_ang     = 1;
stop_ang     = 40;
```

FIGURE E.6.7
Specified input (in bold text) in the Appendix C.5 MATLAB file for Example 6.5.

6.8 Planar Mechanism Static Force Analysis and Modeling in SimMechanics®

As has been noted throughout this chapter, Appendices C.1 through C.6 include user instructions for the planar four-bar, slider-crank, geared five-bar, Watt II, and Stephenson III mechanism MATLAB files, respectively. In these files, the static force and torque equations formulated in this chapter are solved. These MATLAB files provide a means for the user to efficiently conduct planar four-bar, slider-crank, geared five-bar, Watt II, and Stephenson III static force analyses by solving their displacement equations along with their static force and torque equations.

This textbook also utilizes SimMechanics as an alternate approach for simulation-based static force analysis. A library of SimMechanics files is available for download at www.crcpress.com/product/isbn/9781498724937 to conduct static force analyses on the planar four-bar, slider-crank, geared five-bar, Watt II, and Stephenson III mechanisms. With these files, the user specifies the mechanism link dimensions, mass properties, applied loads, driving link parameters, and the static forces and torques at the mechanism locations of interest are measured. The SimMechanics file user instructions for the planar four-bar, slider-crank, geared five-bar, Watt II, and Stephenson III mechanisms are given in Appendices I.1 through I.6, respectively.

While the user can specify input according to any unit type for *length*, *force*, and *torque* in the Appendix C MATLAB files, only two groups of dimensions are available in the Appendix I MATLAB files. The user can specify length, mass, and force quantities in either *inch*, *pound-mass*, and *pound-force*, respectively (the U.S. system), or *meter*, *kilogram*, and *Newton*, respectively (the S.I. system). The user also has the option in the Appendix J files to convert U.S. system input to S.I. system output and vice versa.

Example 6.6

Problem Statement: Repeat Example 6.1 using the Appendix I.1 SimMechanics files.

Known Information: Example 6.1 and Appendix I.1 SimMechanics files.

Solution Approach: Figure E.6.8 includes the input specified (in bold text) in the Appendix I.1 SimMechanics file. Table E.6.16 includes the static torque and forces

```
unit_select = 'SI';

W1 = 0.2013*exp(-i*157.8291*pi/180);
V1 = 0.1583*exp(-i*150.1267*pi/180);
G1 = -0.0876 - i*0.4071;
U1 = 0.3455*exp(i*133.0953*pi/180);
L1 = 0.27*exp(i*4.7572*pi/180);

Fp1 = [0, 4500];
g = -9.81;

R1 = -0.0932 - i*0.038;
R2 = 0.0955 + i*0.0159;
R3 = -0.118 + i*0.1261;

m1 = 8;
m2 = 40;
m3 = 12;

start_ang   = 0;
step_ang    = 1;
stop_ang    = 60;
```

FIGURE E.6.8
Specified input (in bold text) in the Appendix I.1 SimMechanics file for Example 6.6.

TABLE E.6.16

Calculated Stamping Mechanism Static Forces (in N) and Torque (in N-m) (Appendix I.1)

T_{a_0}	F_{a_0}	F_{a_1}	F_{b_0}	F_{b_1}
162.19	73.51, −5341	−74.51, 5419.5	−73.51, 1429.6	73.51, −1311.9

calculated for the planar four-bar stamping mechanism using the Appendix I.1 SimMechanics files.

Example 6.7

Problem Statement: Repeat Example 6.2 using the Appendix I.2 SimMechanics files.
  Known Information: Example 6.2 and Appendix I.2 SimMechanics files.
  Solution Approach: Figure E.6.9 includes the input specified (in bold text) in the Appendix I.2 SimMechanics file. Table E.6.17 includes the static torque and forces calculated for the slider-crank mechanism using the Appendix I.2 SimMechanics files.

Example 6.8

Problem Statement: Repeat Example 6.3 using the Appendix I.4 SimMechanics files.
  Known Information: Example 6.3 and Appendix I.4 SimMechanics files.
  Solution Approach: Figure E.6.10 includes the input specified (in bold text) in the Appendix I.4 SimMechanics file. Table E.6.18 includes the static torque and forces calculated for the geared five-bar mechanism using the Appendix I.4 SimMechanics files.

```
unit_select = ' SI';

LW1 = 0.04;
theta = 45*pi/180;

LU1 = 0;

LV1 = 0.06;

F = [-50, 0];
mu = 0.1;
g = -9.81;

R1 = 0;
R2 = 0.0265 - i*0.0141;

m1 = 0.05;
m2 = 0.025;
m3 = 0.075;

start_ang  = 0;
step_ang   = 1;
stop_ang   = 0;
```

FIGURE E.6.9
Specified input (in bold text) in the Appendix I.2 SimMechanics file for Example 6.7.

TABLE E.6.17

Calculated Slider-Crank Mechanism Static Forces (in N) and Torque (in N-m) (Appendix I.2)

T_{a_0}	F_{a_0}	F_{a_1}	F_{b_1}	F_{normal}	$F_{friction}$
−2.053	47.381, −24.714	−47.381, 25.204	47.381, −25.449	26.815	2.6185

```
unit_select = ' SI';

W1 = 0.5*exp(i*90*pi/180);
V1 = 0.75*exp(i*32.7304*pi/180);
G1 = 0.75;
U1 = 0.75*exp(i*45*pi/180);
L1 = 0.5*exp(i*74.1400*pi/180);
S1 = 0.75*exp(i*149.9847*pi/180);

ratio = 2;
Fp1 = [-2500,-3000];
g = -9.81;

R1 = i*0.0831;
R2 = 0.2558 + i*0.2955;
R3 = -0.3247 + i*0.1876;
R4 = 0.0356 + i*0.0356;

m1 = 22.54;
m2 = 29.785;
m3 = 12.075;
m4 = 75.67;

start_ang   = 0;
step_ang    = 1;
stop_ang    = 60;
```

FIGURE E.6.10

Specified input (in bold text) in the Appendix I.4 SimMechanics file for Example 6.8.

TABLE E.6.18

Calculated Geared Five-Bar Mechanism Static Forces (in N) and Torques (in N-m) (Appendix I.4)

T'_{a_0}	F'_{a_0}	F_{a_1}	F_{c_1}	F_{b_1}
−2206.9	543.08, −1145.2	−543.08, −3173.7	−1956.9, −118.51	−1956.9, −236.97
T_{b_0}	F'_{b_0}			
−1362	1956.9, −3560.7			

Example 6.9

Problem Statement: Repeat Example 6.4 using the Appendix I.5 SimMechanics files.

Known Information: Example 6.4 and Appendix I.5 SimMechanics files.

Solution Approach: Figure E.6.11 includes the input specified (in bold text) in the Appendix I.5 SimMechanics file. Table E.6.19 includes the static torque and forces calculated for the Watt II mechanism using the Appendix I.5 SimMechanics files.

Example 6.10

Problem Statement: Repeat Example 6.5 using the Appendix I.6 SimMechanics files.

Known Information: Example 6.5 and Appendix I.6 SimMechanics files.

Solution Approach: Figure E.6.12 includes the input specified (in bold text) in the Appendix I.6 SimMechanics file. Table E.6.20 includes the static torque and forces calculated for the Stephenson III mechanism using the Appendix I.6 SimMechanics files.

```
unit_select = 'SI';

W1 = 0.5*exp(i*90*pi/180);
V1 = 0.75*exp(i*19.3737*pi/180);
G1 = 0.75;
U1 = 0.75*exp(i*93.2461*pi/180);
L1 = 0.5*exp(i*60.7834*pi/180);

W1s = 0.5*exp(i*45*pi/180);
V1s = 0.75*exp(i*7.9416*pi/180);
G1s = 0.7244 - i*0.1941;
U1s = 0.75*exp(i*60.2717*pi/180);
L1s = 0.5*exp(i*49.3512*pi/180);

Fp1 = [2500, 3000];
Fp1s = [-1500, 2000];
g = -9.81;

R1 = i*0.25;
R2 = 0.3172 + i*0.2284;
R3 = 0.1037 + i*0.3675;
R4 = 0.3562 + i*0.161;
R5 = 0.1860 + i*0.3257;

m1 = 8.05;
m2 = 29.785;
m3 = 33.81;
m4 = 29.785;
m5 = 12.075;

start_ang   = 0;
step_ang    = 1;
stop_ang    = 100;
```

FIGURE E.6.11
Specified input (in bold text) in the Appendix I.5 SimMechanics file for Example 6.9.

TABLE E.6.19

Calculated Watt II Mechanism Static Forces (in N) and Torque (in N-m) (Appendix I.5)

T_{a_0}	F_{a_0}	F_{a_1}	F_{b_0}
1323.5	−412.12, −2721.1	412.12, 2800	−741.85, 211.25

F_{b_1}	$F_{a_1'}$	$F_{b_0'}$	$F_{b_i'}$
2087.9, −92.227	−1346, 212.65	153.97, −1376.7	−153.97, 1495.2

6.9 Summary

In terms of structural force analyses for mechanical systems, a static force analysis is the most basic type of force analysis to consider beyond kinematics. In this type of analysis, loads are considered for each mechanism link according to Newton's first law ($\Sigma \mathbf{F} = \Sigma \mathbf{M} = 0$). In this chapter, static force and moment equations are formulated for the planar four-bar, slider-crank, geared five-bar, Watt II, and Stephenson III mechanisms. These equations form sets of linear simultaneous equations that are solved to determine the static forces and torques present at each mechanism joint. The Appendix C.1 through C.6 MATLAB files provide a means for the user to efficiently conduct planar four-bar, slider-crank, geared five-bar, Watt II, and Stephenson III static force analyses by solving their displacement equations (from Chapter 4) along with their linear simultaneous equation sets.

```
unit_select = 'SI';

W1 = 0.0136*exp(i*-64.4543*pi/180);
V1 = 0.0097*exp(i*57.2740*pi/180);
G1 = 0.0092 - i*0.023;
U1 = 0.019*exp(i*84.2513*pi/180);
L1 = 0.0061*exp(-i*143.6057*pi/180);

V1s = 0.0058*exp(-i*125.7782*pi/180);
G1s = -0.0339 + i*0.0225;
U1s = 0.03*exp(-i*42.1315*pi/180);
L1s = 0.0222*exp(-i*5*pi/180);

Fp1s = [0, -40];
g = -9.81;

R1 = 0.0029 - i*0.0061;
R2 = -0.0058 + i*0.0103;
R3 = 0.001 + i*0.0095;
R4 = 0.0128 - i*0.0055;
R5 = 0.0111 - i*0.01;

m1 = 8;
m2 = 45;
m3 = 8;
m4 = 45;
m5 = 18;

start_ang  = 0;
step_ang   = 1;
stop_ang   = 40;
```

FIGURE E.6.12
Specified input (in bold text) in the Appendix I.6 SimMechanics file for Example 6.10.

TABLE E.6.20

Calculated Gripping Mechanism Static Forces (in N) and Torque (in N-m) (Appendix I.6)

T_{a_0}	F_{a_0}	F_{a_1}	F_{b_0}
3.3492	−1481.1, 983.65	1481.1, −905.17	337.68, 429.88

F_{b_1}	F_{p_1}	$F_{b_0^*}$	$F_{b_1^*}$
−337.68, −351.4	1143.5, −815.12	1143.5, −157.09	−1143.5, 333.67

This textbook also utilizes SimMechanics as an alternate approach for simulation-based static load analyses. Using the Appendix I.1 through I.6 SimMechanics files, the user can conduct static load analyses on the planar four-bar, slider-crank, geared five-bar, Watt II, and Stephenson III mechanisms, respectively, as well as simulating mechanism motion.

References

1. Hibbeler, R. C. 1995. *Engineering Mechanics: Statics*. 7th edn, Chapter 5. Englewood Cliffs, NJ: Prentice-Hall.
2. Beer, F. P. and Johnston, E. R. 1988. *Vector Mechanics for Engineers: Statics and Dynamics*. 5th edn, Chapter 4. New York: McGraw-Hill.

3. Cheng, F. 1997. *Statics and Strength of Materials*. 2nd edn, Chapter 3. New York: Glencoe McGraw-Hill.

4. Spotts, M. F., Shoup, T. E., and Hornberger, L. E. 2004. *Design of Machine Elements*. 8th edn, pp. 709–717. Saddle River, NJ: Prentice-Hall.

Additional Reading

Myszka, D. H. 2005. *Machines and Mechanisms: Applied Kinematic Analysis*. 3rd edn, Chapter 13. Saddle River, NJ: Prentice-Hall.

Waldron, K. J. and Kinzel, G. L. 2004. *Kinematics, Dynamics and Design of Machinery*. 2nd edn, Chapter 13. Saddle River, NJ: Prentice-Hall.

Wilson, C. E. and Sadler, J. P. 2003. *Kinematics and Dynamics of Machinery*. 3rd edn, Chapter 9. Saddle River, NJ: Prentice-Hall.

Problems

1. Figure P.6.1 illustrates a planar four-bar mechanism used to guide a hatch from the closed-hatch position to the opened-hatch position. When the opened-hatch position is reached, a static force of $\mathbf{F}_{P2} = (0, 4500)\text{N}$ is applied at the displaced $\mathbf{p}_1$ (labeled $\mathbf{p}_2$ in Figure P.6.1). The dimensions for the illustrated mechanism are included in Table P.6.1. The masses of the crank, coupler, and follower links are $m_1 = 5$, $m_2 = 63$, and $m_3 = 7$ kg, respectively, and the link center of mass vectors are $\mathbf{R}_1 = -1.0629 +$

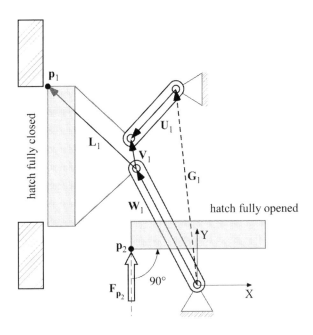

FIGURE P.6.1
Hatch mechanism.

TABLE P.6.1

Hatch Mechanism Dimensions (with Link Lengths in m)

W_1, θ	V_1, ρ	U_1, σ	G_{1x}, G_{1y}	L_1, δ
4.4127,118.7982°	1.0214,101.2268°	2.2807, 225.1319°	−0.7156, 6.4851	4.1345,139.4559°

$i1.9335$, $\mathbf{R}_2 = -2.6808 + i0.5954$, and $\mathbf{R}_3 = -0.8045 - i0.8082$ m. Gravity is -9.81 m/s^2. Considering a crank rotation of $-30°$, determine the crank static torque and joint forces produced at the opened-hatch position (using the Appendix C.1 or I.1 file).

2. Figure P.6.2 illustrates a planar four-bar mechanism used to guide a bucket from the loading bucket position to the unloading position. When the unloading position is reached, a static force of $\mathbf{F}_{p2} = (0, -2250)$ N is applied at the displaced $\mathbf{p}_1$ (labeled $\mathbf{p}_2$ in Figure P.6.2). The dimensions for the illustrated mechanism are included in Table P.6.2. The masses of the crank, coupler, and follower links are $m_1 = 8.75$, $m_2 = 58$, and $m_3 = 10$ kg, respectively, and the link center of mass

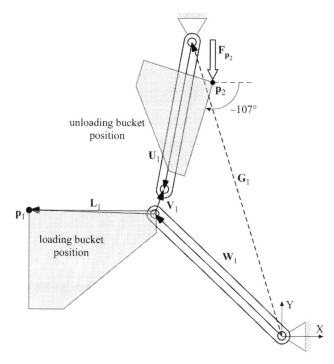

FIGURE P.6.2
Loading–unloading mechanism.

TABLE P.6.2

Loading-Unloading Mechanism Dimensions (with Link Lengths in m)

W_1, θ	V_1, ρ	U_1, σ	G_{1x}, G_{1y}	L_1, δ
4.4109, 136.588°	0.5973, 70.2408°	3.7227, 259.5885°	−2.3295, 7.2548	3.2581, 179.5331°

vectors are $\mathbf{R}_1 = -1.6021 + i1.5157$, $\mathbf{R}_2 = -1.7147 - i1.0699$, and $\mathbf{R}_3 = -0.3364 - i1.8307$. Gravity is $-9.81\,\mathrm{m/s^2}$. Considering a crank rotation of $-30°$, determine the crank static torque and joint forces produced at the unloading bucket position (using the Appendix C.1 or I.1 file).

3. Figure P.6.3 illustrates a planar four-bar mechanism used to guide a digging bucket from the initial position to the final position. When the final digging position is reached, a static force of $\mathbf{F}_{\mathbf{P}_2} = (-3500, -4500)\mathrm{N}$ is applied at the displaced $\mathbf{p}_1$ (labeled $\mathbf{p}_2$ in Figure P.6.3). The dimensions for the illustrated mechanism are included in Table P.6.3. The masses of the crank, coupler, and follower links are $m_1 = 8$, $m_2 = 45$, and $m_3 = 10$ kg, respectively, and the link center of mass vectors are $\mathbf{R}_1 = -1.4003 - i0.7392$, $\mathbf{R}_2 = -4.7688 + i0.7191$, and $\mathbf{R}_3 = -1.3785 - i1.8621$. Gravity is $-9.81\,\mathrm{m/s^2}$. Considering a crank rotation of $-57.4°$, determine the crank static torque and joint forces produced at the opened-hatch position (using the Appendix C.1 or I.1 file).

4. Figure P.6.4 illustrates a planar four-bar mechanism used to guide a component from the initial position to the assembled position. When the assembled position is reached, a static force of $\mathbf{F}_{\mathbf{P}_2} = (5500, 0)\mathrm{N}$ is applied at the displaced $\mathbf{p}_1$ (labeled $\mathbf{p}_2$ in Figure P.6.4). The dimensions for the illustrated mechanism are included in Table P.6.4. The masses of the crank, coupler, and follower links are $m_1 = 40$, $m_2 = 145$, and $m_3 = 20$ kg, respectively, and the link center of mass vectors are $\mathbf{R}_1 = 0.853 - i0.9935$, $\mathbf{R}_2 = 2.3858 + i0.3974$, and $\mathbf{R}_3 = -0.183 - i0.4496$. Gravity is $-9.81\,\mathrm{m/s^2}$. Considering a crank rotation of $66.375°$, determine the crank static torque and joint forces produced at the assembled position (using the Appendix C.1 or I.1 file).

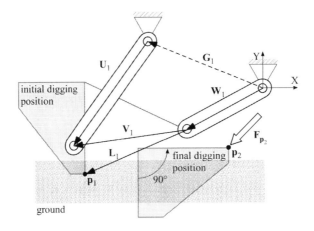

FIGURE P.6.3
Digging mechanism.

TABLE P.6.3

Digging Mechanism Dimensions (with Link Lengths in m)

W_1, θ	V_1, ρ	U_1, σ	G_{1x}, G_{1y}	L_1, δ
3.1669, 207.829°	4.1266, 187.8282°	4.6337, 233.487°	−4.1317, 1.6837	3.9721, 202.3921°

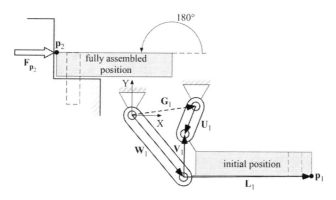

FIGURE P.6.4
Component-assembly mechanism.

TABLE P.6.4

Component Assembly Mechanism Dimensions (with Link Lengths in m)

W_1, θ	V_1, ρ	U_1, σ	G_{1x}, G_{1y}	L_1, δ
2.619, 310.6493°	1.3708, 89.216°	0.9708, −112.147°	2.0908, 0.2827	4.1467, 0.7042°

TABLE P.6.5

Planar Four-Bar Mechanism Configuration (with Link Lengths in m)

W_1, θ	V_1, ρ	U_1, σ	G_{1x}, G_{1y}	L_1, δ
0.2153, 90°	0.246, −16.0138°	0.246, 58.1173°	0.1066, −0.0615	0.246, 12.9412°
F_{p_1} (N)	$R_1 \sim R_3$	$m_1 \sim m_3$		
−150, −150	0	0		

5. As presented in Section 3.4, the total force on the follower link of a planar four-bar mechanism is $F_{follower} = \sqrt{F_{b0x}^2 + F_{b0y}^2} = \sqrt{F_{b1x}^2 + F_{b1y}^2}$ and the transverse and columnar forces on the follower are $F_{follower} \sin(\tau)$ and $F_{follower} \cos(\tau)$, respectively. Knowing this, calculate the minimum and maximum transmission angles (use a 1° crank rotation increment) and the corresponding follower transverse and columnar loads for the planar four-bar mechanism configuration in Table P.6.5 (using the Appendix C.1 or I.1 file). Gravity is to be neglected in this problem.

6. Figure P.6.5 illustrates a planar four-bar mechanism used to guide a wing from the folded position to the extended position. When the extended position is reached, a static force of $\mathbf{F}_{p_2} = (0, -1600)$N is applied at the displaced $\mathbf{p}_1$ (labeled $\mathbf{p}_2$ in Figure P.6.5). The dimensions for the illustrated mechanism are included in Table P.6.6. The masses of the crank, coupler, and follower links are $m_1 = 35$, $m_2 = 100$, and $m_3 = 25$ kg, respectively, and the link center of mass vectors are $\mathbf{R}_1 = 0.2398 + i2.1116$, $\mathbf{R}_2 = 0.8673 + i0.8333$, and $\mathbf{R}_3 = 0.1929 - i1.7563$. Gravity is −9.81 m/s². Considering a crank rotation of −35°, determine the crank static torque and joint forces produced at the extended position (using the Appendix C.1 or I.1 file).

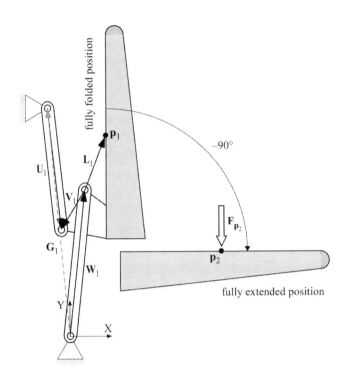

FIGURE P.6.5
Folding-wing mechanism.

TABLE P.6.6

Folding Wing Mechanism Dimensions (with Link Lengths in m)

W_1	V_1	U_1	G_1	L_1
$0.4795 + i4.2231$	$-0.7524 - i1.143$	$0.3857 - i3.5125$	$-0.6586 + i6.5927$	$0.5555 + i0.8439$

7. Figure P.6.6 illustrates a planar four-bar mechanism used to guide a latch from the released position to the applied position. When the applied position is reached, a static force of $\mathbf{F}_{p_2} = (5000,0)\text{N}$ is applied at the displaced $\mathbf{p}_1$ (labeled $\mathbf{p}_2$ in Figure P.6.6). The dimensions for the illustrated mechanism are included in Table P.6.7. The masses of the crank, coupler, and follower links are $m_1 = 8$, $m_2 = 17$, and $m_3 = 5\,\text{kg}$, respectively, and the link center of mass vectors are $\mathbf{R}_1 = 1.3729 - i1.0192$, $\mathbf{R}_2 = 2.2199 + i3.7677$, and $\mathbf{R}_3 = 1.152 - i0.3437$. Gravity is $-9.81\,\text{m/s}^2$. Considering a crank rotation of $70°$, determine the crank static torque and joint forces produced at the applied position (using the Appendix C.1 or I.1 file).

8. Figure P.6.7 illustrates a planar four-bar mechanism used to guide a lower cutting blade from the open position to the close position. When the close position is reached, a static force of $\mathbf{F}_{p_2} = (0,-22)\text{N}$ is applied at the displaced $\mathbf{p}_1$ (labeled $\mathbf{p}_2$ in Figure P.6.7). The dimensions for the illustrated mechanism are included in Table P.6.8. The masses of the crank, coupler, and follower links are $m_1 = 15$, $m_2 = 75$ and $m_3 = 30\,\text{kg}$, respectively, and the link center of mass vectors are $\mathbf{R}_1 = -0.0498 + i0.7196$, $\mathbf{R}_2 = 1.8244 - i1.4446$, and $\mathbf{R}_3 = 1.1707 + i0.5269$. Gravity is $-9.81\,\text{m/s}^2$. Considering a crank rotation of $-45°$, determine the crank static torque and joint forces produced at the close position (using the Appendix C.1 or I.1 file).

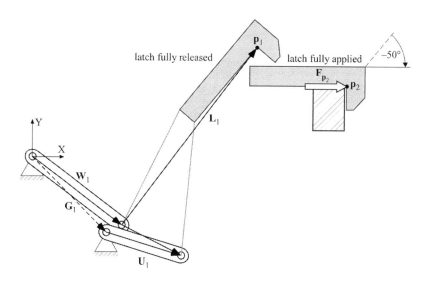

FIGURE P.6.6
Latch mechanism.

TABLE P.6.7

Latch Mechanism Dimensions (with Link Lengths in m)

W_1	V_1	U_1	G_1	L_1
$2.7459 - i2.0384$	$1.8305 - i0.8899$	$2.3039 - i0.6875$	$2.2725 - i2.2408$	$4.1783 + i5.334$

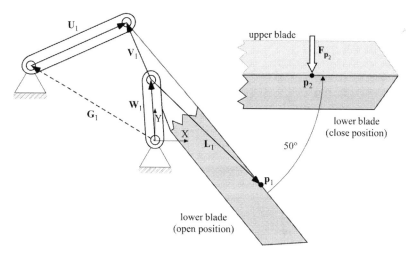

FIGURE P.6.7
Cutting-blade mechanism.

TABLE P.6.8

Cutting Blade Mechanism Dimensions (with Link Lengths in m)

W_1	V_1	U_1	G_1	L_1
$-0.0997 + i1.4392$	$-0.6134 + i1.3743$	$2.3413 + i1.0538$	$-3.0544 + i1.7597$	$2.222 - i1.111$

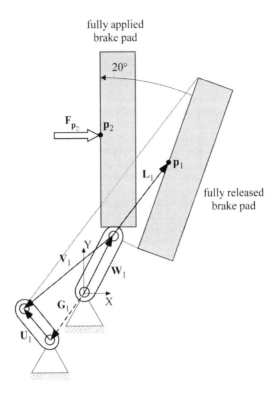

FIGURE P.6.8
Brake-pad mechanism.

TABLE P.6.9

Brake Pad Mechanism Dimensions (with Link Lengths in m)

W_1	V_1	U_1	G_1	L_1
$0.5322 + i0.9695$	$-1.6049 - i1.2391$	$-0.4555 + i0.5396$	$-0.6171 - i0.8093$	$0.899 + i1.2002$

9. Figure P.6.8 illustrates a planar four-bar mechanism used to guide a brake pad from the released position to the applied position. When the applied position is reached, a static force of $\mathbf{F_{p_2}} = (222, 0)\text{N}$ is applied at the displaced $\mathbf{p}_1$ (labeled $\mathbf{p}_2$ in Figure P.6.8). The dimensions for the illustrated mechanism are included in Table P.6.9. The masses of the crank, coupler, and follower links are $m_1 = 8$, $m_2 = 12$, and $m_3 = 5$ kg, respectively, and the link center of mass vectors are $\mathbf{R}_1 = 0.2661 + i0.4847$, $\mathbf{R}_2 = 1.1785 + i1.0948$, and $\mathbf{R}_3 = -0.2278 + i0.2698$. Gravity is -9.81 m/s^2. Considering a crank rotation of $40°$, determine the crank static torque and joint forces produced at the applied position (using the Appendix C.1 or I.1 file).

10. Calculate the static forces and torque for the planar four-bar mechanism in Example 6.1 where $\mathbf{F_{p_1}} = (-1500, 4500)\text{N}$ (using the Appendix C.1 or I.1 file).

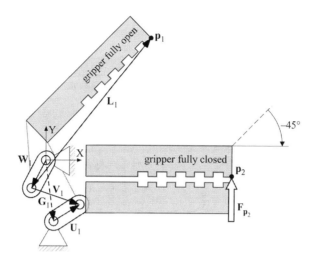

FIGURE P.6.9
Gripper mechanism.

TABLE P.6.10

Gripper Mechanism Dimensions (with Link Lengths in m)

W_1	V_1	U_1	G_1	L_1
$-0.405 - i0.801$	$1.3998 - i0.4425$	$0.7481 + i0.4728$	$0.2466 - i1.7163$	$3.375 + i4.25$

11. Figure P.6.9 illustrates a planar four-bar mechanism used to guide a gripper from the open position to the close position. When the close position is reached, a static force of $\mathbf{F}_{p_2} = (0, 900)$N is applied at the displaced $\mathbf{p}_1$ (labeled $\mathbf{p}_2$ in Figure P.6.9). The dimensions for the illustrated mechanism are included in Table P.6.10. The masses of the crank, coupler, and follower links are $m_1 = 12$, $m_2 = 30$, and $m_3 = 10$ kg, respectively, and the link center of mass vectors are $\mathbf{R}_1 = -0.2025 - i0.4005$, $\mathbf{R}_2 = 1.7259 + i3.1249$, and $\mathbf{R}_3 = 0.3741 + i0.2364$. Gravity is -9.81 m/s^2. Considering a crank rotation of 50°, determine the crank static torque and joint forces produced at the close position (using the Appendix C.1 or I.1 file).

12. Figure P.6.10 illustrates a planar four-bar mechanism used to guide a stamping tool from the released position to the applied position. When the close position is reached, a static force of $\mathbf{F}_{p_2} = (0, 9000)$N is applied at the displaced $\mathbf{p}_1$ (labeled $\mathbf{p}_2$ in Figure P.6.10). The dimensions for the illustrated mechanism are included in Table P.6.11. The masses of the crank, coupler, and follower links are $m_1 = 25, m_2 = 120$, and $m_3 = 25$ kg, respectively, and the link center of mass vectors are $\mathbf{R}_1 = -0.9319 - i0.3798$, $\mathbf{R}_2 = 0.955 + i0.1593$, and $\mathbf{R}_3 = -1.1589 + i0.4722$. Gravity is -9.81 m/s^2. Considering a crank rotation of 60°, determine the crank static torque and joint forces produced at the applied position (using the Appendix C.1 or I.1 file).

13. Repeat Problem 3 where $\mathbf{F}_{p_2} = (0, -5700)$N.

14. Repeat Problem 4 where $\mathbf{F}_{p_2} = (5500, 2000)$N.

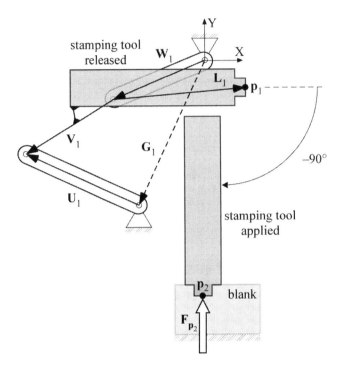

FIGURE P.6.10
Stamping mechanism.

TABLE P.6.11

Stamping Mechanism Dimensions (with Link Lengths in m)

W_1	V_1	U_1	G_1	L_1
$-1.8638 - i0.7595$	$-1.8068 - i0.9841$	$-2.3177 + i0.9444$	$-1.3529 - i2.688$	$2.6903 + i0.2239$

15. Repeat Problem 7 where $\mathbf{F}_{p_2} = (5000, 2500)\text{N}$.

16. Calculate the static forces and torque for the slider-crank mechanism configuration in Table P.6.12 when the crank is rotated 45° (using the Appendix C.2 or I.2 file). The masses of the crank, coupler, and slider links are $m_1 = 2.5$, $m_2 = 7.5$ and $m_3 = 5.5$ kg, respectively, and the link center of mass vectors are $\mathbf{R}_1 = i0.15$ and $\mathbf{R}_2 = 0.477 - i0.15$. Gravity is $-9.81\,\text{m/s}^2$.

17. Calculate the minimum and maximum crank link static torque magnitude $|\mathbf{T}_{a_0}|$ and the corresponding crank displacement angles for the slider-crank mechanism configuration in Table P.6.13 (using the Appendix C.2 or I.2 file). Consider a complete crank rotation cycle at 1° crank rotation increments. The masses of the crank, coupler, and slider links are $m_1 = 0.025$, $m_2 = 0.015$, and $m_3 = 0.015$ kg, respectively,

TABLE P.6.12

Slider-Crank Mechanism Configuration (with Link Lengths in m)

W_1, θ	V_1	U_1	μ	F (N)
0.3, 90°	1	0	0.45	300, 100

TABLE P.6.13

Slider-Crank Mechanism Configuration (with Link Lengths in m)

W_1, θ	V_1	U_1	μ	F (N)
0.03, 90°	0.09	0	0.35	−60, −20

TABLE P.6.14

Slider-Crank Mechanism Configuration (with Link Lengths in m)

W_1, θ	V_1	U_1	μ	F (N)
0.0635, 45°	0.2	−0.035	0.5	400, 0

and the link center of mass vectors are $\mathbf{R}_1 = i0.015$ and $\mathbf{R}_2 = 0.0424 − i0.015$. Gravity is −9.81 m/s².

18. Calculate the static forces and torque for the initial position of the slider-crank mechanism configuration in Table P.6.14 (using the Appendix C.2 or I.2 file). The masses of the crank, coupler, and slider links are $m_1 = 0.025$, $m_2 = 0.035$, and $m_3 = 0.025$ kg, respectively, and the link center of mass vectors are $\mathbf{R}_1 = 0.0225 + i0.0225$ and $\mathbf{R}_2 = 0.0917 − i0.04$. Gravity is −9.81 m/s².

19. Calculate the minimum and maximum crank link static torque magnitude $|\mathbf{T}_{a0}|$ and the corresponding displacement angles for the slider-crank mechanism configuration in Table P.6.15 (using the Appendix C.2 or I.2 file). Consider a complete crank rotation cycle at 1° crank rotation increments. The masses of the crank, coupler, and slider links are $m_1 = 0.012$, $m_2 = 0.01$, and $m_3 = 0.012$ kg, respectively, and the link center of mass vectors are $\mathbf{R}_1 = i0.015$ and $\mathbf{R}_2 = 0.045$. Gravity is −9.81 m/s². Consider a complete crank rotation cycle at 1° crank rotation increments.

20. Calculate the static forces and torque for the slider-crank mechanism in Example 6.2 (using the Appendix C.2 or I.2 file) where $\mathbf{F} = (−100, −25)$N.

21. Calculate the static forces and torques for the initial position of the geared five-bar mechanism configuration in Table P.6.16 (using the Appendix C.3 or I.3 file). The

TABLE P.6.15

Slider-Crank Mechanism Configuration (with Link Lengths in m)

W_1, θ	V_1	U_1	μ	F (N)
0.03, 90°	0.09	0.03	0.25	−111, 0

TABLE P.6.16

Geared Five-Bar Mechanism Configuration (with Link Lengths in m)

W_1, θ	V_1, ρ	U_1, σ	S_1, ψ	G_{1x}, G_{1y}	L_1, δ
0.35, 90°	0.525, 54.7643°	0.35, 60°	0.525, 115.0279°	0.35, 0	0.35, −15.7645°

F_{p1}	Gear Ratio				
0, −1000	−2				

link masses are $m_1 = 12$, $m_2 = 17$, $m_3 = 10$, and $m_4 = 11$ kg, respectively, and the link center of mass vectors are $R_1 = i0.175$, $R_2 = 0.2132 + i0.1112$, $R_3 = -0.1111 + i0.2379$, and $R_4 = 0.0875 + i0.1516$. Gravity is -9.81 m/s^2.

22. Repeat Problem 21 using a gear ratio of +2.

23. Using the geared five-bar mechanism configuration given in Table P.6.17, determine the driving link static torque value corresponding to crank displacement of 120° (using the Appendix C.3 or I.3 file). The link masses are $m_1 = 0.012$, $m_2 = 0.17$, $m_3 = 0.05$, and $m_4 = 0.015$ kg, respectively, and the link center of mass vectors are $R_1 = i0.0127$, $R_2 = -0.0514 + i0.0217$, $R_3 = -0.0685 + i0.0334$, and $R_4 = 0.0049 + i0.0184$. Gravity is -9.81 m/s^2.

24. Using the geared five-bar mechanism configuration given in Table P.6.18, determine the driving link static torque value corresponding to crank displacement of -120° (using the Appendix C.4 or I.4 file). The link masses are $m_1 = 0.012$, $m_2 = 0.17$, $m_3 = 0.025$, and $m_4 = 0.015$ kg, respectively, and the link center of mass vectors are $R_1 = 0.0064 + i0.011$, $R_2 = -0.1054 - i0.0138$, $R_3 = -0.1142 - i0.0046$, and $R_4 = -0.044 + i0.0254$. Gravity is -9.81 m/s^2.

25. Calculate the static forces and torque for the geared five-bar mechanism in Example 6.3 where $F_{p_1} = (1500, -2000)$N (using the Appendix C.4 or I.4 file).

26. Using the Watt II mechanism configuration given in Table P.6.19, determine the driving link static torque value corresponding to crank displacement of 60° (using the Appendix C.5 or I.5 file). The link masses are $m_1 = 4$, $m_2 = 8$, $m_3 = 8$, $m_4 = 8$, and $m_5 = 6$ kg, respectively, and the link center of mass vectors are $R_1 = i0.5$, $R_2 = 0.6344 + i0.4568$, $R_3 = 0.2074 + i0.7349$, $R_4 = 0.7124 + i0.322$, and $R_5 = 0.3719 + i0.6513$. Gravity is -9.81 m/s^2.

27. Using the Watt II mechanism configuration given in Table P.6.19, determine the driving link static torque values corresponding to crank displacements of -90°, -180°, and -270° (using the Appendix C.5 or I.5 file).

TABLE P.6.17

Geared Five-Bar Mechanism Configuration (with Link Lengths in m)

W_1, θ	V_1, ρ	U_1, σ	S_1, ψ	G_1, angle
0.0254, 90°	0.1016, 140.2031°	0.0381, 75°	0.1524, 154.0128°	0.0508, −15°
L_1, δ	F_{p_1} (N)	Gear Ratio		
0.0762, 180°	0,−40	−1.5		

TABLE P.6.18

Geared Five-Bar Mechanism Configuration (with Link Lengths in m)

W_1, θ	V_1, ρ	U_1, σ	S_1, ψ	G_{1x}, G_{1y}
0.0254, 60°	0.1778, 173.6421°	0.1016, 150°	0.2286, 182.2848°	0.1524, 0
L_1, δ	F_{p_1} (N)	Gear Ratio		
0.1524, 203.6421°	20, −40	3		

TABLE P.6.19

Watt II Mechanism Configuration (with all Link Lengths in m)

W_1, θ	V_1, ρ	U_1, σ	G_{1x}, G_{1y}	L_1, δ
1, 90°	1.5, 19.3737°	1.5, 93.2461°	1.5, 0	1, 60.7834°
W_1^*, θ^*	V_1^*, ρ^*	U_1^*, σ^*	G_{1x}^*, G_{1y}^*	L_1^*, δ^*
1, 45°	1.5, 7.9416°	1.5, 60.2717°	1.4489, −0.3882	1, 49.3512°
F_{P_1} **(N)**	$F_{P_i^*}$ **(N)**			
0, −1500	−2500, −1000			

28. Figure P.6.11 includes a Watt II mechanism used in an adjustable chair. The dimensions for the chair in the initial upright position are also included in this figure. A crank rotation of 25° is required to achieve the reclined chair position. At the reclined position, the forces illustrated in Figure P.6.11 are applied at vectors $\mathbf{L}_1$ and $\mathbf{L}_1^*$ (the midpoints of vectors $\mathbf{V}_1$ and $\mathbf{V}_1^*$ respectively). Calculate the static forces and torques at the reclined chair position (using the Appendix C.5 or I.5 file). The link masses are $m_1 = 3$, $m_2 = 7$, $m_3 = 12$, $m_4 = 7$, and $m_5 = 2$ kg, respectively, and the link center of mass vectors are $\mathbf{R}_1 = -0.2201 + i0.7295$, $\mathbf{R}_2 = 0.0602 - i0.4441$, $\mathbf{R}_3 = i0.3279$, $\mathbf{R}_4 = 0.0964 - i0.3749$, and $\mathbf{R}_5 = 0.1039 - i0.2908$. Gravity is −9.81 m/s².

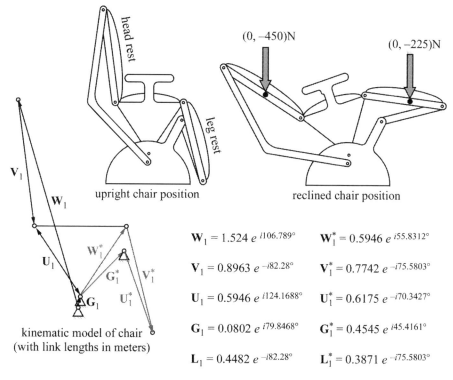

$\mathbf{W}_1 = 1.524 \, e^{\, i106.789°}$ $\mathbf{W}_1^* = 0.5946 \, e^{\, i55.8312°}$

$\mathbf{V}_1 = 0.8963 \, e^{\, -i82.28°}$ $\mathbf{V}_1^* = 0.7742 \, e^{\, -i75.5803°}$

$\mathbf{U}_1 = 0.5946 \, e^{\, i124.1688°}$ $\mathbf{U}_1^* = 0.6175 \, e^{\, -i70.3427°}$

$\mathbf{G}_1 = 0.0802 \, e^{\, i79.8468°}$ $\mathbf{G}_1^* = 0.4545 \, e^{\, i45.4161°}$

$\mathbf{L}_1 = 0.4482 \, e^{\, -i82.28°}$ $\mathbf{L}_1^* = 0.3871 \, e^{\, -i75.5803°}$

FIGURE P.6.11
Adjustable-chair mechanism.

29. Repeat Problem 28 using head rest and leg rest forces of (−150, −450) and (−25, −225) N, respectively.

30. Calculate the static forces and torque for the Watt II mechanism in Example 6.4 where $\mathbf{F}_{p_1} = (−1500, −2000)$N and $\mathbf{F}_{p_i} = (2000, −1000)$N (using the Appendix C.5 or I.5 file).

31. Figure P.6.12 illustrates a Stephenson III mechanism used to guide a seat from the reclined position to the upright position. When the upright position is reached, a static force of $\mathbf{F}_{p_2^*} = (0, −900)$N is applied at the displaced $\mathbf{p}_1^*$ (labeled $\mathbf{p}_2^*$ in Figure P.6.12). The dimensions for the illustrated mechanism are included in Table P.6.20. Considering a crank rotation of −16°, determine the crank static torque and joint forces produced at the upright position (using the Appendix C.6 or I.6 file). The link masses are $m_1 = 3.75$, $m_2 = 12$, $m_3 = 2.5$, $m_4 = 10$, and $m_5 = 2.25$ kg, respectively, and the link center of mass vectors are $\mathbf{R}_1 = −0.3716 + i0.2364$, $\mathbf{R}_2 = 0.3832 + i0.4702$, $\mathbf{R}_3 = −0.0902 − i0.2348$, $\mathbf{R}_4 = 0.3758 − i1.0047$, and $\mathbf{R}_5 = −0.1043 − i0.1201$. Gravity is −9.81 m/s².

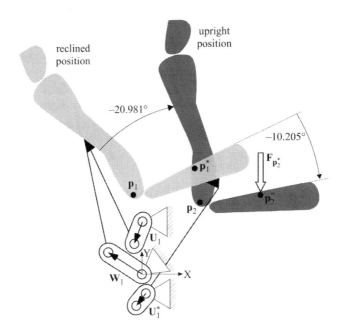

FIGURE P.6.12
Seat mechanism.

TABLE P.6.20

Seat Mechanism Dimensions (with Link Lengths in m)

W_1	V_1	U_1	G_1	L_1
−0.7432 + i0.4727	0.5825 + i0.1724	−0.1804 − i0.4696	0.0197 + i1.1147	0.5671 + i1.2383
L_1^*	V_1^*	U_1^*	G_1^*	
1 + i0.25	0.0552 − i2.3906	−0.2086 − i0.2401	0.068 − i1.5542	

32. Figure P.6.13 illustrates a Stephenson III mechanism used to guide a digging tool from the initial position to the final position. When the final position is reached, a static force of $\mathbf{F}_{\mathbf{p}_2^*} = (0, -9000)$N is applied at the displaced $\mathbf{p}_1^*$ (labeled $\mathbf{p}_2^*$ in Figure P.6.13). The dimensions for the illustrated mechanism are included in Table P.6.21. Considering a crank rotation of 40°, determine the crank static torque and joint forces produced at the final position (using the Appendix C.6 or I.6 file). The link masses are $m_1 = 5$, $m_2 = 10$, $m_3 = 4$, $m_4 = 15$, and $m_5 = 3.75$ kg, respectively, and the link center of mass vectors are $\mathbf{R}_1 = 1.5273 - i0.7343$, $\mathbf{R}_2 = -1.6254 + i0.2655$, $\mathbf{R}_3 = -0.6545 - i0.7694$, $\mathbf{R}_4 = 1.056 - i0.4951$, and $\mathbf{R}_5 = -0.5295 - i0.0657$. Gravity is -9.81 m/s².

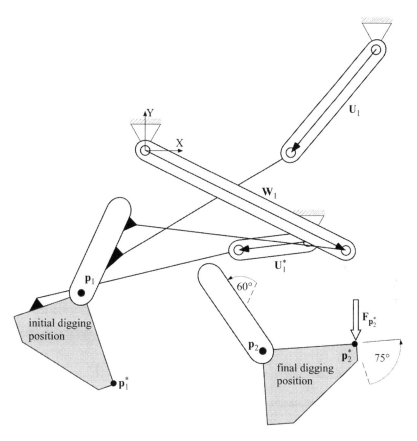

FIGURE P.6.13
Digging mechanism.

TABLE P.6.21

Digging Mechanism Dimensions (with Link Lengths in m)

W_1, θ	V_1, ρ	U_1, σ	G_{1x}, G_{1y}	L_1, δ
3.3893, −25.6775°	1.6644, 120.3504°	2.0202, 229.6156°	3.5225, 1.5065	4.0855, 189.0097°
V_1^*, ρ^*	U_1^*, σ^*	L_1^*, δ^*	G_{1x}^*, G_{1y}^*	
2.4795, 14.6953°	1.0672, 187.0724°	2.25, −70°	−1.0455, −2.8545	

TABLE P.6.22

Stephenson III Mechanism Configuration (with all Link Lengths in m)

W_1, θ	V_1, ρ	U_1, σ	G_{1x}, G_{1y}	L_1, δ
1, 90°	1.5, 19.3737°	1.5, 93.2461°	1.5, 0	1, 60.7834°
L_1^*, δ^*	V_1^*, ρ^*	U_1^*, σ^*	G_{1x}^*, G_{1y}^*	$F_{p_1}^*$ (N)
1, 63.7091°	2, 17.1417°	2, 76.4844°	0.4318, 0.5176	0, 0

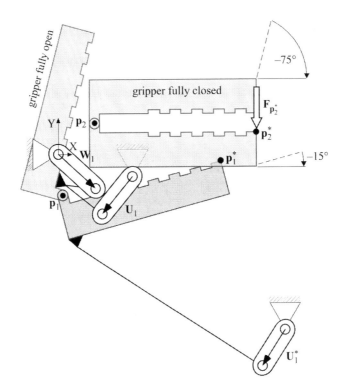

FIGURE P.6.14

Gripping-tool mechanism.

TABLE P.6.23

Gripper Mechanism Dimensions (with Link Lengths in cm)

W_1, θ	V_1, ρ	U_1, σ	G_{1x}, G_{1y}	L_1, δ
1.0283, 318.4178°	0.5195, 286.4362°	0.9418, 230.6143°	1.5138, −0.4529	0.7031, 189.1904°
V_1^*, ρ^*	U_1^*, σ^*	L_1^*, δ^*	G_{1x}^*, G_{1y}^*	
5.2881, 319.9647°	0.8666, 236.6216°	3.2742, 11.8113°	3.0869, −3.0198	

33. Calculate the static forces and torques for the Stephenson III mechanism configuration in Table P.6.22 at a crank displacement angle of $-90°$ (using the Appendix C.6 or I.6 file). The link masses are $m_1 = 4, m_2 = 8, m_3 = 4, m_4 = 8$, and $m_5 = 7$ kg, respectively, and the link center of mass vectors are $R_1 = i0.5$, $R_2 = 0.6344 + i0.4568$, $R_3 = 0.2074 + i07349$, $R_4 = 0.7847 + i0.4953$, and $R_5 = 0.2337 + i0.9723$. Gravity is -9.81 m/s^2.

34. Figure P.6.14 illustrates a Stephenson III mechanism used to guide a gripping tool from the open position to the closed position. When the closed position is reached, a static force of $F_{p_2^*} = (0, -50)$N is applied at the displaced p_1^* (labeled p_2^* in Figure P.6.14). The dimensions for the illustrated mechanism are included in Table P.6.23. Considering a crank rotation of $40°$, determine the crank static torque and joint forces produced at the closed position (using the Appendix C.6 or I.6 file). The link masses are $m_1 = 4, m_2 = 8, m_3 = 4, m_4 = 12$, and $m_5 = 4$ kg, respectively, and the link center of mass vectors are $R_1 = 0.3846 - i0.3412$, $R_2 = -0.6343 + i1.0364$, $R_3 = -0.2988 - i0.3639$, $R_4 = 1.1487 + i0.0597$, and $R_5 = -0.2384 - i0.3618$. Gravity is -9.81 m/s^2.

35. Calculate the static forces and torque for the Stephenson III mechanism in Example 6.5 where $F_{p_1^*} = (0, 0)$N.

7

Dynamic Force Analysis of Planar Mechanisms

CONCEPT OVERVIEW

In this chapter, the reader will gain a central understanding regarding

1. Criteria for dynamic force analysis and its applications
2. Link dynamic loads in 2D space
3. Formulation and solution of linear simultaneous equation sets for dynamic force analysis
4. The effects of gear train inclusion in the dynamic force analysis of five-bar mechanisms
5. The mass moment of inertia, its application in dynamic force analysis, and its calculation in computer-aided design (CAD) software

7.1 Introduction

As explained in Chapter 1, a *dynamic force analysis* (Figure 1.1) is the next type of force analysis to consider beyond a static force analysis when determining the structural forces in mechanical systems. Such an analysis should always be considered when angular velocities and accelerations are substantial (when mechanism motion is truly dynamic). Dynamic force analyses are also more general than static force analyses when mechanism motion is quasi-static. This is because, with a dynamic force analysis, acceleration-based forces and torques (however small in a quasi-static condition) are included.* In a dynamic force analysis, loads such as forces and torques are considered for each mechanism link according to Newton's second law ($\sum\mathbf{F} = m\mathbf{a}$, $\sum\mathbf{M} = I\alpha$) [1–5].†

Unlike the static force equations in Chapter 6, the equation systems presented in this chapter consider the inertia, velocity, and acceleration of each link.‡ Like the Chapter 6 equations, the equation systems presented in this chapter consider in-plane forces and torques, and mechanism links are considered to be rigid. Although link weights are not neglected in the forthcoming equations, as the acceleration of the links exceeds gravitational

---

* In a static force analysis, acceleration-based loads are not included due to Newton's first law.
† In Newton's second law, the variables m and I represent the link *mass* and *mass moment of inertia*, respectively.
‡ While the equations in this chapter do not explicitly consider *time* (only angular *displacement*, *velocity*, and *acceleration*), time can be inversely determined from these quantities (see Section 4.9).

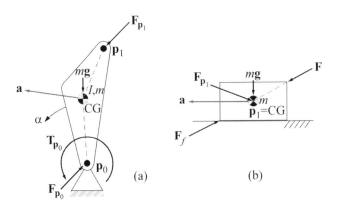

FIGURE 7.1
Dynamic loads on (a) grounded rotating and (b) translating planar links.

acceleration, the effect of gravity (and subsequently the link weights) becomes increasingly negligible. This condition is common in high-speed machinery.

7.2 Dynamic Loading in Planar Space

Figure 7.1 illustrates arbitrarily grounded rotating and translating planar links under loading. Like static loads, dynamic loads can be represented as force vectors applied to link points. For example, force vectors $\mathbf{F}_{\mathbf{p}_0}$, $\mathbf{F}_{\mathbf{p}_1}$, and $m\mathbf{g}$ are applied to link points $\mathbf{p}_0$, $\mathbf{p}_1$, and the link's CG (center of gravity), respectively, in Figure 7.1(a). Also, a torque $\mathbf{T}_{\mathbf{a}_0}$ is applied about point $\mathbf{p}_0$ in Figure 7.1(a). Like static loads, dynamic forces and torques are either applied externally (like force vectors $\mathbf{F}$ and $m\mathbf{g}$ in Figure 7.1(b)) or they are reactions to externally applied loads (like force vectors $\mathbf{F}_{\mathbf{p}_1}$ and $\mathbf{F}_f$ in Figure 7.1(b)).

Link dynamic loads are governed according to Newton's second law. In this law, the sum of all link forces is equal to the product of link mass and linear acceleration (or $\Sigma \mathbf{F} = m\mathbf{a}$), and the sum of all moments is equal to the product of the link mass moment of inertia and rotational acceleration (or $\Sigma M = I\alpha$). Like static loading, under dynamic loading force sums are taken in both directions ($\Sigma F_x = ma_x$ and $\Sigma F_y = ma_y$). The gravitational load or the *weight* of each moving link (denoted by the product $m\mathbf{g}$ in Figure 7.1) is oriented in the negative y-axis direction.

The conditions of Newton's second law must be satisfied for each mechanism link in order for the entire mechanism to be dynamically sound.

7.3 Four-Bar Mechanism Analysis

Figure 7.2a illustrates a planar four-bar mechanism where a force $\mathbf{F}_{\mathbf{p}_1}$ is applied to the coupler link point $\mathbf{p}_1$. To drive the mechanism, a torque $\mathbf{T}_{\mathbf{a}_0}$ is applied about the crank link revolute joint $\mathbf{a}_0$. The user also has the option of specifying a torque $\mathbf{T}_{\mathbf{b}_0}$ about the follower-link revolute joint $\mathbf{b}_0$.[*] The loads on the individual planar four-bar mechanism links are illustrated in Figure 7.2b. Because the joints at $\mathbf{a}_1$ and $\mathbf{b}_1$ are shared among two links, the forces at $\mathbf{a}_1$ and $\mathbf{b}_1$ must be equal but opposite (resulting in $\pm\mathbf{F}_{\mathbf{a}_1}$ and $\pm\mathbf{F}_{\mathbf{b}_1}$ in Figure 7.2b). The remaining force and torque variables, however, remain positive for simplicity.[†] This approach is repeated for the mechanisms in Sections 7.4–7.7.

Taking the sum of the forces and moments for each link according to the conditions $\Sigma\mathbf{F} = m\mathbf{a}$ and $\Sigma\mathbf{M} = I\alpha$ produces two dynamic equations for each link. Expanding these two equations and separating the force equation into two equations, where $\Sigma F_x = ma_{CG_x}$ and $\Sigma F_y = ma_{CG_y}$, ultimately produces three dynamic equations for each link.

Because, in a dynamic condition, moments are taken with respect to each link's center of gravity (also called the *center of mass*), Figure 7.2b includes vectors between the center of gravity and the *load points* (which are the mechanism nodes and the points where external forces are applied) of each mechanism link. Equation 7.1 includes the center of gravity-load point vectors used for the planar four-bar mechanism. To eliminate the

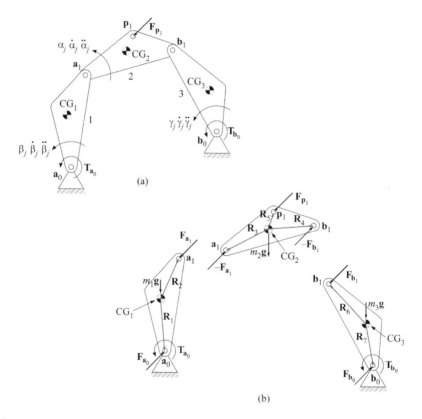

FIGURE 7.2
(a) Planar four-bar mechanism and (b) link dynamic forces and torques.

---

[*] If the user prefers not to consider the coupler link applied force $\mathbf{F}_{p1}$ or the follower link applied torque $\mathbf{T}_{b_0}$, these quantities can be specified as zero in the dynamic load equations.

[†] The signs of the calculated dynamic force and torque variables are not determined by the signs prescribed to them during equation formulation. They are determined by the mechanism position and the applied load values.

need to define unique direction angles for vectors $\mathbf{R}_1$ through $\mathbf{R}_7$, they are expressed in rectangular form.

$$\mathbf{R}_j = R_{jx} + iR_{jy}, \quad j = 1, 2, \ldots, 7 \tag{7.1}$$

As illustrated in Figure 7.2, vectors $\mathbf{R}_1$ through $\mathbf{R}_7$ are defined with respect to the link center of gravity (therefore pointing away from the center of gravity).*

Because the *x*- and *y*-components of the total acceleration of the center of gravity of each link are required in the dynamic load equations, Equation 7.2 includes the acceleration equations for $\mathbf{a}_{CG_1}$ through $\mathbf{a}_{CG_3}$.† These acceleration equations are identical in form to the planar four-bar acceleration equations in Section 4.4.3.

$$
\begin{aligned}
\mathbf{a}_{CG_1} &= i\ddot{\beta}_j(-\mathbf{R}_1)e^{i(\beta_j)} - \dot{\beta}_j^2(-\mathbf{R}_1)e^{i(\beta_j)} \\
\mathbf{a}_{CG_2} &= \mathbf{A}_{a1} + i\ddot{\alpha}_i(-\mathbf{R}_3)e^{i(\alpha_j)} - \dot{\alpha}_j^2(-\mathbf{R}_3)e^{i(\alpha_j)} \\
&= i\ddot{\beta}_j W_1 e^{i(\theta+\beta_j)} - \dot{\beta}_j^2 W_1 e^{i(\theta+\beta_j)} - i\ddot{\alpha}_j \mathbf{R}_3 e^{i(\alpha_j)} + \dot{\alpha}_j^2 \mathbf{R}_3 e^{i(\alpha_j)} \\
\mathbf{a}_{CG_3} &= i\ddot{\gamma}_j(-\mathbf{R}_7)e^{i(\gamma_j)} - \dot{\gamma}_j^2(-\mathbf{R}_7)e^{i(\gamma_j)}
\end{aligned}
\tag{7.2}
$$

Using the first two dynamic load conditions for the crank link (where the moment sum is taken about the center of gravity CG_1) produces

$$
\begin{aligned}
\mathbf{F}_{a_0} + \mathbf{F}_{a_1} + m_1\mathbf{g} &= m_1\mathbf{a}_{CG_1} \\
\mathbf{T}_{a_0} + \mathbf{R}_1 \times \mathbf{F}_{a_0} + \mathbf{R}_2 \times \mathbf{F}_{a_1} &= I_1\ddot{\beta}_j
\end{aligned}
\tag{7.3}
$$

Expanding and separating Equation 7.3 produces

$$
\begin{aligned}
F_{a_0x} + F_{a_1x} &= m_1 a_{CG_1x} \\
F_{a_0y} + F_{a_1y} + m_1 g &= m_1 a_{CG_1y} \\
T_{a_0} - F_{a_0x}R_{1y} + F_{a_0y}R_{1x} - F_{a_1x}R_{2y} + F_{a_1y}R_{2x} &= I_1\ddot{\beta}_j
\end{aligned}
\tag{7.4}
$$

Using the first two dynamic load conditions for the coupler link (where the moment sum is taken about CG_2) produces

$$
\begin{aligned}
-\mathbf{F}_{a_1} - \mathbf{F}_{b_1} + \mathbf{F}_{p_1} + m_2\mathbf{g} &= m_2\mathbf{a}_{CG_2} \\
-\mathbf{R}_3 \times \mathbf{F}_{a_1} - \mathbf{R}_4 \times \mathbf{F}_{b_1} + \mathbf{R}_5 \times \mathbf{F}_{p_1} &= I_2\ddot{\alpha}_j
\end{aligned}
\tag{7.5}
$$

Expanding and separating Equation 7.5 produces

---

* The vectors were established this way for use in the forthcoming moment summation equations.
† To correctly calculate $\mathbf{a}_{CG_1}$ through $\mathbf{a}_{CG_3}$, the signs of $\mathbf{R}_1$, $\mathbf{R}_3$, and $\mathbf{R}_7$ are reversed.

$$-F_{a_{1x}} - F_{b_{1x}} + F_{p_{1x}} = m_2 a_{CG_{2x}}$$

$$-F_{a_{1y}} - F_{b_{1y}} + F_{p_{1y}} + m_2 g = m_2 a_{CG_{2y}} \tag{7.6}$$

$$F_{a_{1x}} R_{3y} - F_{a_{1y}} R_{3x} + F_{b_{1x}} R_{4y} - F_{b_{1y}} R_{4x} - F_{p_{1x}} R_{5y} + F_{p_{1y}} R_{5x} = I_2 \ddot{\alpha}_j$$

Using the first two dynamic load conditions for the follower link (where the moment sum is taken about CG_3) produces

$$\mathbf{F_{b_0}} + \mathbf{F_{b_1}} + m_3 \mathbf{g} = m_3 \mathbf{a}_{CG_3}$$

$$\mathbf{T_{b_0}} + \mathbf{R}_6 \times \mathbf{F_{b_1}} + \mathbf{R}_7 \times \mathbf{F_{b_0}} = I_3 \ddot{\gamma}_j \tag{7.7}$$

Expanding and separating Equation 7.7 produces

$$F_{b_{0x}} + F_{b_{1x}} = m_3 a_{CG_{3x}}$$

$$F_{b_{0y}} + F_{b_{1y}} + m_3 g = m_3 a_{CG_{3y}} \tag{7.8}$$

$$T_{b_0} - F_{b_{1x}} R_{6y} + F_{b_{1y}} R_{6x} - F_{b_{0x}} R_{7y} + F_{b_{0y}} R_{7x} = I_3 \ddot{\gamma}_j$$

Expressing Equations 7.4, 7.6, and 7.8 in a combined matrix form produces

$$\begin{bmatrix} 0 & 1 & 0 & 1 & 0 & 0 & 0 & 0 & 0 \\ 0 & 0 & 1 & 0 & 1 & 0 & 0 & 0 & 0 \\ 1 & -R_{1y\beta_j} & R_{1x\beta_j} & -R_{2y\beta_j} & R_{2x\beta_j} & 0 & 0 & 0 & 0 \\ 0 & 0 & 0 & -1 & 0 & 0 & 0 & -1 & 0 \\ 0 & 0 & 0 & 0 & -1 & 0 & 0 & 0 & -1 \\ 0 & 0 & 0 & R_{3y\alpha_j} & -R_{3x\alpha_j} & 0 & 0 & R_{4y\alpha_j} & -R_{4x\alpha_j} \\ 0 & 0 & 0 & 0 & 0 & 1 & 0 & 1 & 0 \\ 0 & 0 & 0 & 0 & 0 & 0 & 1 & 0 & 1 \\ 0 & 0 & 0 & 0 & 0 & -R_{7y\gamma_j} & R_{7x\gamma_j} & -R_{6y\gamma_j} & R_{6x\gamma_j} \end{bmatrix} \begin{Bmatrix} T_{\mathbf{a_0}} \\ F_{a_{0x}} \\ F_{a_{0y}} \\ F_{a_{1x}} \\ F_{a_{1y}} \\ F_{b_{0x}} \\ F_{b_{0y}} \\ F_{b_{1x}} \\ F_{b_{1y}} \end{Bmatrix}$$

$$= \begin{bmatrix} m_1 a_{CG_{1x}} \\ m_1 a_{CG_{1y}} - m_1 g \\ I_1 \ddot{\beta}_j \\ -F_{p_{1x}} + m_2 a_{CG_{2x}} \\ -F_{p_{1y}} + m_2 a_{CG_{2y}} - m_2 g \\ I_2 \ddot{\alpha}_j + F_{p_{1x}} R_{5y\alpha_j} - F_{p_{1y}} R_{5x\alpha_j} \\ m_3 a_{CG_{3x}} \\ m_3 a_{CG_{3y}} - m_3 g \\ I_3 \ddot{\gamma}_j - T_{\mathbf{b_0}} \end{bmatrix} \tag{7.9}$$

where the x- and y-components of vectors $\mathbf{R}_1$ through $\mathbf{R}_7$ are $R_{nx_{angle\,j}} = R_{nx}\cos\left(angle_j\right) - R_{ny}\sin\left(angle_j\right)$ and $R_{ny_{angle\,j}} = R_{nx}\sin\left(angle_j\right) + R_{ny}\cos\left(angle_j\right)$.* It can be seen in Equations 7.4, 7.6, 7.8, and ultimately in Equation 7.9 that the mass and mass moment of inertia of each link are also required.

Equation 7.9 can be solved using Cramer's rule to determine the unknown forces and torque. The unknown planar four-bar displacement angles α_j and γ_j, angular velocities $\dot{\alpha}_j$ and $\dot{\gamma}_j$, and angular accelerations $\ddot{\alpha}_j$ and $\ddot{\gamma}_j$ are the same quantities calculated from the planar four-bar equations in Sections 4.4.1–4.4.4. Given such solutions, the corresponding dynamic forces and torques can be calculated from Equation 7.9.

Appendix D.1 includes the MATLAB® file user instructions for planar four-bar dynamic force analysis. In this MATLAB file (which is available for download at www.crcpress.com/product/isbn/9781498724937), solutions for Equation 7.9 are calculated.

Example 7.1

Problem Statement: Using the Appendix D.1 MATLAB file, calculate the reaction forces $F_{a_{0x}}$, $F_{a_{0y}}$, $F_{b_{0x}}$, and $F_{b_{0y}}$ over a complete crank rotation range for the planar four-bar mechanism in Tables E.7.1 and E.7.2. For this mechanism $\mathbf{F}_{p_1} = (0,0)\,\text{N}$, $\beta_0 = 1\,\text{rad/s}$, and $\ddot{\beta} = 0\,\text{rad/s}^2$. Also, $T_{b_0} = 0$ and gravity is $-9.81\,\text{m/s}^2$.

Known Information: Tables E.7.1, E.7.2, and Appendix D.1 MATLAB file.

Solution Approach: Figure E.7.1 includes the input specified (in bold text) in the Appendix D.1 MATLAB file. Vectors $\mathbf{R}_2$, $\mathbf{R}_4$, $\mathbf{R}_5$, and $\mathbf{R}_6$ are calculated in the MATLAB file using vector-loop equations (see Appendix D.1). Figure E.7.2 includes the reaction force profiles calculated for the planar four-bar mechanism using the Appendix D.1 MATLAB file.

TABLE E.7.1

Planar Four-Bar Mechanism Dimensions (with Link Lengths in m)

W_1, θ	V_1, ρ	U_1, σ	G_{1x}, G_{1y}	L_1, δ
0.5, 90°	0.75, 19.3737°	0.75, 93.2461°	0.75, 0	0.5, 60.7834°

TABLE E.7.2

Planar Four-Bar Mechanism Dynamic Parameters (with Length in m, Mass in kg, and Inertia in kg–m²)

$\mathbf{R}_1$	$0 - i0.25$	m_1	8.05
$\mathbf{R}_3$	$-0.3172 - i0.2284$	I_1	0.805
$\mathbf{R}_7$	$0.0212 - i0.3744$	m_2	29.785
		I_2	5.635
		m_3	12.075
		I_3	2.415

---

* This method of calculating mechanism forces via matrix manipulation is called the *matrix method*. With this method, link force equations are quickly derived.

```
W1 = 0.5*exp(i*90*pi/180);
V1 = 0.75*exp(i*19.3737*pi/180);
G1 = 0.75;
U1 = 0.75*exp(i*93.2461*pi/180);
L1 = 0.5*exp(i*60.7834*pi/180);

Fp1 = [0, 0];
Tb0 = 0;
g = -9.81;

R1 = -i*0.25;
R3 = -0.3172 - i*0.2284;
R7 = 0.0212 - i*0.3744;

m1 = 8.05;
I1 = 0.805;

m2 = 29.785;
I2 = 5.635;

m3 = 12.075;
I3 = 2.415;

start_ang   = 0;
step_ang    = 1;
stop_ang    = 360;

angular_vel  = 1 * ones(N+1,1);
angular_acc  = 0 * ones(N+1,1);
```

FIGURE E.7.1
Specified input (in bold text) in the Appendix D.1 MATLAB file for Example 7.1.

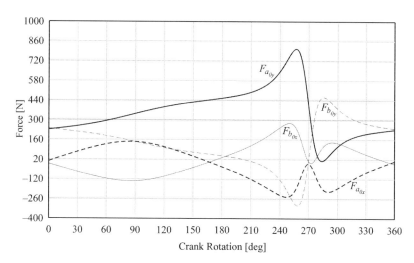

FIGURE E.7.2
$F_{a_{0x}}$, $F_{a_{0y}}$, $F_{b_{0x}}$ and $F_{b_{0y}}$ reaction force profiles for planar four-bar mechanism.

In a mechanism where the crank link rotates continuously, nonconstant force profiles in the mechanism joints (like those illustrated in Figure E.7.1 for the planar four-bar mechanism) will cause the mechanism to oscillate or *vibrate*. Excessive vibration is an undesired effect because it can compromise the design life of a mechanical system as well as its performance. *Vibration analysis* (a study associated with dynamics) is concerned with the oscillatory motions of bodies and the forces associated with them [6].

7.4 Slider-Crank Mechanism Analysis

Figure 7.3a illustrates a slider-crank mechanism where a force $\mathbf{F}$ is applied to the slider link revolute joint $\mathbf{b}_1$ (which is coincident with the slider's center of gravity CG_3). To drive the mechanism, a torque $\mathbf{T}_{a_0}$ is applied about the crank link revolute joint $\mathbf{a}_0$. The loads on the individual slider-crank mechanism links are illustrated in Figure 7.3b. Taking the sum of the forces and moments for each link according to the conditions $\Sigma\mathbf{F} = m\mathbf{a}$ and $\Sigma\mathbf{M} = I\alpha$ produces two dynamic equations for each link. Expanding these two equations and separating the force equation into two equations, where $\Sigma F_x = ma_{CG_x}$ and $\Sigma F_y = ma_{CG_y}$, ultimately produces three dynamic equations for each link.

Equation 7.10 includes the center of gravity-load point vectors used for the slider-crank mechanism.

$$\mathbf{R}_j = R_{jx} + iR_{jy},\ j = 1, 2, \ldots, 4 \tag{7.10}$$

Because the x- and y-components of the total acceleration of the center of gravity of each link are required in the dynamic load equations, Equation 7.11 includes the acceleration equations for $\mathbf{a}_{CG_1}$ through $\mathbf{a}_{CG_3}$. The acceleration equations are identical in form to the slider-crank acceleration equations in Section 4.5.3.

$$\mathbf{a}_{CG_1} = i\ddot{\beta}_i(-\mathbf{R}_1)e^{i(\beta_j)} - \dot{\beta}_j^2(-\mathbf{R}_1)e^{i(\beta_j)}$$

$$\mathbf{a}_{CG_2} = \mathbf{A}_{a1} + i\ddot{\alpha}_j(-\mathbf{R}_3)e^{i(\alpha_j)} - \dot{\alpha}_j^2(-\mathbf{R}_3)e^{i(\alpha_j)}$$

$$= i\ddot{\beta}_j W_1 e^{i(\theta+\beta_j)} - \dot{\beta}_j^2 W_1 e^{i(\theta+\beta_j)} - i\ddot{\alpha}_j \mathbf{R}_3 e^{i(\alpha_j)} + \dot{\alpha}_j^2 \mathbf{R}_3 e^{i(\alpha_j)} \tag{7.11}$$

$$\mathbf{a}_{CG_3} = \mathbf{A}_{b1} = \ddot{G}_j$$

Using the first two dynamic load conditions for the crank link (where the moment sum is taken about the center of gravity CG_1) produces Equation 7.3. Expanding and separating the resulting equations produces Equation 7.4.

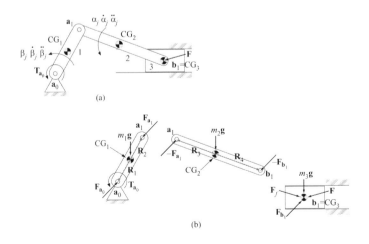

(a)

(b)

FIGURE 7.3
(a) Slider-crank mechanism and (b) link dynamic forces and torque.

Using the first two dynamic load conditions for the coupler link (where the moment sum is taken about CG_2) produces

$$-\mathbf{F}_{\mathbf{a}_1} - \mathbf{F}_{\mathbf{b}_1} + m_2\mathbf{g} = m_2\mathbf{a}_{CG_2}$$

$$-\mathbf{R}_3 \times \mathbf{F}_{\mathbf{a}_1} - \mathbf{R}_4 \times \mathbf{F}_{\mathbf{b}_1} = I_2\ddot{\alpha}_j \tag{7.12}$$

Expanding and separating Equation 7.12 produces

$$-F_{a_{1x}} - F_{b_{1x}} = m_2 a_{CG_{2x}}$$

$$-F_{a_{1y}} - F_{b_{1y}} m_2 g = m_2 a_{CG_{2y}} \tag{7.13}$$

$$F_{a_{1x}} R_{3y} - F_{a_{1y}} R_{3x} + F_{b_{1x}} R_{4y} - F_{b_{1y}} R_{4x} = I_2 \ddot{\alpha}_j$$

Using only the dynamic load condition $\Sigma F = ma$ for the slider link produces

$$\mathbf{F} + \mathbf{F}_{\mathbf{b}_1} + \mathbf{F}_f + m_3\mathbf{g} = m_3\mathbf{a}_{CG_3} \tag{7.14}$$

In Equation 7.14, the x- and y-components of vector $\mathbf{F}_f$ are the *friction force* $\pm\mu F_{\text{normal}}$ and the *normal force* F_{normal}, respectively. Expanding and separating Equation 7.14 produces

$$F_x + F_{b_{1x}} \pm \mu F_{\text{normal}} = m_3 a_{CG_{3x}}$$

$$F_y + F_{b_{1y}} + F_{\text{normal}} + m_3 g = 0 \tag{7.15}$$

Expressing Equations 7.4, 7.13, and 7.15 in a combined matrix form produces

$$\begin{bmatrix} 0 & 1 & 0 & 1 & 0 & 0 & 0 & 0 \\ 0 & 0 & 1 & 0 & 1 & 0 & 0 & 0 \\ 1 & -R_{1y\beta_j} & R_{1x\beta_j} & -R_{2y\beta_j} & R_{2x\beta_j} & 0 & 0 & 0 \\ 0 & 0 & 0 & -1 & 0 & -1 & 0 & 0 \\ 0 & 0 & 0 & 0 & -1 & 0 & -1 & 0 \\ 0 & 0 & 0 & R_{3y\alpha_j} & -R_{3x\alpha_j} & R_{4y\alpha_j} & -R_{4x\alpha_j} & 0 \\ 0 & 0 & 0 & 0 & 0 & 1 & 0 & \pm\mu \\ 0 & 0 & 0 & 0 & 0 & 0 & 1 & 1 \end{bmatrix} \begin{Bmatrix} T_{a_0} \\ F_{a_{0x}} \\ F_{a_{0y}} \\ F_{a_{1x}} \\ F_{a_{1y}} \\ F_{b_{1x}} \\ F_{b_{1y}} \\ F_{\text{normal}} \end{Bmatrix}$$

$$= \begin{bmatrix} m_1 a_{CG_{1x}} \\ m_1 a_{CG_{1y}} - m_1 g \\ I_1 \ddot{\beta}_j \\ m_2 a_{CG_{2x}} \\ m_2 a_{CG_{2y}} - m_2 g \\ I_2 \ddot{\alpha}_j \\ -F_x + m_3 a_{CG_{3x}} \\ -F_y - m_3 g \end{bmatrix} \tag{7.16}$$

where the x- and y-components of vectors $\mathbf{R}_1$ through $\mathbf{R}_4$ are $R_{nx_{angle\,j}} = R_{nx}\cos\left(angle_j\right) - R_{ny}\sin\left(angle_j\right)$ and $R_{nx_{angle\,j}} = R_{nx}\sin\left(angle_j\right) + R_{ny}\cos\left(angle_j\right)$.* It can be seen in Equations 7.4, 7.13, 7.15, and ultimately in Equation 7.16 that the mass and mass moment of inertia of each link are also required.

Equation 7.16 can be solved using Cramer's rule to determine the unknown forces and torque. The unknown slider-crank displacement angles α_j, angular velocities $\dot{\alpha}_j$, and angular accelerations $\ddot{\alpha}_j$ are the same quantities calculated from the slider-crank equations in Sections 4.5.1–4.5.3 (along with the slider accelerations $\ddot{G}_j$). Given such solutions, the corresponding dynamic forces and torques can be calculated from Equation 7.16.

Appendix D.2 includes the MATLAB file user instructions for slider-crank dynamic force analysis. In this MATLAB file (which is available for download at www.crcpress. com/product/isbn/9781498724937), solutions for Equation 7.16 are calculated.

Example 7.2

Problem Statement: Using the Appendix D.2 MATLAB file, calculate the driver torque over a complete crank rotation range for the in-line slider-crank mechanism in Tables E.7.3 and E.7.4. For this mechanism, $\mathbf{F}_{P1} = (0,0)\mathbf{N}$, $\dot{\beta}_0 = 10\,\mathrm{rad/s}$, and $\ddot{\beta} = 0\,\mathrm{rad/s}^2$. Gravity is $-9.81\,\mathrm{m/s}^2$.

Known Information: Tables E.7.3, E.7.4, and Appendix D.2 MATLAB file.

Solution Approach: Figure E.7.3 includes the input specified (in bold text) in the Appendix D.2 MATLAB file. Vectors $\mathbf{R}_2$ and $\mathbf{R}_4$ are calculated in the MATLAB file using vector-loop equations (see Appendix D.2). Figure E.7.4 includes the driver torque profile calculated for the slider-crank mechanism using the Appendix D.2 MATLAB file. After running the Appendix D.2 file twice (once with $\mu = +0.5$ and again with $\mu = -0.5$), only the driver torque where the friction force opposes the slider velocity is retained and assembled to produce Figure E.7.4.

TABLE E.7.3

Slider–Crank Mechanism Dimensions (with Link Lengths in m)

W_1, θ	V_1, ρ	μ
0.5, 90°	0.9014, −33.6901°	±0.5

TABLE E.7.4

Slider–Crank Mechanism Dynamic Parameters (with Length in m, Mass in kg, and Inertia in kg–m²)

$\mathbf{R}_1$	$0 - i0.25$	m_1	8.05
$\mathbf{R}_3$	$-0.3750 + i0.2500$	I_1	0.805
		m_2	14.49
		I_2	4.025
		m_3	30

---

* The correct sign for ±μ in Equation 7.16 will be the sign that produces a friction force that opposes the direction of the slider velocity.

```
LW1 = 0.5;
theta = 90*pi/180;

LU1 = 0;

LV1 = 0.9014;

F = [0, 0];
mu = 0.5;  (NOTE: the file must be run with mu=0.5 and mu=-0.5)
g = -9.81;

R1 = -i*0.25;
R3 = -0.3750 + i*0.25;

m1 = 8.05;
I1 = 0.805;

m2 = 14.49;
I2 = 4.025;

m3 = 30;

start_ang   = 0;
step_ang    = 1;
stop_ang    = 360;

angular_vel  = 10 * ones(N+1,1);
angular_acc  = 0 * ones(N+1,1);
```

FIGURE E.7.3
Specified input (in bold text) in the Appendix D.2 MATLAB file for Example 7.2.

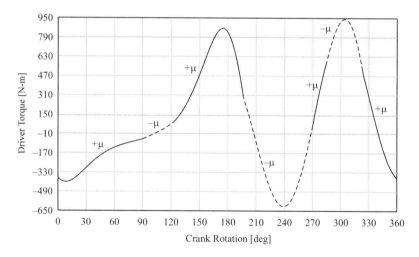

FIGURE E.7.4
T_{a0} profile for slider-crank mechanism.

7.5 Geared Five-Bar Mechanism Analysis

We will begin the formulation of a dynamic force equation system for the geared five-bar mechanism by first formulating a dynamic force equation system for a five-bar mechanism *without* gears. Because such a system has two degrees of freedom, controlled mechanism

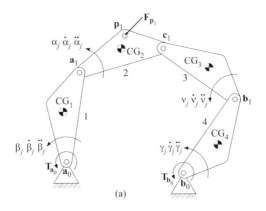

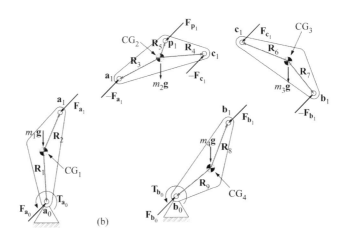

FIGURE 7.4
(a) Geared five-bar mechanism and (b) link dynamic forces and torques.

motion is achieved by independently controlling the rotations of the links containing vectors $\mathbf{W}_1$ and $\mathbf{U}_1$.

Figure 7.4a illustrates a geared five-bar mechanism where a force $\mathbf{F}_{p_1}$ is applied to the intermediate link point $\mathbf{p}_1$. To drive the mechanism, torques $\mathbf{T}_{a_0}$ and $\mathbf{T}_{b_0}$ are applied about the grounded revolute joints $\mathbf{a}_0$ and $\mathbf{b}_0$, respectively. The loads on the individual geared five-bar mechanism links are illustrated in Figure 7.4b. Taking the sum of the forces and moments for each link according to the conditions $\Sigma\mathbf{F} = m\mathbf{a}$ and $\Sigma\mathbf{M} = I\alpha$ produces two dynamic equations for each link. Expanding these two equations and separating the force equation into two equations, where $\Sigma F_x = ma_{CG_x}$ and $\Sigma F_y = ma_{CG_y}$, ultimately produces three dynamic equations for each link.

Equation 7.17 includes the center of gravity-load point vectors used for the geared five-bar mechanism.

$$\mathbf{R}_j = R_{jx} + iR_{jy}, \ j = 1, 2, \ldots, 9 \tag{7.17}$$

Because the x- and y-components of the total acceleration of the center of gravity of each link are required in the dynamic load equations, Equation 7.18 includes the acceleration equations for $\mathbf{a}_{CG_1}$ through $\mathbf{a}_{CG_4}$. The acceleration equations are identical in form to the geared five-bar acceleration equations in Section 4.6.3.

$$\mathbf{a}_{CG_1} = i\ddot{\beta}_j(-\mathbf{R}_1)e^{i(\beta_j)} - \dot{\beta}_j^2(-\mathbf{R}_1)e^{i(\beta_j)}$$

$$\mathbf{a}_{CG_2} = \mathbf{A}_{\mathbf{a}_1} + i\ddot{\alpha}_j(-\mathbf{R}_3)e^{i(\alpha_j)} - \dot{\alpha}_j^2(-\mathbf{R}_3)e^{i(\alpha_j)}$$

$$= i\ddot{\beta}_j W_1 e^{i(\theta+\beta_j)} - \dot{\beta}_j^2 W_1 e^{i(\theta+\beta_j)} - i\ddot{\alpha}_j \mathbf{R}_3 e^{i(\alpha_j)} + \dot{\alpha}_j^2 \mathbf{R}_3 e^{i(\alpha_j)}$$

$$\mathbf{a}_{CG_3} = \mathbf{A}_{\mathbf{b}_1} + i\ddot{v}_j(-\mathbf{R}_7)e^{i(v_j)} - \dot{v}_j^2(-\mathbf{R}_7)e^{i(v_j)} \qquad (7.18)$$

$$= i\ddot{\gamma}_j U_1 e^{i(\sigma+\gamma_j)} - \dot{\gamma}_j^2 U_1 e^{i(\sigma+\gamma_j)} - i\ddot{v}_j \mathbf{R}_7 e^{i(v_j)} + \dot{v}_j^2 \mathbf{R}_7 e^{i(v_j)}$$

$$\mathbf{a}_{CG_4} = i\ddot{\gamma}_j(-\mathbf{R}_9)e^{i(\gamma_j)} - \dot{\gamma}_j^2(-\mathbf{R}_9)e^{i(\gamma_j)}$$

Using the first two dynamic load conditions for the crank link (where the moment sum is taken about the center of gravity CG_1) produces Equation 7.3. Expanding and separating the resulting equations produces Equation 7.4.

Using the first two dynamic load conditions for the intermediate link $\mathbf{a}_1$–$\mathbf{c}_1$ (where the moment sum is taken about CG_2) produces

$$-\mathbf{F}_{\mathbf{a}_1} - \mathbf{F}_{\mathbf{c}_1} + \mathbf{F}_{\mathbf{p}_1} + m_2\mathbf{g} = m_2\mathbf{a}_{CG_2}$$

$$-\mathbf{R}_3 \times \mathbf{F}_{\mathbf{a}_1} - \mathbf{R}_4 \times \mathbf{F}_{\mathbf{c}_1} + \mathbf{R}_5 \times \mathbf{F}_{\mathbf{p}_1} = I_2\ddot{\alpha}_j \qquad (7.19)$$

Expanding and separating Equation 7.19 produces

$$-F_{a_{1x}} - F_{c_{1x}} + F_{p_{1x}} = m_2 a_{CG_{2x}}$$

$$-F_{a_{1y}} - F_{c_{1y}} + F_{p_{1y}} + m_2 g = m_2 a_{CG_{2y}} \qquad (7.20)$$

$$F_{a_{1x}} R_{3y} - F_{a_{1y}} R_{3x} + F_{c_{1x}} R_{4y} - F_{c_{1y}} R_{4x} - F_{p_{1x}} R_{5y} + F_{p_{1y}} R_{5x} = I_2\ddot{\alpha}_j$$

Using the first two dynamic load conditions for the intermediate link $\mathbf{b}_1$–$\mathbf{c}_1$ (where the moment sum is taken about CG_3) produces

$$\mathbf{F}_{\mathbf{c}_1} - \mathbf{F}_{\mathbf{b}_1} + m_3\mathbf{g} = m_3\mathbf{a}_{CG_3}$$

$$\mathbf{R}_6 \times \mathbf{F}_{\mathbf{c}_1} - \mathbf{R}_7 \times \mathbf{F}_{\mathbf{b}_1} = I_3\ddot{v}_j \qquad (7.21)$$

Expanding and separating Equation 7.21 produces

$$F_{c_{1x}} - F_{b_{1x}} = m_3 a_{CG_{3x}}$$

$$F_{c_{1y}} - F_{b_{1y}} + m_3 g = m_3 a_{CG_{3y}} \qquad (7.22)$$

$$-F_{c_{1x}} R_{6y} + F_{c_{1y}} R_{6x} + F_{b_{1x}} R_{7y} - F_{b_{1y}} R_{7x} = I_3\ddot{v}_j$$

Using the first two dynamic load conditions for the crank link $\mathbf{b}_0$–$\mathbf{b}_1$ (where the moment sum is taken about CG_4) produces

$$\mathbf{F}_{\mathbf{b}_0} + \mathbf{F}_{\mathbf{b}_1} + m_4\mathbf{g} = m_4\mathbf{a}_{CG_4}$$

$$\mathbf{T}_{\mathbf{b}_0} + \mathbf{R}_8 \times \mathbf{F}_{\mathbf{b}_1} + \mathbf{R}_9 \times \mathbf{F}_{\mathbf{b}_0} = I_4\ddot{\gamma}_j \qquad (7.23)$$

Expanding and separating Equation 7.23 produces

$$F_{b_0x} + F_{b_1x} = m_4 a_{CG_4x}$$

$$F_{b_0y} + F_{b_1y} + m_4 g = m_4 a_{CG_4y} \tag{7.24}$$

$$T_{b_0} - F_{b_1x} R_{8y} + F_{b_1y} R_{8x} - F_{b_0x} R_{9y} + F_{b_0y} R_{9x} = I_4 \ddot{\gamma}_j$$

Expressing Equations 7.4, 7.20, 7.22, and 7.24 in a combined matrix form produces

$$
\begin{bmatrix}
0 & 1 & 0 & 1 & 0 & 0 & 0 & 0 & 0 & 0 & 0 & 0 \\
0 & 0 & 1 & 0 & 1 & 0 & 0 & 0 & 0 & 0 & 0 & 0 \\
1 & -R_{1y\beta_j} & R_{1x\beta_j} & -R_{2y\beta_j} & R_{2x\beta_j} & 0 & 0 & 0 & 0 & 0 & 0 & 0 \\
0 & 0 & 0 & -1 & 0 & -1 & 0 & 0 & 0 & 0 & 0 & 0 \\
0 & 0 & 0 & 0 & -1 & 0 & -1 & 0 & 0 & 0 & 0 & 0 \\
0 & 0 & 0 & R_{3y\alpha_j} & -R_{3x\alpha_j} & R_{4y\alpha_j} & -R_{4x\alpha_j} & 0 & 0 & 0 & 0 & 0 \\
0 & 0 & 0 & 0 & 0 & 1 & 0 & 0 & 0 & 0 & -1 & 0 \\
0 & 0 & 0 & 0 & 0 & 0 & 1 & 0 & 0 & 0 & 0 & -1 \\
0 & 0 & 0 & 0 & 0 & -R_{6y_{v_j}} & R_{6x_{v_j}} & 0 & 0 & 0 & R_{7y_{v_j}} & -R_{7x_{v_j}} \\
0 & 0 & 0 & 0 & 0 & 0 & 0 & 0 & 1 & 0 & 1 & 0 \\
0 & 0 & 0 & 0 & 0 & 0 & 0 & 0 & 0 & 1 & 0 & 1 \\
0 & 0 & 0 & 0 & 0 & 0 & 0 & 1 & -R_{9y\gamma_j} & R_{9x\gamma_j} & -R_{8y\gamma_j} & R_{8x\gamma_j}
\end{bmatrix}
$$

$$
\begin{Bmatrix}
T_{a_0} \\
F_{a_0x} \\
F_{a_0y} \\
F_{a_1x} \\
F_{a_1y} \\
F_{c_1x} \\
F_{c_1y} \\
T_{b_0} \\
F_{b_0x} \\
F_{b_0y} \\
F_{b_1x} \\
F_{b_1y}
\end{Bmatrix}
=
\begin{Bmatrix}
m_1 a_{CG_1x} \\
m_1 a_{CG_1y} - m_1 g \\
I_1 \ddot{\beta}_j \\
m_2 a_{CG_2x} - F_{p_1x} \\
-F_{p_1y} + m_2 a_{CG_2y} - m_2 g \\
I_2 \ddot{\alpha}_j + F_{p_1x} R_{5y\alpha_j} - F_{p_1y} R_{5x\alpha_j} \\
m_3 a_{CG_3x} \\
m_3 a_{CG_3y} - m_3 g \\
I_3 \ddot{v}_j \\
m_4 a_{CG_4x} \\
m_4 a_{CG_4y} - m_4 g \\
I_4 \ddot{\gamma}_j
\end{Bmatrix}
\tag{7.25}
$$

where the x- and y-components of vectors $\mathbf{R}_1$ through $\mathbf{R}_9$ are $R_{nx_{angle\,j}} = R_{nx} \cos(angle_j)$ $- R_{ny} \sin(angle_j)$ and $R_{ny_{angle\,j}} = R_{nx} \sin(angle_j) + R_{ny} \cos(angle_j)$. It can be seen in Equations 7.4, 7.20, 7.22, 7.24, and ultimately in Equation 7.25 that the mass and mass moment of inertia of each link are also required.

Equation 7.25 can be solved using Cramer's rule to determine the unknown forces and torques. The unknown geared five-bar displacement angles α_j and ν_j, angular velocities $\dot{\alpha}_j$ and $\dot{\nu}_j$, and angular accelerations $\ddot{\alpha}_j$ and $\ddot{\nu}_j$ are the same quantities calculated from the geared five-bar equations in Sections 4.6.1–4.6.4. Given such solutions, the corresponding dynamic forces and torques can be calculated from Equation 7.25. Because this equation (like Equation 6.19 in static force analysis) considers a five-bar mechanism without gears, values for T'_{a_0}, $\mathbf{F}'_{a_0}$, and $\mathbf{F}'_{b_0}$ must be calculated in place of T_{a_0}, $\mathbf{F}_{a_0}$, and $\mathbf{F}_{b_0}$ in Equation 7.25.

Equations 6.20–6.24 (see Section 6.5) are used to calculate the gear-based driver torque T'_{a_0} and gear-based forces $\mathbf{F}'_{a_0}$ and $\mathbf{F}'_{b_0}$ from the Equation 7.25 solutions. Because the mass and inertia of the idler gear are not included in Equation 7.25, the results calculated from this equation for a five-bar mechanism having three gears are approximate solutions.*

Appendices D.3 and D.4 include the MATLAB file user instructions for geared five-bar dynamic force analysis (for two and three gears, respectively). In these MATLAB files (which are available for download at www.crcpress.com/product/isbn/9781498724937), solutions for Equation 7.25 and 6.20–6.24 are calculated.

Example 7.3

Problem Statement: Using the Appendix D.4 MATLAB file, calculate the driver torque over a complete crank rotation range for the geared five-bar mechanism in Tables E.7.5 and E.7.6. The gear ratio is $r = +2$. For this mechanism, $\mathbf{F}_{p_1} = (0, -1000)\,\mathrm{N}$, $\beta_0 = 1\,\mathrm{rad/s}$, and $\ddot{\beta} = 0.25\,\mathrm{rad/s^2}$. Gravity is $-9.81\,\mathrm{m/s^2}$.

Known Information: Tables E.7.5, E.7.6, and Appendix D.4 MATLAB file.

Solution Approach: Figure E.7.5 includes the input specified (in bold text) in the Appendix D.4 MATLAB file. Vectors $\mathbf{R}_2$, $\mathbf{R}_4$, $\mathbf{R}_5$, $\mathbf{R}_6$, and $\mathbf{R}_8$ are calculated in the MATLAB file using vector-loop equations (see Appendix D.4). Figure E.7.6 includes the driver torque profiles calculated for the geared five-bar mechanism using the Appendix D.4 MATLAB file.

TABLE E.7.5

Geared Five–Bar Mechanism Dimensions (with Link Lengths in m)

W_1, θ	V_1, ρ	U_1, σ	S_1, ψ	G_{1x}, G_{1y}	L_1, δ
0.5, 90°	0.75, 32.7304°	0.75, 45°	0.75, 149.9837°	0.75, 0	0.5, 74.14°

TABLE E.7.6

Geared Five–Bar Mechanism Dynamic Parameters (with Length in m, Mass in kg, and Inertia in kg–m²)

$\mathbf{R}_1$	$0 - i0.0831$	m_1	22.54
$\mathbf{R}_3$	$-0.2558 - i0.2955$	I_1	0.505
$\mathbf{R}_7$	$0.3247 - i0.1876$	m_2	29.785
$\mathbf{R}_9$	$-0.0356 - i0.0356$	I_2	5.635
		m_3	12.075
		I_3	2.415
		m_4	75.67
		I_4	5.635

---

* While only approximate solutions are calculated using Equation 7.25 for five-bar mechanisms having three gears, the solutions are most accurate if the mass and inertia of the idler gear are equal to or less than those of the driving gear.

```
W1 = 0.5*exp(i*90*pi/180);
V1 = 0.75*exp(i*32.7304*pi/180);
G1 = 0.75*exp(i*0*pi/180);
U1 = 0.75*exp(i*45*pi/180);
L1 = 0.5*exp(i*74.14*pi/180);
S1 = 0.75*exp(i*149.9847*pi/180);

Fp1 = [0, -1000];
ratio = 2;
g = -9.81;

R1 = -i*0.0831;
R3 = -0.2558 - i*0.2955;
R7 = 0.3247 - i*0.1876;
R9 = -0.0356 - i*0.0356;

m1 = 22.54;
I1 = 0.505;

m2 = 29.785;
I2 = 5.635;

m3 = 12.075;
I3 = 2.415;

m4 = 75.67;
I4 = 5.635;

start_ang    = 0;
step_ang     = 1;
stop_ang     = 360;

angular_vel  = 10 * ones(N+1,1);
angular_acc  = 0 * ones(N+1,1);
```

FIGURE E.7.5
Specified input (in bold text) in the Appendix D.4 MATLAB file for Example 7.3.

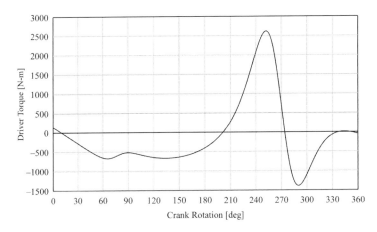

FIGURE E.7.6
T'_{a_0} profile for geared five-bar mechanism.

7.6 Watt II Mechanism Analysis

Figure 7.5a illustrates a Watt II mechanism where forces $\mathbf{F}_{p1}$ and $\mathbf{F}_{p_1^*}$ are applied to the intermediate link points $\mathbf{p}_1$ and $\mathbf{p}_1^*$, respectively. To drive the mechanism, a torque $\mathbf{T}_{a_0}$ is applied

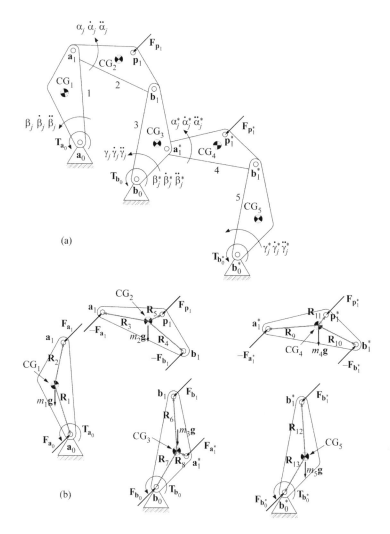

FIGURE 7.5
(a) Watt II mechanism and (b) link dynamic forces and torques.

about the crank link revolute joint. The user also has the option of specifying torques T_{b_0} and $T_{b_0^*}$ about the revolute joints b_0 and b_0^*, respectively.[*] The loads on the individual Watt II mechanism links are illustrated in Figure 7.5b. Taking the sum of the forces and moments for each link according to the conditions $\sum \mathbf{F} = m\mathbf{a}$ and $\sum \mathbf{M} = I\alpha$ produces two dynamic equations for each link. Expanding these two equations and separating the force equation into two equations, where $\sum F_x = ma_{CG_x}$ and $\sum F_y = ma_{CG_y}$, ultimately produces three dynamic equations for each link.

Equation 7.26 includes the center of gravity-load point vectors used for the Watt II mechanism.

$$\mathbf{R}_j = R_{jx} + iR_{jy}, \quad j = 1, 2, ..., 13 \tag{7.26}$$

---

[*] If the user prefers not to consider the applied force $\mathbf{F}_{p_1}$ or the applied torques T_{b_0} or $T_{b_0^*}$, these quantities can be specified as zero in the dynamic load equations.

Because the *x*- and *y*-components of the total acceleration of the center of gravity of each link are required in the dynamic load equations, Equation 7.27 includes the acceleration equations for $\mathbf{a}_{CG_4}$ and $\mathbf{a}_{CG_5}$. These acceleration equations are identical in form to the planar four-bar acceleration equations in Section 4.4.3. The acceleration equations for $\mathbf{a}_{CG_1}$ through $\mathbf{a}_{CG_3}$ are identical to Equation 7.2, since the corresponding links produce a planar four-bar mechanism.

$$
\begin{aligned}
\mathbf{a}_{CG_4} &= \mathbf{A}_{\mathbf{a}_1^*} + i\ddot{\alpha}_j^*(-\mathbf{R}_9)e^{i(\alpha_j^*)} - (\dot{\alpha}_j^*)^2(-\mathbf{R}_9)e^{i(\alpha_j^*)} \\
&= i\ddot{\gamma}_j(-\mathbf{R}_7 + \mathbf{R}_8)e^{i(\gamma_j)} - \dot{\gamma}_j^2(-\mathbf{R}_7 + \mathbf{R}_8)e^{i(\gamma_j)} + i\ddot{\alpha}_j^*(-\mathbf{R}_9)e^{i(\alpha_j^*)} - (\dot{\alpha}_j^*)^2(-\mathbf{R}_9)e^{i(\alpha_j^*)} \quad (7.27) \\
\mathbf{a}_{CG_5} &= i\ddot{\gamma}_j^*(-\mathbf{R}_{13})e^{i(\gamma_j^*)} - (\dot{\gamma}_j^*)^2(-\mathbf{R}_{13})e^{i(\gamma_j^*)}
\end{aligned}
$$

Because the Watt II mechanism includes planar four-bar mechanisms, the dynamic load equations given in Equations 7.4 and 7.6 are used for the crank and coupler links, respectively, of the planar four-bar mechanism loop $\mathbf{a}_0$–$\mathbf{a}_1$–$\mathbf{b}_1$–$\mathbf{b}_0$ in the Watt II mechanism.

Using the first two dynamic load conditions for the follower link of the planar four-bar mechanism loop $\mathbf{a}_0$–$\mathbf{a}_1$–$\mathbf{b}_1$–$\mathbf{b}_0$ (where the moment sum is taken about CG_3) produces

$$
\begin{aligned}
\mathbf{F}_{\mathbf{b}_0} + \mathbf{F}_{\mathbf{b}_1} + \mathbf{F}_{\mathbf{a}_1^*} + m_3\mathbf{g} &= m_3\mathbf{a}_{CG_3} \\
\mathbf{T}_{\mathbf{b}_0} + \mathbf{R}_6 \times \mathbf{F}_{\mathbf{b}_1} + \mathbf{R}_7 \times \mathbf{F}_{\mathbf{b}_0} + \mathbf{R}_8 \times \mathbf{F}_{\mathbf{a}_1^*} &= I_3\ddot{\beta}_j^*
\end{aligned}
\quad (7.28)
$$

Expanding and separating Equation 7.28 produces

$$
\begin{aligned}
F_{b_0x} + F_{b_1x} + F_{a_1^*x} &= m_3 a_{CG_3x} \\
F_{b_0y} + F_{b_1y} + F_{a_1^*y} + m_3 g &= m_3 a_{CG_3y} \\
T_{b_0} - F_{b_1x}R_{6y} + F_{b_1y}R_{6x} - F_{b_0x}R_{7y} + F_{b_0y}R_{7x} - F_{a_1^*x}R_{8y} + F_{a_1^*y}R_{8x} &= I_3\ddot{\beta}_j^*
\end{aligned}
\quad (7.29)
$$

Using the first two dynamic load conditions for the coupler link of the planar four-bar mechanism loop $\mathbf{b}_0 - \mathbf{a}_1^* - \mathbf{b}_1^* - \mathbf{b}_0^*$ (where the moment sum is taken about CG_4) produces

$$
\begin{aligned}
-\mathbf{F}_{\mathbf{a}_1^*} - \mathbf{F}_{\mathbf{b}_1^*} + \mathbf{F}_{\mathbf{p}_1^*} + m_4\mathbf{g} &= m_4\mathbf{a}_{CG_4} \\
-\mathbf{R}_9 \times \mathbf{F}_{\mathbf{a}_1^*} - \mathbf{R}_{10} \times \mathbf{F}_{\mathbf{b}_1^*} + \mathbf{R}_{11} \times \mathbf{F}_{\mathbf{p}_1^*} &= I_4\ddot{\alpha}_j^*
\end{aligned}
\quad (7.30)
$$

Expanding and separating Equation 7.30 produces

$$
\begin{aligned}
-F_{a_1^*x} - F_{b_1^*x} + F_{p_1^*x} &= m_4 a_{CG_4x} \\
-F_{a_1^*y} - F_{b_1^*y} + F_{p_1^*y} + m_4 g &= m_4 a_{CG_4y} \\
F_{a_1^*x}R_{9y} - F_{a_1^*y}R_{9x} + F_{b_1^*x}R_{10y} - F_{b_1^*y}R_{10x} - F_{p_1^*x}R_{11y} + F_{p_1^*y}R_{11x} &= I_4\ddot{\alpha}_j^*
\end{aligned}
\quad (7.31)
$$

Using the first two dynamic load conditions for the follower link of the planar four-bar mechanism loop $\mathbf{b}_0 - \mathbf{a}_1^* - \mathbf{b}_1^* - \mathbf{b}_0^*$ (where the moment sum is taken about $CG5$) produces

$$
\begin{aligned}
\mathbf{F}_{\mathbf{b}_0^*} + \mathbf{F}_{\mathbf{b}_1^*} + m_5\mathbf{g} &= m_5\mathbf{a}_{CG_5} \\
\mathbf{T}_{\mathbf{b}_0^*} + \mathbf{R}_{12} \times \mathbf{F}_{\mathbf{b}_1^*} + \mathbf{R}_{13} \times \mathbf{F}_{\mathbf{b}_0^*} &= I_5\ddot{\gamma}_j^*
\end{aligned}
\quad (7.32)
$$

Expanding and separating Equation 7.32 produces

$$F_{b_0^*x} + F_{b_1^*x} = m_5 a_{CG5x}$$

$$F_{b_0^*y} + F_{b_1^*y} + m_5 g = m_5 a_{CG5y} \tag{7.33}$$

$$T_{b_0} - F_{b_1^*x} R_{12y} + F_{b_1^*y} R_{12x} - F_{b_0^*x} R_{13y} + F_{b_0^*y} R_{13x} = I_5 \ddot{\gamma}_j^*$$

Expressing Equations 7.4, 7.6, 7.29, 7.31 and 7.33 in a combined matrix form produces

$$
\begin{bmatrix}
0 & 1 & 0 & 1 & 0 & 0 & 0 & 0 & 0 & 0 & 0 & 0 & 0 & 0 & 0 \\
0 & 0 & 1 & 0 & 1 & 0 & 0 & 0 & 0 & 0 & 0 & 0 & 0 & 0 & 0 \\
1 & -R_{1y\beta_j} & R_{1x\beta_j} & -R_{2y\beta_j} & R_{2x\beta_j} & 0 & 0 & 0 & 0 & 0 & 0 & 0 & 0 & 0 & 0 \\
0 & 0 & 0 & -1 & 0 & 0 & 0 & -1 & 0 & 0 & 0 & 0 & 0 & 0 & 0 \\
0 & 0 & 0 & 0 & -1 & 0 & 0 & 0 & -1 & 0 & 0 & 0 & 0 & 0 & 0 \\
0 & 0 & 0 & R_{3y\alpha_j} & -R_{3x\alpha_j} & 0 & 0 & R_{4y\alpha_j} & -R_{4x\alpha_j} & 0 & 0 & 0 & 0 & 0 & 0 \\
0 & 0 & 0 & 0 & 0 & 1 & 0 & 1 & 0 & 1 & 0 & 0 & 0 & 0 & 0 \\
0 & 0 & 0 & 0 & 0 & 0 & 1 & 0 & 1 & 0 & 1 & 0 & 0 & 0 & 0 \\
0 & 0 & 0 & 0 & 0 & -R_{7y\gamma_j} & R_{7x\gamma_j} & -R_{6y\gamma_j} & R_{6x\gamma_j} & -R_{8y\gamma_j} & R_{8x\gamma_j} & 0 & 0 & 0 & 0 \\
0 & 0 & 0 & 0 & 0 & 0 & 0 & 0 & 0 & -1 & 0 & 0 & 0 & -1 & 0 \\
0 & 0 & 0 & 0 & 0 & 0 & 0 & 0 & 0 & 0 & -1 & 0 & 0 & 0 & -1 \\
0 & 0 & 0 & 0 & 0 & 0 & 0 & 0 & 0 & R_{9y\alpha_j^*} & -R_{9x\alpha_j^*} & 0 & 0 & -R_{10y\alpha_j^*} & R_{10x\alpha_j^*} \\
0 & 0 & 0 & 0 & 0 & 0 & 0 & 0 & 0 & 0 & 0 & 1 & 0 & 1 & 0 \\
0 & 0 & 0 & 0 & 0 & 0 & 0 & 0 & 0 & 0 & 0 & 0 & 1 & 0 & 1 \\
0 & 0 & 0 & 0 & 0 & 0 & 0 & 0 & 0 & 0 & 0 & -R_{13y\gamma_j^*} & R_{13x\gamma_j^*} & -R_{12y\gamma_j^*} & R_{12x\gamma_j^*}
\end{bmatrix}
$$

$$
\times
\begin{Bmatrix}
T_{a_0} \\
F_{a_0x} \\
F_{a_0y} \\
F_{a_1x} \\
F_{a_1y} \\
F_{b_0x} \\
F_{b_0y} \\
F_{b_1x} \\
F_{b_1y} \\
F_{a_1^*x} \\
F_{a_1^*y} \\
F_{b_0^*x} \\
F_{b_0^*y} \\
F_{b_1^*x} \\
F_{b_1^*y}
\end{Bmatrix}
=
\begin{bmatrix}
m_1 a_{CG1x} \\
m_1 a_{CG1y} - m_1 g \\
I_1 \ddot{\beta}_j \\
m_2 a_{CG2x} - F_{p1x} \\
m_2 a_{CG2y} - m_2 g - F_{p1y} \\
I_2 \ddot{\alpha}_j + F_{p1x} R_{5y\alpha_j} - F_{p1y} R_{5x\alpha_j} \\
m_3 a_{CG3x} \\
m_3 a_{CG3y} - m_3 g \\
I_3 \ddot{\gamma}_j - T_{b_0} \\
m_4 a_{CG4x} - F_{p1^*x} \\
m_4 a_{CG4y} - m_4 g - F_{p1^*y} \\
I_4 \ddot{\alpha}_j^* + F_{p1^*x} R_{11y\alpha_j^*} - F_{p1^*y} R_{11x\alpha_j^*} \\
m_5 a_{CG5x} \\
m_5 a_{CG5y} - m_5 g \\
I_5 \ddot{\gamma}_j^* - T_{b_0^*}
\end{bmatrix}
\tag{7.34}
$$

where the x- and y-components of vectors $\mathbf{R}_1$ through $\mathbf{R}_{13}$ are $R_{nx_{anglej}} = R_{nx}\cos(angle_j) - R_{ny}\sin(angle_j)$ and $R_{nx_{anglej}} = R_{nx}\sin(angle_j) + R_{ny}\cos(angle_j)$. It can be seen in Equations 7.4, 7.6, 7.29, 7.31, 7.33, and ultimately in Equation 7.34 that the mass and mass moment of inertia of each link are also required.

Equation 7.34 can be solved using Cramer's rule to determine the unknown forces and torque. The unknown Watt II displacement angles (α_j, γ_j, α_j^* and γ_j^*), angular velocities ($\dot{\alpha}_j$, $\dot{\gamma}_j$, $\dot{\alpha}_j^*$ and $\dot{\gamma}_j^*$), and angular accelerations ($\ddot{\alpha}_j$, $\ddot{\gamma}_j$, $\ddot{\alpha}_j^*$ and $\ddot{\gamma}_j^*$) are calculated from the planar four-bar equations in Sections 4.4.1–4.4.4. Given such solutions, the corresponding dynamic forces and torques can be calculated from Equation 7.34.

Appendix D.5 includes the MATLAB file user instructions for Watt II dynamic force analysis. In this MATLAB file (which is available for download at www.crcpress.com/ product/isbn/9781498724937), solutions for Equation 7.34 are calculated.

Example 7.4

Problem Statement: Using the Appendix D.5 MATLAB file, calculate the forces at joints a_1, b_1, a_1^*, and b_1^* over a complete crank rotation range (at $-60°$ increments) for the Watt II mechanism in Tables E.7.7 and E.7.8. For this mechanism, $\mathbf{F}_{p_1} = (-500, -500)\,\mathrm{N}$, $\mathbf{F}_{p_1}^* = (-1000, 0)\,\mathrm{N}$, $\dot{\beta}_0 = -1\,\mathrm{rad/s}$, and $\ddot{\beta} = -0.25\,\mathrm{rad/s^2}$. Also $T_{b_0} = T_{b_0}^* = 0$ and gravity is $-9.81\,\mathrm{m/s^2}$.

Known Information: Tables E.7.7, E.7.8, and Appendix D.5 MATLAB file

Solution Approach: Figure E.7.7 includes the input specified (in bold text) in the Appendix D.5 MATLAB file. Vectors $\mathbf{R}_2$, $\mathbf{R}_4$, $\mathbf{R}_5$, $\mathbf{R}_6$, $\mathbf{R}_8$, $\mathbf{R}_{10}$, $\mathbf{R}_{11}$, and $\mathbf{R}_{12}$ are calculated in the MATLAB file using vector-loop equations (see Appendix D.5). Table E.7.9 includes the forces calculated for the Watt II mechanism using the Appendix D.5 MATLAB file.

TABLE E.7.7

Watt II Mechanism Dimensions (with Link Lengths in m)

W_1, θ	V_1, ρ	U_1, σ	G_{1x}, G_{1y}	L_1, δ
0.5, 90°	0.75, 19.3737°	0.75, 93.2461°	0.75, 0	0.5, 60.7834°

W_1^*, θ^*	V_1^*, ρ^*	U_1^*, σ^*	G_{1x}^*, G_{1y}^*	L_1^*, δ^*
0.5, 45°	0.75, 7.9416°	0.75, 60.2717°	0.7244, −0.1941	0.5, 49.3512°

TABLE E.7.8

Watt II Mechanism Dynamic Parameters (with Length in m, Mass in kg, and Inertia in kg–m²)

$\mathbf{R}_1$	$0 - i0.25$	m_1	8.05
$\mathbf{R}_3$	$-0.3172 - i0.2284$	I_1	0.805
$\mathbf{R}_7$	$-0.1037 - i0.3675$	m_2	29.785
$\mathbf{R}_9$	$-0.3562 - i0.161$	I_2	5.635
$\mathbf{R}_{13}$	$-0.1860 - i0.3257$	m_3	33.81
		I_3	5.635
		m_4	29.785
		I_4	5.635
		m_5	12.075
		I_5	2.415

```
W1 = 0.5*exp(i*90*pi/180);
V1 = 0.75*exp(i*19.3737*pi/180);
G1 = 0.75;
U1 = 0.75*exp(i*93.2461*pi/180);
L1 = 0.5*exp(i*60.7834*pi/180);

W1s = 0.5*exp(i*45*pi/180);
V1s = 0.75*exp(i*7.9416*pi/180);
G1s = 0.7244 - i*0.1941;
U1s = 0.75*exp(i*60.2717*pi/180);
L1s = 0.5*exp(i*49.3512*pi/180);

Fp1 = [-500, -500]; Fp1s = [-1000, 0];

Tb0 = 0; Tbs0 = 0; g = -9.81;

R1 = -i*0.25;
R3 = -0.3172 - i*0.2284;
R7 = -0.1037 - i*0.3675;
R9 = -0.3562 - i*0.161;
R13 = -0.1860 - i*0.3257;

m1 = 8.05; I1 = 0.805;
m2 = 29.785; I2 = 5.635;
m3 = 33.81; I3 = 5.635;
m4 = 29.785; I4 = 5.635;
m5 = 12.075; I5 = 2.415;

start_ang   = 0;
step_ang    = -1;
stop_ang    = -360;

angular_vel  = -1 * ones(N+1,1);
angular_acc  = -0.25 * ones(N+1,1);
```

FIGURE E.7.7

Specified input (in bold text) in the Appendix D.5 MATLAB file for Example 7.4.

TABLE E.7.9

Watt II mechanism forces (N)

β (°)	F_{a_1}	F_{b_1}	$F_{a_1^*}$	$F_{b_1^*}$
0	−689, −855	187, 75	−1186, −685	191, 396
−60	164, −488	−610, −304	−1434, −943	436, 621
−120	1087, −3122	−1603, 2282	−1063, −634	4, 353
−180	−54, −1917	−450, 1098	−946, −639	−63, 344
−240	−885, −1616	359, 805	−970, −637	−45, 343
−300	−1189, −1140	651, 364	−1047, −621	34, 345
−360	−669, −820	173, 80	−1153, −672	180, 387

7.7 Stephenson III Mechanism Analysis

Figure 7.6 a illustrates a Stephenson III mechanism where a force $F_{p_1^*}$ is applied to the intermediate link point p_1^*. To drive the mechanism, a torque T_{a_0} is applied about the crank link revolute joint. The user also has the option of specifying torques T_{b_0} and $T_{b_0^*}$ about

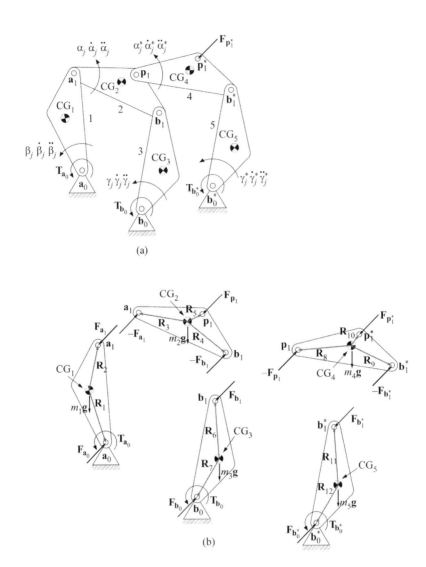

FIGURE 7.6
(a) Stephenson III mechanism and (b) link dynamic forces and torques.

the revolute joints $\mathbf{b}_0$ and $\mathbf{b}_0^*$, respectively.* The forces on the individual Stephenson III mechanism links are illustrated in Figure 7.6b. Taking the sum of the forces and moments for each link according to the conditions $\Sigma\mathbf{F} = m\mathbf{a}$ and $\Sigma\mathbf{M} = I\alpha$ produces two dynamic equations for each link. Expanding these two equations and separating the force equation into two equations, where $\Sigma F_x = ma_{CG_x}$ and $\Sigma F_y = ma_{CG_y}$, ultimately produces three dynamic equations for each link.

Equation 7.35 includes the center of gravity-load point vectors used for the Stephenson III mechanism.

---

* If the user prefers not to consider the applied force $\mathbf{F}_{p1}$ or the applied torques $\mathbf{T}_{b_0}$ or $\mathbf{T}_{b_0^*}$, these quantities can be specified as zero in the dynamic load equations.

$$\mathbf{R}_j = R_{jx} + iR_{jy}, \quad j = 1, 2, \ldots, 12 \tag{7.35}$$

Because the x- and y-components of the total acceleration of the center of gravity of each link are required in the dynamic load equations, Equation 7.36 includes the acceleration equations for $\mathbf{a}_{CG_4}$ and $\mathbf{a}_{CG_5}$. These acceleration equations are identical in form to the Stephenson III acceleration equations in Section 4.8.3. The acceleration equations for $\mathbf{a}_{CG_1}$ through $\mathbf{a}_{CG_3}$ are identical to Equation 7.2, since the corresponding links produce a planar four-bar mechanism.

$$\mathbf{a}_{CG_4} = \mathbf{A}_{p1} + i\ddot{\alpha}_j^* \left(-\mathbf{R}_8\right) e^{i\left(\alpha_j^*\right)} - \left(\dot{\alpha}_j^*\right)^2 \left(-\mathbf{R}_8\right) e^{i\left(\alpha_j^*\right)} = i\ddot{\beta}_j W_1 e^{i\left(\theta + \beta_j\right)} - \dot{\beta}_j^2 W_1 e^{i\left(\theta + \beta_j\right)}$$

$$+ i\ddot{\alpha} L_1 e^{i\left(\delta + \alpha_j\right)} - \left(\dot{\alpha}_j\right)^2 L_1 e^{i\left(\delta + \alpha_j\right)} - i\ddot{\alpha}_j^* \mathbf{R}_8 e^{i\left(\alpha_j^*\right)} + \left(\dot{\alpha}_j\right)^2 \mathbf{R}_8 e^{i\left(\alpha_j^*\right)} \tag{7.36}$$

$$\mathbf{a}_{CG_5} = i\ddot{\gamma}_j^* \left(-\mathbf{R}_{12}\right) e^{i\left(\gamma_j^*\right)} - \left(\dot{\gamma}_j^*\right) \left(-\mathbf{R}_{12}\right) e^{i\left(\gamma_j^*\right)}$$

Because the Stephenson III mechanism includes the planar four-bar mechanism, the dynamic load equations given in Equations 7.4, 7.6, and 7.8 are used for the crank, coupler, and follower links, respectively, of the planar four-bar mechanism included in the Stephenson III mechanism.

Using the first two dynamic load conditions for the intermediate link $\mathbf{p}_1 - \mathbf{b}_1^*$ (where the moment sum is taken about CG_4) produces

$$-\mathbf{F}_{p_1} - \mathbf{F}_{b_1^*} + \mathbf{F}_{p_1^*} + m_4 \mathbf{g} = m_4 \mathbf{a}_{CG_4}$$

$$-\mathbf{R}_8 \times \mathbf{F}_{p_1} - \mathbf{R}_9 \times \mathbf{F}_{b_1^*} + \mathbf{R}_{10} \times \mathbf{F}_{p_1^*} = I_4 \ddot{\alpha}_j^* \tag{7.37}$$

Expanding and separating Equation 7.37 produces

$$-F_{p_1 x} - F_{b_1^* x} + F_{p_1^* x} = m_4 a_{CG_4 x}$$

$$-F_{p_1 y} - F_{b_1^* y} + F_{p_1^* y} + m_4 g = m_4 a_{CG_4 y} \tag{7.38}$$

$$F_{p_1 x} R_{8y} - F_{p_1 y} R_{8x} + F_{b_1^* x} R_{9y} - F_{b_1^* y} R_{9x} - F_{p_1^* x} R_{10y} + F_{p_1^* y} R_{10x} = I_4 \ddot{\alpha}_j^*$$

Using the first two dynamic load conditions for the grounded link $\mathbf{b}_0^* - \mathbf{b}_1^*$ (where the moment sum is taken about CG_5) produces

$$\mathbf{F}_{b_0^*} + \mathbf{F}_{b_1^*} + m_5 \mathbf{g} = m_5 \mathbf{a}_{CG_5}$$

$$\mathbf{T}_{b_0^*} + \mathbf{R}_{11} \times \mathbf{F}_{b_1^*} + \mathbf{R}_{12} \times \mathbf{F}_{b_0^*} = I_5 \ddot{\gamma}_j^* \tag{7.39}$$

Expanding and separating Equation 7.39 produces

$$F_{b_0^* x} + F_{b_1^* x} = m_5 a_{CG_5 x}$$

$$F_{b_0^* y} + F_{b_1^* y} + m_5 g = m_5 a_{CG_5 y} \tag{7.40}$$

$$T_{b_0} - F_{b_1^* x} R_{11y} + F_{b_1^* y} R_{11x} - F_{b_0^* x} R_{12y} + F_{b_0^* y} R_{12x} = I_5 \ddot{\gamma}_j^*$$

Expressing Equations 7.4, 7.6, 7.8, 7.38, and 7.40 in a combined matrix form produces

$$
\begin{bmatrix}
0 & 1 & 0 & 1 & 0 & 0 & 0 & 0 & 0 & 0 & 0 & 0 & 0 & 0 & 0 \\
0 & 0 & 1 & 0 & 1 & 0 & 0 & 0 & 0 & 0 & 0 & 0 & 0 & 0 & 0 \\
1 & -R_{1y\beta_j} & R_{1x\beta_j} & -R_{2y\beta_j} & R_{2x\beta_j} & 0 & 0 & 0 & 0 & 0 & 0 & 0 & 0 & 0 & 0 \\
0 & 0 & 0 & -1 & 0 & 0 & 0 & -1 & 0 & 1 & 0 & 0 & 0 & 0 & 0 \\
0 & 0 & 0 & 0 & -1 & 0 & 0 & 0 & -1 & 0 & 1 & 0 & 0 & 0 & 0 \\
0 & 0 & 0 & R_{3y\alpha_j} & -R_{3x\alpha_j} & 0 & 0 & R_{4y\alpha_j} & -R_{4x\alpha_j} & -R_{5y\alpha_j} & R_{5x\alpha_j} & 0 & 0 & 0 & 0 \\
0 & 0 & 0 & 0 & 0 & 1 & 0 & 1 & 0 & 0 & 0 & 0 & 0 & 0 & 0 \\
0 & 0 & 0 & 0 & 0 & 0 & 1 & 0 & 1 & 0 & 0 & 0 & 0 & 0 & 0 \\
0 & 0 & 0 & 0 & 0 & -R_{7y\gamma_j} & R_{7x\gamma_j} & -R_{6y\gamma_j} & R_{6x\gamma_j} & 0 & 0 & 0 & 0 & 0 & 0 \\
0 & 0 & 0 & 0 & 0 & 0 & 0 & 0 & 0 & -1 & 0 & 0 & 0 & -1 & 0 \\
0 & 0 & 0 & 0 & 0 & 0 & 0 & 0 & 0 & 0 & -1 & 0 & 0 & 0 & -1 \\
0 & 0 & 0 & 0 & 0 & 0 & 0 & 0 & 0 & R_{8y\alpha_j^*} & -R_{8x\alpha_j^*} & 0 & 0 & R_{9y\alpha_j^*} & -R_{9x\alpha_j^*} \\
0 & 0 & 0 & 0 & 0 & 0 & 0 & 0 & 0 & 0 & 0 & 1 & 0 & 1 & 0 \\
0 & 0 & 0 & 0 & 0 & 0 & 0 & 0 & 0 & 0 & 0 & 0 & 1 & 0 & 1 \\
0 & 0 & 0 & 0 & 0 & 0 & 0 & 0 & 0 & 0 & 0 & -R_{12y\gamma_j^*} & R_{12x\gamma_j^*} & -R_{11y\gamma_j^*} & R_{11x\gamma_j^*}
\end{bmatrix}
$$

$$
\times
\begin{Bmatrix}
T_{a_0} \\
F_{a_{0x}} \\
F_{a_{0y}} \\
F_{a_{1x}} \\
F_{a_{1y}} \\
F_{b_{0x}} \\
F_{b_{0y}} \\
F_{b_{1x}} \\
F_{b_{1y}} \\
F_{p_{1x}} \\
F_{p_{1y}} \\
F_{b_{0x}^*} \\
F_{b_{0y}^*} \\
F_{b_{1x}^*} \\
F_{b_{1y}^*}
\end{Bmatrix}
=
\begin{bmatrix}
m_1 a_{CG1x} \\
m_1 a_{CG1y} - m_1 g \\
I_1 \ddot{\beta}_j \\
m_2 a_{CG2x} \\
m_2 a_{CG2y} - m_2 g \\
I_2 \ddot{\alpha}_j \\
m_3 a_{CG3x} \\
m_3 a_{CG3y} - m_3 g \\
I_3 \ddot{\gamma}_j - T_{b_0} \\
m_4 a_{CG4x} - F_{p_{1x}}^* \\
m_4 a_{CG4y} - m_4 g - F_{p_{1y}}^* \\
I_4 \ddot{\alpha}_j^* + F_{p_{1x}}^* R_{10y\alpha_j^*} - F_{p_{1y}}^* R_{10x\alpha_j^*} \\
m_5 a_{CG5x} \\
m_5 a_{CG5y} - m_5 g \\
I_5 \ddot{\gamma}_j^* - T_{b_0}^*
\end{bmatrix}
\tag{7.41}
$$

where the x- and y-components of vectors $\mathbf{R}_1$ through $\mathbf{R}_{12}$ are $R_{nx_{angle\,j}} = R_{nx} \cos(angle_j) - R_{ny} \sin(angle_j)$ and $R_{ny_{angle\,j}} = R_{nx} \sin(angle_j) + R_{ny} \cos(angle_j)$. It can be seen in Equations 7.4, 7.6, 7.8, 7.38, 7.40, and ultimately in Equation 7.41 that the mass and mass moment of inertia of each link are also required.

Equation 7.41 can be solved using Cramer's rule to determine the unknown forces and torques. The unknown Stephenson III displacement angles α_j^* and γ_j^*, angular velocities

$\dot{\alpha}_j^*$ and $\dot{\gamma}_{j}^*$, and angular accelerations $\ddot{\alpha}_j^*$ and $\ddot{\gamma}_j^*$ are the same quantities calculated from the Stephenson III equations in Sections 4.8.1–4.8.4. Given such solutions (along with the solutions for the planar four-bar mechanism), the corresponding dynamic forces and torques can be calculated from Equation 7.41.

Appendix D.6 includes the MATLAB file user instructions for Stephenson III dynamic force analysis. In this MATLAB file (which is available for download at www.crcpress.com/product/isbn/9781498724937), solutions for Equation 7.41 are calculated.

Example 7.5

Problem Statement: Using the Appendix D.6 MATLAB file, calculate the forces at joints $\mathbf{a}_1$, $\mathbf{b}_1$, $\mathbf{p}_1$, and $\mathbf{b}_1^*$ over a complete crank rotation range (at $-60°$ increments) for the Stephenson III mechanism in Tables E.7.10 and E.7.11. In this mechanism, $\mathbf{F}_{\mathbf{p}_1^*} = (0, -1000)\,\mathrm{N}$, $\dot{\beta}_0 = -1\,\mathrm{rad/s}$, and $\ddot{\beta} = -0.25\,\mathrm{rad/s^2}$. Also $T_{\mathbf{b}_0} = T_{\mathbf{b}_0^*} = 0$ and gravity is $-9.81\,\mathrm{m/s^2}$.

Known Information: Tables E.7.10, E.7.11, and Appendix D.6 MATLAB file.

Solution Approach: Figure E.7.8 includes the input specified (in bold text) in the Appendix D.6 MATLAB file. Vectors $\mathbf{R}_2$, $\mathbf{R}_4$, $\mathbf{R}_5$, $\mathbf{R}_6$, $\mathbf{R}_9$, $\mathbf{R}_{10}$, and $\mathbf{R}_{11}$ are calculated in the MATLAB file using vector-loop equations (see Appendix D.6). Table E.7.12 includes the forces calculated for the Stephenson III mechanism using the Appendix D.6 MATLAB file.

TABLE E.7.10

Stephenson III Mechanism Dimensions (with Link Lengths in m)

W_1, θ	V_1, ρ	U_1, σ	G_{1x}, G_{1y}	L_1, δ
0.5, 90°	0.75, 19.3737°	0.75, 93.2461°	0.75, 0	0.5, 60.7834°

V_1^*, ρ^*	L_1^*, δ^*	U_1^*, σ^*	G_{1x}^*, G_{1y}^*	
1, 17.1417°	0.5, 63.7091°	1, 76.4844°	0.2159, 0.2588	

TABLE E.7.11

Stephenson III Mechanism Dynamic Parameters (with Length in m, Mass in kg, and Inertia in kg–m²)

$\mathbf{R}_1$	$0 - i0.25$	m_1	8.05
$\mathbf{R}_3$	$-0.3172 - i0.2284$	I_1	0.805
$\mathbf{R}_7$	$0.0212 - i0.3744$	m_2	29.785
$\mathbf{R}_8$	$-0.3923 - i0.2477$	I_2	5.635
$\mathbf{R}_{12}$	$-0.1169 - i0.4862$	m_3	12.075
		I_3	2.415
		m_4	43.47
		I_4	16.1
		m_5	15.925
		I_5	5.635

```
W1 = 0.5*exp(i*90*pi/180);
V1 = 0.75*exp(i*19.3737*pi/180);
G1 = 0.75;
U1 = 0.75*exp(i*93.2461*pi/180);
L1 = 0.5*exp(i*60.7834*pi/180);

V1s = 1*exp(i*17.1417*pi/180);
G1s = 0.2159 + i*0.2588;
U1s = exp(i*76.4844*pi/180);
L1s = 0.5*exp(i*63.7091*pi/180);

Fp1s = [0, -1000];

Tb0 = 0; Tbs0 = 0; g = -9.81;

R1 = -i*0.25;
R3 = -0.3172 - i*0.2284;
R7 = 0.0212 - i*0.3744;
R8 = -0.3923 - i*0.2477;
R12 = -0.1169 - i*0.4862;

m1 = 8.05; I1 = 0.805;
m2 = 29.785; I2 = 5.635;
m3 = 12.075; I3 = 2.415;
m4 = 43.47; I4 = 16.1;
m5 = 15.295; I5 = 5.635;

start_ang    = 0;
step_ang     = -1;
stop_ang     = -360;

angular_vel  = -1 * ones(N+1,1);
angular_acc  = -0.25 * ones(N+1,1);
```

FIGURE E.7.8

Specified input (in bold text) in the Appendix D.5 MATLAB file for Example 7.4.

TABLE E.7.12

Stephenson III Mechanism Forces (N)

β (°)	F_{a_1}	F_{b_1}	F_{p_1}	$F_{b_1'}$
0	90, −722	34, −527	126, −968	−124, −442
−60	1219, 135	−551, −1208	613, −781	−469, −661
−120	2160, −3768	−1977, 1528	200, −1900	−262, 360
−180	725, −2221	−650, 291	79, −1611	−99, 147
−240	−177, −1802	30, 52	−121, −1439	83, 0
−300	−466, −1217	280, −245	−148, −1185	104, −211
−360	123, −645	31, −517	151, −922	−127, −438

7.8 Mass Moment of Inertia and Computer-Aided Design Software

The *mass moment of inertia* is a measure of a solid object's resistance to change in rotational speed about a particular axis of rotation. This is the quantity used in the dynamic load condition $\Sigma\mathbf{M} = I\alpha$. The mass moment of inertia of a rotating body is generally defined as

$$I = \int_m r^2 \, dm \tag{7.42}$$

or more specifically as

$$I_x = \int_m \left(y^2 + z^2\right) dm$$

$$I_y = \int_m \left(z^2 + x^2\right) dm \tag{7.43}$$

$$I_z = \int_m \left(x^2 + y^2\right) dm$$

where dm is the mass of an infinitesimally small part of the body and r, x, y, and z are the distances between dm and the axis of rotation (see Figure 7.7).*

Mass moment of inertia equations are available for a range of primitive solid shapes and a variety of rotation conditions [6]. Although such equations provide the user with a convenient means to calculate mass moments of inertia, the user is limited to the geometry and rotation condition considered in the equations.

With the development of *computer-aided design* (or *CAD*) software, specifically in the area of *solid modeling* (creating virtual representations of solid geometry), the user can

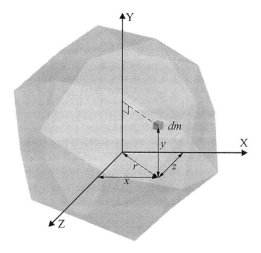

FIGURE 7.7
Arbitrary body with mass moment of inertia variables.

* Because only planar mechanisms are considered in this chapter, only link rotations about the z-axis and the subsequent moment of inertia about the z-axis (I_z in Equation 7.43) are considered.

now readily calculate the mass moment of inertia of any solid shape produced and under any rotation condition established.* With more accurate CAD-produced inertia values for mechanism components, the accuracy of calculated quantities such as mechanism dynamic forces and torques will also be improved. Mass moment of inertia calculation using CAD software is the preferred approach among engineers today—particularly when the mechanism geometry considered cannot be accurately represented by primitive geometry.

7.9 Planar Mechanism Dynamic Force Analysis and Modeling in Simmechanics®

As noted throughout this chapter, Appendices D.1–D.6 include user instructions for the planar four-bar, slider-crank, geared five-bar, Watt II, and Stephenson III mechanism MATLAB files, respectively. In these files, the dynamic force and torque equations formulated in this chapter are solved. These MATLAB files provide means for the user to efficiently conduct planar four-bar, slider- crank, geared five-bar, Watt II, and Stephenson III dynamic force analyses by solving their displacement, velocity, and acceleration equations along with their dynamic force and torque equations.

This textbook also utilizes SimMechanics as an alternate approach for simulation-based dynamic force analysis. A library of SimMechanics files is available for download at www.crcpress.com/ product/isbn/9781498724937 to conduct dynamic force analyses on the planar four-bar, slider-crank, geared five-bar, Watt II, and Stephenson III mechanisms. With these files, the user specifies the mechanism link dimensions, mass properties, applied loads, and driving link parameters, and the dynamic forces and torques at the mechanism locations of interest are measured. Additionally, the motion of the mechanism itself is simulated. The SimMechanics file user instructions for the planar four-bar, slider-crank, geared five-bar, Watt II, and Stephenson III mechanisms are given in Appendices J.1–J.6, respectively.

While the user can specify input according to any unit type for *length, force,* and *torque* in the Appendix D MATLAB files, only two groups of dimensions are available in the Appendix J MATLAB files. The user can specify length, mass, and force quantities in either *inch, pound-mass,* and *pound-force* respectively (the U.S. system), or *meter, kilogram,* and *Newton* respectively (the S.I. system). The user also has the option in the Appendix J files to convert U.S. system input to S.I. system output and vice versa.

Example 7.6

Problem Statement: Figure E.7.2 (in Example 7.1) includes plots for mechanism forces $F_{a_{0x}}$, $F_{a_{0y}}$, $F_{b_{0x}}$, and $F_{b_{0y}}$. Using the Appendix J.1 SimMechanics files, measure the maximum and minimum values (considering value magnitudes and not directions) for the planar four-bar mechanism forces in Example 7.1. Consider 1° crank rotation increments.

---

* Commercial CAD software producers include (but are by no means limited to) *Autodesk, PTC, Dassault Systemes,* and *Siemens PLM Software.*

Known Information: Example 7.1 and Appendix J.1 SimMechanics files.

Solution Approach: Figure E.7.9 includes the input specified (in bold text) in the Appendix J.1 MATLAB file. Table E.7.13 includes the maximum and minimum force values calculated using the Appendix D.1 MATLAB file and the Appendix J.1 SimMechanics files.

```
unit_select = 'SI';

W1 = 0.5*exp(i*90*pi/180);
V1 = 0.75*exp(i*19.3737*pi/180);
G1 = 0.75;
U1 = 0.75*exp(i*93.2461*pi/180);
L1 = 0.5*exp(i*60.7834*pi/180);

Fp1 = [0, 0];
Tb0 = 0;
g = -9.81;

R1 = -i*0.25;
R3 = -0.3172 - i*0.2284;
R7 = 0.0212 - i*0.3744;

m1 = 8.05;
I1 = 0.805;

m2 = 29.785;
I2 = 5.635;

m3 = 12.075;
I3 = 2.415;

start_ang    = 0;
step_ang     = 1;
stop_ang     = 360;

angular_vel  = 1;
angular_acc  = 0;
```

FIGURE E.7.9

Specified input (in bold text) in the Appendix J.1 MATLAB file for Example 7.6.

TABLE E.7.13

Minimum and Maximum Planar Four-Bar Mechanism Forces (N)

	β (°)	Appendix D.1 MATLAB File	β (°)	Appendix J.1 SimMechanics Files
$F_{a_{0x}\,max}$	247	−245.22	246.97	−245.23
$F_{a_{0x}\,min}$	168	−0.14	167.99	−0.11
$F_{a_{0y}\,max}$	257	806.31	256.98	806.31
$F_{a_{0y}\,min}$	284	10.19	283.98	10.19
$F_{b_{0x}\,max}$	250	279.36	250	279.36
$F_{b_{0x}\,min}$	166	−0.07	165.97	−0.13
$F_{b_{0y}\,max}$	286	463.81	286	463.81
$F_{b_{0y}\,min}$	214	−0.61	214	−0.60

Example 7.7

Problem Statement: The minimum and maximum torque values in Figure E.7.2 (in Example 7.2) appear at crank displacement angles of 239° and 304°, respectively. Using the Appendix J.2 SimMechanics files, measure the maximum and minimum torque values for the slider-crank mechanism in Example 7.2.

Known Information: Example 7.2 and Appendix J.2 SimMechanics files.

Solution Approach: Figure E.7.10 includes the input specified (in bold text) in the Appendix J.1 MATLAB file. Table E.7.14 includes the maximum and minimum torque values calculated using the Appendix D.2 MATLAB file and the Appendix J.2 SimMechanics files. While the Appendix D.2 file requires the user to run it twice (once using +μ and again using –μ) and then to filter the results (only keeping data rows where the sliding friction direction and the slider velocity direction are opposite), only a single run required for the Appendix J.2 files. This is because the latter includes filter (a decision operator that considers the sliding friction direction and the slider velocity direction) to exclude all invalid result data.

```
LW1 = 0.5;
theta = 90*pi/180;

LU1 = 0;

LV1 = 0.9014;

F = [0, 0];
mu = 0.5;
g = -9.81;

R1 = -i*0.25;
R3 = -0.3750 + i*0.25;

m1 = 8.05;
I1 = 0.805;

m2 = 14.49;
I2 = 4.025;

m3 = 30;

start_ang  = 0;
step_ang   = 1;
stop_ang   = 360;

angular_vel  = 10;
angular_acc  = 0;
```

FIGURE E.7.10
Specified input (in bold text) in the Appendix J.2 MATLAB file for Example 7.7.

TABLE E.7.14

Slider-Crank Mechanism Torque (in N-m)

Appendix D.2 MATLAB file	Appendix J.2 SimMechanics Files
$T_{a_0\min} = -612.53$ at $\beta = 239°$	$T_{a_0\min} = -612.51$ at $\beta = 239°$
$T_{a_0\max} = 940.87$ at $\beta = 304°$	$T_{a_0\max} = 940.83$ at $\beta = 304°$

Example 7.8

Problem Statement: Figure E.7.6 (in Example 7.3) includes plots for the driver torque T'_{a_0}. Using the Appendix J.4 SimMechanics files, measure the maximum and minimum values (considering value directions) for this torque in Example 7.1. Consider $1°$ crank rotation increments.

Known Information: Example 7.3 and Appendix J.4 SimMechanics files.

Solution Approach: Figure E.7.11 includes the input specified (in bold text) in the Appendix J.4 MATLAB file. Table E.7.15 includes the maximum and minimum torque values calculated using the Appendix D.4 MATLAB file and the Appendix J.4 SimMechanics files. Unlike the Appendix D.4 file, the mass (m_5) and inertia (I_5) of the idler gear are used in the Appendix J.4 files (see Appendix J.4). This makes the results from the latter more accurate.

```
unit_select = 'SI';

W1 = 0.5*exp(i*90*pi/180);
V1 = 0.75*exp(i*32.7304*pi/180);
G1 = 0.75*exp(i*0*pi/180);
U1 = 0.75*exp(i*45*pi/180);
L1 = 0.5*exp(i*74.14*pi/180);
S1 = 0.75*exp(i*149.9847*pi/180);

Fp1 = [0, -1000];
ratio = 2;
g = -9.81;

R1 = -i*0.0831;
R3 = -0.2558 - i*0.2955;
R7 = 0.3247 - i*0.1876;
R9 = -0.0356 - i*0.0356;

m1 = 22.54; I1 = 0.505;

m2 = 29.785; I2 = 5.635;

m3 = 12.075; I3 = 2.415;

m4 = 75.67; I4 = 5.635;

m5 = 22.54; I5 = 0.505;

start_ang   = 0;
step_ang    = 1;
stop_ang    = 360;

angular_vel  = 10;
angular_acc  = 0;
```

FIGURE E.7.11
Specified input (in bold text) in the Appendix J.4 MATLAB file for Example 7.8.

TABLE E.7.15

Geared Five-Bar (with 3 Gears) Mechanism Driver Torque (in N-m)

Appendix D.4 MATLAB File	Appendix J.4 SimMechanics Files (Where $m_5 = m_1$ and $I_5 = I_1$)
$T'_{a_0\text{max}} = 2618.4$ at $\beta = 252°$	$T'_{a_0\text{max}} = 2618.4$ at $\beta = 251.99°$
$T'_{a_0\text{min}} = -1386.5$ at $\beta = 290°$	$T'_{a_0\text{min}} = -1386.5$ at $\beta = 290.01°$

Example 7.9

Problem Statement: Repeat Example 7.4 using the Appendix J.5 SimMechanics files.

Known Information: Example 7.4 and Appendix J.5 SimMechanics files.

Solution Approach: Figure E.7.12 includes the input specified (in bold text) in the Appendix J.5 MATLAB file. Table E.7.16 includes the forces calculated for the Watt II mechanism using the Appendix J.5 SimMechanics files.

```
unit_select = 'SI';

W1 = 0.5*exp(i*90*pi/180);
V1 = 0.75*exp(i*19.3737*pi/180);
G1 = 0.75;
U1 = 0.75*exp(i*93.2461*pi/180);
L1 = 0.5*exp(i*60.7834*pi/180);

W1s = 0.5*exp(i*45*pi/180);
V1s = 0.75*exp(i*7.9416*pi/180);
G1s = 0.7244 - i*0.1941;
U1s = 0.75*exp(i*60.2717*pi/180);
L1s = 0.5*exp(i*49.3512*pi/180);

Fp1 = [-500, -500]; Fp1s = [-1000, 0];

Tb0 = 0; Tbs0 = 0; g = -9.81;

R1 = -i*0.25;
R3 = -0.3172 - i*0.2284;
R7 = -0.1037 - i*0.3675;
R9 = -0.3562 - i*0.161;
R13 = -0.1860 - i*0.3257;

m1 = 8.05; I1 = 0.805;
m2 = 29.785; I2 = 5.635;
m3 = 33.81; I3 = 5.635;
m4 = 29.785; I4 = 5.635;
m5 = 12.075; I5 = 2.415;

start_ang  = 0;
step_ang   = -60;
stop_ang   = -360;

angular_vel  = -1;
angular_acc  = -0.25;
```

FIGURE E.7.12
Specified input (in bold text) in the Appendix J.5 MATLAB file for Example 7.9.

TABLE E.7.16

Watt II Mechanism Forces (N)

β (°)	F_{a_1}	F_{b_1}	$F_{a_1^*}$	$F_{b_1^*}$
0	−688.85, −854.54	186.89, 74.82	−1185.5, −685.08	190.79, 395.58
−60.03	164.28, −488	−610.03, −303.93	−1433.4, −943.03	436.41, 621.4
−119.98	1087, −3123	−1603.6, 2283	−1063.3, −633.55	4.4, 353.01
−180.01	−54.46, −1916.7	−449.73, 1097.5	−945.63, −638.55	−63.3, 343.6
−239.99	−884.78, −1616.4	358.55, 805.15	−970.38, −636.66	−44.95, 342.65
−200.01	−1189.4, −1140.2	650.49, 363.57	−1047, −621.43	34.28, 345.36
−359.94	−670.23, −820.15	173.52, 79.92	−1153.1, −671.78	179.86, 386.55

Example 7.10

Problem Statement: Repeat Example 7.5 using the Appendix J.6 SimMechanics files.

Known Information: Example 7.5 and Appendix J.6 SimMechanics files.

Solution Approach: Figure E.7.13 includes the input specified (in bold text) in the Appendix J.6 MATLAB file. Table E.7.17 includes the forces calculated for the Stephenson III mechanism using the Appendix J.6 SimMechanics files.

```
unit_select = 'SI';

W1 = 0.5*exp(i*90*pi/180);
V1 = 0.75*exp(i*19.3737*pi/180);
G1 = 0.75;
U1 = 0.75*exp(i*93.2461*pi/180);
L1 = 0.5*exp(i*60.7834*pi/180);

V1s = 1*exp(i*17.1417*pi/180);
G1s = 0.2159 + i*0.2588;
U1s = exp(i*76.4844*pi/180);
L1s = 0.5*exp(i*63.7091*pi/180);

Fp1s = [0, -1000];

Tb0 = 0; Tb0s = 0; g = -9.81;

R1 = -i*0.25;
R3 = -0.3172 - i*0.2284;
R7 = 0.0212 - i*0.3744;
R8 = -0.3923 - i*0.2477;
R12 = -0.1169 - i*0.4862;

m1 = 8.05; I1 = 0.805;
m2 = 29.785; I2 = 5.635;
m3 = 12.075; I3 = 2.415;
m4 = 43.47; I4 = 16.1;
m5 = 15.295; I5 = 5.635;

start_ang   = 0;
step_ang    = -60;
stop_ang    = -360;

angular_vel  = -1;
angular_acc  = -0.25;
```

FIGURE E.7.13

Specified input (in bold text) in the Appendix J.6 MATLAB file for Example 7.10.

TABLE E.7.17

Stephenson III Mechanism Forces (N)

β (°)	F_{a_1}	F_{b_1}	F_{p_1}	$F_{b_1^*}$
0	90.14, −721.61	33.53, −526.57	125.62, −968.46	−124.04, −441.54
−59.99	1218.3, 135.01	−550.79, −1207.7	613.34, −780.76	−468.53, −660.51
−119.99	2160.3, −3768.2	−1976.8, 1528.7	200.02, −1899.6	−262.28, 359.84
−180.02	724.37, −2221	−649.45, 291.22	79.12, −1610.6	−98.84, 146.98
−240.01	−177.29, −1802.3	29.73, 51.89	−121.33, −1439.2	83.41, −0.14
−300.03	−466.26, −1216.6	279.81, −245.28	−147.58, −1185.3	104.4, −210.7
−359.94	122.58, −645.1	31.49, −516.87	150.76, −921.74	−126.66, −438.17

7.10 Summary

In terms of structural force analyses for mechanical systems, a dynamic force analysis is the next type of force analysis to consider beyond a static force analysis. In this type of analysis, loads are considered for each mechanism link according to Newton's second law ($\sum \mathbf{F} = m\mathbf{a}$, $\sum \mathbf{M} = I\alpha$). In this chapter, a system of dynamic force and moment equations are formulated for the planar four-bar, slider-crank, geared five-bar, Watt II, and Stephenson III mechanisms. These equations form sets of linear simultaneous equations that are solved to determine the dynamic forces and torques present at each mechanism joint. The Appendix D.1–D.6 MATLAB files provide a means for the user to efficiently conduct planar four-bar, slider-crank, geared five-bar, Watt II, and Stephenson III dynamic force analyses by solving their displacement, velocity, and acceleration from Chapter 4 along with their linear simultaneous equation sets.

This textbook also utilizes SimMechanics as an alternate approach for simulation-based dynamic load analyses. Using the Appendix J.1–J.6 SimMechanics files, the user can conduct dynamic load analyses on the planar four-bar, slider-crank, geared five-bar, Watt II, and Stephenson III mechanisms, respectively, as well as simulate mechanism motion.

References

1. Beer, F. P. and Johnston, E. R. 1988. *Vector Mechanics for Engineers: Statics and Dynamics.* 5th edn, Chapter 16. New York: McGraw-Hill.
2. Myszka, D. H. 2005. *Machines and Mechanisms: Applied Kinematic Analysis.* 3rd edn, Chapter 14. Saddle River, NJ: Prentice-Hall.
3. Norton, R. L. 2008. *Design of Machinery.* 4th edn, Chapter 11. New York: McGraw-Hill.
4. Wilson, C. E. and Sadler, J. P. 2003. *Kinematics and Dynamics of Machinery.* 3rd edn, Chapter 10. Saddle River, NJ: Prentice-Hall.
5. Waldron, K. J. and Kinzel, G. L. 2004. *Kinematics, Dynamics and Design of Machinery.* 2nd edn, Chapter 14. Saddle River, NJ: Prentice-Hall.
6. Lindeburg, M. R. 2001. *Mechanical Engineering Reference Manual for the PE Exam.* 11th edn, Chapter 58. Belmont, CA: Professional Publications.

Additional Reading

Hibbeler, R. C. 1998. *Engineering Mechanics: Dynamics.* 8th edn, Chapter 17. Upper Saddle River, NJ: Prentice-Hall.
McGill, D. J. and King, W. W. 1995. *Engineering Mechanics: An Introduction to Dynamics.* Chapter 2. Boston: PWS Publishing.

Problems

1. Figure P.7.1 illustrates a planar four-bar mechanism used to guide a hatch from the opened position to the closed position over a crank rotation of 31.7°. The dimensions

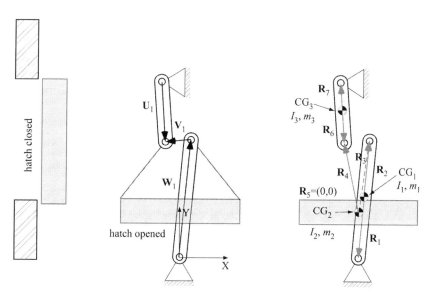

FIGURE P.7.1
Hatch mechanism.

TABLE P.7.1

Hatch Mechanism Dimensions (with Link Lengths in m)

W_1	V_1	U_1	G_1	L_1
$0.1658 + i3.2649$	$-0.7423 - i0.1494$	$-0.0462 - i1.6889$	$-0.5302 + i4.8044$	$-0.3024 - i1.9839$

TABLE P.7.2

Hatch Mechanism Dynamic Parameters (with Length in m, Mass in kg, and Inertia in kg-m²)

R_1	$-0.0829 - i1.6324$	m_1	2.8811
R_3	$0.3024 + i1.9839$	I_1	10.8216
R_7	$0.0231 + i0.8445$	m_2	56.0143
		I_2	281.0267
		m_3	1.6092
		I_3	1.7066

for the mechanism at the opened-hatch position are included in Table P.7.1 and the link dynamic parameters are included in Table P.7.2. Gravity is −9.81 m/s². Plot the driver torque over a crank rotation range of 31.7° with an initial crank angular velocity of 0 rad/s and an angular acceleration of 0.175 rad/s² (using the Appendix D.1 or J.1 files).

2. Repeat Problem 1 where a constant follower-link torque of $T_{b_0} = 10,000$ N-m is also applied.

3. Figure P.7.2 illustrates a planar four-bar mechanism used to guide a cutting blade from the opened position to the closed position over a crank rotation of −45°.

The dimensions for the mechanism at the opened-blade position are included in Table P.7.3 and the link dynamic parameters are included in Table P.7.4. Gravity is $-9.81\,\text{m/s}^2$. Plot the force components at the grounded crank revolute joint $\left(F_{a_{0x}}, F_{a_{0y}}\right)$ over a crank rotation range of $-45°$ with an initial crank angular velocity of $0\,\text{rad/s}$ and an angular acceleration of $-0.55\,\text{rad/s}^2$ (using the Appendix D.1 or J.1 files).

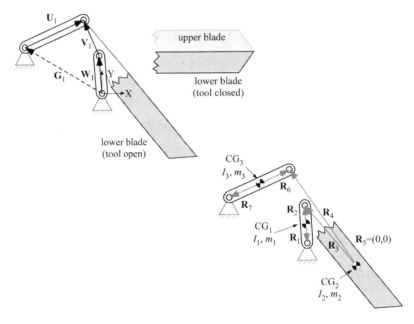

FIGURE P.7.2
Cutting mechanism.

TABLE P.7.3

Cutting Mechanism Dimensions (with Link Lengths in m)

W_1	V_1	U_1	G_1	L_1
$-0.0899 + i1.2955$	$-0.5523 + i1.237$	$2.1075 + i0.9489$	$-2.7497 + i1.5836$	$1.8547 - i2.0853$

TABLE P.7.4

Cutting Mechanism Dynamic Parameters (with Length in m, Mass in kg and Inertia in kg–m²)

R_1	$0.045 - i0.6478$	m_1	1.2985
R_3	$-1.8547 + i2.0853$	I_1	0.8477
R_7	$-1.0538 - i0.4744$	m_2	4.6682
		I_2	40.9753
		m_3	2.1139
		I_3	4.0685

4. Repeat Problem 3 where a constant follower-link torque of $T_{b_0} = 250$ N-m is also applied.

5. Figure P.7.3 illustrates a planar four-bar mechanism used to guide a brake pad from the released position to the applied position over a crank rotation of $40°$. The dimensions for the mechanism at the released-hatch position are included in Table P.7.5 and the link dynamic parameters are included in Table P.7.6. A force of $\mathbf{F}_{p_1} = (400, 0)$N is constantly applied and gravity is -9.81 m/s². Plot the driver torque over a crank rotation range of $40°$ with an initial crank angular velocity of 0 rad/s and an angular acceleration of 1 rad/s² (using the Appendix D.1 or J.1 files).

6. For Problem 5, plot the force components at the grounded follower-revolute joint $(F_{b_{0x}}, F_{b_{0y}})$.

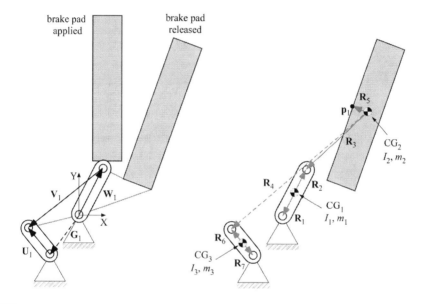

FIGURE P.7.3
Brake mechanism.

TABLE P.7.5

Brake Mechanism Dimensions (with Link Lengths in m)

$\mathbf{W_1}$	$\mathbf{V_1}$	$\mathbf{U_1}$	$\mathbf{G_1}$	$\mathbf{L_1}$
$0.5967 + i1.0893$	$-1.801 - i1.3932$	$-0.5121 + i0.6055$	$-0.6922 - i0.9094$	$1.0539 + i1.3509$

TABLE P.7.6

Brake Mechanism Dynamic Parameters (with Length in m, Mass in kg, and Inertia in kg–m²)

$\mathbf{R_1}$	$-0.2983 - i0.5446$	m_1	1.2526
$\mathbf{R_3}$	$-1.3766 - i1.2335$	I_1	0.7543
$\mathbf{R_7}$	$0.256 - i0.3028$	m_2	24.2732
		I_2	106.9483
		m_3	0.8911
		I_3	0.2439

7. Figure P.7.4 illustrates a planar four-bar mechanism used to guide a latch from the released position to the applied position over a crank rotation of 70°. The dimensions for the mechanism at the released-latch position are included in Table P.7.7 and the link dynamic parameters are included in Table P.7.8. Gravity is −9.81 m/s². Plot the driver torque over a crank rotation range of 70° with an initial crank angular velocity of 0 rad/s and an angular acceleration of 0.25 rad/s² (using the Appendix D.1 or J.1 files).

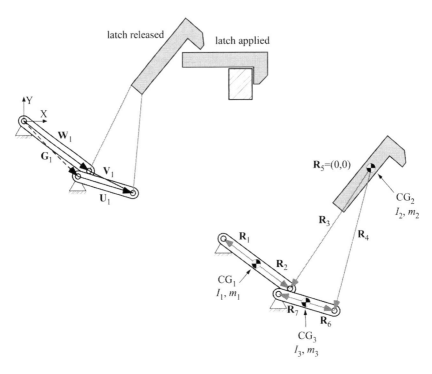

FIGURE P.7.4
Latch mechanism.

TABLE P.7.7

Latch Mechanism Dimensions (with Link Lengths in m)

W_1	V_1	U_1	G_1	L_1
$3.0887 - i2.2929$	$2.0591 - i1.001$	$2.5916 - i0.7733$	$2.5562 - i2.5206$	$3.7416 + i5.3422$

TABLE P.7.8

Latch Mechanism Dynamic Parameters (with Length in m, Mass in kg, and Inertia in kg–m²)

R_1	$-1.5444 + i1.1465$	m_1	3.3504
R_3	$-3.7416 - i5.3422$	I_1	17.2769
R_7	$-1.2958 + i0.3867$	m_2	12.8341
		I_2	566.2676
		m_3	2.4303
		I_3	6.3265

8 Repeat Problem 7 where a constant follower-link torque of $T_{b_0} = -100$ N-m is also applied.

9 Figure P.7.5 illustrates a planar four-bar mechanism used to guide a wiper blade. The dimensions for the mechanism are included in Table P.7.9 and the link dynamic parameters are included in Table P.7.10. Gravity is -9.81 m/s^2. Plot the force components at the grounded crank revolute joint ($F_{a_{0x}}$, $F_{a_{0y}}$) over a complete crank rotation range at a constant crank angular velocity of 5.25 rad/s (using the Appendix D.1 or J.1 files).

10. Plot the force components at the grounded follower-revolute joint ($F_{b_{0x}}$, $F_{b_{0y}}$) for Problem 9 at a constant crank angular velocity of 6.25 rad/s.

11. Figure P.7.6 illustrates a planar four-bar mechanism used to guide a gripper component from the open position to the closed position over a crank rotation of 50°. The dimensions for the mechanism at the open gripper position are included in Table P.7.11 and the link dynamic parameters are included in Table P.7.12. A force

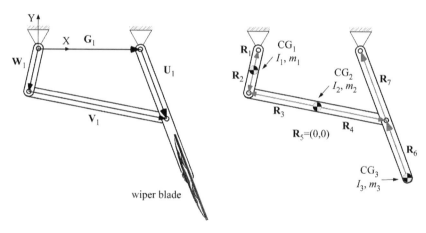

FIGURE P.7.5
Wiper-blade mechanism.

TABLE P.7.9

Wiper Mechanism Dimensions (with Link Lengths in m)

W_1	V_1	U_1	G_1	L_1
$-0.0626 - i0.2727$	$0.6258 - i0.2874$	$0.0632 - i0.5601$	$0.5 + i0$	$0.3129 - i0.1437$

TABLE P.7.10

Wiper Mechanism Dynamic Parameters (with Length in m, Mass in kg, and Inertia in kg–m^2)

R_1	$0.0313 + i0.1364$	m_1	0.0177
R_3	$-0.3129 + i0.1437$	I_1	0.0081
R_7	$-0.1023 + i0.9063$	m_2	0.0386
		I_2	0.0064
		m_3	0.0499
		I_3	0.252

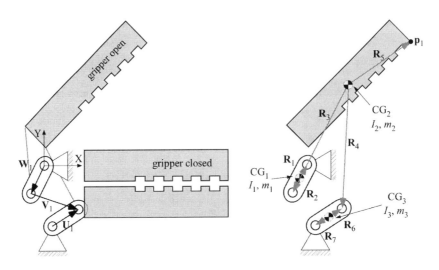

FIGURE P.7.6
Gripper mechanism.

TABLE P.7.11

Gripper Mechanism Dimensions (with Link Lengths in m)

W_1	V_1	U_1	G_1	L_1
$-0.324 - i0.6408$	$1.1198 - i0.354$	$0.5985 + i0.3782$	$0.1973 - i1.373$	$2.7357 + i3.415$

TABLE P.7.12

Gripper Mechanism Dynamic Parameters (with Length in m, Mass in kg, and Inertia in kg–m^2)

R_1	$0.162 + i0.3204$	m_1	0.8308
R_3	$-1.2711 - i2.4568$	I_1	0.1924
R_7	$-0.2993 - i0.1891$	m_2	8.8936
		I_2	76.9644
		m_3	0.8227
		I_3	0.1868

of $\mathbf{F}_{p_1} = (0, 500)\,\mathrm{N}$ is constantly applied. Gravity is $-9.81\,\mathrm{m/s^2}$. Determine the maximum driver torque magnitude $|\mathbf{T}_{a_0}|$ produced given a crank rotation increment of $1°$, an initial crank angular velocity of $0\,\mathrm{rad/s}$ and an angular acceleration of $0.1\,\mathrm{rad/s^2}$ (using the Appendix D.1 or J.1 files).

12. Repeat Problem 11 where a force of $\mathbf{F}_{p_1} = (-150, 400)\,\mathrm{N}$ is constantly applied.

13. Figure P.7.7 illustrates a planar four-bar mechanism used to guide a component from its initial position to its assembled position over a crank rotation of $66°$. The dimensions for the mechanism at the initial component position are included in Table P.7.13 and the link dynamic parameters are included in Table P.7.14. Gravity is $-9.81\,\mathrm{m/s^2}$. Determine the maximum driver torque magnitude $|\mathbf{T}_{a_0}|$ produced given a crank rotation increment of $0.1°$, an initial crank angular velocity of $0\,\mathrm{rad/s}$ and an angular acceleration of $0.55\,\mathrm{rad/s^2}$ (using the Appendix D.1 or J.1 files).

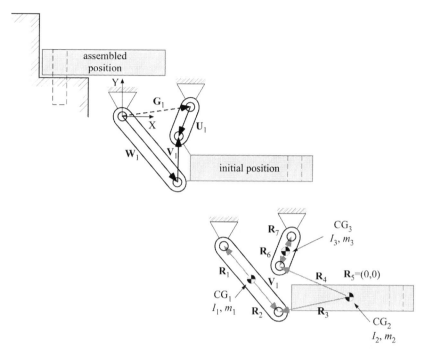

FIGURE P.7.7
Assembly mechanism.

TABLE P.7.13

Assembly Mechanism Dimensions (with Link Lengths in m)

W_1	V_1	U_1	G_1	L_1
$1.5356 - i1.7885$	$0.0169 + i1.2337$	$-0.3294 - i0.8093$	$1.8819 + i0.2545$	$1.7054 + i0.8903$

TABLE P.7.14

Assembly Mechanism Dynamic Parameters (with Length in m, Mass in kg, and Inertia in kg–m²)

R_1	$-0.7678 + i0.8943$	m_1	2.1510
R_3	$-1.7054 - i0.8903$	I_1	4.2995
R_7	$0.1647 + i0.4047$	m_2	9.3356
		I_2	43.7912
		m_3	0.9563
		I_3	0.3091

14. Repeat Problem 13 where the minimum and maximum values of $|\mathbf{F}_{b_0}|$ are determined (instead of the maximum driver torque magnitude).

15. Figure P.7.8 illustrates a planar four-bar mechanism used to guide a digging bucket from its initial position to its final position over a crank rotation of −58°. The dimensions for the mechanism at the initial component position are included in Table P.7.15 and the link dynamic parameters are included in Table P.7.16. A force of $\mathbf{F}_{p_1} = (-3182, -3182)\,\text{N}$ is constantly applied. Gravity is −9.81 m/s². Determine

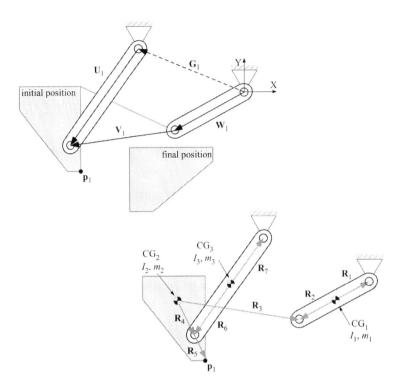

FIGURE P.7.8
Digging mechanism.

TABLE P.7.15

Digging Mechanism Dimensions (with Link Lengths in m)

W_1	V_1	U_1	G_1	L_1
$-2.5208 - i1.3307$	$-3.6797 - i0.5059$	$-2.4816 - i3.3521$	$-3.7189 + i1.5155$	$-3.3055 - i1.3619$

TABLE P.7.16

Digging Mechanism Dynamic Parameters (with Length in m, Mass in kg, and Inertia in kg–m²)

R_1	$1.2604 + i0.6654$	m_1	2.5478
R_3	$4.1869 - i0.3915$	I_1	7.3408
R_7	$1.2408 + i1.676$	m_2	36.419
		I_2	677.1258
		m_3	3.6112
		I_3	21.8099

the maximum force magnitude $\left|\mathbf{F}_{a_1}\right|$ produced given a crank rotation increment of $-0.1°$, an initial crank angular velocity of $0\,\text{rad/s}$ and an angular acceleration of $-0.35\,\text{rad/s}^2$ (using the Appendix D.1 or J.1 files).

16. Repeat Problem 15 where the minimum and maximum values of $\left|\mathbf{F}_{b_1}\right|$ are determined (instead of the maximum force magnitude $\left|\mathbf{F}_{a_1}\right|$).

17. Figure P.7.9 illustrates a planar four-bar mechanism used to guide a bucket from its loading position to its unloading position over a crank rotation of $-30°$. The dimensions for the mechanism at the initial component position are included in Table P.7.17 and the link dynamic parameters are included in Table P.7.18. Gravity is $-9.81\,\text{m/s}^2$. Determine the maximum driver torque magnitude $|\mathbf{T}_{a_0}|$ produced given a crank rotation increment of $-1°$, an initial crank angular velocity of $0\,\text{rad/s}$ and an angular acceleration of $-0.025\,\text{rad/s}^2$ (using the Appendix D.1 or J.1 files).

18. Repeat Problem 17 where the maximum force magnitude $|\mathbf{F}_{b_0}|$ is determined (instead of the maximum driver torque magnitude).

19. Plot the driving link torque for the planar four-bar mechanism in Example 7.1 over a complete crank rotation range.

20. Repeat Example 7.1 using $\mathbf{F}_{p_1} = (-1500, 4500)\,\text{N}$, an initial crank angular velocity of $1.25\,\text{rad/s}$, and an angular acceleration of $0.2\,\text{rad/s}^2$.

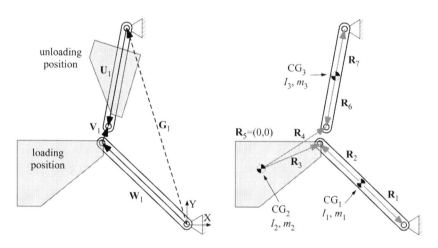

FIGURE P.7.9
Load–unload mechanism.

TABLE P.7.17

Load–Unload Mechanism Dimensions (with Link Lengths in m)

W_1	V_1	U_1	G_1	L_1
$-2.8841 + i2.7285$	$0.1817 + i0.5059$	$-0.6055 - i3.2954$	$-2.0968 + i6.5299$	$-1.7775 - i0.8574$

TABLE P.7.18

Load–Unload Mechanism Dynamic Parameters (with Length in m, Mass in kg, and Inertia in kg–m²)

R_1	$1.442 - i1.3642$	m_1	3.4494
R_3	$1.7775 + i0.8574$	I_1	18.9215
R_7	$0.3028 + i1.6477$	m_2	36.419
		I_2	174.9659
		m_3	2.9503
		I_3	11.6282

21. Plot $F_{a_{0x}}$ and $F_{a_{0y}}$ (vs. the crank angular displacement) for the slider-crank mechanism in Example 7.2 over a complete crank rotation range. For this problem, let $\mu = \pm 0.1$.

22. Plot $F_{a_{0x}}$ and $F_{a_{0y}}$ (vs. the crank angular displacement) for the slider-crank mechanism in Example 7.2 over a complete crank rotation range, using a material with half the density of the original material. For this problem, let $\mu = \pm 0.1$.

23. Repeat Example 7.2 using F = (–100, –25) N, an initial crank angular velocity of 10 rad/s, and an angular acceleration of 2.5 rad/s². For this problem, let $\mu = \pm 0.1$.

24. Table P.7.19 includes the dimensions for an offset slider-crank mechanism and Table P.7.20 includes dynamic parameters for this mechanism. For this problem, $F = (0, -100)$ N, the initial crank angular velocity is 0 rad/s and the angular acceleration is –5 rad/s². Gravity is –9.81 m/s². Calculate the driver torque over a complete crank rotation range (using the Appendix D.2 or J.2 files).

25. Plot $F_{a_{0y}}$ (vs. the crank angular displacement) for the slider-crank mechanism in Problem 24 over a complete crank rotation range, using $m_3 = 25$ kg and $m_3 = 10$ kg.

26. Repeat Problem 24 using $F = (-150, -50)$ N, an initial crank angular velocity of –15 rad/s and an angular acceleration of 0 rad/s².

27. Table P.7.21 includes the dimensions for a geared five-bar mechanism and P.7.22 includes dynamic parameters for this mechanism. The initial crank angular velocity is 1.25 rad/s and the angular acceleration is 0.15 rad/s². Plot the driving and driven link torque over a complete crank rotation range (using the Appendix D.3 or J.3 files).

TABLE P.7.19

Slider–Crank Mechanism Dimensions (with Link Lengths in m)

W_1, θ	V_1	U_1	μ
0.5, 90°	0.9014	0.25	±0.15

TABLE P.7.20

Slider–Crank Mechanism Dynamic Parameters (with Length in m, Mass in kg, and Inertia in kg–m²)

R_1	$0 - i0.25$	m_1	8.05
R_3	$-0.433 + i0.125$	I_1	0.805
		m^2	14.49
		I_2	4.025
		m_3	50

TABLE P.7.21

Geared Five-Bar Mechanism Configuration (with Link Lengths in m)

W_1, θ	V_1, ρ	U_1, σ	S_1, ψ	G_{1x}, G_{1y}	L_1, δ
0.35, 90°	0.525, 54.7643°	0.35, 60°	0.525, 115.0279°	0.35, 0	0.35, –15.7645°

F_{P_1} (N)	Gravity	Gear Ratio			
0, –1000	–9.81 m/s²	–2			

TABLE P.7.22

Geared Five–Bar Mechanism Dynamic Parameters (with Length in m, Mass in kg, and Inertia in kg–m²)

R_1	$0 - i0.0666$		m_1	22.54
R_3	$-0.2132 - i0.1112$		I_1	0.505
R_7	$0.1111 - i0.2379$		m_2	29.785
R_9	$-0.0074 - i0.0128$		I_2	5.635
			m_3	12.075
			I_3	2.415
			m_4	75.67
			I_4	5.635

28. Plot the force magnitudes $\left|\mathbf{F}'_{a_0}\right|$ and $\left|\mathbf{F}'_{b_0}\right|$ (versus the crank angular displacement) for the geared five-bar mechanism in Problem 27 over a complete crank rotation range,

29. Repeat Example 7.3 using $\mathbf{F}_{P_1} = (-1000, -750)\,\mathrm{N}$, an initial crank angular velocity of 1.25 rad/s, and an angular acceleration of 0.1 rad/s².

30. Plot $F_{C_{1x}}$ and $F_{C_{1y}}$ (vs. the crank angular displacement) for the geared five-bar mechanism in Problem 29 over a complete crank rotation range.

31. Calculate the forces $\mathbf{F}_{a_0}$, $\mathbf{F}_{b_0}$, and $\mathbf{F}_{b_0^*}$ for the Watt II mechanism configuration in Table P.7.23 over a $-90°$ crank rotation range at $-15°$ increments (using the Appendix D.5 or J.5 files). Table P.7.24 includes the dynamic parameters for this mechanism. The initial crank angular velocity and acceleration are -1.25 rad/s and -0.1 rad/s², respectively.

32. Repeat Example 7.4 using $\mathbf{F}_{P_1} = (0, -750)\,\mathrm{N}$, $\mathbf{F}_{P_1^*} = (0, -1500)\,\mathrm{N}$, and $T_{b_0} = T_{b_0^*} = 350\,\mathrm{N}$-m. The initial crank angular velocity and acceleration are -2.5 rad/s and 0 rad/s², respectively.

33. Figure P.7.10 illustrates a Stephenson III mechanism used to guide a digging bucket from its initial position to its final position over a crank rotation of 40°. The dimensions for the mechanism at the open gripper position are included in Table P.7.25 and the link dynamic parameters are included in Table P.7.26. A force of $\mathbf{F}_{P_1^*} = (-2500, 0)\,\mathrm{N}$ is constantly applied. Gravity is $-9.81\,\mathrm{m/s^2}$. Determine the maximum driver torque magnitude $\left|\mathbf{T}_{a_0}\right|$ produced given a crank rotation increment of 1°, an initial crank angular velocity of 0 rad/s and an angular acceleration of 0.25 rad/s² (using the Appendix D.6 or J.6 files).

TABLE P.7.23

Watt II Mechanism Configuration (with Link Lengths in m)

W_1, θ	V_1, ρ	U_1, σ	G_{1x}, G_{1y}	L_1, δ
1, 90°	1.5, 19.3737°	1.5, 93.2461°	1.5, 0	1, 60.7834°
W_1^*, θ^*	V_1^*, ρ^*	U_1^*, σ^*	G_{1x}^*, G_{1y}^*	L_1^*, δ^*
1, 45°	1.5, 7.9416°	1.5, 60.2717°	1.4489, −0.3882	1, 49.3512°
F_{P_1} (N)	$F_{P_1^*}$ (N)	T_{b_0} (N-m)	$T_{b_0^*}$ (N-m)	Gravity
0, −1500	−2500, −1000	0	0	−9.81 m/s²

TABLE P.7.24

Watt II Mechanism Dynamic Parameters (with Length in m, Mass in kg, and Inertia in kg–m²)

R_1	$0 - i0.5$	m_1	8.05
R_3	$-0.6344 - i0.4568$	I_1	0.805
R_7	$-0.2074 - i0.7349$	m_2	29.785
R_9	$-0.7123 - i0.322$	I_2	5.635
R_{13}	$-0.3719 - i0.6513$	m_3	33.81
		I_3	5.635
		m_4	29.785
		I_4	5.635
		m_5	12.075
		I_5	2.415

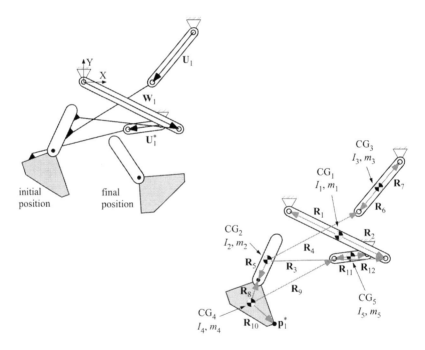

FIGURE P.7.10

Stephenson III digging mechanism.

TABLE P.7.25

Stephenson III Digging Mechanism Dimensions (with Link Lengths in m)

W_1	V_1	U_1	G_1	L_1
$3.0546 - i1.4686$	$-0.841 + i1.4363$	$-1.3089 - i1.5388$	$3.5225 + i1.5065$	$-4.0351 - i0.6398$
V_1^*	L_1^*	U_1^*	G_1^*	$T_{b_0}, T_{b_0}^*$ (N-m)
$2.3984 + i0.629$	$0.4974 - i1.3665$	$-1.0591 - i0.1314$	$-1.0455 - i2.8545$	0

TABLE P.7.26

Stephenson III Mechanism Dynamic Parameters (with Length in m, Mass in kg, and Inertia in kg–m²)

R_1	$-1.5273 + i0.7343$	m_1	2.9817
R_3	$3.7288 - i0.0171$	I_1	12.0162
R_7	$0.6545 + i0.7694$	m_2	2.5865
R_8	$0.2166 + i0.6934$	I_2	36.6452
R12	$0.5295 + i0.0657$	m_3	1.8797
		I_3	2.799
		m_4	9.1046
		I_4	80.2505
		m_5	1.1125
		I_5	0.5096

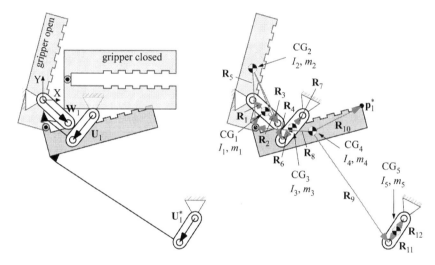

FIGURE P.7.11
Stephenson III gripper mechanism.

TABLE P.7.27

Stephenson III Gripper Mechanism Dimensions (with Link Lengths in m)

W_1, θ	V_1, ρ	U_1, σ	G_{1x}, G_{1y}	L_1, δ
1.0283, 318.4178°	0.5195, 286.4362°	0.9418, 230.6143°	1.5138, −0.4529	0.7031, 189.1904°
V_1^*, ρ^*	U_1^*, σ^*	L_1^*, δ^*	G_{1x}^*, G_{1y}^*	$T_{b_0}, T_{b_0}^*$ (N-m)
5.2881, 319.9647°	0.8666, 236.6216°	3.2742, 11.8113°	3.0869, −3.0198	25, 75

34. Figure P.7.11 illustrates a Stephenson III mechanism used to guide a gripping tool from its open position to its closed position over a crank rotation of 40°. The dimensions for the mechanism at the open gripper position are included in Table P.7.27 and the link dynamic parameters are included in Table P.7.28. A force of $\mathbf{F}_{p_1}^* = (0, -2500)$N is constantly applied. Gravity is −9.81 m/s². Calculate the forces $\mathbf{F}_{a_1}$, $\mathbf{F}_{p_1}$, $\mathbf{F}_{b_0}$ and $\mathbf{F}_{b_0}^*$ for the Stephenson III mechanism configuration over

a 40° crank rotation range at 5° increments (using the Appendix D.6 or J.6 files). The initial crank angular velocity and acceleration are $0\,\text{rad/s}$ and $0.025\,\text{rad/s}^2$, respectively.

35. Calculate the forces $\mathbf{F}_{a_0}$, $\mathbf{F}_{p_1}$, $\mathbf{F}_{b_0}$ and $\mathbf{F}_{b_0^*}$ for the Stephenson III mechanism configuration in Table P.7.29 over a 360° crank rotation range at 60° increments (using the Appendix D.6 or J.6 files). Table P.7.30 includes the dynamic parameters for this mechanism. The initial crank angular velocity and acceleration are $0\,\text{rad/s}$ and $0.125\,\text{rad/s}^2$, respectively. Also $\mathbf{F}_{p_1^*} = (0,0)\text{N}$ and gravity is $-9.81\,\text{m/s}^2$.

TABLE P.7.28

Stephenson III Mechanism Dynamic Parameters (with Length in m, Mass in kg, and Inertia in kg–m²)

R_1	$-0.3846 + i0.3412$	m_1	1.0811
R_3	$0.8113 - i1.4891$	I_1	0.4637
R_7	$0.2988 + i0.3640$	m_2	9.1255
R_8	$-1.6013 + i0.1172$	I_2	35.7589
$R12$	$0.2384 + i0.3618$	m_3	1.0111
		I_3	0.3719
		m_4	9.1255
		I_4	64.0933
		m_5	0.9507
		I_5	0.3027

TABLE P.7.29

Stephenson III Mechanism Dimensions (with Link Lengths in m)

W_1, θ	V_1, ρ	U_1, σ	G_{1x}, G_{1y}	L_1, δ
1, 90°	1.5, 19.3737°	1.5, 93.2461°	1.5, 0	1, 60.7834°

V_1^*, ρ^*	L_1^*, δ^*	U_1^*, σ^*	G_{1x}^*, G_{1y}^*	$T_{b_0}, T_{b_0^*}$ (N-m)
2, 17.1417°	1, 63.7091°	2, 76.4844°	0.4318, 0.5176	−150

TABLE P.7.30

Stephenson III Mechanism Dynamic Parameters (with Length in m, Mass in kg, and Inertia in kg–m²)

R_1	$0 - i0.5$	m_1	16.1
R_3	$-0.6344 - i0.4568$	I_1	1.61
R_7	$0.0425 - i0.7488$	m_2	59.57
R_8	$-0.7847 - i0.4953$	I_2	11.27
R_{12}	$-0.2337 - i0.9723$	m_3	24.15
		I_3	4.83
		m_4	86.94
		I_4	32.2
		m_5	30.59
		I_5	11.27

8

Design and Kinematic Analysis of Gears

CONCEPT OVERVIEW

In this chapter, the reader will gain a central understanding regarding

1. Purposes, designs, and functions of *spur, planetary, rack and pinion, helical, bevel,* and *worm* gears
2. Criteria for optimal gear operation and its relationship with gear design variables and design equations
3. Equations and solution methods for the kinematics of *spur, planetary, rack and pinion, helical, bevel,* and *worm* gears

8.1 Introduction

Gears are mechanical components used to transmit motion from one shaft to another. In Chapters 3 and 4, gears have been introduced in the planar five-bar mechanism. By including a gear pair, as illustrated in Figure 3.4, or a *gear train* (three or more gears), as illustrated in Figure 4.13, to interconnect the driving links of this mechanism, the rotation of link a_0–a_1 is transmitted to link b_0–b_1.

A simple design to transmit motion between shafts can include cylinders, where friction maintains the rolling contact between the cylinders (Figure 8.1a). With this design, as long as the contact between the cylinders is pure rolling (e.g., no slip), the velocity relationships given in Equation 8.1 hold true. As a result of the friction force being the limiting factor for the torque capacity of rolling cylinders, motion transmission through rolling cylinders is limited to low-torque applications in practice.

Replacing the rolling cylinders with gears (Figure 8.1b) maintains the velocity relationships in Equation 8.1 while substantially increasing torque capacity, since the teeth in the driving gear (also called the *pinion*) interface or *mesh* with the teeth of the driven gear.* With gears, torque capacity is limited to the bending strength of the gear teeth (as opposed to contact friction in rolling cylinders) [1]. The ratio of the driving and driven gear angular velocities (in Equation 8.1) is called the *velocity ratio* (*VR*).

Although rotation can be transmitted through other mechanical components (e.g., belt-pulley systems and chain-sprocket systems), gears are commonly used, appearing in mechanical systems of all sizes. The advantages of gears over belt-pulley, chain-sprocket, and even linkage systems include higher torque, speed and power capacities, no slip, greater durability, efficiency, and suitability for confined spaces. As you will see

---

* In Figure 8.1b, variables r_{p1} and r_{p2} are the *pitch circle radii* (the *pitch circle diameter* is presented in Section 8.3.1) of the driving and driven gears, respectively, and are analogous to r_1 and r_2 of the rolling cylinders in Figure 8.1a.

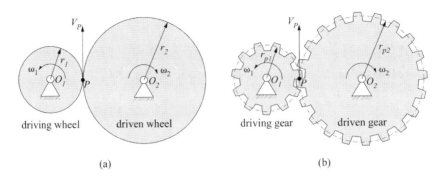

FIGURE 8.1
(a) Rolling cylinder pair and (b) gear pair.

throughout this chapter, gear types and gear systems vary primarily by the shaft orientations they accommodate, the input–output rotation ratios possible, and the overall gear system design.

$$V_P = r_1\omega_1 = r_2\omega_2$$

or

$$\frac{\omega_1}{\omega_2} = \frac{r_2}{r_1} = VR \tag{8.1}$$

8.2 Gear Types

Spur gears (Figure 8.1b) are used to transmit motion between parallel shafts. Figure 8.2a illustrates the two types of spur-gear designs: *external* and *internal*. In an external gear, the gear teeth point away from the gear center, while the teeth point toward the center in an internal gear. An internal gear with an infinite radius forms a *rack gear*, which, when included with an external gear, is called a *rack and pinion gear*. In rack and pinion gears

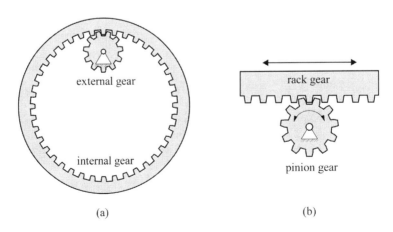

FIGURE 8.2
(a) External and internal gears and (b) rack and pinion gear.

(Figure 8.2b), the rotational motion of the pinion gear produces the translational motion of the rack gear.

Helical gears can be used to transmit motion between parallel shafts (Figure 8.3a) and nonparallel shafts that do not intersect (Figure 8.3b). The helical tooth profiles of these gears give them a greater load-bearing capacity (for increased power transmission), greater wear resistance, quieter operation, greater operating speeds, and smoother operation than spur gears.*

Bevel gears (Figure 8.4) are used to transmit motion between intersecting shafts. Bevel gears are typically used for shafts that intersect at 90° (although they are not limited to this angle). Like helical gears, *worm gears* (Figure 8.5) are used to transmit motion between orthogonal shafts that do not intersect. Worm gears are used when large reductions in velocity ratios are required.

(a)   (b)

FIGURE 8.3
Helical gears on (a) parallel and (b) nonparallel shafts.

FIGURE 8.4
Bevel gear.

---

* The helical tooth profile results in a greater contact area for helical gear teeth than for a spur gear of the same thickness and radius. The enhanced capacities noted for the helical gear are the result of the increased contact area. This statement assumes that the variables common to both gear types have identical values.

FIGURE 8.5
Worm gear.

8.3 SPUR-Gear Nomenclature and Relationships of Mating Gears

8.3.1 Spur-Gear Nomenclature

Figure 8.6 illustrates the principal features of a spur-gear tooth. The *pitch circle* represents the size of the rolling cylinder that would replace the gear (as illustrated in Figure 8.1)*. The diameter of a pitch circle is called the *pitch diameter* (represented by the variable d_p).[†]

The *number of teeth* (represented by the variable N) is the total number of teeth on the gear. This quantity is always an integer (gear teeth fractions are not used).

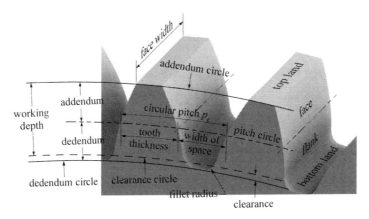

FIGURE 8.6
Spur-gear tooth features.

---

* The point *P* in Figure 8.1 is the point of contact between the two pitch circles and is called the *pitch point*.
† While the pitch diameter cannot be directly measured (due to its location within the gear), gears are commonly referenced by their pitch diameters.

The *circular pitch* (represented by the variable p_c) is the distance measured along the pitch circle from one point on a gear tooth to the corresponding point on an adjacent gear tooth. Given the variables d_p and N, an equation for the circular pitch is expressed as

$$p_c = \frac{\pi d_p}{N} \tag{8.2}$$

To enable proper operation for mating gears, the circular pitch values for the gears must be identical.

The gear-tooth profile is constructed from the *base circle* of the gear and the diameter of a base circle is called the *base circle diameter* (represented by the variable d_b). Section 8.4.5 includes details on how a gear-tooth profile is generated from a base circle.

The *face width* (represented by the variable F) is the length of a gear tooth in the direction parallel to the shaft axis.

The *addendum* (represented by the variable a) is the radial distance from the pitch circle to the top of a gear tooth and the *dedendum* (represented by the variable b) is the radial distance from the pitch circle to the bottom of a gear tooth. The sum of the addendum and dedendum is called the *whole depth* (represented by the variable h_T). The amount that the addendum exceeds the dedendum is called the *clearance* (represented by the variable c).[*]

The *diametral pitch* (represented by the variable P_d) is the number of gear teeth per inch of pitch diameter and can be expressed as

$$P_d = \frac{N}{d_p} \tag{8.3}$$

Diametral pitch (also called *pitch*) is an often-referenced parameter for gear-tooth size specifications in *US customary units* (or simply *US units*). Figure 8.7 includes several standard gear-tooth sizes and their corresponding diametral pitch values. To enable proper operation for mating gears, the diametral pitch values for the gears must be identical.

The *module* (represented by the variable m) is the ratio of pitch diameter to the number of gear teeth and is expressed as

$$m = \frac{d_p}{N} \tag{8.4}$$

The module is an often-referenced parameter in the *International System of Units* (or simply *SI units*) and has a unit of millimeters. It can be seen from Equations 8.3 and 8.4 that the module is the reciprocal of the diametral pitch.

After substituting Equations 8.4 and 8.3 into Equation 8.2, the circular pitch becomes

$$p_c = \frac{\pi d_p}{N} = \frac{\pi}{P_d} = \pi m \tag{8.5}$$

Most of the gear-tooth features identified in Figure 8.6 are standardized with respect to the diametral pitch.[†] Table 8.1 includes several gear-tooth feature equations as given and

---

[*] In mating gears, the clearance is the gap between the top of a tooth of one gear and the bottom of a tooth of the other gear.

[†] Gear tooth features are also standardized with respect to the circular pitch.

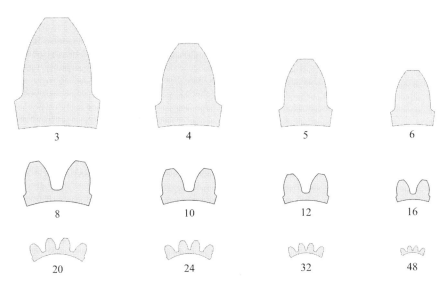

FIGURE 8.7
Spur-gear tooth size variation with diametral pitch.

TABLE 8.1

Formulas for Spur-Gear Tooth Proportions (20° and 25° Pressure
Angle Involute Full-Depth Teeth ANSI Coarse Pitch)

Tooth Feature to Calculate	Diametral Pitch P_d Known
Addendum, a	$a = 1/P_d$
Dedendum (preferred), b	$b = 1.25/P_d$
Dedendum (shaved or ground teeth), b	$b = 1.35/P_d$
Working depth, h_k	$h_k = 2/P_d$
Whole depth (preferred), h_t	$h_t = 2.25/P_d$
Whole depth (shaved or ground teeth), h_t	$h_t = 2.35/P_d$
Fillet radius, r_f	$r_f = 0.3/P_d$
Clearance (preferred), c	$c = 0.25/P_d$
Clearance (shaved or ground teeth), c	$c = 0.35/P_d$
Circular tooth thickness, t	$t = \pi/2P_d$

Source: ANSI B6.1-1968 (R1974). Coarse-pitch spur-gear tooth forms. New York: American National Standards Institute [3].

certified by the *American Gear Manufacturers Association* (AGMA) and the *American National Standards Institute* (ANSI), respectively [2].

Example 8.1

Problem Statement: Using Equations 8.2–8.4, calculate the pitch diameter, diametral pitch, and module for a spur gear having 20 teeth and a circular pitch of 0.5. Assuming this gear is an ANSI coarse pitch with a 25° pressure angle (Table 8.1), calculate the working depth and circular tooth thickness of this gear also.

Known Information: N and p_c.

Solution Approach: Figure E.8.1 includes the calculation procedure in the MATLAB® command window.

```
>> N = 20;
>> PC = 0.5;
>> dp = N*PC/pi

dp =

      3.1831

>> Pd = N/dp

Pd =

      6.2832

>> m = dp/N

m =

      0.1592

>> hk = 2/Pd

hk =

      0.3183

>> t = pi/(2*Pd)

t =

      0.2500

>>
```

FIGURE E.8.1
Example 8.1 solution calculation procedure in MATLAB.

8.3.2 Pressure Angle and Involute Tooth Profile

The *pressure angle* (represented by the variable ϕ) is the angle between the line tangent to both pitch circles of mating gears (called the *pitch line* in Figure 8.8) and the line perpendicular to both gear-tooth surfaces at the contact point (called the *pressure line* in Figure 8.8).* The pressure line is also tangent to both the base circles of the mating gears. The relative gear-tooth shape is influenced in part by the pressure angle. To enable proper operation of the mating gears, the pressure angle (like the diametral pitch) values for the gears must be identical.

It can be explained from Figure 8.8 how the pressure angle affects the relative gear-tooth shape. Because the pressure line is tangent to both base circles, any increase in the pressure angle decreases the size of the base circles. Conversely, any decrease in the pressure angle increases their size. The radial distance between the base and pitch circles influences the gear-tooth shape because the specific gear-tooth profile (explained later in this section) lies along this radial distance only. Most gears today are standardized at pressure angles of 20° and 25°. Although reducing the pressure angle increases the possibility of poor gear-tooth engagement (due to interference), it also results in more efficient torque transmission and smaller radial load transfer to supporting shafts.

---

* Because forces are transmitted in the direction perpendicular to the surfaces of the contacting bodies (and the pressure line is perpendicular to both contacting gear-tooth surfaces), the forces acting on a gear tooth are transmitted along the pressure line. The point of contact between the gear surfaces that lies on the pitch circles is called the *pitch point* and is represented by the variable P in Figure 8.8.

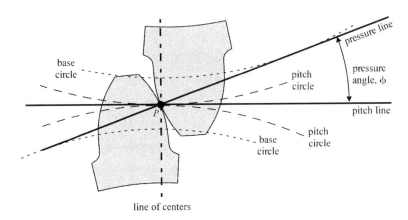

FIGURE 8.8
Pressure angle.

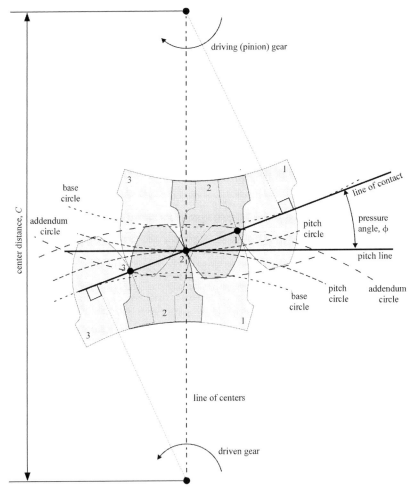

FIGURE 8.9
Gear mating process.

Smooth gear motion is achieved when the velocity of the driven gear from gear-tooth engagement to gear-tooth disengagement is constant.[*] A requirement for this condition is that the path of gear-tooth contact is a straight line and that this line must intersect the point of contact of both pitch circles. Figure 8.9 illustrates three moments of engagement of two gears: initial tooth engagement (1), an intermediate moment of engagement (2), and tooth disengagement (3). The points of contact at each of these moments of engagement lie along the *line of contact* (or *contact line*) and this line intersects the point of contact between the pitch circles. Therefore, a constant velocity ratio is ensured for the gears in Figure 8.9.

These requirements for a constant velocity ratio are expressed in *the fundamental law of gearing. In this law, to maintain a constant velocity ratio, the gear-tooth profile must be designed in such a way that the common normal to both contacting tooth surfaces (the pressure line) passes through the pitch point on the line of contact.*

The *involute* of a circle is one of the possible curve types that is adequate for a gear-tooth profile.[†] An involute curve is produced by unwinding a taut cord from the base circle (having a diameter represented by the variable d_b), and tracing the path produced by a point on the cord. A gear-tooth profile is formed from a section of the involute curve. Figure 8.10 illustrates the involute curve.

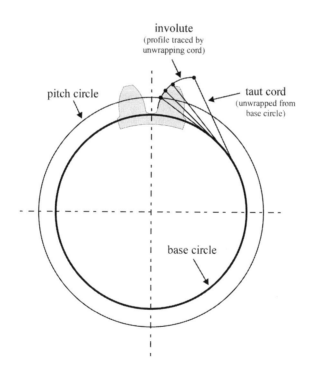

FIGURE 8.10
Involute gear tooth.

---

[*] This condition also means that the velocity ratio of the driving and driven gears is constant.

[†] Another possible curve is called a *cycloid*, which is the planar path traced by a point on a circle as it rolls on a fixed surface. An *epicycloid* is produced by a circle rolling over a fixed circle, and a *hypocycloid* is produced by a circle rolling within a fixed circle.

The pressure line and line of contact are identical for gears with involute profiles. The pressure angle (or the inclination of the line of contact) is determined from the involute curve section used for the gear tooth. As previously noted in this section, an increase in the pressure angle decreases the size of the base circles and a decrease in the angle increases their size. The relationship between the pressure angle (ϕ), the pitch diameter (d_p), and the base circle diameter (d_b) can be expressed as

$$d_b = d_p \cos \phi \tag{8.6}$$

8.3.3 Gear Center Distance and Contact Ratio

The *center distance* (represented by variable C) is the center-to-center distance between two mating gears.* In terms of pitch radii and pitch diameters, the equation for center distance for external gears can be expressed as

$$C_{\text{external}} = r_{p1} + r_{p2} = \frac{d_{p1} + d_{p2}}{2} \tag{8.7}$$

Substituting the pitch diameters in Equation 8.7 with Equation 8.3 produces

$$C_{\text{external}} = \frac{N_1 + N_2}{2P_d} \tag{8.8}$$

For a gear pair consisting of an internal and an external gear (Figure 8.2a, for example), the center distance can be expressed as

$$C_{\text{ext-int}} = r_{p1} - r_{p2} = \frac{d_{p1} - d_{p2}}{2} = \frac{N_1 - N_2}{2P_d} \tag{8.9}^{\dagger}$$

The *contact ratio* (represented by the variable m_p) is the average number of gear teeth in contact at any instant in time. A contact ratio of 1.2 (generally the minimum considered in design) means that one pair of gear teeth is always in contact while another pair is in contact only 20% of the time. By increasing the contact ratio, more gear teeth are in contact from the moment of engagement to the moment of disengagement. This results in greater duration and power transmission (since loads are shared among more teeth), as well as smoother operation.

A contact ratio equation can be expressed as

$$m_p = \frac{Z}{p_b} \tag{8.10}$$

where the variable p_b is the *base pitch* (the distance measured along the base circle from one point on a gear tooth to the corresponding point on an adjacent gear tooth) and the variable Z is the length of the line of contact (from engagement to disengagement). A base pitch equation can be expressed as

---

* This distance is also the center distance between the two shafts supporting the gears.
† For external-internal gear pairs, the center distance is the (positive) difference between the pitch radii.

$$p_b = \frac{\pi d_{p1} \cos\phi}{N_1} = \frac{\pi d_{p2} \cos\phi}{N_2} \tag{8.11}$$

and a contact line length equation can be expressed as

$$Z = \sqrt{\left(r_{p2} + a_2\right)^2 - \left(r_{p2}\cos\phi\right)^2} - r_{p2}\sin\phi + \sqrt{\left(r_{p1} + a_1\right)^2 - \left(r_{p1}\cos\phi\right)^2} - r_{p1}\sin\phi \tag{8.12}$$

Substituting Equations 8.11 and 8.12 into Equation 8.10 produces a contact ratio equation in terms of gear-tooth geometry.

Example 8.2

Problem Statement: Given an external gear pair where $N_1=25$, $N_2=35$ and the circular pitch for the pinion is $p_{c1}=0.25$, calculate the center distance, base pitch, contact line length, and contact ratio for the gear pair. Assume the gears are ANSI coarse pitch with 20° pressure angles (Table 8.1).

Known Information: N_1, N_2, p_{c1}, and ϕ.

Solution Approach: Figure E.8.2 includes the calculation procedure in MATLAB's command window.

```
>> N1 = 25;
>> pc1 = 0.25;
>> N2 = 35;
>> phi = 20*pi/180;
>> dp1 = pc1*N1/pi;
>> Pd = N1/dp1;
>> dp2 = N2/Pd;
>> C = (dp1 + dp2)/2

C =

    2.3873

>> pb = pi*dp1*cos(phi)/N1

pb =

    0.2349

>> a = 1/Pd;
>> Z =...
sqrt((0.5*dp2 + a)^2 - (0.5*dp2*cos(phi))^2) - 0.5*dp2*sin(phi)...
+ sqrt((0.5*dp1 + a)^2 - (0.5*dp1*cos(phi))^2) - 0.5*dp1*sin(phi)

Z =

    0.3874

>> mp = Z/pb

mp =

    1.6491

>>
```

FIGURE E.8.2
Example 8.2 solution calculation procedure in MATLAB.

8.3.4 Gear-Tooth Interference and Undercutting

As illustrated in Figure 8.9, the contact between mating gear teeth should only occur along the line of contact. Gear-tooth contact at any other point is known as *interference* and violates the required constant velocity ratio condition for gears.* To avoid interference, it is important that gear-tooth contact only occurs between their involute portions.

Since the involute profile of a gear exists above its base circle (see Figure 8.11), it is important to eliminate and/or minimize the noninvolute gear tooth portion below the base circle to avoid interference. One way to meet this requirement (and subsequently avoid interference) is by avoiding gears having too few teeth. Inequality Eq. (8.13) determines the maximum number of gear teeth to avoid interference given the number of pinion teeth and pressure angle (in this equation $a_1 = k/P_{d1}$).[†‡] Table 8.2 includes sample interference-avoiding gear tooth combinations calculated from Inequality Eq. (8.13).

$$N_2 < \frac{N_1^2 \sin^2 \phi - 4k^2}{4k - 2N_1 \sin^2 \phi} \tag{8.13}$$

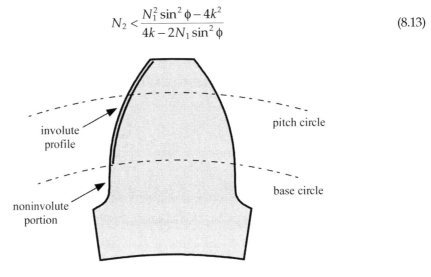

FIGURE 8.11
Gear-tooth involute portion.

TABLE 8.2

Sample Gear-Tooth Combinations to Avoid Interference

$\phi=20°$		$\phi=25°$	
Number of Pinion Teeth, N_1	**Max. Number of Gear Teeth, N_2**	**Number of Pinion Teeth, N_1**	**Max. Number of Gear Teeth, N_2**
Less than 13	Interference	Less than 9	Interference
13	16	9	13
14	26	10	32
15	45	11	249

---

* Interference also produces excessive gear noise, vibration and wear.
[†] Since $a = 1/P_d$ in Table 8.1, $k = 1$ in Inequality Eq. (8.13) for the gear types represented in Table 8.1.
[‡] Inequality Eq. (8.13) can also be expressed as $a < \sqrt{r_b^2 + c^2 \sin^2 \phi}$ where a is the addendum, r_b is the base circle radius, C is the center-to-center gear distance and ϕ is the pressure angle. Interference occurs when this condition is violated.

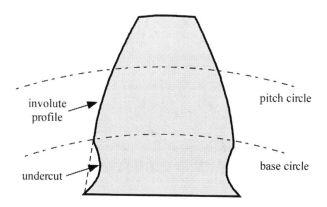

FIGURE 8.12
Undercut gear tooth.

In addition to avoiding gears having too few teeth, interference can also be avoided by removing gear-tooth material between the base and dedendum circles. This procedure is called *undercutting*. Figure 8.12 illustrates an undercut gear tooth. Because undercutting removes gear-tooth material (subsequently compromising gear-tooth strength), it should be avoided if possible.

8.3.5 Backlash

Backlash is the clearance measured along the pitch circle between the nondriving surfaces of mating gear teeth (in general, it is the amount of play between mating teeth). Figure 8.13 illustrates mating gear teeth with corresponding backlash labeled. A limited amount of backlash is necessary to prevent the mating gear teeth from binding with each other. Backlash also helps to enable gear-tooth lubrication because it provides clearance for lubricant flow. Although it is important to include clearance between mating gear teeth for

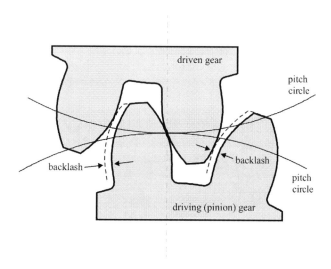

FIGURE 8.13
Backlash in mating gears.

proper gear operation, excessive backlash can produce inaccurate gear motion as well as large dynamic loads. AGMA provides tables of recommended backlash ranges [4].

8.4 Helical-Gear Nomenclature

As mentioned in Section 8.2, helical gears can be used to transmit motion between parallel and nonparallel shafts (also called *crossed shafts*) that do not intersect.[*] It was also noted in that section that helical gears have a greater load-bearing capacity and wear resistance, quieter and smoother operation than spur gears. These advantages are the result of the teeth of a helical gear lying at an angle (labeled the *helix angle* in Figure 8.15) with respect to the gear's center axis. This angled tooth profile increases the contact between mating gear teeth for greater load distribution (resulting in the advantages given).[†]

There are two designations for helical gears: *right-hand* and *left-hand*. As illustrated in Figure 8.14, the teeth in a right-hand helical gear slope downward-right and the teeth in a left-hand helical gear slope downward-left. This designation holds true whether the gears are driving gears or driven gears. Considering these designations, the helical gear illustrated in Figure 8.15 is a left-hand helical gear.

As you will notice in this section, the forthcoming formulas for helical gears are similar to those of spur gears, except for the inclusion of the *helix angle* (represented by the

Left-Hand Gear   Right-Hand Gear
(teeth slope downward-left)   (teeth slope downward-right)

FIGURE 8.14
Left-hand and right-hand helical gears.

---

[*] When used on nonparallel, nonintersecting shafts, the term *crossed helical gears* is used to describe the system.
[†] Although the angled tooth profile makes helical gears more difficult to manufacture than spur gears, they are often preferred over spur gears due to the advantages given.

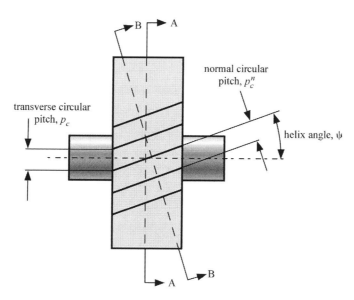

FIGURE 8.15
Helical-gear geometry.

variable ψ). In fact, by considering a spur gear to be a special type of helical gear (one where $\psi = 0°$), the forthcoming equations are directly applicable to spur gears.

As illustrated in Figure 8.15, there are two directions to consider with helical gears: the *normal* direction (labeled B-B) and the *transverse* direction (labeled A-A). There are several helical-gear variables that are given in both the normal and transverse directions. For example, in Figure 8.15, there is a circular pitch that is given in the transverse direction, as well one given in the normal direction. Other helical-gear variables that are given in both directions include the diametral pitch, module, and pressure angle.

The *normal circular pitch* (represented by variable p_c^n) is the distance measured along the pitch circle from one point on a gear tooth to the corresponding point on an adjacent gear tooth *in the normal direction* (or normal to the gear tooth). The equation for the normal circular pitch is expressed as

$$p_c^n = p_c \cos \psi \tag{8.14}$$

The *normal diametral pitch* (represented by the variable P_d^n) is the number of gear teeth per inch of pitch diameter *in the normal direction* and can be expressed as

$$P_d^n = \frac{\pi}{p_c^n} \tag{8.15}$$

The *normal module* (represented by the variable m^n) is the ratio of the pitch diameter to the number of gear teeth *in the normal direction* and can be expressed (in a form identical to Equation 8.5) as

$$p_c^n = \frac{\pi}{p_d^n} = \pi m^n \tag{8.16}$$

and as

$$m = \frac{m^n}{\cos \psi} \tag{8.17}$$

The *normal pressure angle* (represented by the variable ϕ^n) is the angle the pitch line makes with the pressure line *in the normal direction* and can be expressed as

$$\tan \phi^n = \tan \phi \cos \psi \tag{8.18}$$

As previously noted, Equations 8.14 through 8.18 are directly applicable for spur gears at $\psi = 0°$. Therefore, at a helix angle of zero, these equations become identical to their respective spur-gear equations in Sections 8.3.1 and 8.3.2.

Just as spur gear-tooth features are standardized with respect to the diametral pitch, helical gear-tooth features are standardized with respect to the normal diametral pitch. Table 8.3 includes several gear-tooth feature equations as given and certified by AGMA and ANSI, respectively [5].

TABLE 8.3

Formulas for Helical Gear-Tooth Proportions (14.5°, 20°, and 25° Pressure Angle Involute ANSI Fine Pitch)

Tooth Feature to Calculate	Normal Diametral Pitch P_d^n Known
Addendum, a	$a = \dfrac{1}{P_d^n}$
Dedendum, b	$b = \left[\dfrac{1.200}{P_d^n} \right] + 0.002 \text{(min)}$
Working depth, h_k	$h_k = \dfrac{2.000}{P_d^n}$
Whole depth, h_t	$h_t = \left[\dfrac{2.200}{P_d^n} \right] + 0.200 \text{(min)}$
Clearance (standard), c	$c = \left[\dfrac{0.200}{P_d^n} \right] + 0.002 \text{(min)}$
Clearance (shaved or ground teeth), c	$c = \left[\dfrac{0.350}{P_d^n} \right] + 0.002 \text{(min)}$
Normal circular tooth thickness, t_n	$t_n = \dfrac{\pi}{2 P_d^n}$
Pitch diameter, d_p	$d_p = \dfrac{N}{\left[P_d^n \cos \Psi \right]}$
Center distance (external), C_{external}	$C_{\text{extrnel}} = \dfrac{[N_1 + N_2]}{\left[2 P_d^n \cos \Psi \right]}$
Center distance (ext–int), $C_{\text{ext–int}}$	$C_{\text{ext–int}} = \dfrac{[N_1 - N_2]}{\left[2 P_d^n \cos \Psi \right]}$

Source: ANSI B6.7-1977. Fine-pitch helical-gear tooth forms. New York: American National Standards Institute [6].

```
>> N = 20;
>> pc = 0.5;
>> psi = 35*pi/180;
>> phi = 25*pi/180;
>> pcn = pc*cos(psi);
>> Pdn = pi/pcn

Pdn =

     7.6704

>> mn = 1/Pdn;
>> m = mn/cos(psi)

m =

     0.1592

>> phin = atan(tan(phi)*cos(psi))*180/pi

phin =

     20.9057

>> hk = 2/Pdn

hk =

     0.2607

>> dp = N/(Pdn*cos(psi))

dp =

     3.1831

>>
```

FIGURE E.8.3
Example 8.3 solution calculation procedure in MATLAB.

Example 8.3

Problem Statement: Calculate the normal diametral pitch, module, and normal pressure angle for a helical gear having 20 teeth, a circular pitch of 0.5, and a helix angle of 35°. Assuming this gear is an ANSI fine pitch with a 25° pressure angle (Table 8.3), calculate the working depth and pitch diameter of this gear also.

   Known Information: N, p_c, ψ, and ϕ.

   Solution Approach: Figure E.8.3 includes the calculation procedure in MATLAB's command window.

8.5 Gear Kinematics

8.5.1 Spur Gears and Gear Trains

By adhering to the fundamental law of gearing, a constant velocity ratio is maintained. In terms of the variables given in Figure 8.16, Equation 8.1 can be expressed as

$$V_P = r_{p1}\omega_1 = r_{p2}\omega_2$$

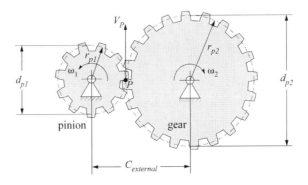

FIGURE 8.16
Mating-gear pair in motion.

or

$$VR = \frac{\omega_1}{\omega_2} = \frac{r_{p2}}{r_{p1}} \tag{8.19}$$

where the velocity variable V_p in Figure 8.16 is called the pitch-line velocity: the velocity of the pitch point (point P) of the mating-gear pair.

The velocity ratio can also be expressed in terms of the pitch diameters or the number of gear teeth, since $d_p = 2r_p$ and $N = d_p P_d$ from Equation 8.3. When expressed in terms of these variables, Equation 8.19 becomes

$$VR = \frac{\omega_1}{\omega_2} = \frac{r_{p2}}{r_{p1}} = \frac{d_{p2}}{d_{p1}} = \frac{N_2}{N_1} \tag{8.20}$$

While Equation 8.20 includes the gear angular velocity ratio ω_1/ω_2, it can also include the ratios of gear angular displacement θ_1/θ_2 or angular acceleration α_1/α_2.

A group of mating-gear pairs is called a *gear train* (Figure 8.17). Gear trains are often used when large velocity reductions are required because the amount of velocity reduction possible in a single-gear pair for practical use is limited.* Rather than achieve a large velocity reduction in a single-gear pair, such a reduction is achieved over multiple stages using multiple gear pairs.

The velocity ratio for a gear train is called a *train value*. This value is the ratio (of quantities including those in Equation 8.20) of the initial driving gear to the final output gear in the gear train. Because a gear train is comprised of multiple gear pairs, the train value is also the product of the individual velocity ratios of each gear pair in the gear train. If we define the velocity ratio of each gear pair as VR_i (where $i = 1, 2, 3 \ldots$), the equation for the train value can be expressed as

$$\frac{\omega_{input}}{\omega_{output}} = (VR_1)(VR_2)(VR_3)\ldots \tag{8.21}$$

---

* The velocity reduction achieved in a single-gear pair is limited by the maximum number of gear teeth and the maximum gear size.

FIGURE 8.17
Gear train.

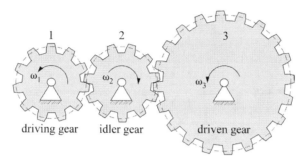

FIGURE 8.18
Gear train with idler gear.

When calculating a train value, it is important to note that if the input and output gears in an individual gear pair rotate in opposite directions, the corresponding velocity ratio is negative and the velocity ratio is positive when they rotate in the same direction.

Because the driven gear will always rotate in the opposite direction to the driving gear in a gear pair, it is necessary to include an *idler gear* to enable both gears to rotate in the same direction. Making this gear (Figure 8.18) identical and equivalent in size and number of teeth, respectively, to the driving gear minimizes both the additional space needed to include it and its rotation speed. Using Equation 8.21, it can be determined that the train value for the gear train in Figure 8.18 is indeed positive (and subsequently consistent with rotation direction of Gears 1 and 3) since it is the product of two negative velocity ratios—the velocity ratios for gear pairs 1–2 and 2–3.

In addition to gear kinematic motion, knowing the forces and torques acting on spur gears are important in gear design and operation. Figure 8.19 illustrates the normal force F_n exerted by the driving gear on the driven gear. Because this force is normal to the contact surfaces of the mating gear teeth at point P, it acts along the pressure line (the line of contact) and subsequently is orientated at the pressure angle ϕ. The radial and tangential components of F_n are the radial force F_r and tangent force F_t, and are expressed as

$$F_r = F_n \sin \phi \tag{8.22}$$

$$F_t = F_n \cos \phi \tag{8.23}$$

The radial force is directed toward the gear center and acts to deflect the gear shaft (acting to move the driven gear out of contact with the driving gear). The tangential force is tangent to the pitch circles at point P and acts to rotate the driven gear.

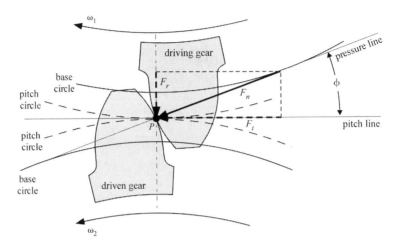

FIGURE 8.19
Gear-tooth force and force components.

The torque T produced about the center of a gear is given by

$$T = F_n\left(\frac{d_b}{2}\right) = F_n\left(\frac{d_p}{2}\right)\cos\phi = F_t\left(\frac{d_p}{2}\right) \tag{8.24}$$

The power (in horsepower or hp) transmitted by a rotating gear is given by

$$hp = \frac{Tn}{63,025} = \frac{F_t V_P}{33,000} \tag{8.25}$$

where the gear rotation speed (in revolutions/min) is represented by the variable n.[*] The power (in kilowatts or kW) transmitted by a rotating gear is given by

$$kW = \frac{T\omega}{1,000,000} = \frac{F_t V_P}{1,000,000} \tag{8.26}$$

where the gear rotation speed (in rad/s) is represented by the variable ω.[†]

Example 8.4

Problem Statement: Given a train of four gears where $N_1 = 20$, $N_2 = 35$, $N_3 = 55$, $N_4 = 80$, calculate the train value. Assuming the rotational speed of the input gear is $\omega_1 = 7\,\text{rad/s}$ and the output gear torque is $T_4 = 150$ N-mm, calculate the power transmitted by the output gear.

Known Information: N_1, N_2, N_3, N_4, ω_1, and T_4.

Solution Approach: Figure E.8.4 includes the calculation procedure in MATLAB's command window.

---

[*] In Equation 8.25, T is given in inch-pounds, F_t is given in pounds, and V_P is given in feet per minute.

[†] In Equation 8.26, T is given in newton-millimeters, F_t is given in newtons and V_P is given in millimeters per second.

```
>> N1 = 20;
>> N2 = 35;
>> N3 = 55;
>> N4 = 80;
>> omega1 = 7;
>> T4 = 150;
>> tv = -N2/N1*-N3/N2*-N4/N3

tv =

     -4

>> omega4 = omega1*N1/N4;
>> kW = T4*omega4/1000000

kW =

   2.6250e-04

>>
```

FIGURE E.8.4
Example 8.4 solution calculation procedure in MATLAB.

8.5.2 Planetary Gear Trains

The gear train illustrated in Figure 8.20a is called a planetary gear train.* The name planetary is used to describe this gear system because it consists of a gear (called a planet gear) that can rotate about a center gear (called a sun gear). Figure 8.20a illustrates the most basic planetary gear train. In this system, the sun gear is typically connected to an input shaft and the ring gear is typically connected to an output shaft. The planet gear mates with both the sun and ring gears and the carrier constrains the planet gear (to rotation about the sun gear). Although this textbook considers the most basic planetary gear train, multiple variations of the planetary gear train have been identified and used in practice.† Figure 8.20b illustrates a simple line diagram (called a skeleton diagram) of the planetary

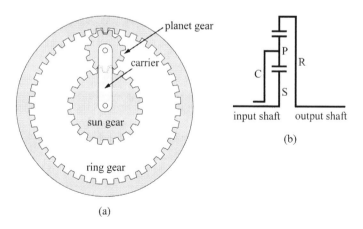

(a)

FIGURE 8.20
(a) A simple planetary gear train and (b) skeleton diagram.

---

* Planetary gear trains are also called *epicyclic gear trains*.
† Planetary gear train variations include various combinations of sun, planet, and ring gears.

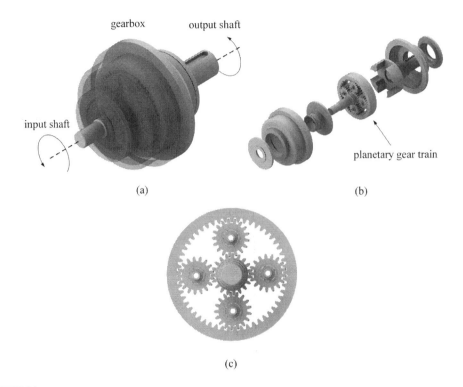

FIGURE 8.21
(a) Speed reduction system with (b, c) planetary gear train.

gear train. In this diagram, the planet, sun, ring, and carrier are labeled by the letters P, S, R, and C, respectively.

Planetary gear trains have several distinct advantages over conventional gear trains. For example, large velocity reductions can be achieved with a planetary gear train in a workspace that is more compact than a conventional gear train. Gear ratio changes are also achieved by constraining different members (gears or carriers) in the planetary gear train. In transmission systems (where planetary gears are commonly utilized), gear ratios are changed by *computer-operated* controls (in automatic transmissions) or *manually-operated* controls (in manual transmissions).

Planetary gear trains are commonly used in transmission systems. Figures 8.21b and c illustrate a planetary gear train used in a speed reduction system (Figure 8.21a). Within this speed reduction system (as the name implies), the input rotational speed of the driving shaft is reduced and delivered to the output shaft.

Given the number of teeth in each gear, there is a three-step procedure for calculating the rotations of each member of the planetary gear train.* The steps of this method are as follows:

Step 1: Assume any initially fixed member is unconstrained, the motion of the carrier is fixed, and calculate the rotations of the remaining gears, given a rotation and rotation direction of any single gear.

Step 2: Assume all members are unconstrained and include the full rotation of all members in the direction opposite to the given rotation direction in Step 1.

---

* This three-step procedure is called the *tabular method*.

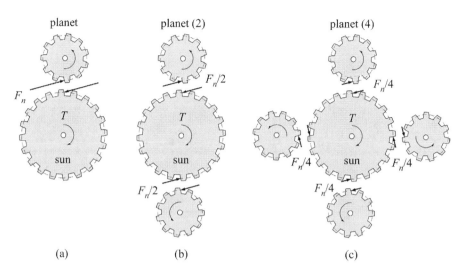

planet   planet (2)   planet (4)

F_n T $F_n/2$ T $F_n/4$ T

sun   sun $F_n/4$ sun $F_n/4$

$F_n/2$ $F_n/4$

(a)   (b)   (c)

FIGURE 8.22
Sun gear conjugating with (a) one, (b) two, and (c) four planet gears.

Step 3: Take the sum of the rotations from Steps 1 and 2 for each member of the gear train (this procedure is called *superposition*). The rotations calculated in this step are the true rotations of the gear train under the initial conditions.

This three-step procedure can be completed in a table of three rows and n columns, where each row is a step in the procedure and each column is a member of the gear train. The rotational velocities of each gear can be calculated from the known velocity ratios.

Unlike the simple planetary gear train illustrated in Figure 8.20, the planetary gear train in Figure 8.22 includes four planet gears. The inclusion of additional planet gears decreases the normal force transmitted between each sun–planet pair. By reducing F_n, the torque capacity of the sun gear is increased. This effect is illustrated in Figure 8.22. While a single sun–planet pair produces a normal force of F_n, this force is reduced by 50% with the inclusion of two sun–planet pairs (Figure 8.22b). Likewise, F_n is reduced by 75% with the inclusion of four sun–planet pairs (Figure 8.22c). As a result, the sun gears in Figures 8.22b and c have double and quadruple the torque capacity, respectively, of the sun gear in Figure 8.22a.*

Example 8.5

Problem Statement: Consider the simple planetary gear train design in Figure 8.20. In this example, $N_{sun}=20$, $N_{planet}=10$, $N_{ring}=40$. Also, the sun gear is fixed, the input is the carrier, and the output is the ring gear.[†] Calculate the rotations of the sun, planet, ring, and carrier. Also calculate the output rotational speed if the input rotational speed is $\omega_{carrier}=35\,\text{rad/s}$.

    Known Information: N_{sun}, N_{planet}, N_{ring}, and ω_{sun}.

    Solution Approach: Equations 8.20 and 8.21 also hold true for gear angular displacements. Knowing this, the angular displacements for the planet and ring gears (Step 1) can be calculated. Assuming a full counterclockwise sun gear rotation, Figure E.8.5 includes the Step 1 calculation procedure in MATLAB's command window for the planet and ring gears.

---

* The sun–planet force transmitted becomes F_n/N_{planet} and the sun torque capacity becomes $(N_{planet})T$, where F_n and T are the force transmitted and torque capacity, respectively, and N_{planet} is the total number of planet gears.
† If the carrier is fixed, this problem can be solved using Equations 8.20 and 8.21 (tabular method not needed).

Table E.8.1 includes the results calculated for Step 1. In accordance with Step 2, a full clockwise rotation is specified for all links and Step 3 includes the sum of the results from Steps 1 and 2. Table E.8.1 also includes the values specified and calculated for Steps 2 and 3, respectively.

With the planetary gear rotations calculated from Step 3 and the known rotation speed of the carrier, the output rotation speed (the rotation speed of the carrier) can be calculated using Equations 8.20 and 8.21. Figure E.8.6 includes this calculation procedure in MATLAB's command window.*

```
>> Ns = 20;
>> Np = 10;
>> Nr = 40;
>> rot_s = 1;
>> rot_p = rot_s*-Ns/Np

rot_p =

    -2

>> rot_r = rot_p*Np/Nr

rot_r =

   -0.5000

>>
```

FIGURE E.8.5
Example 8.5 solution calculation procedure (for Step 1) in MATLAB.

TABLE E.8.1

Planetary Gear Rotation Analysis Table for Example 8.5

	Sun Gear	Planet Gear	Ring Gear	Carrier
Step 1 (fixed carrier)	1	−2	−0.5	0
Step 2 (full counterrotation)	−1	−1	−1	−1
Step 3 (Step 1+Step 2)	0	−3	−1.5	−1

```
>> rot_p = -3;
>> rot_r = -1.5;
>> rot_c = -1;
>> omega_c = 35;
>> omega_p = omega_c*rot_p/rot_c

omega_p =

   105

>> omega_r = omega_p*rot_r/rot_p

omega_r =

   52.5000

>>
```

FIGURE E.8.6
Example 8.5 solution calculation procedure (for rotational speed) in MATLAB.

---

* In this problem, positive and negative rotation values correspond to counterclockwise and clockwise rotations, respectively.

8.5.3 Rack and Pinion Gears

As noted in Section 8.2, in the rack and pinion gear, the rotational motion of the pinion gear produces the translational motion of the rack gear. In Equation 8.27, the translation of the rack is represented by the variable Δs and the rotation of the pinion (in radians) is represented by the variable $\Delta\theta$. The variables r_p and d_p are the pitch radius and diameter of the pinion, respectively.

$$\Delta s = r_p \, \Delta\theta = \frac{d_p}{2} \, \Delta\theta \tag{8.27}$$

The rack and pinion gear is kinematically identical to a gear system consisting of an external gear (the pinion gear) and an internal gear of infinite radius (the rack gear). Because of this, the pitch-line velocity V_P in a spur-gear pair is identical to the velocity of the rack in a rack and pinion gear. Therefore, for the rack and pinion gear, Equation 8.19 can be expressed as

$$V_P = V_{\text{rack}} = r_p \, \omega = \frac{d_p}{2} \, \omega \tag{8.28}$$

Where the variables V_{rack} and ω are the velocity of the rack and the angular velocity of the pinion, respectively.

Example 8.6

Problem Statement: Calculate the translation and velocity of a rack in a rack and pinion gear where the pinion rotation, speed, and diameter are $\pi/6$ rad, 7 rad/s, and 25 mm, respectively.

Known Information: $\Delta\theta$, ω, and d_p.

Solution Approach: Figure E.8.7 includes the solution calculation procedure in MATLAB's command window.

```
>> theta = pi/6;
>> omega = 7;
>> dp = 25;
>> s = theta*dp/2

s =

    6.5450

>> Vrack = omega*dp/2

Vrack =

   87.5000

>>
```

FIGURE E.8.7
Example 8.6 solution calculation procedure in MATLAB.

8.5.4 Helical Gears

In addition to spur gears, Equations 8.19 and 8.20 are also applicable for the kinematic analysis of helical gears. Figure 8.23 illustrates the normal force F_n exerted on the driven helical gear. There are three orientation angles associated with F_n. It has a pressure angle of ϕ in the transverse direction (direction A-A in Figure 8.15), a pressure angle of ϕ^n in the normal direction (direction B-B in Figure 8.15), and a helix angle of ψ, since it is normal to the helical gear teeth in contact at the pressure point.

As illustrated in Figure 8.23, F_n has three components: a radial component F_r, a tangential component F_t, and an axial (or thrust) component F_a. These forces are expressed as

$$F_r = F_n \sin \phi^n = F_t \tan \phi \tag{8.29}$$

$$F_t = F_n \cos \phi^n \cos \psi \tag{8.30}$$

$$F_a = F_n \cos \phi^n \sin \Psi = F_t \tan \Psi \tag{8.31}$$

Like spur-gear forces, F_t contributes to torque and F_r contributes to shaft bending for helical gears. The axial force component F_a is in the direction parallel to the gear shaft axis and acts to move the gear along its shaft axis.[*] As illustrated in Figure 8.24, for pinion gears having a counterclockwise rotation (using the right-hand rule), F_a acts leftward in left-hand pinion gears and rightward in right-hand pinion gears. For a clockwise pinion gear rotation, the direction of F_a is reversed for both gears in Figure 8.24. For the gears that are directly driven by these pinions, both the rotation direction and F_a direction are the opposite of those in their pinions (assuming only external gears and parallel shafts are used).

The torque and horsepower equations given in Section 8.5.1 for spur gears (Equations 8.24–8.26) are also applicable to helical gears.

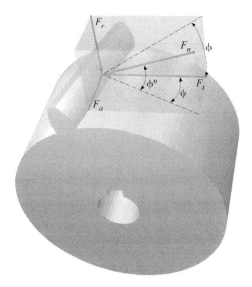

FIGURE 8.23
Helical gear-tooth force and force components.

---

[*] Helical gears are often fitted with *thrust bearings*—a particular type of bearing designed to handle axial loads.

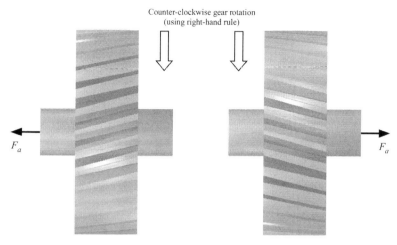

Counter-clockwise gear rotation
(using right-hand rule)

F_a

F_a

Left-Hand Pinion Gear

Right-Hand Pinion Gear

FIGURE 8.24
Direction of axial force in helical gears.

Example 8.7

Problem Statement: Calculate the torque and power transmitted by a helical gear having 40 teeth, a circular pitch of 25 mm, and a helix angle of 35°. Assume this gear is an ANSI fine pitch with a 25° pressure angle (Table 8.3). This gear also has a rotational speed of 7 rad/s and a normal force of 25 N.

Known Information: N, p_c, ψ, ϕ, ω, and F_n.

Solution Approach: Figure E.8.8 includes the calculation procedure in MATLAB's command window.

```
>> N = 40;
>> pc = 25;
>> psi = 35*pi/180;
>> phi = 25*pi/180;
>> omega = 7;
>> Fn = 25;
>> pcn = pc*cos(psi);
>> Pdn = pi/pcn;
>> dp = N/(Pdn*cos(psi));
>> phin = atan(tan(phi)*cos(psi));
>> Ft = Fn*cos(phin)*cos(psi);
>> T = Ft*dp/2

T =

   3.0447e+03

>> kW = T*omega/1000000

kW =

   0.0213

>>
```

FIGURE E.8.8
Example 8.7 solution calculation procedure in MATLAB.

8.5.5 Bevel Gears

In addition to spur and helical gears, Equations 8.19 and 8.20 are also applicable to the kinematic analysis of bevel gears. In fact, the rolling cylinder description used to describe the motion of spur gears become *rolling conical wheels* for bevel gears.

Figure 8.25 includes several bevel-gear design features. As noted in Section 8.2, bevel gears are designed to transmit motion between nonparallel shafts that intersect. The *shaft angle* (represented by the variable Σ in Figure 8.25), is the angle between the center axes of the pinion and gear; although, in many bevel applications, this angle is 90°, bevel gears are not limited to a right shaft angle orientation.*

The shaft angle is the sum of the *pinion-pitch angle* and the *gear-pitch angle* (Figure 8.25). The pitch angle is the angle of the cone upon which the bevel gear is constructed. The gear- and pinion-pitch angles are represented by the variables γ_{pinion} and γ_{gear}, respectively. The equations for γ_{pinion} and γ_{gear} can be expressed as

$$\tan \gamma_{pinion} = \frac{\sin \Sigma}{\cos \Sigma + \dfrac{N_{gear}}{N_{pinion}}} \tag{8.32}$$

$$\tan \gamma_{gear} = \frac{\sin \Sigma}{\cos \Sigma + \dfrac{N_{pinion}}{N_{gear}}} \tag{8.33}$$

and the shaft angle can be expressed as

$$\Sigma = \gamma_{pinion} + \gamma_{gear} \tag{8.34}$$

It can be observed that Equations 8.32 and 8.33 include the velocity ratio N_{gear}/N_{pinion}.

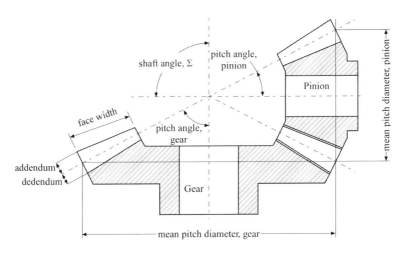

FIGURE 8.25
Mating bevel-gear pair (cross-section view).

---

* *Miter gears* are mating bevel gears having equal numbers of teeth, diametral pitches (giving them a 1:1 gear ratio), and right shaft angle orientations.

Example 8.8

Problem Statement: Calculate the pitch angles for a bevel-gear pair where the pinion and gear have 30 and 45 teeth, respectively, and the shaft angle is 90°.

Known Information: N_{pinion}, N_{gear}, and Σ.

Solution Approach: Figure E.8.9 includes the calculation procedure in MATLAB's command window.

The forces acting on bevel gears are identical in in terms of force components to those acting helical gears. Figure 8.26 illustrates the forces exerted in a mating pair of straight-tooth bevel gears. Equations 8.35–8.37 are the tangential, radial and axial force equations for a straight-tooth pinion bevel gear.

```
>> Np = 30;
>> Ng = 45;
>> sigma = 90*pi/180;
>> gamma_p = atan(sin(sigma)/(cos(sigma) + Ng/Np))*180/pi

gamma_p =

   33.6901

>> gamma_g = atan(sin(sigma)/(cos(sigma) + Np/Ng))*180/pi

gamma_g =

   56.3099

>>
```

FIGURE E.8.9
Example 8.8 solution calculation procedure in MATLAB.

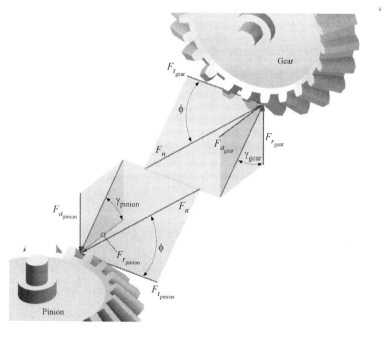

FIGURE 8.26
Bevel gear pair force and force components.

$$F_{t_{\text{pinion}}} = F_n \cos \phi \tag{8.35}$$

$$F_{r_{\text{pinion}}} = F_n \sin \phi \cos \gamma_{\text{pinion}} = F_{t_{\text{pinion}}} \tan \phi \cos \gamma_{\text{pinion}} \tag{8.36}$$

$$F_{a_{\text{pinion}}} = F_n \sin \phi \sin \gamma_{\text{pinion}} = F_{t_{\text{pinion}}} \tan \phi \sin \gamma_{\text{pinion}} \tag{8.37}$$

Equations 8.38–8.40 are the tangential, radial and axial force equations for the driven bevel gear.

$$F_{t_{\text{gear}}} = F_n \cos \phi = F_{t_{\text{pinion}}} \tag{8.38}$$

$$F_{r_{\text{gear}}} = F_n \sin \phi \cos \gamma_{\text{gear}} = F_{t_{\text{gear}}} \tan \phi \cos \gamma_{\text{gear}} \tag{8.39}$$

$$F_{a_{\text{gear}}} = F_n \sin \phi \sin \gamma_{\text{gear}} = F_{t_{\text{gear}}} \tan \phi \sin \gamma_{\text{gear}} \tag{8.40}$$

As illustrated in Figure 8.26, the radial force on the pinion and the axial force on the gear are equal in magnitude while the axial force on the pinion and the radial force on the gear are equal in magnitude. As a result, Equations 8.41 and 8.42 hold true for 90° shaft orientations. Equations 8.43 and 8.44 result from the 90° shaft orientation condition $\gamma_{\text{pinion}} + \gamma_{\text{gear}} = 90°$.

$$F_{r_{\text{pinion}}} = F_{a_{\text{gear}}} = F_n \sin \phi \cos \gamma_{\text{pinion}} = F_n \sin \phi \sin \gamma_{\text{gear}} \tag{8.41}$$

$$F_{a_{\text{pinion}}} = F_{r_{\text{gear}}} = F_n \sin \phi \sin \gamma_{\text{pinion}} = F_n \sin \phi \cos \gamma_{\text{gear}} \tag{8.42}$$

$$\sin \gamma_{\text{pinion}} = \sin \left(90° - \gamma_{\text{gear}}\right) = \cos \gamma_{\text{gear}} \tag{8.43}$$

$$\cos \gamma_{\text{pinion}} = \cos \left(90° - \gamma_{\text{gear}}\right) = \sin \gamma_{\text{gear}} \tag{8.44}$$

Because gear torque can be defined as the product of the tangential force and the pitch radius (Equation 8.24), equations for the pinion and gear torque can be expressed as

$$T_{\text{pinion}} = F_{t_{\text{pinion}}} r_{m_{\text{pinion}}} \tag{8.45}$$

$$T_{\text{gear}} = F_{t_{\text{gear}}} r_{m_{\text{gear}}} = F_{t_{\text{pinion}}} r_{m_{\text{gear}}} \tag{8.46}$$

where $r_{m_{\text{pinion}}}$ and $r_{m_{\text{gear}}}$ are the *mean pitch radii* of the pinion and gear (Figure 8.25).[*]

Example 8.9

Problem Statement: Consider the bevel gear pair in Example 8.8 where $F_n = 400\,\text{N}$, $r_{m_{\text{pinion}}} = 76.2\,\text{mm}$, $\phi = 25°$ and $\omega_{\text{pinion}} = 5.25\,\text{rad/s}$. For this gear pair, calculate the tangential, radial and axial forces on the gear, the pinion and gear torque and the power transmitted.

Known Information: Example 8.8, F_n, $r_{m_{\text{pinion}}}$, ϕ, and ω_{pinion}.

Solution Approach: Figure E.8.10 includes the calculation procedure in MATLAB's command window.

---

[*] The mean pitch radius for a bevel gear can be calculated using $r_m = (d_i + d_o)/4$ where d_i and d_o are the bevel gear's inside and outside pitch diameters respectively.

```
>> Np = 30;
>> Ng = 45;
>> gamma_p = 33.6901*pi/180;
>> gamma_g = 56.3099*pi/180;
>> Fn = 400;
>> rm_p = 76.2;
>> phi = 25*pi/180;
>> omega_p = 5.25;
>> Ft_g = Fn*cos(phi)

Ft_g =

   362.5231

>> Fr_g = Fn*sin(phi)*cos(gamma_g)

Fr_g =

   93.7707

>> Fa_g = Fn*sin(phi)*sin(gamma_g)

Fa_g =

   140.6558

>> Ft_p = Ft_g;
>> T_p = Ft_p*rm_p

T_p =

   2.7624e+04

>> rm_g = rm_p*Ng/Np;
>> T_g = Ft_g*rm_g

T_g =

   4.1436e+04

>> kW = T_p*omega_p/1000000

kW =

    0.1450

>>
```

FIGURE E.8.10
Example 8.9 solution calculation procedure in MATLAB.

8.5.6 Worm Gears

Like helical gears, worm gears are used to transmit motion between nonparallel, nonintersecting shafts and, like planetary gears, worm gears can produce large velocity ratios in a compact work-space. Although the shafts do not intersect with worm gears, they are orthogonal to each other. The worm gear is comprised of two mating gears: the larger *gear* and the smaller *worm*. In practice, the worm typically transmits motion to the gear.

Due to the structural resistance and friction resistance between the worm and the gear, the gear is typically incapable of driving the worm. This means that a stationary worm will lock the motion of the gear. The *self-locking* capacity of worm gears can be advantageous in applications where a system is to be locked in position when the power is turned off (such as a loaded hoist, crane, or jack) for safety.

The worm resembles a screw. In fact, the teeth of the worm are called *treads*. The *number of worm teeth* (or the number of threads) is represented by the variable N_{worm}. The gear is

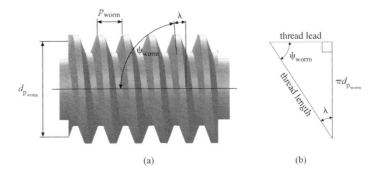

FIGURE 8.27
a) Worm geometry and features (single thread) and (b) trigonometric relationship from worm geometry (Figure 8.27b is not drawn to scale with Figure 8.27a)

often a helical gear, although the involute profile of the gear is often concave (rather than the usual convex involute profile) to better enclose the worm threads.[*]

Figure 8.27 includes the geometry of the worm. The *worm-pitch diameter, worm pitch* and *worm lead angle* are represented by variables $d_{p_{worm}}$, p_{worm}, and λ, respectively. The worm lead angle and the *worm helix angle* ψ_{worm} are complementary (therefore $\lambda + \psi_{worm} = 90°$). The worm-pitch diameter is the diameter of the circle that remains tangent to the pitch diameter of the gear. The worm pitch is the axial distance between one point on a thread to the corresponding point on an adjacent thread. If we can recall the definitions of these two terms for spur gears, we can see that the worm-pitch diameter and pitch are defined in a similar manner as for spur gears. The worm illustrated in Figure 8.27a is a *single thread*. By this we mean that the thread is a single helix. Worms that include two helixes have *double thread* (and *triple thread* for three helixes and so on).

Figure 8.27b illustrates the relationship between the *thread length, thread lead*, and *pitch circumference* (or $\pi d_{p_{worm}}$) of a worm. The thread lead is the axial distance that a thread advances in one revolution of the worm and can be expressed as N_{worm} and P_{worm}.[†]

From the trigonometric relationship illustrated in Figure 8.27b, a relationship that includes the worm geometry features can be expressed as

$$\tan \lambda = \cot \psi_{worm} = \frac{N_{worm} p_{worm}}{\pi d_{p_{worm}}} \tag{8.47}$$

For a mating worm-gear pair, the worm pitch and the circular pitch of the gear must be identical. Therefore, from Equation 8.5,

$$p_{worm} = p_{c_{gear}} = \frac{\pi}{P_d} \tag{8.48}$$

Because worm gears are used for shafts that are aligned at 90° angles, the lead angle of the worm and the helix angle of the gear must be identical.

---

[*] The technique of cutting concave gear teeth to better enclose the worm threads is called *enveloping worm-gear teeth* and produces a line of contact rather than a point of contact (for better force transfer). In *double enveloping*, the worm thread is cut concave along its length for even greater contact and force transfer than with enveloping.

[†] From this relationship, it can be observed that the leads for single, double, and triple thread worms are one, two, and three times the worm pitch, respectively.

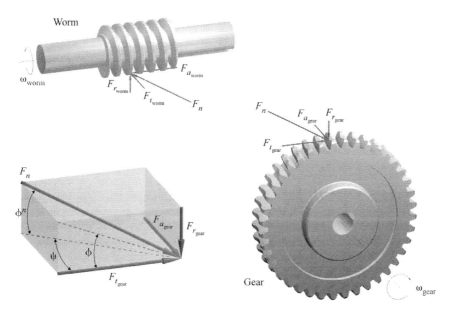

FIGURE 8.28
Worm-gear pair force and force components.

The velocity ratio of a worm-gear pair is defined in the same manner as spur, rack and pinion, helical, and bevel gears. Considering the notation for a worm-gear pair, the velocity ratio becomes

$$\frac{\omega_{\text{gear}}}{\omega_{\text{worm}}} = \frac{N_{\text{worm}}}{N_{\text{gear}}}$$

(8.49)

As noted before, the worm drives the gear.

The forces acting on worms and gears are identical in terms of force components to those acting on helical gears. Figure 8.28 illustrates the forces exerted in a mating worm-gear pair. Equations 8.50–8.52 are the tangential, radial, and axial forces acting on the gear (where ψ is the helix angle of the gear). These equations are identical in form to the helical-gear equations for the same forces (Equations 8.29–8.31).

$$F_{t_{\text{gear}}} = F_n \cos\phi^n \cos\psi$$

(8.50)

$$F_{r_{\text{gear}}} = F_n \sin\phi^n = F_{t_{\text{gear}}} \tan\phi$$

(8.51)

$$F_{a_{\text{gear}}} = F_n \cos\phi^n \sin\psi = F_{t_{\text{gear}}} \tan\psi$$

(8.52)

Because the helix angle of the gear (ψ) and the lead angle of the worm (λ) are identical for shafts aligned at 90°, for this condition, Equations 8.50 and 8.52 become

$$F_{t_{\text{gear}}} = F_n \cos\phi^n \cos\lambda$$

(8.53)

$$F_{a_{\text{gear}}} = F_n \cos\phi^n \sin\lambda = F_{t_{\text{gear}}} \tan\lambda$$

(8.54)

Because torque in gears can be defined as the product of the tangential force and the pitch radius (Equation 8.24), equations for the torque on the worm and gear can be expressed as

$$T_{\text{worm}} = F_{t_{\text{worm}}} r_{p_{\text{worm}}} \tag{8.55}$$

$$T_{\text{gear}} = F_{t_{\text{gear}}} r_{p_{\text{gear}}} \tag{8.56}$$

where $r_{p_{\text{worm}}}$ and $r_{p_{\text{gear}}}$ are the pitch radii for the worm and gear, respectively.

For ANSI fine-pitch worms and gears, the pitch diameters for the worm and gear are [7]

$$d_{p_{\text{worm}}} = \frac{N_{\text{worm}} p_{\text{worm}}}{\pi \tan \lambda} \tag{8.57}$$

$$d_{p_{\text{gear}}} = \frac{N p_{c\,\text{gear}}}{\pi} \tag{8.58}$$

From Figure 8.28, it can be observed that $F_{a_{\text{gear}}} = F_{t_{\text{worm}}}$ (as well as $F_{t_{\text{gear}}} = F_{a_{\text{worm}}}$ and $F_{r_{\text{gear}}} = F_{r_{\text{worm}}}$). Making this substitution in Equation 8.55, as well as observing that $F_{a_{\text{gear}}} = F_{t_{\text{gear}}} \tan \lambda$ in Equation 8.54, the worm torque equation can be expressed as

$$T_{\text{worm}} = F_{a_{\text{gear}}} r_{p_{\text{worm}}} = F_{t_{\text{gear}}} (\tan \lambda) r_{p_{\text{worm}}} \tag{8.59}$$

Example 8.10

Problem Statement: Calculate the torque and power transmitted by the gear in a worm-gear pair. The worm has a single thread and rotates at 15 rad/s and the gear has 40 teeth and a circular pitch of 15 mm. A normal force of 45 N is applied to the gear. Assume the worm and gear are ANSI fine pitch with helix and pressure angles of 40° and 20°, respectively.

Known Information: N_{worm}, ω_{worm}, N_{gear}, $p_{c_{\text{gear}}}$, ψ, ϕ, and F_n.

Solution Approach: Figure E.8.11 includes the calculation procedure in MATLAB's command window.

```
>> Nw = 1;
>> Ng = 40;
>> omega_w = 15;
>> Fn = 45;
>> phi = 20*pi/180;
>> psi = 40*pi/180;
>> pc_g = 15;
>> phin = atan(tan(phi)*cos(psi));
>> Ft_g = Fn*cos(phin)*cos(psi);
>> dp_g = Ng*pc_g/pi;
>> omega_g = omega_w*Nw/Ng;
>> T_g = Ft_g*dp_g/2

T_g =

    3.1709e+03

>> kW = T_g*omega_g/1000000

kW =

    0.0012

>>
```

FIGURE E.8.11
Example 8.9 solution calculation procedure in MATLAB.

8.6 Summary

Gears are mechanical components used to transmit motion from one shaft to another. Although rotation can be transmitted through other mechanical components (e.g., belt-pulley systems and chain-sprocket systems), gears are more commonly used, appearing in mechanical systems of all sizes. Gears offer the advantages of higher torque, speed, and power capacities, no slip, greater durability, efficiency, and suitability for confined spaces.

Six types of gears are considered in this textbook. Spur gears are used to transmit motion between parallel shafts. A group of two or more mating-gear pairs form a gear train. Gear trains are often used to achieve large velocity ratios.

Planetary gear trains have several distinct advantages over conventional gear trains. For example, large velocity reductions can be achieved with a planetary gear train in a workspace that is more compact than a conventional gear train. Gear ratio changes are also achieved by constraining different members (gears or carriers) in the planetary gear train.

In rack and pinion gears, the rotational motion of the pinion gear produces the translational motion of the rack gear. Helical gears can be used to transmit motion between parallel shafts and nonparallel shafts that do not intersect. Bevel gears are used to transmit motion between intersecting shafts.

Like helical gears, worm gears are used to transmit motion between nonparallel shafts that do not intersect. Although the shafts do not intersect with worm gears, they are orthogonal to each other. Worm gears are used when large reduction ratios are required.

For optimal gear operation, the gear-tooth profile must be designed in such a way that the common normal to both contacting tooth surfaces (the pressure line) passes through the pitch point on the line of contact. These requirements, which are expressed in the fundamental law of gearing, ensure a constant velocity ratio (ratio of the driving and driven gear speeds).

References

1. Spotts, M. F., Shoup, T. E., and Hornberger, L. E. 2004. *Design of Machine Elements*. 8th edn, pp. 585–609. Upper Saddle River, NJ: Pearson Prentice-Hall.
2. Oberg, E., Jones, F. D., Horton, H. L., and Ryffel, H. H. 2000. *Machinery's Handbook*. 26th edn, pp. 2004–2005. New York: Industrial Press.
3. ANSI B6.1-1968 (R1974). Coarse-pitch spur-gear tooth forms. New York: American National Standards Institute.
4. Oberg, E., Jones, F. D., Horton, H. L., and Ryffel, H. H. 2000. *Machinery's Handbook*. 26th edn, pp. 2036–2041. New York: Industrial Press.
5. Ibid, p. 2008.
6. ANSI B6.7-1977. Fine-pitch helical-gear tooth forms. New York: American National Standards Institute.
7. Oberg, E., Jones, F. D., Horton, H. L., and Ryffel, H. H. 2000. *Machinery's Handbook*. 26th edn, p. 2065. New York: Industrial Press.

Additional Reading

Myszka, D. H. 2005. *Machines and Mechanisms: Applied Kinematic Analysis*. 3rd edn, Chapter 10. Saddle River, NJ: Prentice-Hall.

Norton, R. L. 2008. *Design of Machinery*. 4th edn, Chapter 9. New York: McGraw-Hill.

Waldron, Kenneth J. and Kinzel, Gary L. 2004. *Kinematics, Dynamics and Design of Machinery*. 2nd edn, Chapters 10–12. Saddle River, NJ: Prentice-Hall.

Wilson, C. E. and J. P. Sadler 2003. *Kinematics and Dynamics of Machinery*. 3rd edn. Chapters 6–8. Saddle River, NJ: Prentice-Hall.

Problems

1. From among the gear types presented in Chapter 8, list the gear type(s) that can accommodate shafts that are (a) *parallel*, (b) *intersecting*, and (c) *orthogonal and nonintersecting*.

2. Using Equations 8.2–8.4, calculate the pitch diameter, diametral pitch, and module for a spur gear having 45 teeth and a circular pitch of 0.6. Assuming this gear is an ANSI coarse pitch with a 20° pressure angle (Table 8.1), calculate the fillet radius, addendum, working depth, and circular tooth thickness of this gear also.

3. Using Equations 8.2–8.4, calculate the pitch diameter, diametral pitch, and module for a spur gear having 45 teeth and a circular pitch of 0.75. Assuming this gear is an ANSI coarse pitch with a 25° pressure angle (Table 8.1), calculate the clearance, dedendum, and whole depth (all preferred) of this gear also.

4. Given an external gear pair where $N_1 = 45$, $N_2 = 75$ and the circular pitch for the pinion is $p_{c_1} = 0.43$, calculate the center distance, base pitch, contact line length, and contact ratio for the gear pair. Assume the gears are ANSI coarse pitch with 25° pressure angles (Table 8.1).

5. Given an external–internal gear pair where $N_1 = 30$, $N_2 = 85$ and the circular pitch for the pinion is $p_{c_1} = 0.45$, calculate the center distance, base pitch, contact line length, and contact ratio for the gear pair. Assume the gears are ANSI coarse pitch with 20° pressure angles (Table 8.1).

6. Using Equation 8.13, calculate the maximum number of gear teeth for a pinion having 16 and 17 teeth and a 20° pressure angle (see footnote regarding variable k in Inequality (8.13)).

7. Determine which single gear from the gear train illustrated in Figure P.8.1 should be replaced and explain why?

8. What is the purpose of the idler gear? Describe its design. Also, what general advantages do helical gears have over spur gears and why?

9. Calculate the normal diametral pitch, module, and normal pressure angle for a helical gear having 25 teeth, a circular pitch of 0.4 and a helix angle of 45°. Assuming this gear is an ANSI fine pitch with a 14.5° pressure angle (Table 8.3), calculate the dedendum, center distance (external, where $N_2 = 35$), working depth, and pitch diameter of this gear also.

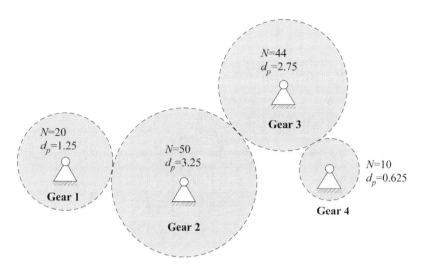

FIGURE P.8.1
Spur gear train.

10. Calculate the normal diametral pitch, module, and normal pressure angle for a helical gear having 40 teeth, a circular pitch of 0.3 and a helix angle of 55°. Assuming this gear is an ANSI fine pitch with a 20° pressure angle (Table 8.3), calculate the addendum, center distance (internal–external, where $N_2 = 125$), whole depth, and clearance (standard) of this gear also.

11. Calculate each of the quantities in Table 8.1 for a gear module of 0.185.

12. Calculate each of the quantities in Table 8.3 for a gear module and helix angle of 0.125 and 45°, respectively. Let $N_1 = 20$ and $N_2 = 45$ for an external gear pair, and $N_1 = 20$ and $N_2 = 90$ for an external–internal gear pair.

13. Given a train of four gears where $N_1 = 25$, $N_2 = 35$, $N_3 = 60$, and $N_4 = 80$, calculate the train value. Assuming an input gear pitch radius, rotational speed, and torque of $r_{p1} = 17$ mm, $\omega_1 = 10$ rad/s, and $T_1 = 125$ N-mm, respectively, calculate the power transmitted by each gear.

14. Given a train of six gears where $N_1 = 24$, $N_2 = 40$, $N_3 = 55$, $N_4 = 75$, $N_5 = 115$, and $N_6 = 135$, calculate the train value. Assuming an input gear pitch radius, rotational speed, and torque of $r_{p1} = 15$ mm, $\omega_1 = 8.5$ rad/s, and $T_1 = 230$ N-mm, respectively, calculate the power transmitted by each gear.

15. A five-bar mechanism that includes a gear pair is illustrated in Figure P.8.2. Gear 1 is attached to the input link and gear 2 is attached to the output link. The input link rotates twice as much as the output link, its gear has 33 teeth, a diametral pitch of 3.117, and a pressure angle of 25°. If gear 1 transmits a torque of 121 N-mm, what is the torque transmitted by gear 2?

16. A five-bar mechanism that includes a gear train is illustrated in Figure P.8.3. Gear 1 is attached to the input link, gear 2 is an idler gear, and gear 3 is attached to the output link. The input link rotates three times as much as the output link, its gear has 54 teeth, a diametral pitch of 1.57, and a pressure angle of 20°. If gear 1 transmits a torque of 75 N-mm, what is the torque transmitted by gear 3?

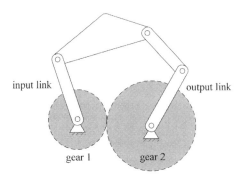

FIGURE P.8.2
Five-bar mechanism with a gear pair.

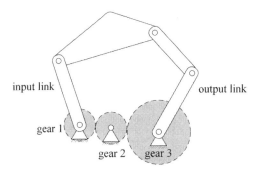

FIGURE P.8.3
Five-bar mechanism with a gear train.

17. What general advantages do planetary gear trains have over conventional gear trains?

18. Consider the simple planetary gear train design in Figure P.8.4. In this example, $N_{sun} = 21$, $N_{planet} = 17$, $N_{ring} = 55$. Also, the carrier is fixed, the input is the sun gear and the output is the ring gear. Calculate the rotations of the sun, planet, ring, and carrier. Also, assuming an input gear pitch radius, rotational speed, and torque of $r_{p_{sun}} = 28\,\text{mm}$, $\omega_{sun} = 25\,\text{rad/s}$, and $T_{sun} = 85\,\text{N-mm}$, respectively, calculate the power transmitted by each gear.

19. Consider the simple planetary gear train design in Figure P.8.4. In this example, $N_{sun} = 25$, $N_{planet} = 16$, $N_{ring} = 62$. Also, the sun gear is fixed, the input is the carrier and the output is the ring gear. Calculate the rotations of the sun, planet, ring, and carrier.

20. Consider the simple planetary gear train design in Figure P.8.4. In this example, $N_{sun} = 23$, $N_{planet} = 14$, $N_{ring} = 77$. Also, the ring gear is fixed, the input is the sun gear and the output is the carrier. Calculate the rotations of the sun, planet, ring, and carrier. Also, assuming an input gear pitch radius, rotational speed, and torque of $r_{p_{sun}} = 28\,\text{mm}$, $\omega_{sun} = 41\,\text{rad/s}$, and $T_{sun} = 65\,\text{N-mm}$, respectively, calculate the power transmitted by each gear.

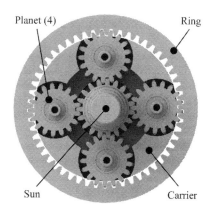

Planet (4)

Ring

Sun

Carrier

FIGURE P.8.4
Planetary gear train (see Figure 8.21).

21. Consider the simple planetary gear train design in Figure P.8.4. In this example, $N_{sun} = 23$, $N_{planet} = 18$, $N_{ring} = 51$. Also, the ring gear is fixed, the input is the carrier and the output is the sun gear. Calculate the rotations of the sun, planet, ring, and carrier.

22. Consider the simple planetary gear train design in Figure P.8.4. In this example, $N_{sun} = 18$, $N_{planet} = 16$, $N_{ring} = 56$. Also, the carrier is fixed, the input is the ring gear and the output is the sun gear. Calculate the rotations of the sun, planet, ring, and carrier. Also, assuming an input gear pitch radius, rotational speed, and torque of $r_{p_{ring}} = 85\,mm$, $\omega_{ring} = 27\,rad/s$, and $T_{ring} = 270\,N\text{-}mm$, respectively, calculate the power transmitted by each gear.

23. Consider the simple planetary gear train design in Figure P.8.4. In this example, $N_{sun} = 32$, $N_{planet} = 19$, $N_{ring} = 89$. Also, the sun gear is fixed, the input is the ring gear and the output is the carrier. Calculate the rotations of the sun, planet, ring, and carrier. Also, assuming an input gear pitch radius, rotational speed, and torque of $r_{p_{ring}} = 105\,mm$, $\omega_{ring} = 32\,rad/s$, and $T_{ring} = 171\,N\text{-}mm$, respectively, calculate the power transmitted by each gear.

24. Describe the types of input–output motion produced by rack and pinion gears.

25. Calculate the translation and velocity of the rack in a rack and pinion gear where the pinion rotation, speed, and diameter are $3\pi/4\,rad$, $4.55\,rad/s$, and $56\,mm$, respectively.

26. Calculate the angular displacement and velocity of the pinion in a rack and pinion gear where the rack displacement, rack speed, and pinion diameter are $132\,mm$, $24\,mm/s$, and $73\,mm$, respectively.

27. Calculate the radial, axial, and tangential forces transmitted by a helical gear having 49 teeth and a helix angle of $39°$. Assume this gear is an ANSI fine pitch with a $14.5°$ pressure angle. This gear also has a normal force of $42\,N$.

28. Calculate the torque and power transmitted by a helical gear having 37 teeth, a circular pitch of $45\,mm$ and a helix angle of $50°$. Assume this gear is an ANSI fine pitch with a $20°$ pressure angle (Table 8.3). This gear also has a rotational speed of $18\,rad/s$ and a normal force of $25\,N$.

29. Calculate the pinion and pitch angles for a bevel gear pair where the pinion and gear have 38 and 60 teeth respectively and the shaft angle is 90°.

30. For the bevel gears in Problem 8.29, calculate the pitch radius and angular velocity of the gear if the pinion has an angular velocity of −5.85 rad/s.

31. For a bevel gear pair where $\gamma_{pinion} = 55°$, $\gamma_{gear} = 35°$, $F_n = 175\,N$, and $\phi = 20°$, calculate the tangential, radial, and axial forces on the gear and pinion.

32. For a bevel gear pair where $N_{pinion} = 55$, $N_{gear} = 80$, $\gamma_{pinion} = 35°$, $\gamma_{gear} = 45°$, $F_n = 110N$, $r_{m_{pinion}} = 75\,mm$, $\phi = 25°$, and $\omega_{pinion} = 7.25\,rad/s$, calculate the torque and the power transmitted by the gear.

33. While planetary gear trains and worm gears can both achieve large velocity reductions, what additional capability is provided by worm gears?

34. Calculate the radial, axial, and tangential forces transmitted by the gear in a worm-gear pair. The worm has 2 threads and the gear has 82 teeth. A normal force of 18 N is applied to the gear. Assume the worm and gear are ANSI fine pitch with helix and pressure angles of 34° and 20°, respectively.

35. Calculate the torque and power transmitted by the gear in a worm-gear pair. The worm has 3 threads and rotates at 29 rad/s and the gear has 65 teeth and a circular pitch of 21 mm. A normal force of 38 N is applied to the gear. Assume the worm and gear are ANSI fine pitch with helix and pressure angles of 45° and 25°, respectively.

9

Design and Kinematic Analysis of Disk Cams

CONCEPT OVERVIEW

In this chapter, the reader will gain a central understanding regarding

1. Purposes and functions of *radial cam* systems and follower types
2. Components of follower motion and follower motion types
3. Radial cam follower kinematics
4. Criteria for optimal radial cam operation and its relationship with follower motion types
5. Radial cam design
6. Criteria for optimal radial cam operation and its relationship with radial cam system design equations

9.1 Introduction

A *disk cam* (also called a *radial cam, flat-faced cam,* or simply *cam*) is a mechanical component used to convert rotary motion into oscillating rotary or translation motion.* In its most basic form, a *cam system* includes a rotating disk member (the cam) that compels the motion of an oscillating member called the follower. The names *disk cam* and *flat-faced cam* refer to the flat, disk-like shape of the cam geometry. Since the rotating disk cam compels motion by pushing components away from its center of rotation—or in a radial direction—the name *radial cam* is also used.

Among other applications, cam systems are commonly used in the valve trains of internal combustion engines, particularly in automotive engines. Figure 9.1 illustrates a cam system used in an overhead valve train. The rotating cams produce an oscillating translation motion in the rod components which subsequently produce an oscillation rotation motion in the rocker components. Ultimately, the oscillating motion of the rocker components produce an oscillating translation motion in the valve components, causing them to open and close in a precisely timed manner. The cam system illustrated in Figure 9.1 enables proper fuel entry into the engine and proper exit of the combustion products from the engine.

---

* Other common (nonradial) cam types not covered in this text include *cam slots, cylindrical cams, yokes,* and *wedge cams.*

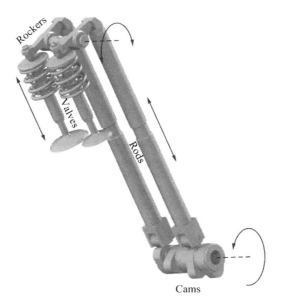

FIGURE 9.1
Overhead valve train and actuation mechanism.

9.2 Follower Types

The types of followers used in disk cam systems vary in terms of their shape and the oscillating motion they exhibit. Figure 9.2 includes five follower types.

The *knife-edge follower* (Figure 9.2a) translates when in contact with the rotating cam. Although this type of follower is simple in design, its sharp edge produces high contact stresses.* The *flat-faced follower* (Figure 9.2b) also translates when in contact with the rotating cam. While the contact stresses with this follower type are less concentrated than those with the knife-edge follower due to its larger contact surface, greater sliding friction forces are produced with the flat-faced follower due to its larger contact surface. The translating *roller follower* (Figure 9.2c) includes an additional rolling wheel component (pinned to the follower) that rolls over the cam surface.† Though more complex in design that the knife-edge and flat-faced followers, the rolling wheel of the roller follower produces lower contact stresses and friction forces. Figures 9.2d and 9.2e are variations of the flat-faced slider and roller followers, respectively. In these follower variations, a rotational motion is produced when in contact with a rotating cam.

---

* Because contact stress is proportional to the rate of wear, an increase in contact stress results in an increase in component wear rate and subsequently component life.
† The rods in Figure 9.1 are translating roller followers (Figure 9.2c).

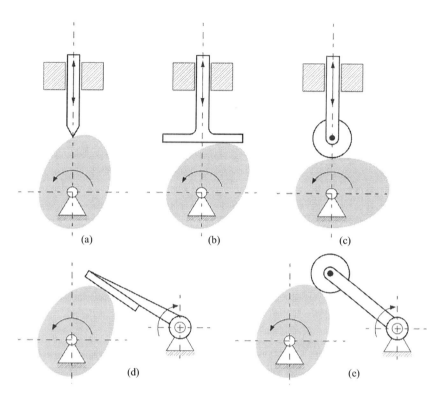

FIGURE 9.2
Translating (a) knife-edge, (b) flat-faced, and (c) roller followers, and rotating (d) flat-faced and (e) roller followers.

9.3 Follower Motion

9.3.1 Rise, Fall, and Dwell

No matter the type of follower used, a follower displacement profile is comprised of three distinct displacement profiles: *rise, fall,* and *dwell* profiles. These profiles are determined by the shape of the cam and its rotation direction.

Figure 9.3 illustrates a disk cam rotating counterclockwise direction. This figure also includes the follower displacement achieved by the rotating cam. As the cam rotates, there is a steady increase in its radii of curvature (or an increasing rate of change) in the cam region bounded by radii r_1 and r_2. As the cam passes through this region, the follower will be displaced radially away from the cam's center of rotation (the positive y-direction in Figure 9.3). Positive follower displacement is called *rise*.

The radius of curvature is constant in the cam region bounded by radii r_2 and r_3. Because of the constant curvature, as the cam rotates through this region, the follower will experience no displacement at all. Zero follower displacement is called *dwell*.

There is a steady decrease in radii of curvature (or a decreasing rate of change) in the cam region bounded by radii r_3 and r_4. As the cam rotates through this region, the follower will be displaced radially toward the cam's center of rotation (the negative y-direction in Figure 9.3). Negative follower displacement is called *fall*.

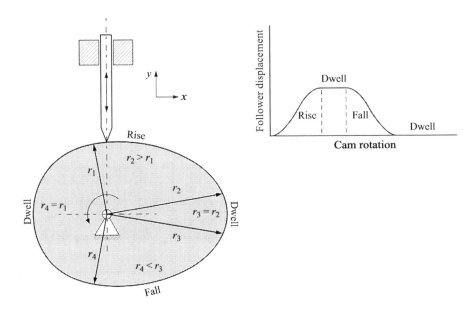

FIGURE 9.3
Disk cam and follower displacement curve with rise, fall, and dwell regions labeled.

The radius of curvature is also constant in the in the cam region bounded by radii r_4 and r_1. Because of the constant curvature, as the cam rotates through this region, the follower will again experience dwell.*

Figure 9.4 illustrates a simple follower displacement diagram. This diagram includes a *rise-dwell-fall-dwell* follower displacement interval sequence (identical to the displacement sequence in Figure 9.3). As with all disk cams, the follower displacement sequence is achieved over a complete (360°) cam rotation cycle. The variables β_1 through β_4 are the cam rotation ranges associated with each rise, fall, and dwell interval. The variable h_1 is the follower displacement range for the corresponding rise and fall intervals.

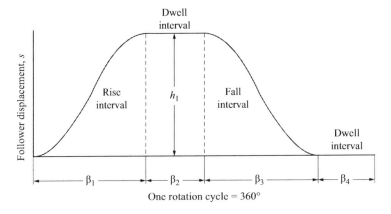

FIGURE 9.4
Simple follower displacement diagram.

---

* For rotating followers, dwell would correspond to zero rotational displacement and rise and fall would correspond to counterclockwise and clockwise rotations, respectively (or vice versa if the user prefers).

9.3.2 Displacement, Velocity, Acceleration, and Jerk

As we have demonstrated in prior chapters (Chapters 2 and 4, for example), differentiating a displacement equation produces an equation for velocity and differentiating a velocity equation produces an equation for acceleration. Determining the follower displacement, velocity, and acceleration is necessary in cam design. It is also necessary to know the rate of change in follower acceleration in cam design. This quantity is called *jerk*. Differentiating an acceleration equation produces an equation for jerk.

For a cam to operate beyond low-speed applications, its profile must be designed so that its follower displacement, velocity, and acceleration profiles are continuous throughout a complete cam rotation cycle. For this condition to be true, the follower jerk profile must be finite throughout a complete cam rotation cycle. In Sections 9.3.2–9.3.7, we will evaluate different types of follower motion with respect to the cam design conditions of continuous displacement, velocity, and acceleration profiles and a finite jerk profile.*

9.3.3 Constant Velocity Motion

Constant velocity motion is the most basic type of follower motion. Here, the follower rise and fall profiles are defined as linear functions.

Table 9.1 includes the follower rise and fall displacement, velocity, acceleration, and jerk equations for constant velocity motion [1]. When expressing these quantities in terms of cam rotation, the variable θ represents the cam rotation increment over the rotation interval range β.

In Table 9.1, the follower equations are also given in terms of time. When expressed this way, the cam rotation interval range β becomes the time-interval range T and the rotation increment θ becomes the time increment t. The quantity $\dot{\theta}$ in the velocity equations represents the cam rotation speed.[†]

Although constant velocity motion is the simplest type of follower motion, it violates the fundamental law of cam design (given in Section 9.3.2). Figure 9.5 illustrates the follower displacement, velocity, acceleration, and jerk profiles for constant velocity motion. Because the slope is constant in the displacement profile (Figure 9.5a), its derivative becomes a stepped velocity profile (Figure 9.5b). Such a velocity profile violates the continuous velocity condition for cam design.

TABLE 9.1

Follower Displacement, Velocity, Acceleration, and Jerk
Equations: Constant Velocity Motion

	Rise	Fall
	For $0 < \theta < \beta$ or $0 < t < T$	
Displacement, s	$s = h\theta/\beta = ht/T$	$s = \dfrac{h}{\beta}(1-\theta) = \dfrac{h}{T}(1-t)$
Velocity, $\dot{s}$	$\dot{s} = \dfrac{h\dot{\theta}}{\beta} = \dfrac{h}{T}$	$\dot{s} = -\dfrac{h\dot{\theta}}{\beta} = -\dfrac{h}{T}$
Acceleration, $\ddot{s}$	$\ddot{s} = 0$	$\ddot{s} = 0$
Jerk, $\dddot{s}$	$\dddot{s} = 0$	$\dddot{s} = 0$

---

* These cam design conditions are also called the *fundamental law of cam design*. Violating these cam design conditions does not mean a cam design cannot be used, only that it will be limited to low-speed applications.
† The quantity $\dot{\theta}$ is often expressed in units of degrees/time, radians/time, or revolutions/time.

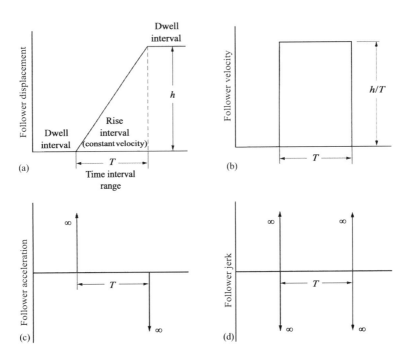

FIGURE 9.5
Follower (a) displacement, (b) velocity, (c) acceleration, and (d) jerk profiles for constant velocity motion.

Because the slopes at the start and end of the stepped velocity profile are infinite, these infinite slopes appear as infinite spikes in the acceleration profile (Figure 9.5c).* Such an acceleration profile violates the continuous acceleration condition for cam design. While an infinite acceleration will not appear in actual cam use, these quantities will appear as excessive accelerations (impulse or shock loads). Such loads result in excessive cam wear and damage and will ultimately shorten the life of the cam substantially.

Being the derivative of an acceleration profile having infinite spikes, the follower jerk profile (Figure 9.5d) also includes infinite spikes. Such a jerk profile violates the finite jerk condition for cam design.

Appendix E.1 includes the MATLAB® file user instructions for generating constant velocity motion-based displacement, velocity, acceleration, and jerk diagrams. In this MATLAB file (which is available for download at www.crcpress.com/product/isbn/9781498724937), the constant velocity motion equations in Table 9.1 are used.

Example 9.1

Problem Statement: Using the Appendix E.1 MATLAB file, plot the follower displacement and velocity profiles under constant velocity motion for the following 6 intervals (with $\pi/3$ rad for each interval): rise (25 mm)-dwell-rise (30 mm)-dwell-fall (55 mm)-dwell. Assume the cam rotates at a constant speed of 45 rpm.

Known Information: The follower displacement sequence, cam rotation speed, and the Appendix E.1 MATLAB file.

---

* In mathematics, these spikes are idealized unit impulses and are known as *Dirac delta functions.*

```
Event = ['R','D','R','D','F','D'];
Beta = [60,60,60,60,60,60];
Si = [25,25,55,55,0,0];

dTheta = 45*pi/30;
Rbase = 1; (NOTE: this value is not relevant since "Rbase" is not
used in this example)
```

FIGURE E.9.1
Specified input (in bold text) in the Appendix E.1 MATLAB file for Example 9.1.

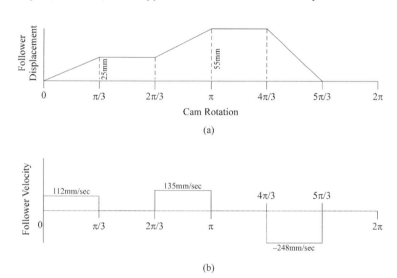

(a)

(b)

FIGURE E.9.2
Constant velocity follower (a) displacement and (b) velocity profiles.

Solution Approach: Given the crank rotation speed and the constant cam rotation value for each interval, we can determine that the each interval has a time period of $T = 0.2222\,\text{s}$. Figure E.9.1 includes the input specified (in bold text) in the Appendix E.1 MATLAB file. Figure E.9.2 illustrates the follower displacement and velocity profiles calculated from the Appendix E.1 MATLAB file.

9.3.4 Constant Acceleration Motion

Unlike the first-order (or linear) follower displacement profile associated with constant velocity motion, *constant acceleration motion* includes a second-order (or parabolic) follower displacement profile.* Table 9.2 includes the follower rise and fall displacement, velocity, acceleration, and jerk equations for constant acceleration motion [1]. As shown in this table, the range (be it a time range or rotation range) is divided so that profiles for each half of the range are plotted using separate equations.

The parabolic follower displacement profile (Figure 9.6a), when differentiated, produces a triangular velocity profile (Figure 9.6b). This velocity profile, when differentiated, produces a stepped acceleration profile (Figure 9.6c). In comparison to the acceleration profile for constant velocity motion (Figure 9.5c), the acceleration profile for constant acceleration motion is an improvement because it has a finite height. In addition, for a given angle of

---

* Constant acceleration motion is also called *parabolic motion*.

TABLE 9.2

Follower Displacement, Velocity, Acceleration, and Jerk Equations: Constant Acceleration Motion

	Rise	Fall
	For $0 < \theta < \beta/2$ or $0 < t < T/2$	
Displacement, s	$s = 2\,h(\theta/\beta)^2 = 2\,h(t/T)^2$	$s = h - 2\,h(\theta/\beta)^2 = h - 2\,h(t/T)^2$
Velocity, $\dot{s}$	$\dot{s} = \dfrac{4h\dot{\theta}\theta}{\beta^2} = \dfrac{4ht}{T^2}$	$\dot{s} = -\dfrac{4h\dot{\theta}\theta}{\beta^2} = -\dfrac{4ht}{T^2}$
Acceleration, $\ddot{s}$	$\ddot{s} = 4h\left(\dfrac{\dot{\theta}}{\beta}\right)^2 = \dfrac{4h}{T^2}$	$\ddot{s} = -4h\left(\dfrac{\dot{\theta}}{\beta}\right)^2 = -\dfrac{4h}{T^2}$
Jerk, $\dddot{s}$	$\dddot{s} = 0$	$\dddot{s} = 0$
	For $\beta/2 < \theta < \beta$ or $T/2 < t < T$	
Displacement, s	$s = h - 2h[1-\theta/\beta]^2 = h - 2h[1-t/T]^2$	$s = 2h[1-\theta/\beta]^2 = 2h[1-t/T]^2$
Velocity, $\dot{s}$	$\dot{s} = \dfrac{4h\dot{\theta}}{\beta}\left(1 - \dfrac{\theta}{\beta}\right) = \dfrac{4ht}{T}\left(1 - \dfrac{t}{T}\right)$	$\dot{s} = -\dfrac{4h\dot{\theta}}{\beta}\left(1 - \dfrac{\theta}{\beta}\right) = -\dfrac{4ht}{T}\left(1 - \dfrac{t}{T}\right)$
Acceleration, $\ddot{s}$	$\ddot{s} = -4h\left(\dfrac{\dot{\theta}}{\beta}\right)^2 = -\dfrac{4h}{T^2}$	$\ddot{s} = 4h\left(\dfrac{\dot{\theta}}{\beta}\right)^2 = \dfrac{4h}{T^2}$
Jerk, $\dddot{s}$	$\dddot{s} = 0$	$\dddot{s} = 0$

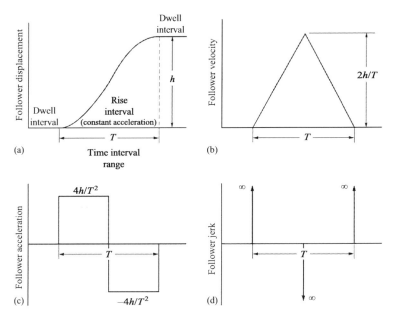

FIGURE 9.6
Follower (a) displacement, (b) velocity, (c) acceleration, and (d) jerk profiles for constant acceleration motion.

rotation and rise, constant acceleration motion produces the smallest acceleration among the motion types presented in this chapter [2]. However, because the acceleration profile changes abruptly, shock loads will be produced in the cam system. This discontinuous acceleration profile also violates the continuous acceleration condition for cam design.

Being the derivative of a stepped acceleration profile, the follower jerk profile (Figure 9.6d) includes infinite spikes. Such a jerk profile violates the finite jerk condition for cam design.

Appendix E.2 includes the MATLAB file user instructions for generating constant acceleration motion-based displacement, velocity, acceleration, and jerk diagrams. In this MATLAB file (which is available for download at www.crcpress.com/product/isbn/9781498724937), the constant acceleration motion equations in Table 9.2 are used.

Example 9.2

Problem Statement: Using the Appendix E.2 MATLAB file, plot the follower displacement, velocity and acceleration profiles under constant acceleration motion for the follower displacement interval and cam rotation speed data given in Example 9.1.

Known Information: Example 9.1 and Appendix E.2 MATLAB file.

Solution Approach: Figure E.9.3 includes the input specified (in bold text) in the Appendix E.2 MATLAB file. Figure E.9.4 illustrates the follower displacement, velocity, and acceleration profiles calculated from the Appendix E.2 MATLAB file.

```
Event = ['R','D','R','D','F','D'];
Beta = [60,60,60,60,60,60];
Si = [25,25,55,55,0,0];

dTheta = 45*pi/30;
Rbase = 1; (NOTE: this value is not relevant since "Rbase" is not
used in this example)
```

FIGURE E.9.3
Specified input (in bold text) in the Appendix E.2 MATLAB file for Example 9.2.

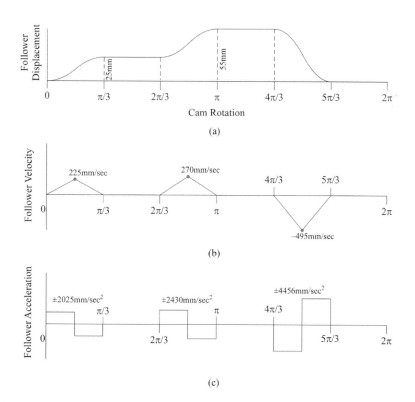

FIGURE E.9.4
Constant acceleration follower (a) displacement, (b) velocity, and (c) acceleration profiles.

9.3.5 Simple Harmonic Motion

In *simple harmonic motion*, the follower displacement equations include periodic functions.*
Unlike nonperiodic functions, periodic functions can be differentiated indefinitely without ever producing a constant solution (thus preventing infinite spikes in cam motion profiles). Table 9.3 includes the follower rise and fall displacement, velocity, acceleration, and jerk equations for simple harmonic motion [3].

In simple harmonic motion, a smooth follower displacement profile is produced (Figure 9.7a). This profile, when differentiated, produces a continuous velocity profile (Figure 9.7b). This velocity profile, when differentiated, produces a continuous acceleration profile, having steps at its start and end (Figure 9.7c). The acceleration profile from simple harmonic motion is an improvement over the profile from constant velocity motion (Figure 9.5c), since it is smooth and continuous within its interval range. Simple harmonic motion also produces a finite follower jerk profile (Figure 9.7d), thus satisfying the finite jerk condition for cam design. However, in simple harmonic motion, the acceleration profile changes abruptly at its start and end, producing shock loads in the cam system and violating the continuous acceleration condition for cam design.

TABLE 9.3

Follower Displacement, Velocity, Acceleration, and Jerk Equations: Simple Harmonic Motion

	Rise	Fall
	For $0 < \theta < \beta$ or $0 < t < T$	
Displacement, s	$s = \dfrac{h}{2}\left[1 - \cos\left(\dfrac{\pi\theta}{\beta}\right)\right]$	$s = \dfrac{h}{2}\left[1 + \cos\left(\dfrac{\pi\theta}{\beta}\right)\right]$
	$= \dfrac{h}{2}\left[1 - \cos\left(\dfrac{\pi t}{T}\right)\right]$	$= \dfrac{h}{2}\left[1 + \cos\left(\dfrac{\pi t}{T}\right)\right]$
Velocity, $\dot{s}$	$\dot{s} = \dfrac{\pi h \dot{\theta}}{2\beta}\left[\sin\left(\dfrac{\pi\theta}{\beta}\right)\right]$	$\dot{s} = -\dfrac{\pi h \dot{\theta}}{2\beta}\left[\sin\left(\dfrac{\pi\theta}{\beta}\right)\right]$
	$= \dfrac{\pi h}{2T}\left[\sin\left(\dfrac{\pi t}{T}\right)\right]$	$= -\dfrac{\pi h}{2T}\left[\sin\left(\dfrac{\pi t}{T}\right)\right]$
Acceleration, $\ddot{s}$	$\ddot{s} = \dfrac{\pi^2 h \dot{\theta}^2}{2\beta^2}\left[\cos\left(\dfrac{\pi\theta}{\beta}\right)\right]$	$\ddot{s} = -\dfrac{\pi^2 h \dot{\theta}^2}{2\beta^2}\left[\cos\left(\dfrac{\pi\theta}{\beta}\right)\right]$
	$= \dfrac{\pi^2 h}{2T^2}\left[\cos\left(\dfrac{\pi t}{T}\right)\right]$	$= -\dfrac{\pi^2 h}{2T^2}\left[\cos\left(\dfrac{\pi t}{T}\right)\right]$
Jerk, $\dddot{s}$	$\dddot{s} = -\dfrac{\pi^3 h \dot{\theta}^3}{2\beta^3}\left[\sin\left(\dfrac{\pi\theta}{\beta}\right)\right]$	$\dddot{s} = \dfrac{\pi^3 h \dot{\theta}^3}{2\beta^3}\left[\sin\left(\dfrac{\pi\theta}{\beta}\right)\right]$
	$= -\dfrac{\pi^3 h}{2T^3}\left[\sin\left(\dfrac{\pi t}{T}\right)\right]$	$= \dfrac{\pi^3 h}{2T^3}\left[\sin\left(\dfrac{\pi t}{T}\right)\right]$

---

* *Sine* and *cosine* functions (also sine and cosine-based functions) are examples of periodic functions—functions that repeat their values at regular intervals or periods.

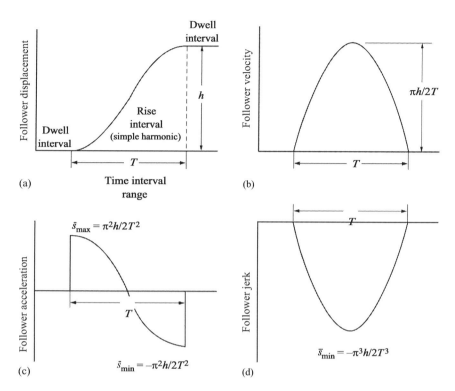

FIGURE 9.7
Follower (a) displacement, (b) velocity, (c) acceleration, and (d) jerk profiles for simple harmonic motion.

Appendix E.3 includes the MATLAB file user instructions for generating simple harmonic motion-based displacement, velocity, acceleration, and jerk diagrams. In this MATLAB file (which is available for download at www.crcpress.com/product/isbn/9781498724937), the simple harmonic motion equations in Table 9.3 are used.

Example 9.3

Problem Statement: Using the Appendix E.3 MATLAB file, plot the follower displacement, velocity, acceleration, and jerk profiles under simple harmonic motion for the follower displacement interval and cam rotation speed data given in Example 9.1.

Known Information: Example 9.1 and Appendix E.3 MATLAB file.

Solution Approach: Figure E.9.5 includes the input specified (in bold text) in the Appendix E.3 MATLAB file. Figure E.9.6 illustrates the follower displacement, velocity, acceleration, and jerk profiles calculated from the Appendix E.3 MATLAB file.

```
Event = ['R','D','R','D','F','D'];
Beta = [60,60,60,60,60,60];
Si = [25,25,55,55,0,0];

dTheta = 45*pi/30;
Rbase = 1; (NOTE: this value is not relevant since "Rbase" is not
used in this example)
```

FIGURE E.9.5
Specified input (in bold text) in the Appendix E.3 MATLAB file for Example 9.3.

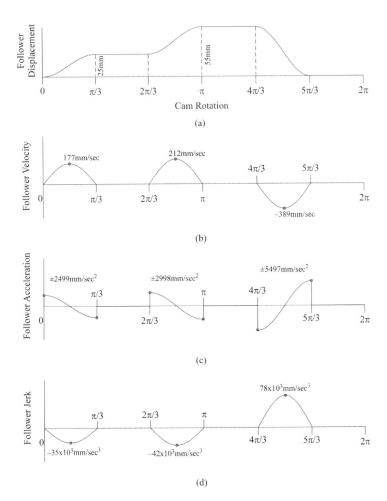

FIGURE E.9.6
Simple harmonic motion follower (a) displacement, (b) velocity, (c) acceleration, and (d) jerk profiles.

9.3.6 Cycloidal Motion

Like simple harmonic motion, the follower displacement equations also include periodic functions in *cycloidal motion*. Table 9.4 includes the follower rise and fall displacement, velocity, acceleration, and jerk equations for cycloidal motion [4].

While smooth and continuous follower displacement and velocity profiles are produced in cycloidal motion (Figures 9.8a and b), the main advantages with this type of motion is that its acceleration profile is smooth and continuous (Figure 9.8c) and its jerk profile (Figure 9.8d) is finite. Because the acceleration profile includes no abrupt changes, the shock loads resulting from such abrupt changes in acceleration are not produced. Under cycloidal motion, the cam design conditions for continuous displacement, velocity, and acceleration profiles and a finite jerk profile are all satisfied. As a result, cycloidal motion-based cam designs are well suited for high-speed applications.

Appendix E.4 includes the MATLAB file user instructions for generating cycloidal motion-based displacement, velocity, acceleration, and jerk diagrams. In this MATLAB file (which is available for download at www.crcpress.com/product/isbn/9781498724937), the cycloidal motion equations in Table 9.4 are used.

TABLE 9.4

Follower Displacement, Velocity, Acceleration, and Jerk Equations: Cycloidal Motion

	Rise	Fall
	For $0 < \theta < \beta$ or $0 < t < T$	
Displacement, s	$s = h\left[\dfrac{\theta}{\beta} - \dfrac{1}{2\pi}\sin\left(\dfrac{2\pi\theta}{\beta}\right)\right]$	$s = h\left[1 - \dfrac{\theta}{\beta} + \dfrac{1}{2\pi}\sin\left(\dfrac{2\pi\theta}{\beta}\right)\right]$
	$= h\left[\dfrac{t}{T} - \dfrac{1}{2\pi}\sin\left(\dfrac{2\pi t}{T}\right)\right]$	$= h\left[1 - \dfrac{t}{T} + \dfrac{1}{2\pi}\sin\left(\dfrac{2\pi t}{T}\right)\right]$
Velocity, $\dot{s}$	$\dot{s} = \dfrac{h\dot{\theta}}{\beta}\left[1 - \cos\left(\dfrac{2\pi\theta}{\beta}\right)\right]$	$\dot{s} = -\dfrac{h\dot{\theta}}{\beta}\left[1 - \cos\left(\dfrac{2\pi\theta}{\beta}\right)\right]$
	$= \dfrac{h}{T}\left[1 - \cos\left(\dfrac{2\pi t}{T}\right)\right]$	$= -\dfrac{h}{T}\left[1 - \cos\left(\dfrac{2\pi t}{T}\right)\right]$
Acceleration, $\ddot{s}$	$\ddot{s} = \dfrac{2\pi h\dot{\theta}^2}{\beta^2}\left[\sin\left(\dfrac{2\pi\theta}{\beta}\right)\right]$	$\ddot{s} = -\dfrac{2\pi h\dot{\theta}^2}{\beta^2}\left[\sin\left(\dfrac{2\pi\theta}{\beta}\right)\right]$
	$= \dfrac{2\pi h}{T^2}\left[\sin\left(\dfrac{2\pi t}{T}\right)\right]$	$= -\dfrac{2\pi h}{T^2}\left[\sin\left(\dfrac{2\pi t}{T}\right)\right]$
Jerk, $\dddot{s}$	$\dddot{s} = \dfrac{4\pi^2 h\dot{\theta}^3}{\beta^3}\left[\cos\left(\dfrac{2\pi\theta}{\beta}\right)\right]$	$\dddot{s} = -\dfrac{4\pi^2 h\dot{\theta}^3}{\beta^3}\left[\cos\left(\dfrac{2\pi\theta}{\beta}\right)\right]$
	$= \dfrac{4\pi^2 h}{T^3}\left[\cos\left(\dfrac{2\pi t}{T}\right)\right]$	$= -\dfrac{4\pi^2 h}{T^3}\left[\cos\left(\dfrac{2\pi t}{T}\right)\right]$

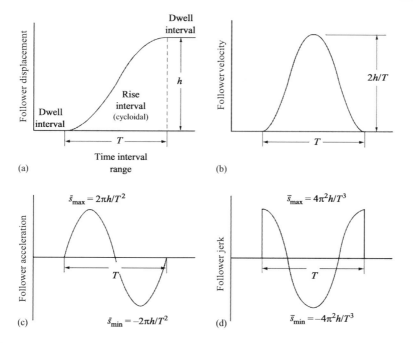

FIGURE 9.8
Follower (a) displacement, (b) velocity, (c) acceleration, and (d) jerk profiles for cycloidal motion.

Example 9.4

Problem Statement: Using the Appendix E.4 MATLAB file, plot the follower displacement, velocity, acceleration, and jerk profiles under cycloidal motion for the follower displacement interval and cam rotation speed data given in Example 9.1.

Known Information: Example 9.1 and Appendix E.4 MATLAB file.

Solution Approach: Figure E.9.7 includes the input specified (in bold text) in the Appendix E.4 MATLAB file. Figure E.9.8 illustrates the follower displacement, velocity, acceleration, and jerk profiles calculated from the Appendix E.4 MATLAB file.

```
Event = ['R','D','R','D','F','D'];
Beta = [60,60,60,60,60,60];
Si = [25,25,55,55,0,0];

dTheta = 45*pi/30;
Rbase = 1; (NOTE: this value is not relevant since "Rbase" is not
used in this example)
```

FIGURE E.9.7
Specified input (in bold text) in the Appendix E.4 MATLAB file for Example 9.4.

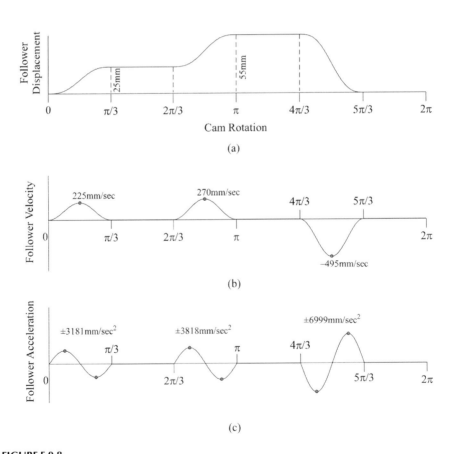

FIGURE E.9.8
Cycloidal motion follower (a) displacement, (b) velocity, (c) acceleration, and (d) jerk profiles.

(*Continued*)

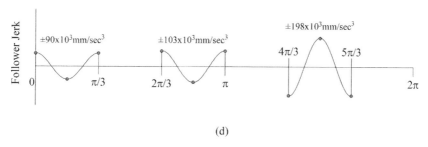

(d)

FIGURE E.9.8 (CONTINUED)
Cycloidal motion follower (a) displacement, (b) velocity, (c) acceleration, and (d) jerk profiles.

9.3.7 Polynomial Motion

Follower motion functions can also be formulated using polynomials. The primary advantage of polynomial-based follower motion functions is that boundary conditions for follower displacement, velocity, acceleration, and jerk can be prescribed. With capability, the user can ensure that the profiles for these quantities are finite and continuous, and thus satisfactory according to the cam design conditions in Section 9.3.2. Equation 9.1 includes the general form of the polynomial function.* In this equation, the variable x becomes θ/β if cam rotational displacement is considered or t/T if time is considered.

$$s = C_0 + C_1 x + C_2 x^2 + C_3 x^3 + C_4 x^4 + C_5 x^5 \ldots C_n x^n \tag{9.1}$$

One type of polynomial used to define follower motion is called the *3-4-5 polynomial*. This function begins as a polynomial of order 5 (also called a fifth-order polynomial). Expressing Equation 9.1 as a fifth-order polynomial where the term θ/β is used in place of x, the resulting polynomial becomes

$$s = C_0 + C_1\left(\frac{\theta}{\beta}\right) + C_2\left(\frac{\theta}{\beta}\right)^2 + C_3\left(\frac{\theta}{\beta}\right)^3 + C_4\left(\frac{\theta}{\beta}\right)^4 + C_5\left(\frac{\theta}{\beta}\right)^5 \tag{9.2}$$

To formulate the coefficients C_0 through C_5 in Equation 9.2 for the rise function, the following boundary conditions are specified:

$$\text{At } \theta = 0, s = \dot{s} = \ddot{s} = 0 \quad \text{and} \quad \text{at} \theta = \beta, s = h, \dot{s} = \ddot{s} = 0$$

To formulate the coefficients in Equation 9.2 for the fall function, the following boundary conditions are specified:

$$\text{At } \theta = 0, s = h, \dot{s} = \ddot{s} = 0 \quad \text{and} \quad \text{at } \theta = \beta, s = \dot{s} = \ddot{s} = 0$$

Table 9.5 includes the resulting follower rise and fall displacement, velocity, acceleration, and jerk equations for 3-4-5 polynomial motion.[†] The name "3-4-5 polynomial" reflects the exponent orders that appear in these displacement functions. Table 9.5 also includes the follower velocity, acceleration, and jerk functions.

---

* The greatest exponent value given in a polynomial determines its order. For example, the function $s = C_0 + C_1 x + C_2 x^2 + C_3 x^3$ is a polynomial of order 3 (or a third-order polynomial) because its greatest exponent is 3.
† Using the given boundary conditions, the coefficients C_0, C_1, and C_2 become zero in the 3–4–5 polynomial.

TABLE 9.5

Follower Displacement, Velocity, Acceleration, and Jerk Equations: 3-4-5 Polynomial Motion

	Rise	Fall
	For $0 < \theta < \beta$ or $0 < t < T$	
Disp., s	$s = h\left[10\left(\dfrac{\theta}{\beta}\right)^3 - 15\left(\dfrac{\theta}{\beta}\right)^4 + 6\left(\dfrac{\theta}{\beta}\right)^5\right]$	$s = h\left[1 - 10\left(\dfrac{\theta}{\beta}\right)^3 + 15\left(\dfrac{\theta}{\beta}\right)^4 - 6\left(\dfrac{\theta}{\beta}\right)^5\right]$
	$= h\left[10\left(\dfrac{t}{T}\right)^3 - 15\left(\dfrac{t}{T}\right)^4 + 6\left(\dfrac{t}{T}\right)^5\right]$	$= h\left[1 - 10\left(\dfrac{t}{T}\right)^3 + 15\left(\dfrac{t}{T}\right)^4 - 6\left(\dfrac{t}{T}\right)^5\right]$
Vel., $\dot{s}$	$\dot{s} = \dfrac{h\dot{\theta}}{\beta}\left[30\left(\dfrac{\theta}{\beta}\right)^2 - 60\left(\dfrac{\theta}{\beta}\right)^3 + 30\left(\dfrac{\theta}{\beta}\right)^4\right]$	$\dot{s} = \dfrac{h\dot{\theta}}{\beta}\left[-30\left(\dfrac{\theta}{\beta}\right)^2 + 60\left(\dfrac{\theta}{\beta}\right)^3 - 30\left(\dfrac{\theta}{\beta}\right)^4\right]$
	$= \dfrac{h}{T}\left[30\left(\dfrac{t}{T}\right)^2 - 60\left(\dfrac{t}{T}\right)^3 + 30\left(\dfrac{t}{T}\right)^4\right]$	$= \dfrac{h}{T}\left[-30\left(\dfrac{t}{T}\right)^2 + 60\left(\dfrac{t}{T}\right)^3 - 30\left(\dfrac{t}{T}\right)^4\right]$
Acc., $\ddot{s}$	$\ddot{s} = \dfrac{h\dot{\theta}^2}{\beta^2}\left[60\left(\dfrac{\theta}{\beta}\right) - 180\left(\dfrac{\theta}{\beta}\right)^2 + 120\left(\dfrac{\theta}{\beta}\right)^3\right]$	$\ddot{s} = \dfrac{h\dot{\theta}^2}{\beta^2}\left[-60\left(\dfrac{\theta}{\beta}\right) + 180\left(\dfrac{\theta}{\beta}\right)^2 + 120\left(\dfrac{\theta}{\beta}\right)^3\right]$
	$= \dfrac{h}{T^2}\left[60\left(\dfrac{t}{T}\right) - 180\left(\dfrac{t}{T}\right)^2 + 120\left(\dfrac{t}{T}\right)^3\right]$	$= \dfrac{h}{T^2}\left[-60\left(\dfrac{t}{T}\right) + 180\left(\dfrac{t}{T}\right)^2 + 120\left(\dfrac{t}{T}\right)^3\right]$
Jerk, $\dddot{s}$	$\dddot{s} = \dfrac{h\dot{\theta}^3}{\beta^3}\left[60 - 360\left(\dfrac{\theta}{\beta}\right) + 360\left(\dfrac{\theta}{\beta}\right)^2\right]$	$\dddot{s} = \dfrac{h\dot{\theta}^3}{\beta^3}\left[-60 + 360\left(\dfrac{\theta}{\beta}\right) - 360\left(\dfrac{\theta}{\beta}\right)^2\right]$
	$= \dfrac{h}{T^3}\left[60 - 360\left(\dfrac{t}{T}\right) + 360\left(\dfrac{t}{T}\right)^2\right]$	$= \dfrac{h}{T^3}\left[-60 + 360\left(\dfrac{t}{T}\right) - 360\left(\dfrac{t}{T}\right)^2\right]$

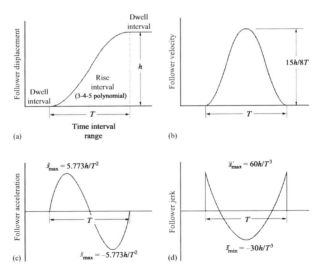

FIGURE 9.9

Follower (a) displacement, (b) velocity, (c) acceleration, and (d) jerk profiles for 3–4–5 polynomial motion.

The follower displacement, velocity, acceleration, and jerk profiles produced from the 3-4-5 polynomial all satisfy the continuity and finite conditions for cam design (Figure 9.9). In addition to cycloidal motion-based cam designs, 3-4-5 polynomial motion-based cam designs are also well suited for high-speed applications.

Appendix E.5 includes the MATLAB file user instructions for generating 3-4-5 polynomial motion-based displacement, velocity, acceleration, and jerk diagrams. In this MATLAB file (which is available for download at www.crcpress.com/product/isbn/9781498724937), the 3-4-5 polynomial motion equations in Table 9.5 are used.

Example 9.5

Problem Statement: Using the Appendix E.5 MATLAB file, plot the follower displacement, velocity, acceleration, and jerk profiles under 3-4-5 polynomial motion for the follower displacement interval and cam rotation speed data given in Example 9.1.

Known Information: Example 9.1 and Appendix E.5 MATLAB file.

Solution Approach: Figure E.9.9 includes the input specified (in bold text) in the Appendix E.5 MATLAB file. Figure E.9.10 illustrates the follower displacement, velocity, acceleration, and jerk profiles calculated from the Appendix E.5 MATLAB file.

```
Event = ['R','D','R','D','F','D'];
Beta = [60,60,60,60,60,60];
Si = [25,25,55,55,0,0];

dTheta = 45*pi/30;
Rbase = 1; (NOTE: this value is not relevant since "Rbase" is not
used in this example)
```

FIGURE E.9.9
Specified input (in bold text) in the Appendix E.5 MATLAB file for Example 9.5.

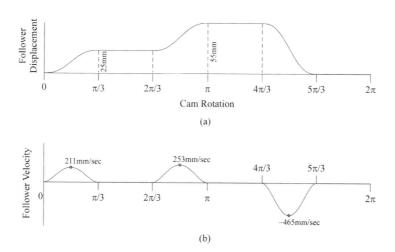

(a)

(b)

FIGURE E.9.10
Polynomial (3-4-5) motion follower (a) displacement, (b) velocity, (c) acceleration, and (d) jerk profiles.

(Continued)

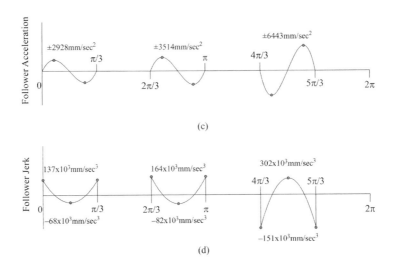

(c)

(d)

FIGURE E.9.10 (CONTINUED)
Polynomial (3-4-5) motion follower (a) displacement, (b) velocity, (c) acceleration, and (d) jerk profiles.

Another type of polynomial used to define follower motion is called the *4-5-6-7 polynomial*. This function begins as a polynomial of order 7 (also called a seventh-order polynomial). Expressing Equation 9.1 as a seventh-order polynomial where the term θ/β is used in place of x, the resulting polynomial becomes

$$s = C_0 + C_1\left(\frac{\theta}{\beta}\right) + C_2\left(\frac{\theta}{\beta}\right)^2 + C_3\left(\frac{\theta}{\beta}\right)^3 + C_4\left(\frac{\theta}{\beta}\right)^4 + C_5\left(\frac{\theta}{\beta}\right)^5 + C_6\left(\frac{\theta}{\beta}\right)^6 + C_7\left(\frac{\theta}{\beta}\right)^7 \quad (9.3)$$

To formulate the coefficients C_0 through C_7 in Equation 9.3 for the rise function, the following boundary conditions are specified:

$$\text{At } \theta = 0, s = \dot{s} = \ddot{s} = \dddot{s} = 0 \quad \text{and} \quad \text{at } \theta = \beta, s = h, \dot{s} = \ddot{s} = \dddot{s} = 0$$

To formulate the coefficients in Equation 8.3 for the dwell function, the following boundary conditions are specified:

$$\text{At } \theta = 0, s = h, \dot{s} = \ddot{s} = \dddot{s} = 0 \quad \text{and} \quad \text{at } \theta = \beta, s = \dot{s} = \ddot{s} = \dddot{s} = 0$$

Table 9.6 includes the resulting follower rise and fall displacement, velocity, acceleration, and jerk equations for 4-5-6-7 polynomial motion.[*] The name "4-5-6-7 polynomial" reflects the exponent orders that appear in these displacement functions. Table 9.6 also includes the follower velocity, acceleration, and jerk functions.

The follower displacement, velocity, acceleration, and jerk profiles produced from the 4-5-6-7 polynomial all satisfy the continuity and finite conditions for cam design (Figure 9.10). In addition to cycloidal and 3-4-5 polynomial motion-based cam designs, 4-5-6-7 polynomial motion-based cam designs are also well suited for high-speed applications.

---

[*] Using the given boundary conditions, the coefficients C_0, C_1, C_2, and C_3 become zero in the 4-5-6-7 polynomial.

TABLE 9.6

Follower Displacement, Velocity, Acceleration, and Jerk Equations: 4–5–6–7 Polynomial Motion

	Rise	Fall

For $0 < \theta < B$ or $0 < t < T$

Disp., s

$$s = h\left[35\left(\frac{\theta}{\beta}\right)^4 - 84\left(\frac{\theta}{\beta}\right)^5 + 70\left(\frac{\theta}{\beta}\right)^6 - 20\left(\frac{\theta}{\beta}\right)^7\right]$$

$$s = h\left[1 - 35\left(\frac{\theta}{\beta}\right)^4 + 84\left(\frac{\theta}{\beta}\right)^5 - 70\left(\frac{\theta}{\beta}\right)^6 + 20\left(\frac{\theta}{\beta}\right)^7\right]$$

$$= h\left[35\left(\frac{t}{T}\right)^4 - 84\left(\frac{t}{T}\right)^5 + 70\left(\frac{t}{T}\right)^6 - 20\left(\frac{t}{T}\right)^7\right]$$

$$= h\left[1 - 35\left(\frac{t}{T}\right)^4 + 84\left(\frac{t}{T}\right)^5 - 70\left(\frac{t}{T}\right)^6 + 20\left(\frac{t}{T}\right)^7\right]$$

Vel., $\dot{s}$

$$\dot{s} = \frac{h\dot{\theta}}{\beta}\left[140\left(\frac{\theta}{\beta}\right)^3 - 420\left(\frac{\theta}{\beta}\right)^4 + 420\left(\frac{\theta}{\beta}\right)^5 - 140\left(\frac{\theta}{\beta}\right)^6\right]$$

$$\dot{s} = \frac{h\dot{\theta}}{\beta}\left[140\left(\frac{\theta}{\beta}\right)^3 + 420\left(\frac{\theta}{\beta}\right)^4 - 420\left(\frac{\theta}{\beta}\right)^5 + 140\left(\frac{\theta}{\beta}\right)^6\right]$$

$$= \frac{h}{T}\left[140\left(\frac{t}{T}\right)^3 - 420\left(\frac{t}{T}\right)^4 + 420\left(\frac{t}{T}\right)^5 - 140\left(\frac{t}{T}\right)^6\right]$$

$$= \frac{h}{T}\left[-140\left(\frac{t}{T}\right)^3 + 420\left(\frac{t}{T}\right)^4 - 420\left(\frac{t}{T}\right)^5 + 140\left(\frac{t}{T}\right)^6\right]$$

Acc., $\ddot{s}$

$$\ddot{s} = \frac{h\dot{\theta}^2}{\beta^2}\left[420\left(\frac{\theta}{\beta}\right)^2 - 1680\left(\frac{\theta}{\beta}\right)^3 + 2100\left(\frac{\theta}{\beta}\right)^4 - 840\left(\frac{\theta}{\beta}\right)^5\right]$$

$$\ddot{s} = \frac{h\dot{\theta}^2}{\beta^2}\left[-420\left(\frac{\theta}{\beta}\right)^2 + 1680\left(\frac{\theta}{\beta}\right)^3 - 2100\left(\frac{\theta}{\beta}\right)^4 + 840\left(\frac{\theta}{\beta}\right)^5\right]$$

$$= \frac{h}{T^2}\left[420\left(\frac{t}{T}\right)^2 - 1680\left(\frac{t}{T}\right)^3 + 2100\left(\frac{t}{T}\right)^4 - 840\left(\frac{t}{T}\right)^5\right]$$

$$= \frac{h}{T^2}\left[-420\left(\frac{t}{T}\right)^2 + 1680\left(\frac{t}{T}\right)^3 - 2100\left(\frac{t}{T}\right)^4 + 840\left(\frac{t}{T}\right)^5\right]$$

Jerk., $\dddot{s}$

$$\dddot{s} = \frac{h\dot{\theta}^3}{\beta^3}\left[840\left(\frac{\theta}{\beta}\right) - 5040\left(\frac{\theta}{\beta}\right)^2 + 8400\left(\frac{\theta}{\beta}\right)^3 - 4200\left(\frac{\theta}{\beta}\right)^4\right]$$

$$\dddot{s} = \frac{h\dot{\theta}^3}{\beta^3}\left[-840\left(\frac{\theta}{\beta}\right) + 5040\left(\frac{\theta}{\beta}\right)^2 - 8400\left(\frac{\theta}{\beta}\right)^3 + 4200\left(\frac{\theta}{\beta}\right)^4\right]$$

$$= \frac{h}{T^3}\left[840\left(\frac{t}{T}\right) - 5040\left(\frac{t}{T}\right)^2 + 8400\left(\frac{t}{T}\right)^3 - 4200\left(\frac{t}{T}\right)^4\right]$$

$$= \frac{h}{T^3}\left[-840\left(\frac{t}{T}\right) + 5040\left(\frac{t}{T}\right)^2 - 8400\left(\frac{t}{T}\right)^3 + 4200\left(\frac{t}{T}\right)^4\right]$$

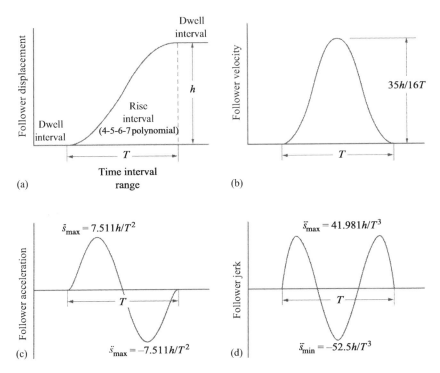

FIGURE 9.10
Follower (a) displacement, (b) velocity, (c) acceleration, and (d) jerk profiles for 4–5–6–7 polynomial motion.

Appendix E.6 includes the MATLAB file user instructions for generating 4-5-6-7 polynomial motion-based displacement, velocity, acceleration, and jerk diagrams. In this MATLAB file (which is available for download at www.crcpress.com/product/isbn/9781498724937), the 4-5-6-7 polynomial motion equations in Table 9.6 are used.

Example 9.6

Problem Statement: Using the Appendix E.6 MATLAB file, plot the follower displacement, velocity, acceleration, and jerk profiles under 4-5-6-7 polynomial motion for the follower displacement interval and cam rotation speed data given in Example 9.1.

Known Information: Example 9.1 and Appendix E.6 MATLAB file.

Solution Approach: Figure E.9.11 includes the input specified (in bold text) in the Appendix E.6 MATLAB file. Figure E.9.12 illustrates the follower displacement, velocity, acceleration, and jerk profiles calculated from the Appendix E.6 MATLAB file.

```
Event = ['R','D','R','D','F','D'];
Beta = [60,60,60,60,60,60];
Si = [25,25,55,55,0,0];

dTheta = 45*pi/30;
Rbase = 1; (NOTE: this value is not relevant since "Rbase" is not
used in this example)
```

FIGURE E.9.11
Specified input (in bold text) in the Appendix E.6 MATLAB file for Example 9.6.

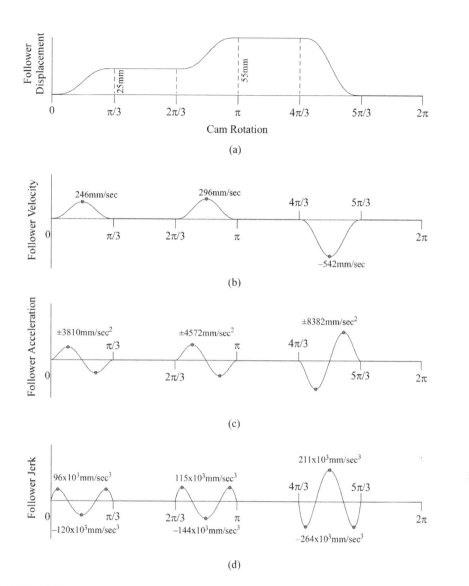

FIGURE E.9.12
Polynomial (4-5-6-7) motion follower (a) displacement, (b) velocity, (c) acceleration, and (d) jerk profiles.

9.4 Disk Cam Design and Pressure Angle

The actual shape of a disk cam is designed from the follower displacement profile. Figure 9.11 illustrates the disk cam design features and nomenclature. The *cam profile* is produced by wrapping the follower displacement profile over the *base circle*—the smallest circle centered at the cam's axis of rotation. Because the size of the base circle will ultimately determine the size of the cam produced, its size is typically restricted to the

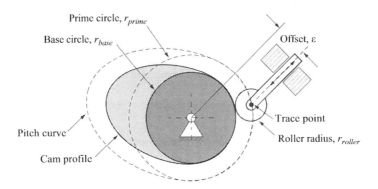

FIGURE 9.11
Disk cam nomenclature.

workspace restrictions for the cam application.* The radius of the base circle is represented by variable r_{base} in Figure 9.11.

The x and y coordinates of the disk cam profile can be formulated as

$$x = (r_{\text{base}} + s)\cos\theta$$
$$y = (r_{\text{base}} + s)\sin\theta \tag{9.4}$$

The *prime circle* (defined by variable r_{prime}) is a circle drawn through a point on the follower (called the *trace point*) while the follower is at a position of zero radial displacement (also called a *home position*). For the roller follower illustrated in Figure 9.11, the trace point is drawn through the center of the roller. As a result, the prime circle radius for a roller follower is the sum of the base circle radius and the radius of the follower roller (or $r_{\text{prime}} = r_{\text{base}} + r_{\text{roller}}$). The path traced by the follower over the rotating cam (at the trace point) is called the *pitch curve*.

As noted in Chapter 8, for two contacting surfaces, the contact force between the surface pair is oriented in the direction normal to the common tangent of the surface pair. For a cam and follower, as for gears, the orientation angle of the contact force is called the *pressure angle*. In a cam with a translating follower, this angle is formed by the common normal and the axis of translation of the follower. Figure 9.12 illustrates the contact force and pressure angle of a disk cam and offset roller follower.

As shown in Figure 9.12, the contact force can have two components: components acting *along* the direction of and *normal* to the direction of follower motion. The contact force component acting in the direction of follower motion is necessary for effective follower operation. The contact force component acting normal to the follower motion direction produces side-thrust loads on the cam and follower.[†] Designers often attempt to minimize the latter contact force component. In practice, pressure angles do not exceed ±30° for translating followers and ±45° for rotating followers [5]. Techniques to reduce the pressure angle in radial cams include increasing the size of the base circle, decreasing the follower offset (represented by the variable ε in Figures 9.11 and 9.12), and modifying the follower displacement profile.

---

* For a given cam shape, a larger base circle will typically result in improved force transmission than a smaller base circle.
† These side-thrust loads contribute to the lateral deflection of the follower.

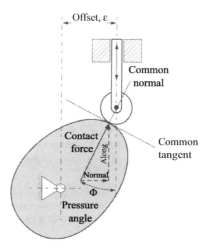

FIGURE 9.12
Offset cam system and pressure angle.

A general pressure-angle equation for a disk cam and translating roller follower can be expressed as

$$\phi = \delta - \tan^{-1}\left(\frac{\varepsilon}{r_{base} + s + r_{roller}}\right) \tag{9.5}$$

where

$$\delta = \tan^{-1}\left[\left(\frac{\dot{s}}{\dot{\theta}}\right)\left(\frac{r_{base} + s + r_{roller}}{\varepsilon^2 + \left(r_{base} + s + r_{roller}\right)^2 - \varepsilon\left(\dot{s}/\dot{\theta}\right)}\right)\right] \tag{9.6}$$

In these equations, the variables s, $\dot{s}$ and $\dot{\theta}$ are the displacement and velocity of the follower and the rotational velocity of the cam, respectively (all previously defined in this chapter). To consider a cam and in-line follower using Equations 9.5 and 9.6, the offset variable ε should be zero. For proper operation in a cam with a translating follower, $r_{prime} \geq \varepsilon$ (meaning $r_{base} + r_{roller} \geq \varepsilon$).

In a cam with a rotating follower, the pressure angle is formed by the common normal and the velocity vector of the follower arm (which is perpendicular to the follower arm).[*] Figure 9.13 illustrates the contact force and pressure angle of a disk cam and rotating roller follower.

A general pressure-angle equation for a disk cam and rotating roller follower (Figure 9.14) can be expressed as

$$\phi = \delta - \frac{\pi}{2} + \tan^{-1}\left[\left(\frac{\dot{s}}{\dot{\theta}}\right)\left(\frac{1}{r_{base} + s + r_{roller} - \left(\frac{\dot{s}}{\dot{\theta}}\right)\cos\delta}\right)\right] \tag{9.7}$$

---

[*] While the velocity vector is always perpendicular to the follower arm, its direction depends on the rotation direction of the follower arm (clockwise or counterclockwise).

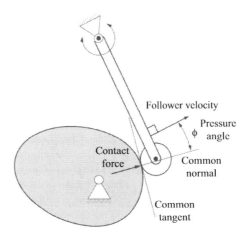

FIGURE 9.13
Cam system with rotating roller follower and pressure angle.

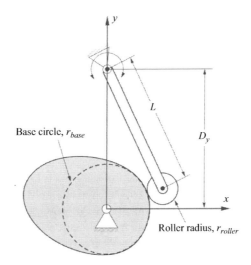

FIGURE 9.14
Disk cam with rotating roller follower nomenclature.

where

$$\delta = \cos^{-1}\left(\frac{L^2 + (r_{\text{base}} + s + r_{\text{roller}})^2 - D_y^2}{2L(r_{\text{base}} + s + r_{\text{roller}})}\right) \tag{9.8}$$

The variable L in Equation 9.8 is the length of the follower's rotating arm (between the arm's center of rotation and the center of the roller). The variable D_y in this equation is the distance along the y-axis between the arm's center of rotation and the cam's center of rotation. As illustrated in Figure 9.14, both centers of rotation are along the y-axis.

```
Event = ['R','D','R','D','F','D'];
Beta = [60,60,60,60,60,60];
Si = [25,25,55,55,0,0];

dTheta = 1; (NOTE: this value is not relevant since "dTheta" is
not used in this example)
Rbase = 100;
```

FIGURE E.9.13
Specified input (in bold text) in the Appendix E.6 MATLAB file for Example 9.7.

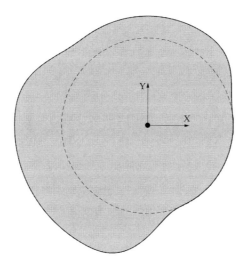

FIGURE E.9.14
Disk cam profile from 4-5-6-7 polynomial motion (with dashed base circle).

Example 9.7

Problem Statement: Using the Appendix E.6 MATLAB file, plot the radial cam profile for the follower displacement interval given in Example 9.1. Use a 100 mm base circle radius.

Known Information: Example 9.1 and Appendix E.6 MATLAB file.

Solution Approach: Figure E.9.13 includes the input specified (in bold text) in the Appendix E.6 MATLAB file. Figure E.9.14 illustrates the disk cam profile produced from Equation 9.4 (which is utilized in the Appendix E.6 MATLAB file).

Example 9.8

Problem Statement: Using the follower displacement and velocity data produced in Example 9.6, plot the pressure angle profile for a translating follower. Assume a zero offset, a base circle radius of 100 mm and a roller radius of 5 mm.

Known Information: Displacement and velocity results from Example 9.6, Equations 9.5 and 9.6, r_{base} and r_{roller}.

Solution Approach: From Example 9.6, we know that the cam rotates at a constant speed of 45 rpm (or 4.71 rad/s). Because a zero offset condition is assumed, the pressure angle equation is reduced to Equation 9.6 (where $\phi = \delta$). Figure E.9.15 illustrates the resulting pressure angle profile.

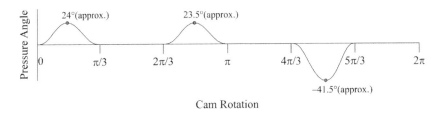

FIGURE E.9.15
Pressure angle profile for 4-5-6-7 polynomial cam and translating follower.

```
>> rb = 100;
>> rr = 5;
>> s = 27.5;
>> ds = -541.41;
>> dtheta = 4.71;
>> L = 275;
>> Dy = 250;
>> delta = acos((L^2 + (rb + s + rr)^2 - Dy^2)/...
(2*L*(rb + s + rr)));
>> phi = (delta - pi/2 + atan((ds/dtheta)*...
(1/(rb + s + rr - (ds/dtheta)*cos(delta)))))*180/pi

phi =

  -57.3323

>>
```

FIGURE E.9.16
Example 9.9 solution calculation procedure in MATLAB.

Example 9.9

Problem Statement: Using the follower displacement and velocity data produced in Example 9.6, calculate the pressure angle for a rotating follower at a cam rotation of $3\pi/2$ radians. Assume a base circle radius of 100 mm, a roller radius of 5 mm, a follower arm length of 275 mm and a center-to-center y-axis distance of 250 mm.

Known Information: Displacement and velocity results from Example 9.6, Equations 9.7 and 9.8, r_{base}, r_{roller}, L and D_y.

Solution Approach: From Example 9.6, we know that can cam rotates at a constant speed of 45 rpm (or 4.71 rad/s). From Example 9.6, it can be determined that at a cam rotation of $3\pi/2$ radians (or 270°), $s = 27.5$ mm and $\dot{s} = -541.41$ mm/s. Figure E.9.16 includes the calculation procedure for a single pressure angle value in MATLAB's command window.

9.5 Summary

A disk cam is a mechanical component used to convert rotation motion into oscillating rotation or translation motion. In its most basic form, a cam system includes a rotating disk member (the cam) that compels the motion of an oscillating member called the follower. Among other applications, cam systems are commonly used in the valve trains of internal combustion engines, particularly in automotive engines.

The types of followers used in disk cam systems vary in terms of their shape, motion and contact stress. Common follower types include the knife-edge, flat-faced, and roller followers. Followers are designed to exhibit either translating motion or rotational motion when in contact with a rotating cam.

A follower displacement profile is comprised of three distinct displacement profiles: the rise, fall, and dwell profiles. These profiles are determined by the rate of change in the radius of curvature of the cam and its rotation direction. The first, second, and third derivatives of a follower displacement function produce follower velocity, acceleration, and jerk functions. In cam design for high-speed applications, the follower displacement, velocity, and acceleration profiles must be continuous and the follower jerk profile must be finite over a complete cam rotation cycle. These conditions are expressed in the fundamental law of cam design.

Common types of follower motion include constant velocity, constant acceleration, simple harmonic, cycloidal motion, and polynomial motion. Of these follower motion types, both cycloidal and polynomial motion fully satisfy the conditions for continuous and finite follower motion profiles and are subsequently suitable for high-speed applications. In the Appendix E.1 through E.6 MATLAB files, the user can produce follower displacement, velocity, acceleration, and jerk profiles for constant velocity, constant acceleration, simple harmonic, cycloidal motion, and polynomial motion (both 3-4-5 and 4-5-6-7 polynomials).

For two contacting surfaces, the contact force between the surface pair is oriented in the direction normal to the common tangent of the surface pair. For a cam and follower, as for gears, the orientation angle of the contact force is called the pressure angle. The contact force can have two components: components acting *along* the direction of and *normal* to the direction of follower motion. The contact force component acting in the direction of follower motion is necessary for effective follower operation. The contact force component acting normal to the follower motion direction produces side-thrust loads on the cam and follower. Designers often attempt to minimize the latter contact force component. In practice, pressure angles do not exceed 30° for translating followers and 45° for rotating followers.

References

1. Oberg, E., Jones, F. D., Horton, H. L., and Ryffel, H. H. 2000. *Machinery's Handbook*. 26th edn, p. 2165. New York: Industrial Press.
2. Ibid., pp. 2165–2166.
3. Ibid., p. 2166.
4. Ibid., p. 2167.
5. Ibid., pp. 2176–2178.

Additional Reading

Myszka, D. H. 2005. *Machines and Mechanisms: Applied Kinematic Analysis*. 3rd edn, Chapter 9. Saddle River, NJ: Prentice-Hall.

Norton, R. L. 2008. *Design of Machinery*. 4th edn, Chapter 8. New York: McGraw-Hill.

Waldron, K. J. and Gary, L. K. 2004. *Kinematics, Dynamics and Design of Machinery*. 2nd edn, Chapter 8. Saddle River, NJ: Prentice-Hall.

Wilson, C. E. and Sadler, J. P. 2003. *Kinematics and Dynamics of Machinery*. 3rd edn, Chapter 5. Saddle River, NJ: Prentice-Hall.

Problems

1. Of the six follower motion types presented in this chapter, which type produces the smallest acceleration?

2. Of the six follower motion types presented in this chapter, which types fully satisfy the fundamental law of cam design?

3. In comparison to cams produced from follower motion types that fully satisfy the fundamental law of cam design, what is the primary limitation of cams produced from follower motion types that do not fully satisfy this law?

4. What are the structural effects in a cam follower system that does not fully satisfy the fundamental law of cam design?

5. Describe a primary similarity and difference between the motion in a cam follower system and a rack and pinion gear.

6. Plot constant velocity motion-based follower displacement and velocity profiles for the follower displacement data given in Table P.9.1. Consider a cam rotation speed of 75 rad/s and a cam base circle radius of 15 mm.

7. Plot constant velocity motion-based follower displacement and velocity profiles for the follower displacement data given in Table P.9.2. Consider a cam rotation speed of 85 rad/s and a cam base circle radius of 15 mm.

8. Plot the cam profiles using the follower displacement data from Problems 6 and 7.

9. Plot constant acceleration motion-based follower displacement, velocity, and acceleration profiles for the follower displacement data given in Table P.9.1. Consider a cam rotation speed of 70 rad/s and a cam base circle radius of 15 mm.

10. Plot constant acceleration motion-based follower displacement, velocity, and acceleration profiles for the follower displacement data given in Table P.9.2. Consider a cam rotation speed of 80 rad/s and a cam base circle radius of 15 mm.

TABLE P.9.1

Follower Dwell-Rise-Dwell-Fall Displacement Data

Event	Rotation Range	Total Follower Displacement (mm)
Dwell	90°	0
Rise	60°	4
Dwell	30°	4
Fall	180°	0

TABLE P.9.2

Follower Rise-Dwell-Fall-Dwell-Fall-Dwell
Displacement Data

Event	Rotation Range	Total Follower Displacement (mm)
Rise	60°	3
Dwell	60°	3
Fall	60°	1.5
Dwell	60°	1.5
Fall	60°	0
Dwell	60°	0

11. Plot the cam profiles using the follower displacement data from Problems 9 and 10.

12. Plot simple harmonic motion-based follower displacement, velocity, acceleration, and jerk profiles for the follower displacement data given in Table P.9.3. Consider a cam rotation speed of 55 rad/s and base circle radius of 20 mm.

13. Plot simple harmonic motion-based follower displacement, velocity, acceleration, and jerk profiles for the follower displacement data given in Table P.9.4. Consider a cam rotation speed of 55 rad/s and base circle radius of 20 mm.

TABLE P.9.3

Follower Rise-Dwell-Fall-Dwell-Rise-Dwell-Fall-
Dwell Displacement Data

Event	Rotation Range	Total Follower Displacement (mm)
Rise	15°	3.5
Dwell	65°	3.5
Fall	30°	0
Dwell	50°	0
Rise	60°	2
Dwell	90°	2
Fall	35°	0
Dwell	15°	0

TABLE P.9.4

Follower Dwell-Rise-Dwell-Rise-Dwell-Fall
Displacement Data

Event	Rotation Range	Total Follower Displacement (mm)
Dwell	15°	0
Rise	45°	1.5
Dwell	30°	1.5
Rise	90°	4
Dwell	45°	4
Fall	135°	0

14. Plot the cam profiles using the follower displacement data from Problem 12.

15. Plot cycloidal motion-based follower displacement, velocity, acceleration, and jerk profiles for the follower displacement data given in Table P.9.3. Consider a cam rotation speed of 25 rad/s and base circle radius of 20 mm.

16. Plot cycloidal motion-based follower displacement, velocity, acceleration, and jerk profiles for the follower displacement data given in Table P.9.4. Consider a cam rotation speed of 50 rad/s and base circle radius of 20 mm.

17. Plot the cam profiles using the follower displacement data from Problem 16.

18. Plot 3-4-5 polynomial motion-based follower displacement, velocity, acceleration, and jerk profiles for the follower displacement data given in Table P.9.5. Consider a cam rotation speed of 40 rad/s and base circle radius of 25 mm.

19. Plot 3-4-5 polynomial motion-based follower displacement, velocity, acceleration, and jerk profiles (and the corresponding cam profile) for the follower displacement data given in Table P.9.6. Consider a cam rotation speed of 40 rad/s and base circle radius of 25 mm.

20. Plot the cam profiles using the follower displacement data from Problem 18.

21. Plot 4-5-6-7 polynomial motion-based follower displacement, velocity, acceleration, and jerk profiles (and the corresponding cam profile) for the follower displacement data given in Table P.9.5. Consider a cam rotation speed of 35 rad/s and base circle radius of 25 mm.

TABLE P.9.5

Follower Rise-Dwell-Rise-Dwell-Fall-Dwell-Fall-Dwell Displacement Data

Event	Rotation Range	Total Follower Displacement (mm)
Rise	60°	2.5
Dwell	30°	2.5
Rise	60°	5
Dwell	30°	5
Fall	80°	3.5
Dwell	50°	3.5
Fall	25°	0
Dwell	25°	0

TABLE P.9.6

Follower Dwell-Rise-Dwell-Fall Displacement Data

Event	Rotation Range	Total Follower Displacement (mm)
Dwell	30°	0
Rise	150°	5
Dwell	30°	5
Fall	150°	0

22. Plot 4-5-6-7 polynomial motion-based follower displacement, velocity, acceleration, and jerk profiles for the follower displacement data given in Table P.9.6. Consider a cam rotation speed of 35 rad/s and base circle radius of 25 mm.

23. Plot the cam profiles using the follower displacement data from Problem 22.

24. Using the follower displacement and velocity data produced in Problem 6, plot the pressure angle profile for a translating follower. Assume no offset and a roller radius of 3.75 mm.

25. Using the follower displacement and velocity data produced in Problem 12, plot the pressure angle profile for a translating follower. Assume no offset and a roller radius of 5 mm.

26. Using the follower displacement and velocity data produced in Problem 18, plot the pressure angle profile for a translating follower. Assume no offset and a roller radius of 3.9 mm.

27. Using the follower displacement and velocity data produced in Problem 10, plot the pressure angle profile for a translating follower. Assume a 5 mm offset and a roller radius of 2.75 mm.

28. Using the follower displacement and velocity data produced in Problem 16, plot the pressure angle profile for a translating follower. Assume a 15 mm offset and a roller radius of 1.75 mm.

29. Using the follower displacement and velocity data produced in Problem 22, plot the pressure angle profile for a translating follower. Assume a 3 mm offset and a roller radius of 3.95 mm.

30. Using the follower displacement and velocity data produced in Problem 6, plot the pressure angle profile for a rotating follower. Assume a roller radius of 3.5 mm, a follower arm length of 275 mm and a center-to-center y-axis distance of 280 mm.

31. Using the follower displacement and velocity data produced in Problem 12, plot the pressure angle profile for a rotating follower. Assume a roller radius of 5.75 mm, a follower arm length of 300 mm and a center-to-center y-axis distance of 280 mm.

32. Using the follower displacement and velocity data produced in Problem 18, plot the pressure angle profile for a rotating follower. Assume a roller radius of 1.75 mm, a follower arm length of 290 mm and a center-to-center y-axis distance of 280 mm.

33. Using the follower displacement and velocity data produced in Problem 10, plot the pressure angle profile for a rotating follower. Assume a roller radius of 3.5 mm, a follower arm length of 295 mm and a center-to-center y-axis distance of 300 mm.

34. Using the follower displacement and velocity data produced in Problem 16, plot the pressure angle profile for a rotating follower. Assume a roller radius of 5.75 mm, a follower arm length of 285 mm and a center-to-center y-axis distance of 300 mm.

35. Using the follower displacement and velocity data produced in Problem 22, plot the pressure angle profile for a rotating follower. Assume a roller radius of 2.25 mm, a follower arm length of 275 mm and a center-to-center y-axis distance of 300 mm.

10

Kinematic Analysis of Spatial Mechanisms

CONCEPT OVERVIEW

In this chapter, the reader will gain a central understanding regarding

1. Advantages and disadvantages of spatial mechanisms and the ongoing aim of spatial mechanism modeling
2. Intermediate and total spatial kinematics of the *RRSS*, *RSSR*, and *4R spherical* mechanisms
3. Planar mechanism kinematic analysis using spatial mechanism kinematic models

10.1 Introduction

Planar mechanisms are restricted to motion in two-dimensional or planar space. *Spatial* mechanisms can exhibit three-dimensional or spatial motion. Spatial mechanism motion is predominantly determined by the degrees of freedom of the mechanism joints used and spatial orientation of the joints.

The cylindrical joint (Figure 3.9) can enable spatial mechanism motion because this joint has a translational degree of freedom (DOF) along the z-axis in addition to the planar rotational DOF about the z-axis. The spherical joint (Figure 3.9) can enable spatial mechanism motion because this joint has rotational DOFs about the x- and y-axes in addition to the planar rotational DOF about the z-axis. Even revolute and prismatic joints (Figure 3.9), though possessing only a single DOF, will enable spatial mechanism motion when positioned to have joint axes (or lines of action in the case of prismatic joints) that are skewed to the x–y plane.

Because spatial mechanisms have the capacity to exhibit 3D motion, they offer a greater variety of possible motions and are structurally more general than planar mechanisms. However, because the equations for spatial mechanism analysis are often much larger in scale and greater in complexity than those for planar mechanisms, their real-world applications are often limited. In practice, it is not uncommon to find complicated planar mechanism solutions when, in fact, a simpler spatial mechanism solution is possible. It is, therefore, an ongoing task to devise simple methods of calculation, to produce design aids with diagrams, and to set design standards for spatial mechanisms [1].

This textbook considers three types of four-bar spatial mechanisms: the *revolute-revolute-spherical-spherical* or *RRSS*, the *revolute-revolute-revolute-revolute spherical* or *4R spherical*, and the *revolute-spherical-spherical-revolute* or *RSSR* mechanisms (Figures 3.7a, b, and c, respectively)

[2, 3]. The RRSS, 4R spherical, and RSSR are among the more basic four-bar spatial mechanisms in terms of the types of joints used and the required linkage assembly conditions for motion.

10.2 RRSS Mechanism Analysis

10.2.1 Displacement Equations

The revolute-revolute-spherical-spherical mechanism or RRSS mechanism (Figure 10.1) is a spatial four-link mechanism with a mobility of 2 (with $L = 4$ and $J_1 = J_3 = 2$). Although one DOF of the RRSS mechanism enables it to be crank driven, the other DOF is produced by the follower link. The follower link (link $\mathbf{b}_0$–$\mathbf{b}_1$ in Figure 10.1) is bounded by spherical joints. As a result, this link can rotate about its own axis of symmetry. This DOF is called a *passive degree of freedom*. Passive DOFs generally do not contribute to the overall kinematic motion of a mechanism.*

Suh and Radcliffe presented displacement equations for the RRSS mechanism [4]. The following follower-link constant-length equation forms the basis for the RRSS displacement equations:

$$\left(\mathbf{b} - \mathbf{b}_0\right)^T \left(\mathbf{b} - \mathbf{b}_0\right) - \left(\mathbf{b}_1 - \mathbf{b}_0\right)^T \left(\mathbf{b}_1 - \mathbf{b}_0\right) = 0 \tag{10.1}$$

The rotation of the crank and coupler links (links $\mathbf{a}_0$–$\mathbf{a}_1$ and $\mathbf{a}_1$–$\mathbf{b}_1$, respectively, in Figure 10.1) and the rotation of the crank moving pivot joint axis $\mathbf{ua}_1$ about the crank fixed pivot joint axis $\mathbf{ua}_0$ (by rotation angle θ) are given by

$$\mathbf{a} = \left[\mathbf{R}_{\theta,\mathbf{ua}_0}\right]\left(\mathbf{a}_1 - \mathbf{a}_0\right) + \mathbf{a}_0 \tag{10.2}$$

$$\mathbf{ua} = \left[R_{\theta,\mathbf{ua}_0}\right]\mathbf{ua}_1 \tag{10.3}$$

$$\mathbf{b}_1' = \left[R_{\theta,\mathbf{ua}_0}\right]\left(\mathbf{b}_1 - \mathbf{a}_0\right) + \mathbf{a}_0 \tag{10.4}$$

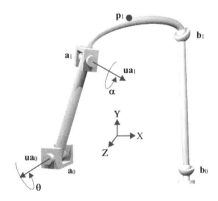

FIGURE 10.1
RRSS mechanism displacement variables.

---

* The passive DOFs associated with the RSSR and RRSS mechanisms generally do not contribute to mechanism motion because the effects of these DOFs are typically highly localized.

The global displacement of the follower moving pivot $\mathbf{b}_1$ is expressed as

$$\mathbf{b} = \left[R_{\alpha,\mathbf{ua}}\right]\left(\mathbf{b}_1' - \mathbf{a}\right) + \mathbf{a} \tag{10.5}$$

The rotational displacement matrices $[R_{\theta,\mathbf{ua}_0}]$ and $[R_{\alpha,\mathbf{ua}}]$ are identical in form to Matrix 2.28. Substituting Equation 10.5 into Equation 10.1 produces

$$E\cos(\alpha) + F\sin(\alpha) + G = 0 \tag{10.6}$$

In Equation 10.6,

$$E = \left(\mathbf{a} - \mathbf{b}_0\right)^T \left\{\left[I - Q_{\mathbf{ua}}\right]\left(\mathbf{b}_1' - \mathbf{a}\right)\right\} \tag{10.7}*$$

$$F = \left(\mathbf{a} - \mathbf{b}_0\right)^T \left\{\left[P_{\mathbf{ua}}\right]\left(\mathbf{b}_1' - \mathbf{a}\right)\right\} \tag{10.8}$$

$$G = \left(\mathbf{a} - \mathbf{b}_0\right)^T \left\{\left[Q_{\mathbf{ua}}\right]\left(\mathbf{b}_1' - \mathbf{a}\right)\right\} + \frac{1}{2}\left\{\begin{matrix}\left(\mathbf{b}_1' - \mathbf{a}\right)^T\left(\mathbf{b}_1' - \mathbf{a}\right) + \left(\mathbf{a} - \mathbf{b}_0\right)^T\left(\mathbf{a} - \mathbf{b}_0\right) \\ -\left(\mathbf{b}_1 - \mathbf{b}_0\right)^T\left(\mathbf{b}_1 - \mathbf{b}_0\right)\end{matrix}\right\} \tag{10.9}$$

$$\left[P_{\mathbf{ua}}\right] = \begin{bmatrix} 0 & -ua_z & ua_y \\ ua_z & 0 & -ua_x \\ -ua_y & ua_x & 0 \end{bmatrix} \tag{10.10}$$

and

$$\left[Q_{\mathbf{ua}}\right] = \begin{bmatrix} ua_x^2 & ua_x ua_y & ua_x ua_z \\ ua_x ua_y & ua_y^2 & ua_y ua_z \\ ua_x ua_z & ua_y ua_z & ua_z^2 \end{bmatrix} \tag{10.11}$$

The two coupler-link rotation-angle solutions (α) for Equation 10.6 are

$$\alpha_{1,2} = 2\tan^{-1}\frac{-F \pm \sqrt{E^2 + F^2 - G^2}}{G - E} \tag{10.12}†$$

With α known, the displaced RRSS moving pivot $\mathbf{b}$ from Equation 10.5 and also the displaced RRSS coupler point $\mathbf{p}$ from Equation 10.13 can be calculated.

As noted in Section 4.10, two sets of link displacement angles are calculated (for a given crank displacement angle) in algebraic four-bar mechanism displacement equations. The ± term in Equation 10.12 corresponds to two coupler displacement angle solutions—one solution for each mechanism assembly configuration.

The global displacement of an arbitrary RRSS coupler point $\mathbf{p}_1$ is expressed as

$$\mathbf{p} = \left[R_{\alpha,\mathbf{ua}}\right]\left(\mathbf{p}_1' - \mathbf{a}\right) + \mathbf{a} \tag{10.13}$$

---

* Matrix I in Equations 10.7 and 10.33 is a 3×3 identity matrix.
† This coupler-link displacement angle is not measured with respect to the initial coupler position (like the planar four-bar mechanism for example), but with respect to the crank link.

where

$$\mathbf{p}_1' = \left[R_{\theta,\mathbf{ua}_0}\right](\mathbf{p}_1 - \mathbf{a}_0) + \mathbf{a}_0 \tag{10.14}$$

The RRSS variables $\mathbf{a}_0$, $\mathbf{a}_1$, $\mathbf{ua}_0$, $\mathbf{ua}_1$, $\mathbf{b}_0$, $\mathbf{b}_1$, and $\mathbf{p}_1$ are 3×1 vectors containing x-, y-, and z-components.

10.2.2 Velocity Equations

Figure 10.2 includes the velocity variables for the RRSS mechanism. The following equation, the derivative of the follower-link constant-length equation, forms the basis for the RRSS velocity equations:

$$\left(\dot{\mathbf{b}}\right)^T (\mathbf{b} - \mathbf{b}_0) = 0 \tag{10.15}$$

The global velocity of the RRSS moving pivot $\mathbf{b}_1$ (variable $\dot{\mathbf{b}}$ in Equation 10.15) is expressed as

$$\dot{\mathbf{b}} = \dot{\mathbf{a}}\left[P_{\mathbf{ua}}\right](\mathbf{b} - \mathbf{a}) + \dot{\mathbf{b}}' = \left[V_{\dot{\alpha},\mathbf{ua}}\right](\mathbf{b} - \mathbf{a}) + \dot{\mathbf{b}}' \tag{10.16}$$

where

$$\dot{\mathbf{b}}' = \dot{\theta}\left[\mathbf{P}_{\mathbf{ua}_0}\right](\mathbf{b} - \mathbf{a}_0) = \left[V_{\dot{\theta},\mathbf{ua}_0}\right](\mathbf{b} - \mathbf{a}_0) \tag{10.17}$$

The rotational velocity matrices $\left[V_{\dot{\theta},\mathbf{ua}_0}\right]$ and $\left[V_{\dot{\alpha},\mathbf{ua}}\right]$ are identical in form to Matrix 2.33.
   In an identical form, the global velocity of an arbitrary RRSS coupler point $\mathbf{p}_1$ is expressed as

$$\dot{\mathbf{p}} = \left[V_{\dot{\alpha},\mathbf{ua}}\right](\mathbf{p} - \mathbf{a}) + \dot{\mathbf{p}}' \tag{10.18}$$

where

$$\dot{\mathbf{p}}' = \left[V_{\dot{\theta},\mathbf{ua}_0}\right](\mathbf{p} - \mathbf{a}_0) \tag{10.19}$$

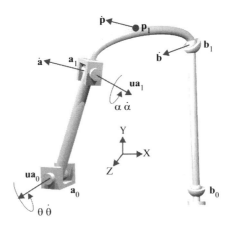

FIGURE 10.2
RRSS mechanism velocity variables.

Substituting Equations 10.16 and 10.17 into Equation 10.15 and solving for the coupler angular velocity. $\dot{\alpha}\cdot$ produces the coupler angular velocity equation

$$\dot{\alpha} = -\dot{\theta}\frac{(\mathbf{b}-\mathbf{b}_0)^T\left\{[P_{\mathbf{ua}_0}](\mathbf{b}-\mathbf{a}_0)\right\}}{(\mathbf{b}-\mathbf{b}_0)^T\left\{[P_{\mathbf{ua}}](\mathbf{b}-\mathbf{a})\right\}} \tag{10.20}$$

With $\dot{\alpha}$ known, the displaced RRSS moving pivot velocity $\dot{\mathbf{b}}$ from Equation 10.16 and also the displaced RRSS coupler point velocity $\dot{\mathbf{p}}$ from Equation 10.18 can be calculated.

Lastly, the global velocity of the RRSS moving pivot $\mathbf{a}_1$ is expressed as

$$\dot{\mathbf{a}} = \dot{\theta}\left[P_{\mathbf{ua}_0}\right](\mathbf{a}-\mathbf{a}_0) = \left[V_{\dot{\theta},\mathbf{ua}_0}\right](\mathbf{a}-\mathbf{a}_0) \tag{10.21}$$

which is identical in form to Equations 10.17 and 10.19.

10.2.3 Acceleration Equations

Figure 10.3 includes the acceleration variables for the RRSS mechanism. The following equation, the second derivative of the follower-link constant-length equation, forms the basis for the RRSS acceleration equations:

$$\left(\ddot{\mathbf{b}}\right)^T(\mathbf{b}-\mathbf{b}_0) + \left(\dot{\mathbf{b}}\right)^T\left(\dot{\mathbf{b}}\right) = 0 \tag{10.22}$$

The global acceleration of the RRSS moving pivot $\mathbf{b}_1$ (variable $\ddot{\mathbf{b}}$ in Equation 10.22) is expressed as

$$\ddot{\mathbf{b}} = \ddot{\mathbf{b}}' + \left\{\ddot{\alpha}[P_{\mathbf{ua}}] + \dot{\alpha}\left[\dot{P}_{\mathbf{ua}}\right] + \dot{\alpha}^2[P_{\mathbf{ua}}][P_{\mathbf{ua}}]\right\}(\mathbf{b}-\mathbf{a}) + 2\dot{\theta}[P_{\mathbf{ua}_0}]\left\{\dot{\alpha}[P_{\mathbf{ua}}](\mathbf{b}-\mathbf{a})\right\}$$

$$= \ddot{\mathbf{b}}' + \left[A_{\dot{\alpha},\ddot{\alpha},\mathbf{ua}}\right](\mathbf{b}-\mathbf{a}) + 2\left[V_{\dot{\theta},\mathbf{ua}_0}\right]\left\{\left[V_{\dot{\alpha},\mathbf{ua}}\right](\mathbf{b}-\mathbf{a})\right\} \tag{10.23}$$

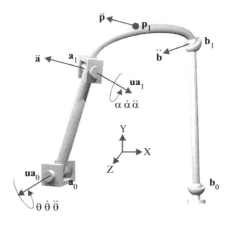

FIGURE 10.3
RRSS mechanism acceleration variables.

where

$$\ddot{\mathbf{b}}' = \left\{\ddot{\theta}\left[P_{\mathbf{ua}_0}\right] + \dot{\theta}\left[\dot{P}_{\mathbf{ua}_0}\right] + \dot{\theta}^2\left[P_{\mathbf{ua}_0}\right]\left[P_{\mathbf{ua}_0}\right]\right\}(\mathbf{b} - \mathbf{a}_0) = \left[A_{\dot{\theta},\ddot{\theta},\mathbf{ua}_0}\right](\mathbf{b} - \mathbf{a}_0) \qquad (10.24)$$

The matrices $\left[\dot{P}_{\mathbf{ua}}\right]$ and $\left[\dot{P}_{\mathbf{ua}_0}\right]$ that appear in Equations 10.23 and 10.24 become zero, since $\mathbf{ua}_0$ and $\mathbf{ub}_0$, being grounded, do not exhibit motion [5]. The rotational acceleration matrices $\left[A_{\dot{\theta},\ddot{\theta},\mathbf{ua}_0}\right]$ and $\left[A_{\dot{\alpha},\ddot{\alpha},\mathbf{ua}}\right]$ are identical in form to Matrix 2.37.

In an identical form, the global acceleration of an arbitrary RRSS coupler point $\mathbf{p}_1$ is expressed as

$$\ddot{\mathbf{p}} = \ddot{\mathbf{p}}' + \left[A_{\dot{\alpha},\ddot{\alpha},\mathbf{ua}}\right](\mathbf{p} - \mathbf{a}) + 2\left[V_{\dot{\theta},\mathbf{ua}_0}\right]\left\{\left[V_{\dot{\alpha},\mathbf{ua}}\right](\mathbf{p} - \mathbf{a})\right\} \qquad (10.25)$$

where

$$\ddot{\mathbf{p}}' = \left[A_{\dot{\theta},\ddot{\theta},\mathbf{ua}_0}\right](\mathbf{p} - \mathbf{a}_0) \qquad (10.26)$$

Substituting Equations 10.23 and 10.24 into Equation 10.22 and solving for the coupler angular acceleration $\ddot{\alpha}$ produces the coupler angular acceleration equation

$$\ddot{\alpha} = -\frac{(\mathbf{b} - \mathbf{b}_0)^T\left(\ddot{\mathbf{b}}' + \left\{\dot{\alpha}^2\left[P_{\mathbf{ua}}\right]\left[P_{\mathbf{ua}}\right]\right\}(\mathbf{b} - \mathbf{a}) + 2\left[V_{\dot{\theta},\mathbf{ua}_0}\right]\left\{\left[V_{\dot{\alpha},\mathbf{ua}}\right](\mathbf{b} - \mathbf{a})\right\}\right) + (\dot{\mathbf{b}})^T(\dot{\mathbf{b}})}{(\mathbf{b} - \mathbf{b}_0)^T\left\{\left[P_{\mathbf{ua}}\right](\mathbf{b} - \mathbf{a})\right\}} \qquad (10.27)$$

With $\ddot{\alpha}$ known, the displaced RRSS moving point acceleration $\ddot{\mathbf{b}}$ from Equation 10.23 and also the displaced RRSS coupler point velocity $\ddot{\mathbf{p}}$ from Equation 10.25 can be calculated.

Lastly, the global acceleration of the RRSS moving pivot $\mathbf{a}_1$ is expressed as

$$\ddot{\mathbf{a}} = \left[A_{\dot{\theta},\ddot{\theta},\mathbf{ua}_0}\right](\mathbf{a} - \mathbf{a}_0) \qquad (10.28)$$

which is identical in form to Equations 10.24 and 10.26.

Appendix F.1 includes the MATLAB® file user instructions for RRSS displacement, velocity, and acceleration analysis. This MATLAB file (which is available for download at www.crcpress.com/product/isbn/9781498724937) utilizes the RRSS displacement, velocity, and acceleration equations presented in Sections 10.2.1–10.2.3.

Although the equations presented in Sections 10.2.1–10.2.3 are presented for the spatial RRSS mechanism, the same equations can also be used for the displacement, velocity, and acceleration analysis of a planar four-bar mechanism (specifically, a planar RRSS mechanism). To accomplish this, the user should specify planar mechanism values for the RRSS mechanism variables (e.g., $\mathbf{ua}_0 = \mathbf{ua}_1 = (0, 0, 1)$ and $a_{0z} = a_{1z} = b_{0z} = b_{1z} = p_{1z} = 0$).*

Example 10.1

Problem Statement: Figure E.10.1 illustrates an RRSS mechanism use to guide a solar panel. Using the Appendix F.1 MATLAB file, calculate the solar panel positions (the

---

* Due to the passive DOF of the follower link, rotation angles for this link cannot be directly calculated in the RRSS mechanism equations, even when restricted to planar motion (unlike the planar four-bar equations in Section 4.3).

displaced values of points p_1, q_1, and r_1) achieved by the RRSS mechanism in Table E.10.1 at crank displacement angles of 50°, 100°, and 200°.

Known Information: Table E.10.1 and Appendix F.1 MATLAB file.

Solution Approach: Figure E.10.2 includes the input specified (in bold text) in the Appendix F.1 MATLAB file. Table E.10.2 includes the spatial coordinates of **p**, **q**, and **r** calculated using the Appendix F.1 MATLAB file for the three crank displacement angles. The dimensions in Table E.10.1 and the results in Table E.10.2 are for the initial RRSS mechanism configuration (which corresponds to the first branch in the Appendix F.1 MATLAB file).*

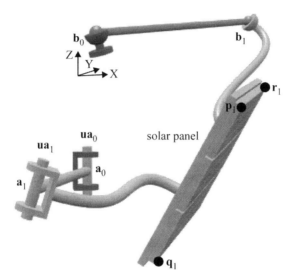

FIGURE E.10.1
RRSS mechanism used to guide a solar panel.

TABLE E.10.1

RRSS Mechanism Dimensions

Variable	Value
a_0	−0.0576, 0.2890, −1.4112
a_1	0.1452, −2.5421, −1.18
ua_0	−0.0003, 0.0814, 0.9967
ua_1	0.304, 0.0992, 0.9475
b_0	0.0851, 0.457, 0.5096
b_1	1.7725, 5.1566, 0.6499
p_1	1.7321, 0, −1
q_1	1.2321, 0, −1.866
r_1	1.9486, 0, −1.125

---

* The Appendix F.1 MATLAB file produces results for both the initial (branch 1) and second (branch 2) RRSS mechanism configurations.

```
a0  = [-0.0576,  0.289, -1.4112]';
a1  = [0.1452, -2.5421, -1.18]';
ua0 = [-0.0003,  0.0814,  0.9967]';
ua1 = [0.304,   0.0992,  0.9475]';
b0  = [0.0851,  0.457,   0.5096]';
b1  = [1.7725,  5.1566,  0.6499]';
p1  = [1.7321,  0,  -1]';
q1  = [1.2321,  0,  -1.866]';
r1  = [1.9486,  0,  -1.125]';

start_ang   = 0;
step_ang    = 50;
stop_ang    = 200;

angular_vel  = 0 * ones(N+1,1);
angular_acc  = 0 * ones(N+1,1);
```

FIGURE E.10.2
Specified input (in bold text) in the Appendix F.1 MATLAB file for Example 10.1.

TABLE E.10.2

Solar Panel Positions Achieved by RRSS Mechanism

θ	P	q	r
50°	1.0926, 1.3848, −1.0402	0.7287, 0.9589, −1.8685	1.2045, 1.5598, −1.1793
100°	−0.1671, 1.7348, −1.0237	−0.1141, 1.1570, −1.8382	−0.2507, 1.9244, −1.1637
200°	−1.7270, −0.0051, −0.9672	−1.2325, 0.0396, −1.8352	−1.9224, −0.1086, −1.0839

10.3 RSSR Mechanism Analysis

10.3.1 Displacement Equations

The revolute-spherical-spherical-revolute or RSSR mechanism (Figure 10.4) is a spatial four-bar mechanism having two degrees of freedom (with $L = 4$ and $J_1 = J_3 = 2$), the rotation of the coupler link (link $\mathbf{a}_1$–$\mathbf{b}_1$) about its own axis of symmetry being a passive DOF.

Suh and Radcliffe presented displacement equations for the RSSR mechanism [6]. The following coupler-link constant-length equation forms the basis for the RSSR displacement equations:

$$(\mathbf{a} - \mathbf{b})^T (\mathbf{a} - \mathbf{b}) - (\mathbf{a}_1 - \mathbf{b}_1)^T (\mathbf{a}_1 - \mathbf{b}_1) = 0 \tag{10.29}$$

The rotation of the crank and follower links (links $\mathbf{a}_0$–$\mathbf{a}_1$ and $\mathbf{b}_0$–$\mathbf{b}_1$, respectively, in Figure 10.4) about their fixed pivot joint axes by crank and follower rotation angles θ and ϕ, respectively, are given by

$$\mathbf{a} = \left[R_{\theta, \mathbf{ua}_0} \right] (\mathbf{a}_1 - \mathbf{a}_0) + \mathbf{a}_0 \tag{10.30}$$

and

$$\mathbf{b} = \left[R_{\phi, \mathbf{ub}_0} \right] (\mathbf{b}_1 - \mathbf{b}_0) + \mathbf{b}_0 \tag{10.31}$$

The rotational displacement matrices $\left[R_{\theta, \mathbf{ua}_0} \right]$ and $\left[R_{\phi, \mathbf{ub}_0} \right]$ are identical in form to Matrix 2.28.

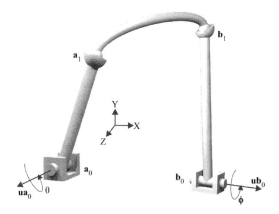

FIGURE 10.4
RSSR mechanism displacement variables.

Substituting Equations 10.30 and 10.31 into Equation 10.29 produces

$$E\cos(\phi) + F\sin(\phi) + G = 0 \tag{10.32}$$

In Equation 10.32,

$$E = (\mathbf{a} - \mathbf{b}_0)^T \{[I - Q_{\mathbf{ub}_0}](\mathbf{b}_1 - \mathbf{b}_0)\} \tag{10.33}$$

$$F = (\mathbf{a} - \mathbf{b}_0)^T \{[P_{\mathbf{ub}_0}](\mathbf{b}_1 - \mathbf{b}_0)\} \tag{10.34}$$

$$G = (\mathbf{a} - \mathbf{b}_0)^T \{[Q_{\mathbf{ub}_0}](\mathbf{b}_1 - \mathbf{b}_0)\} + \frac{1}{2} \left[\begin{aligned} (\mathbf{a}_1 - \mathbf{b}_1)^T(\mathbf{a}_1 - \mathbf{b}_1) - (\mathbf{a} - \mathbf{b}_0)^T(\mathbf{a} - \mathbf{b}_0) \\ -(\mathbf{b}_1 - \mathbf{b}_0)^T(\mathbf{b}_1 - \mathbf{b}_0) \end{aligned} \right] \tag{10.35}$$

In Equations 10.33–10.35, matrices $[Q_{\mathbf{ub}_0}]$ and $[P_{\mathbf{ub}_0}]$ are identical in form to Equations 10.10 and 10.11. The two follower-link rotation-angle solutions (ϕ) for Equation 10.32 are

$$\phi_{1,2} = 2\tan^{-1}\frac{-F \pm \sqrt{E^2 + F^2 - G^2}}{G - E} \tag{10.36}*$$

With ϕ known, the displaced RSSR moving pivot **b** from Equation 10.32 can be calculated. The variables $\mathbf{a}_0$, $\mathbf{a}_1$, $\mathbf{ua}_0$, $\mathbf{b}_0$, $\mathbf{b}_1$, and $\mathbf{ub}_0$ are 3×1 vectors containing x-, y-, and z-components.

10.3.2 Velocity Equations

Figure 10.5 includes the velocity variables for the RSSR mechanism. The following equation, the derivative of the coupler-link constant-length equation, forms the basis for the RSSR velocity equations:

$$(\dot{\mathbf{a}} - \dot{\mathbf{b}})^T(\mathbf{a} - \mathbf{b}) = 0 \tag{10.37}$$

---

* Like Equation 10.12, there are two ϕ solutions corresponding to the two RSSR assembly configurations (the open and crossed configurations) for every θ value.

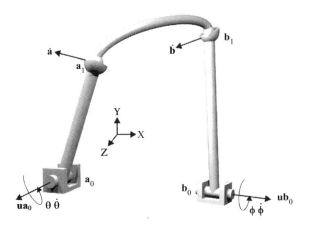

FIGURE 10.5
RSSR mechanism velocity variables.

The global velocity of the RSSR moving pivot $\mathbf{b}_1$ (variable $\dot{\mathbf{b}}$ in Equation 10.37) is expressed as

$$\dot{\mathbf{b}} = \dot{\phi}\left[P_{\mathbf{ub}_0} \right](\mathbf{b} - \mathbf{b}_0) = \left[V_{\dot{\phi}, \mathbf{ub}_0} \right](\mathbf{b} - \mathbf{b}_0) \tag{10.38}$$

and the global velocity of the RSSR moving pivot $\mathbf{a}_1$ (variable $\dot{\mathbf{a}}$ in Equation 10.38) is expressed as

$$\dot{\mathbf{a}} = \dot{\theta}\left[P_{\mathbf{ua}_0} \right](\mathbf{a} - \mathbf{a}_0) = \left[V_{\dot{\theta}, \mathbf{ua}_0} \right](\mathbf{a} - \mathbf{a}_0) \tag{10.39}$$

The rotational velocity matrices $\left[V_{\dot{\theta}, \mathbf{ua}_0} \right]$ and $\left[V_{\dot{\phi}, \mathbf{ub}_0} \right]$ are identical in form to Matrix 2.33.

Substituting Equations 10.38 and 10.39 into Equation 10.37 and solving for the follower angular velocity $\dot{\phi}$ produces the follower angular velocity equation

$$\dot{\phi} = \frac{(\dot{\mathbf{a}})^T (\mathbf{a} - \mathbf{b})}{(\mathbf{a} - \mathbf{b})^T \left\{ \left[P_{\mathbf{ub}_0} \right] (\mathbf{b} - \mathbf{b}_0) \right\}} \tag{10.40}$$

With $\dot{\phi}$ known, the displaced RSSR moving pivot velocity $\dot{\mathbf{b}}$ from Equation 10.38 can be calculated.

10.3.3 Acceleration Equations

Figure 10.6 includes the acceleration variables for the RSSR mechanism. The following equation, the second derivative of the coupler-link constant-length equation, forms the basis for the RSSR acceleration equations:

$$\left(\ddot{\mathbf{a}} - \ddot{\mathbf{b}} \right)^T (\mathbf{a} - \mathbf{b}) + \left(\dot{\mathbf{a}} - \dot{\mathbf{b}} \right)^T \left(\dot{\mathbf{a}} - \dot{\mathbf{b}} \right) = 0 \tag{10.41}$$

The global acceleration of the RSSR moving pivot $\mathbf{b}_1$ (variable $\ddot{\mathbf{b}}$ in Equation 10.41) is expressed as

$$\ddot{\mathbf{b}} = \left\{ \ddot{\phi}\left[P_{\mathbf{ub}_0} \right] + \dot{\phi}^2 \left[P_{\mathbf{ub}_0} \right]\left[P_{\mathbf{ub}_0} \right] \right\}(\mathbf{b} - \mathbf{b}_0) = \left[A_{\ddot{\phi}, \dot{\phi}, \mathbf{ub}_0} \right](\mathbf{b} - \mathbf{b}_0) \tag{10.42}$$

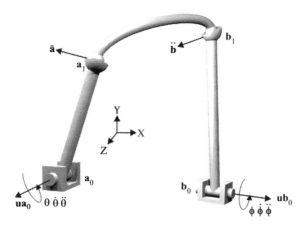

FIGURE 10.6
RSSR mechanism acceleration variables.

and the global acceleration of the RSSR moving pivot $\mathbf{a}_1$ (variable $\ddot{\mathbf{a}}$ in Equation 10.41) is expressed as

$$\ddot{\mathbf{a}} = \left\{ \ddot{\theta} \left[P_{\mathbf{ua}_0} \right] + \dot{\theta}^2 \left[P_{\mathbf{ua}_0} \right] \left[P_{\mathbf{ua}_0} \right] \right\} (\mathbf{a} - \mathbf{a}_0) = \left[A_{\dot{\theta}, \ddot{\theta}, \mathbf{ua}_0} \right] (\mathbf{a} - \mathbf{a}_0) \tag{10.43}*$$

The rotational acceleration matrices $\left[A_{\dot{\theta}, \ddot{\theta}, \mathbf{ua}_0} \right]$ and $\left[A_{\dot{\phi}, \ddot{\phi}, \mathbf{ub}_0} \right]$ are identical in form to Matrix 2.37.

Substituting Equations 10.42 and 10.43 into Equation 10.41 and solving for the follower angular acceleration $\ddot{\phi}$ produces the follower angular acceleration equation

$$\ddot{\phi} = \frac{(\mathbf{a} - \mathbf{b})^T \left\{ \ddot{\mathbf{a}} - \dot{\phi}^2 \left[P_{\mathbf{ub}_0} \right] \left[P_{\mathbf{ub}_0} \right] (\mathbf{b} - \mathbf{b}_0) \right\} + \left(\dot{\mathbf{a}} - \dot{\mathbf{b}} \right)^T \left(\dot{\mathbf{a}} - \dot{\mathbf{b}} \right)}{(\mathbf{a} - \mathbf{b})^T \left\{ \left[P_{\mathbf{ub}_0} \right] (\mathbf{b} - \mathbf{b}_0) \right\}} \tag{10.44}$$

With $\ddot{\phi}$ known, the displaced RSSR moving pivot acceleration $\ddot{\mathbf{b}}$ from Equation 10.42 can be calculated.

Appendix F.2 includes the MATLAB file user instructions for RSSR displacement, velocity, and acceleration analysis. This MATLAB file (which is available for download at www.crcpress.com/product/isbn/9781498724937) utilizes the RSSR displacement, velocity, and acceleration equations presented in Sections 10.3.1–10.3.3.

Like the RRSS mechanism equations in Section 10.2, the RSSR equations can also be used for the displacement, velocity, and acceleration analysis of a planar four-bar mechanism (specifically, a planar RSSR mechanism). To accomplish this, the user should specify planar mechanism values for the RSSR mechanism variables (e.g., $\mathbf{ua}_0 = \mathbf{ub}_0 = (0,0,1)$ and $a_{0z} = a_{1z} = b_{0z} = b_{1z} = 0$).[†]

---

* In Equation 10.42 (and similarly in Equation 10.43), the term $\ddot{\phi} \left[P_{\mathbf{ub}_0} \right] (\mathbf{b} - \mathbf{b}_0) \cdot$ corresponds to the tangential acceleration and the term $\dot{\phi}^2 \left[P_{\mathbf{ub}_0} \right] \left[P_{\mathbf{ub}_0} \right] (\mathbf{b} - \mathbf{b}_0)$ corresponds to the normal acceleration.

† Due to the passive DOF of the coupler link, rotation angles for this link cannot be directly calculated in the RSSR mechanism equations, even when restricted to planar motion (unlike the planar four-bar equations in Section 4.3).

Example 10.2

Problem Statement: Figure E.10.3 illustrates an RSSR mechanism used in a steering system. All of the gear pairs shown have 1:1 gear ratios. Using the Appendix F.2 MATLAB file, calculate the wheel rotation angles achieved by the RSSR mechanism in Table E.10.3 over a steering wheel displacement angle range of 50° (at 10° increments).

   Known Information: Table E.10.3 and Appendix F.2 MATLAB file.

   Solution Approach: Figure E.10.4 includes the input specified (in bold text) in the Appendix F.2 MATLAB file. Table E.10.4 includes the wheel rotation angles calculated using the Appendix F.2 MATLAB file over the steering wheel displacement angles. The dimensions in Table E.10.3 and the results in Table E.10.4 are for the initial RSSR mechanism configuration (which corresponds to the first branch in the Appendix F.2 MATLAB file).*

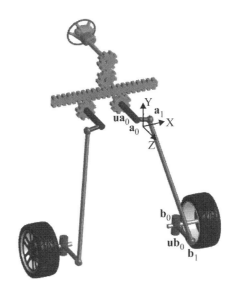

FIGURE E.10.3
RSSR mechanism used in a steering system.

TABLE E.10.3

RSSR Mechanism Dimensions

Variable	Value
a_0	0, 0, −0.4023
a_1	0.3356, −0.0708, −0.4023
ua_0	0, 0, 1
b_0	1, −2.3885, 0
b_1	1, −2.3885, 0.73
ub_0	0, −1, 0

---

* The Appendix F.2 MATLAB file produces results for both the initial (branch 1) and second (branch 2) RSSR mechanism configurations.

```
a0  = [0, 0, -0.4023]';
a1  = [0.3356, -0.0708, -0.4023]';
ua0 = [0, 0, 1]';
b0  = [1, -2.3885, 0]';
b1  = [1, -2.3885, 0.73]';
ub0 = [0, -1, 0]';

start_ang  = 0;
step_ang   = 10;
stop_ang   = 50;

angular_vel  = 0 * ones(N+1,1);
angular_acc  = 0 * ones(N+1,1);
```

FIGURE E.10.4
Specified input (in bold text) in the Appendix F.2 MATLAB file for Example 10.2.

TABLE E.10.4

Wheel Rotation Angles Achieved by RSSR Mechanism

θ (°)	φ (°)
10	15.041
20	29.998
30	44.907
40	60.009
50	76.05

10.4 Four-Revolute Spherical Mechanism Analysis

The revolute-revolute-revolute-revolute spherical mechanism or four-revolute spherical mechanism or simply 4R spherical mechanism (Figure 10.7) is a single-DOF four-bar mechanism that exhibits *spherical motion*—a unique type of spatial motion [7].* Contrary to Gruebler's equation, from which a mobility of −2 is calculated for the 4R spherical mechanism (with $L = J_1 = 4$), it actually has a mobility of 1.[†] Because the links of the 4R spherical mechanism are circular arcs (with all links having a common center) and the mechanism joint axes all intersect at that common center, the links of this mechanism have spherical surface workspaces.

The RRSS kinematic equations introduced in Section 10.2 can be applied directly to the 4R spherical mechanism (Figure 10.8) [8]. This application is particularly useful for calculating 4R spherical mechanism link positions, velocities, and accelerations. To apply the RRSS kinematic equations, the user should specify the appropriate 4R spherical mechanism values for $\mathbf{a}_0$, $\mathbf{a}_1$, $\mathbf{b}_0$, $\mathbf{b}_1$, $\mathbf{p}_1$, $\mathbf{ua}_0$, and $\mathbf{ua}_1$. The variables $\mathbf{a}_0$, $\mathbf{a}_1$, $\mathbf{b}_0$, and $\mathbf{b}_1$ should all lie on the surface of a sphere and variables $\mathbf{ua}_0$ and $\mathbf{ua}_1$ should intersect at the center of the sphere. If it is assumed that the 4R spherical mechanism should lie on a sphere of unit radius ($r_{sphere} = 1$), then $\mathbf{a}_0 = \mathbf{ua}_0$ and $\mathbf{a}_1 = \mathbf{ua}_1$.

---

* While planar mechanisms have *planar-surface* workspaces and most spatial mechanisms have *volume* workspaces, spherical mechanisms have *spatial-surface* workspaces.
† Mechanisms like the 4R spherical mechanism, having true DOFs that are contrary to Gruebler's equation, are called *paradoxes* or *maverick mechanisms*.

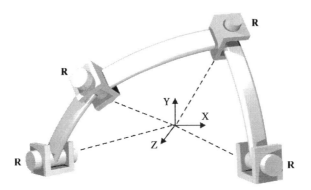

FIGURE 10.7
4R spherical mechanism.

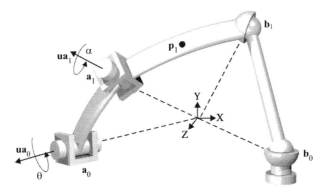

FIGURE 10.8
RRSS mechanism configured as a 4R spherical mechanism.

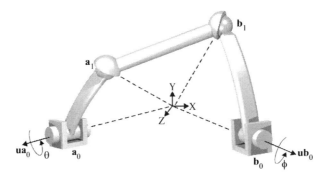

FIGURE 10.9
RSSR mechanism configured as a 4R spherical mechanism.

   The RSSR kinematic equations introduced in Section 10.3 can be applied directly to the 4R spherical mechanism (Figure 10.9) [9]. This application is particularly useful for calculating 4R spherical mechanism link angular positions, velocities, and accelerations. To apply the RSSR kinematic equations, the user should specify the appropriate 4R spherical mechanism values for $\mathbf{a}_0$, $\mathbf{a}_1$, $\mathbf{b}_0$, $\mathbf{b}_1$, $\mathbf{ua}_0$, and $\mathbf{ub}_0$. The variables $\mathbf{a}_0$, $\mathbf{a}_1$, $\mathbf{b}_0$, and $\mathbf{b}_1$ should all lie on the surface of a sphere and variables ua_0 and ub_0 should intersect at the center of the sphere. If it is assumed that the 4R spherical mechanism should lie on a sphere of unit radius ($r_{sphere} = 1$), then $a_0 = ua_0$ and $b_0 = ub_0$.

Example 10.3

Problem Statement: Figure E.10.5 illustrates a 4R Spherical mechanism used in a folding wing system. Using the Appendix F.1 MATLAB file, calculate the location and velocity of point p_1 achieved by the 4R Spherical mechanism in Table E.10.5 over a crank displacement angle range of 168° (at 28° increments). The rotational speed of the crank is 0.5 revolutions/min.

    Known Information: Table E.10.5 and Appendix F.1 MATLAB file.

    Solution Approach: Figure E.10.6 includes the input specified (in bold text) in the Appendix F.1 MATLAB file. Table E.10.6 includes the p_1 location and velocity values calculated using the Appendix F.1 MATLAB file. The dimensions in Table E.10.5 and the results in Table E.10.6 are for the initial RRSS mechanism configuration (which corresponds to the first branch in the Appendix F.1 MATLAB file).

Example 10.4

Problem Statement: Figure E.10.7 illustrates the initial and final positions of a 4R Spherical mechanism used to displace a camera (affixed to the follower link). The mechanism

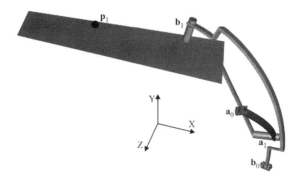

FIGURE E.10.5
4R Spherical mechanism used in a folding wing system.

TABLE E.10.5

4R Spherical Mechanism Dimensions (with Link Lengths in m)

Variable	Value
a_0	0.2612, 0.7274, 0.6346
a_1	−0.0151, 0.9879, 0.1552
b_0	0.1793, 0.9837, 0.0172
b_1	−0.1761, 0.2162, 0.9603
p_1	−0.2081, 0.1675, 0.8227

```
a0 = [0.2612, 0.7274, 0.6346]';
a1 = [-0.0151, 0.9879, 0.1552]';
ua0 = a0;
ua1 = a1;
b0 = [0.1793, 0.9837, 0.0172]';
b1 = [-0.1761, 0.2162, 0.9603]';
p1 = [-0.2081, 0.1675, 0.8227]';
q1 = p1;
r1 = p1;

start_ang   = 0;
step_ang    = 28;
stop_ang    = 168;

angular_vel = (0.5*pi/30) * ones(N+1,1);
angular_acc = 0 * ones(N+1,1);
```

FIGURE E.10.6

Specified input (in bold text) in the Appendix F.1 MATLAB file for Example 10.3.

TABLE E.10.6

Wing Point Locations and Velocities Achieved by 4R Spherical
Mechanism

θ (°)	p (m)	(m/s)
28	0.1998, 0.0772, 0.838	0.0261, −0.0059, −0.0057
56	0.4058, 0.0319, 0.7632	0.0195, −0.0041, −0.0102
84	0.5745, −0.0011, 0.6466	0.0168, −0.003, −0.0149
112	0.7174, −0.0252, 0.4827	0.0136, −0.0021, −0.0203
140	0.8209, −0.04, 0.2697	0.0082, −0.001, −0.0251
168	0.8637, −0.0434, 0.023	0.0007, 0.0003, −0.0271

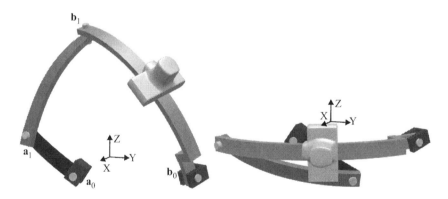

FIGURE E.10.7

4R Spherical camera rotation mechanism at (left) initial and (right) final positions.

dimensions are $a_0 = (1,0,0)$, $a_1 = (0.9083, -0.3824, 0.1695)$, $b_0 = (\sin 45°, \cos 45°, 0)$, and $b_1 = (-0.0166, -0.1928, 0.9811)$. Given a 180° crank displacement, calculate the resulting follower angular displacement using the Appendix F.2 MATLAB file.

Known Information: Appendix F.2 MATLAB file.

Solution Approach: Figure E.10.8 includes the input specified (in bold text) in the Appendix F.2 MATLAB file. A follower displacement of 80.255° was calculated. The 4R Spherical dimensions and results given in this example problem are for the initial RSSR mechanism configuration (which corresponds to the first branch in the Appendix F.2 MATLAB® file).

```
a0 = [1, 0, 0]';
a1 = [0.9083, -0.3824, 0.1695]';
ua0 = a0;
b0 = [sin(45*pi/180), cos(45*pi/180), 0]';
b1 = [-0.0166, -0.1928, 0.9811]';
ub0 = b0;

start_ang   = 0;
step_ang    = 180;
stop_ang    = 180;

angular_vel  = 0 * ones(N+1,1);
angular_acc  = 0 * ones(N+1,1);
```

FIGURE E.10.8
Specified input (in bold text) in the Appendix F.2 MATLAB file for Example 10.4.

10.5 Planar Four-Bar Kinematic Analysis Using RRSS and RSSR Kinematic Equations

As noted in this chapter, the kinematic equations for the spatial RRSS and RSSR mechanisms can also be applied to the planar four-bar mechanism. To consider four-bar mechanism motion in the x-y plane, the z-components of $\mathbf{a}_0$, $\mathbf{a}_1$, $\mathbf{b}_0$, $\mathbf{b}_1$, and $\mathbf{p}_1$ should be specified as zero (therefore $a_{0z} = a_{1z} = b_{0z} = b_{1z} = p_{1z} = 0$).* Also, the joint axis vectors should be along the z-axis (therefore, $\mathbf{ua}_0 = \mathbf{ua}_1 = \mathbf{ub}_0 = (0,0,1)$).

Slider-crank mechanisms can also be modeled using the RRSS and RSSR kinematic equations. While an infinite follower length cannot be specified for a planar four-bar mechanism (to perfectly replicate a slider-crank mechanism), a follower length can be specified to produce an acceptable maximum slider error.[†] For example, a planar four-bar mechanism having a crank length of 1, a coupler length of 2.23, and a follower length of 100,000 will produce a maximum slider error in the order of 10^{-5}.

Example 10.5

Problem Statement: Using the Appendix F.1 MATLAB file, repeat Example 4.2 using the displacement equations for the spatial RRSS mechanism. When expressed as x–y coordinates, the mechanism dimensions in Table E.4.2 become $\mathbf{a}_0 = (0,0,0)$, $\mathbf{a}_1 = (1.3266, 1.4428, 0)$, $\mathbf{b}_0 = (0.6075, -0.6909, 0)$, $\mathbf{b}_1 = (1.5312, 1.1839, 0)$, and $\mathbf{p}_1 = (2.0642, 0.509, 0)$.

Known Information: Example 4.2 and Appendix F.1 MATLAB file.

Solution Approach: Figure E.10.9 includes the input specified (in bold text) in the Appendix F.1 MATLAB file. Figure E.10.10 includes the level-luffing crane and coupler curve calculated. The crane dimensions and results given in this figure are for the initial RRSS mechanism configuration (which corresponds to the first branch in the Appendix F.1 MATLAB file).

Example 10.6

Problem Statement: Using the Appendix F.1 MATLAB file, calculate the minimum and maximum slider velocities for the slider-crank mechanism in Example 4.6 using the displacement equations for the spatial RRSS mechanism. When expressed as x–y coordinates,

---

* If additional coupler points are used (e.g., coupler points $\mathbf{q}_1$ or $\mathbf{r}_1$), their z-components should also be zero.
† A slider-crank mechanism is kinematically identical to a planar four-bar mechanism having an infinite follower length (see Chapter 3).

```
a0 = [0, 0, 0]';
a1 = [1.3266, 1.4428, 0]';
ua0 = [0, 0, 1]';
ua1 = [0, 0, 1]';
b0 = [0.6075, -0.6909, 0]';
b1 = [1.5312, 1.1839, 0]';
p1 = [2.0642, 0.509, 0]';
q1 = p1;
r1 = p1;

start_ang  = 10;
step_ang   = -1;
stop_ang   = -32;

angular_vel  = 0 * ones(N+1,1);
angular_acc  = 0 * ones(N+1,1);
```

FIGURE E.10.9
Specified input (in bold text) in the Appendix F.1 MATLAB file for Example 10.5.

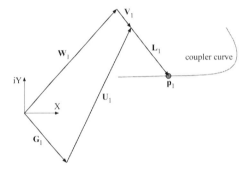

FIGURE E.10.10
Level-luffing crane mechanism with calculated coupler curve.

the mechanism dimensions in Table E.4.5 become $\mathbf{a}_0 = (0, 0, 0)$, $\mathbf{a}_1 = (0.7071, 0.7071, 0)$, and $\mathbf{b}_1 = (2.03, 0, 0)$. Let $\mathbf{b}_0 = (2.03, -500000, 0)$.

Known Information: Example 4.6 and Appendix F.1 MATLAB file.

Solution Approach: Figure E.10.11 includes the input specified (in bold text) in the Appendix F.1 MATLAB file. Maximum and minimum slider velocities of ±122.69 cm/s were calculated. The mechanism dimensions and results given in this example are for the initial RRSS mechanism configuration (which corresponds to the first branch in the Appendix F.1 MATLAB file).

```
a0 = [0, 0, 0]';
a1 = [0.7071, 0.7071, 0]';
ua0 = [0, 0, 1]';
ua1 = [0, 0, 1]';
b0 = [2.03, -500000, 0]';
b1 = [2.03, 0, 0]';
p1 = [0, 0, 0]';
q1 = p1;
r1 = p1;

start_ang  = 0;
step_ang   = 1;
stop_ang   = 720;

angular_vel  = 100 * ones(N+1,1);
angular_acc  = 0 * ones(N+1,1);
```

FIGURE E.10.11
Specified input (in bold text) in the Appendix F.1 MATLAB file for Example 10.6.

Example 10.7

Problem Statement: Using the Appendix F.2 MATLAB file, plot the follower angular displacement versus the crank angular displacement (for both mechanism branches) for the mechanism in Example 4.2. When expressed as x–y coordinates, the mechanism dimensions in Table E.4.2 become $\mathbf{a}_0 = (0,0,0)$, $\mathbf{a}_1 = (1.3266, 1.4428, 0)$, $\mathbf{b}_0 = (0.6075, -0.6909, 0)$, and $\mathbf{b}_1 = (1.5312, 1.1839, 0)$.

Known Information: Example 4.2 and Appendix F.2 MATLAB file

Solution Approach: Figure E.10.12 includes the input specified (in bold text) in the Appendix F.2 MATLAB file. Figure E.10.13 includes the follower versus crank angular displacement plots for the level-luffing crane mechanism calculated using the Appendix F.2 MATLAB file.

```
a0  = [0, 0, 0]';
a1  = [1.3266, 1.4428, 0]';
ua0 = [0, 0, 1]';
b0  = [0.6075, -0.6909, 0]';
b1  = [1.5312, 1.1839, 0]';
ub0 = [0, 0, 1]';

start_ang  = 10;
step_ang   = -1;
stop_ang   = -32;

angular_vel  = 0 * ones(N+1,1);
angular_acc  = 0 * ones(N+1,1);
```

FIGURE E.10.12
Specified input (in bold text) in the Appendix F.2 MATLAB file for Example 10.7.

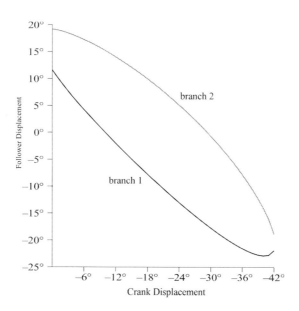

FIGURE E.10.13
Follower versus crank angular displacement plots for level-luffing crane mechanism.

10.6 Spatial Mechanism Kinematic Analysis and Modeling in Simmechanics®

It has been noted throughout this chapter that Appendices F.1 and F.2 include user instructions for the RRSS and RSSR mechanism MATLAB files, respectively. In these files, the displacement, velocity, and acceleration equations formulated in this chapter are used. The Appendix F.1 and F.2 MATLAB files provide means for the user to efficiently conduct RRSS and RSSR kinematic analyses by calculating solutions from their displacement, velocity, and acceleration equations.

This textbook also utilizes SimMechanics as an alternate approach for simulation-based kinematic analysis. A library of SimMechanics files is available for download at www.crcpress.com/product/isbn/9781498724937 to conduct RRSS and RSSR mechanism displacement, velocity, and acceleration analyses. With these files, the user specifies the mechanism link dimensions and driving link parameters (e.g. crank displacements, velocities, and accelerations) and measures the calculated displacements, velocities, and accelerations of the mechanism locations of interest. Additionally, the motion of the mechanism itself is simulated. The SimMechanics file user instructions for the RRSS and RSSR mechanisms are given in Appendices K.1and K.2, respectively.

Due to the passive DOF in the follower S-S link of the RRSS mechanism and the coupler S-S link in the RSSR mechanism, these links are not visible in the SimMechanics animation window. Also, the links of 4R spherical mechanisms are depicted as straight lines (rather than circular arcs) in the SimMechanics animation window.

Example 10.8

Problem Statement: Repeat Example 10.1 using the Appendix K.1 SimMechanics files.

Known Information: Example 10.1 and Appendix K.1 SimMechanics files.

Solution Approach: Figure E.10.14 includes the input specified (in bold text) in the Appendix K.1 SimMechanics file. Table E.10.7 includes the solutions calculated using the Appendix K.1 SimMechanics files.

```
a0  = [-0.0576, 0.289, -1.4112]';
a1  = [0.1452, -2.5421, -1.18]';
ua0 = [-0.0003, 0.0814, 0.9967]';
ua1 = [0.304, 0.0992, 0.9475]';
b0  = [0.0851, 0.457, 0.5096]';
b1  = [1.7725, 5.1566, 0.6499]';
p1  = [1.7321, 0, -1]';
q1  = [1.2321, 0, -1.866]';
r1  = [1.9486, 0, -1.125]';

start_ang  = 0;
step_ang   = 50;
stop_ang   = 200;

angular_vel  = 0;
angular_acc  = 0;
```

FIGURE E.10.14
Specified input (in bold text) in the Appendix K.1 SimMechanics files for Example 10.1.

TABLE E.10.7

Solar Panel Positions Achieved by RRSS Mechanism (Example 10.1)

θ	p	q	r
49.997°	1.0928, 1.3848, −1.0402	0.7289, 0.9589, −1.8685	1.2047, 1.5597, −1.1793
99.997°	−0.167, 1.735, −1.0239	−0.1141, 1.1572, −1.8383	−0.2505, 1.9246, −1.1638
200°	−1.7271, −0.0052, −0.9672	−1.2325, 0.0395, −1.8352	−1.9225, −0.1086, −1.0839

```
a0  = [0, 0, -0.4023]';
a1  = [0.3356, -0.0708, -0.4023]';
ua0 = [0, 0, 1]';
b0  = [1, -2.3885, 0]';
b1  = [1, -2.3885, 0.73]';
ub0 = [0, -1, 0]';

start_ang   = 0;
step_ang    = 10;
stop_ang    = 50;

angular_vel  = 0;
angular_acc  = 0;
```

FIGURE E.10.15
Specified input (in bold text) in the Appendix K.2 SimMechanics files for Example 10.2.

TABLE E.10.8

Wheel Rotation Angles Achieved by RSSR
Mechanism (Example 10.2)

θ (°)	φ (°)
9.9969	15.037
19.997	29.993
29.997	44.902
39.997	60.004
50	76.05

Example 10.9

Problem Statement: Repeat Example 10.2 using the Appendix K.2 SimMechanics files.
Known Information: Example 10.2 and Appendix K.2 SimMechanics files.
Solution Approach: Figure E.10.15 includes the input specified (in bold text) in the Appendix K.2 SimMechanics file. Table E.10.8 includes the solutions calculated using the Appendix K.2 SimMechanics files.

Example 10.10

Problem Statement: Repeat Example 10.3 using the Appendix K.1 SimMechanics files.
Known Information: Example 10.3 and Appendix K.1 SimMechanics files.
Solution Approach: Figure E.10.16 includes the input specified (in bold text) in the Appendix K.1 SimMechanics file. Table E.10.9 includes the solutions calculated the Appendix K.1 SimMechanics files.

```
a0 = [0.2612, 0.7274, 0.6346]';
a1 = [-0.0151, 0.9879, 0.1552]';
ua0 = a0;
ua1 = a1;
b0 = [0.1793, 0.9837, 0.0172]';
b1 = [-0.1761, 0.2162, 0.9603]';
p1 = [-0.2081, 0.1675, 0.8227]';
q1 = p1;
r1 = p1;

start_ang  = 0;
step_ang   = 28;
stop_ang   = 168;

angular_vel = 0.5*pi/30;
angular_acc = 0;
```

FIGURE E.10.16
Specified input (in bold text) in the Appendix K.1 SimMechanics files for Example 10.3.

TABLE E.10.9

Wing Point Locations and Velocities Achieved by 4R Spherical
Mechanism (Example 10.3)

θ (°)	p (m)	$\dot{p}$ (m/s)
28.004	0.1998, 0.0772, 0.838	0.0261, −0.0059, −0.0057
55.993	0.4057, 0.0319, 0.7633	0.0195, −0.0041, −0.0102
84.0105	0.5746, −0.0011, 0.6466	0.0168, −0.003, −0.0149
112	0.7174, −0.0252, 0.4827	0.0136, −0.0021, −0.0203
139.99	0.8209, −0.0399, 0.2698	0.0082, −0.001, −0.0251
168	0.8636, −0.0433, 0.0231	0.0007, 0.0003, −0.0271

10.7 Summary

Just as planar mechanisms are configured so that mechanism motion is restricted to two-dimensional space, spatial mechanisms are configured to exhibit motion in three-dimensional space. Spatial mechanism motion is determined in part by the mobility of the mechanism joints used and the spatial orientation of the joints. The four-bar spatial mechanisms considered in this chapter are the RRSS, RSSR, and 4R spherical mechanisms.

Because spatial mechanisms can exhibit spatial motion, they offer a greater variety of possible motions and are structurally more general than planar mechanisms. However, because spatial mechanism analysis equations are often much greater in scale and complexity than planar mechanism equations, their applications in practice are often limited. It is therefore an ongoing task to devise simple methods of calculation, to produce design aids with diagrams, and to set design standards for spatial mechanisms.

Suh and Radcliffe presented displacement, velocity, and acceleration equations for the RRSS and RSSR mechanisms. These displacement equations can be directly applied to the 4R spherical mechanism and even the planar four-bar mechanism. In the Appendix F.1 and F.2 MATLAB files, the displacement, velocity, and acceleration equations for the RRSS and RSSR mechanisms are used.

This textbook also utilizes SimMechanics as an alternate approach for simulation-based kinematic analyses. Using the Appendix K.1 and K.2 SimMechanics files, the user can conduct displacement, velocity, and acceleration analyses on the RRSS, RSSR, and 4R spherical mechanisms, respectively, as well as simulate mechanism motion.

References

1. Capellen, W. M. 1966. Kinematics—A survey in retrospect and prospect. *Mechanism and Machine Theory*, 1: 211–228.
2. Russell, K., Shen, Q., and Sodhi, R. 2014. *Mechanism Design: Visual and Programmable Approaches.* Chapter 7. Boca Raton: CRC Press.
3. Suh, C. H. and Radcliffe, C. W. 1978. *Kinematics and Mechanisms Design.* Chapter 4. New York: John Wiley.
4. Ibid, pp. 83–85.
5. Ibid, p. 85.
6. Ibid, pp. 79–80.
7. Waldron, K. J. and Kinzel, G. L. 2004. *Kinematics, Dynamics and Design of Machinery.* 2nd edn, pp. 340–347. Saddle River, NJ: Prentice-Hall.
8. Russell, K., Shen, Q., and Sodhi, R. 2014. *Mechanism Design: Visual and Programmable Approaches.* pp. 137–141. Boca Raton: CRC Press.
9. Ibid, pp. 141–143.

Problems

1. The general fixed and moving pivot variables for the RRSS mechanism in spatial motion are

$$\mathbf{a}_0 = \left(a_{0x}, a_{0y}, a_{0z}\right), \quad \mathbf{ua}_0 = \left(ua_{0x}, ua_{0y}, ua_{0z}\right), \quad \mathbf{a}_1 = \left(a_{1x}, a_{1y}, a_{1z}\right), \quad \mathbf{ua}_1 = \left(ua_{1x}, ua_{1y}, ua_{1z}\right),$$

$$\mathbf{b}_0 = \left(b_{0x}, b_{0y}, b_{0z}\right), \quad \mathbf{b}_1 = \left(b_{1x}, b_{1y}, b_{1z}\right).$$

Express $\mathbf{a}_0$, $\mathbf{ua}_0$, $\mathbf{a}_1$, $\mathbf{ua}_1$, $\mathbf{b}_0$, and $\mathbf{b}_1$ for an RRSS mechanism restricted to motion in the X–Z plane.

2. The general fixed and moving pivot variables for the RSSR mechanism in spatial motion are

$$\mathbf{a}_0 = \left(a_{0x}, a_{0y}, a_{0z}\right), \quad \mathbf{ua}_0 = \left(ua_{0x}, ua_{0y}, ua_{0z}\right), \quad \mathbf{a}_1 = \left(a_{1x}, a_{1y}, a_{1z}\right),$$

$$\mathbf{b}_0 = \left(b_{0x}, b_{0y}, b_{0z}\right), \quad \mathbf{ub}_0 = \left(ub_{0x}, ub_{0y}, ub_{0z}\right), \quad \mathbf{b}_1 = \left(b_{1x}, b_{1y}, b_{1z}\right).$$

Express $\mathbf{a}_0$, $\mathbf{ua}_0$, $\mathbf{a}_1$, $\mathbf{b}_0$, $\mathbf{ub}_0$, and $\mathbf{b}_1$ for an RSSR mechanism restricted to motion in the Y–Z plane.

3. From the dimensions provided for the hatch mechanism in Problem 4.1, it can be determined that $\mathbf{a}_0 = (0,0,0)$, $\mathbf{a}_1 = (-2.1257, 3.8669, 0)$, $\mathbf{b}_0 = (-0.7156, 6.4851, 0)$, $\mathbf{b}_1 = (-2.3246, 4.8688, 0)$, and $\mathbf{p}_1 = (-5.2675, 6.5545, 0)$. Tabulate the coordinates of coupler points $\mathbf{p}_1$ and $\mathbf{q}_1$ (where $\mathbf{q}_1 = (-5.2675, 2, 0)$) over the $-30°$ crank rotation range at $-5°$ rotation increments (using the Appendix F.1 or K.1 MATLAB files).

4. From the dimensions provided for the loading–unloading mechanism in Problem 4.2, it can be determined that $\mathbf{a}_0 = (0,0,0)$, $\mathbf{a}_1 = (-2.1631, 2.0464, 0)$, $\mathbf{b}_0 = (-1.5726, 4.8975, 0)$, $\mathbf{b}_1 = (-2.0268, 2.4259, 0)$, and $\mathbf{p}_1 = (-4.3625, 2.0643, 0)$. Tabulate the coordinates of coupler points $\mathbf{p}_1$ and $\mathbf{q}_1$ (where $\mathbf{q}_1 = (-3, 2.0643, 0)$) over the $-22.7°$ crank rotation range at $-4.54°$ rotation increments (using the Appendix F.1 or K.1 MATLAB files).

5. For the mechanism in Problem 3, plot the velocity magnitude of coupler point $\mathbf{p}_1$ versus the crank displacement angle over the $-30°$ crank rotation range at $-1°$ rotation increments (using the Appendix F.1 or K.1 MATLAB files). Consider an initial crank angular velocity and an angular acceleration value of $-1.25\,\mathrm{rad/s}$ and $0\,\mathrm{rad/s^2}$, respectively.

6. From the dimensions provided for the component assembly mechanism in Problem 4.6, it can be determined that $\mathbf{a}_0 = (0,0,0)$, $\mathbf{a}_1 = (1.9963, -2.3251, 0)$, $\mathbf{b}_0 = (2.4465, 0.3309, 0)$, $\mathbf{b}_1 = (2.0182, -0.7213, 0)$, and $\mathbf{p}_1 = (6.8484, -2.2654, 0)$. Tabulate the coordinates of coupler points $\mathbf{p}_1$ and $\mathbf{q}_1$ (where $\mathbf{q}_1 = (2, -2.2654, 0)$) over the $66.375°$ crank rotation range at $7.375°$ rotation increments (using the Appendix F.1 or K.1 MATLAB files).

7. From the dimensions provided for the digging mechanism in Problem 4.7, it can be determined that $\mathbf{a}_0 = (0,0,0)$, $\mathbf{a}_1 = (-3.6552, -1.9295, 0)$, $\mathbf{b}_0 = (-5.3924, 2.1975, 0)$, $\mathbf{b}_1 = (-8.9907, -2.6631, 0)$, and $\mathbf{p}_1 = (-8.4482, -3.9043, 0)$. Tabulate the coordinates of coupler points $\mathbf{p}_1$ and $\mathbf{q}_1$ (where $\mathbf{q}_1 = (-8.4482, -1, 0)$) over the $-57.4°$ crank rotation range at $-8.2°$ rotation increments (using the Appendix F.1 or K.1 MATLAB files).

8. For the mechanism in Problem 7, tabulate the velocity and acceleration values of coupler points $\mathbf{p}_1$ and $\mathbf{q}_1$ over the $-57.4°$ crank rotation range at $-8.2°$ rotation increments (using the Appendix F.1 or K.1 MATLAB files). Consider an initial crank angular velocity and an angular acceleration value of $-1\,\mathrm{rad/s}$ and $0.499\,\mathrm{rad/s^2}$, respectively.

9. For the slider-crank mechanism in Problem 4.16, a maximum slider displacement of $12.675\,\mathrm{cm}$ is achieved at a crank displacement of $273°$. At this crank displacement, a slider velocity and acceleration of 0.0829 cm/s and -15.609 cm/s^2, respectively, are also achieved. Expressing the dimensions of this slider-crank mechanism as $\mathbf{a}_0 = (0,0,0)$, $\mathbf{a}_1 = (0, 3.175, 0)$, $\mathbf{b}_1 = (9.2225, 0.7938, 0)$, and $\mathbf{b}_0 = (9.2225, -1000000, 0)$, calculate the slider displacement, velocity, and acceleration at the same crank displacement (using the Appendix F.1 or K.1 MATLAB files).

10. For the slider-crank mechanism in Example 4.6, a maximum slider displacement of $2.5\,\mathrm{cm}$ is achieved at a crank displacement of $315°$. At this crank displacement, a slider velocity and acceleration of 0.0000 cm/s and $-16,667$ cm/s^2, respectively, are also achieved. Expressing the dimensions of this slider-crank mechanism as $\mathbf{a}_0 = (0, 0, 0)$, $\mathbf{a}_1 = (0.7071, 0.7071, 0)$, $\mathbf{b}_1 = (2.03, 0, 0)$, and $\mathbf{b}_0 = (2.03, -1000000, 0)$, calculate the slider displacement, velocity, and acceleration at the same crank displacement (using the Appendix F.2 or K.2 MATLAB files).

11. From the dimensions provided for the leveling crane mechanism in Problem 4.3, it can be determined that $\mathbf{a}_0 = (0,0,0)$, $\mathbf{a}_1 = (4.7137, 7.3718, 0)$, $\mathbf{b}_0 = (2.4762, -2.8162, 0)$, and $\mathbf{b}_1 = (5.102, 6.1836, 0)$. Tabulate the follower displacement angles over the $-35°$ crank rotation range at $-5°$ rotation increments (using the Appendix F.2 or K.2 MATLAB files).

12. From the dimensions provided for the wiper blade mechanism in Problem 4.15, it can be determined that $\mathbf{a}_0 = (0,0,0)$, $\mathbf{a}_1 = (1.4859, -2.3708, 0)$, $\mathbf{b}_0 = (1.25, 0, 0)$, and $\mathbf{b}_1 = (1.8708, -1.6203, 0)$. Tabulate the follower angular displacement, velocity, and acceleration values over the $45°$ crank rotation range at $5°$ rotation increments (using the Appendix F.2 or K.2 MATLAB files).

13. Using the planar RRSS mechanism dimensions in Table P.10.1, plot the paths traced by coupler point $\mathbf{p}_1$ (for both mechanism branches) over a complete crank rotation (using the Appendix F.1 or K.1 MATLAB files). Consider crank rotation increments of $1°$.

14. Using the planar RRSS mechanism dimensions in Table P.10.1, produce plots of the coupler displacement angle versus the crank displacement angle (for both mechanism branches) over a complete crank rotation (using the Appendix F.1 or K.1 MATLAB files). Consider crank rotation increments of $1°$.

15. Using the planar RSSR mechanism dimensions in Table P.10.2, produce plots of the follower displacement angle versus the crank displacement angle (for both mechanism branches) over a complete crank rotation (using the Appendix F.2 or K.2 MATLAB files). Consider crank rotation increments of $1°$.

TABLE P.10.1

Planar RRSS Mechanism Dimensions

Variable	Values
$\mathbf{a}_0$	0, 0, 0
$\mathbf{a}_1$	0, 1.125, 0
$\mathbf{ua}_0$	0, 0, 1
$\mathbf{ua}_1$	0, 0, 1
$\mathbf{b}_0$	1.4063, 0, 0
$\mathbf{b}_1$	1.3779, 1.4059, 0
$\mathbf{p}_1$	0.5135, 2.126, 0

TABLE P.10.2

Planar RSSR Mechanism Dimensions

Variable	Values
$\mathbf{a}_0$	0, 0, 0
$\mathbf{a}_1$	0, 1.55, 0
$\mathbf{ua}_0$	0, 0, 1
$\mathbf{b}_0$	2.325, 0, 0
$\mathbf{b}_1$	2.1227, 2.2464, 0
$\mathbf{ub}_0$	0, 0, 1

TABLE P.10.3

RRSS Mechanism Dimensions (mm)

Variable	Value
$\mathbf{a}_0$	0, 0, 0
$\mathbf{a}_1$	0, 22.86, 0
$\mathbf{ua}_0$	0, 0, 1
$\mathbf{ua}_1$	0, $\sin(\pi/6)$, $\cos(\pi/6)$
$\mathbf{b}_0$	45.72, 0, 0
$\mathbf{b}_1$	45.72, 45.72, −11.43
$\mathbf{p}_1$	12.7, 31.75, 0

16. For the RRSS mechanism in Table P.10.3, tabulate the location, velocity, and acceleration of coupler point $\mathbf{p}_1$ (for the 1st mechanism branch only) over a complete crank rotation (using the Appendix F.1 or K.1 MATLAB files). Consider a crank rotation increment of 30° with an initial angular velocity and angular acceleration of 0.75 rad/s and 0.25 rad/s², respectively.

17. For the RRSS mechanism dimensions in Example 10.1, calculate the velocities of coupler points $\mathbf{p}_1$, $\mathbf{q}_1$, and $\mathbf{r}_1$ (for the first mechanism branch only) at the end of the 200° crank rotation range, considering an initial rotational velocity of 1.8 rad/s and rotational acceleration of −0.45 rad/s² (using the Appendix F.1 or K.1 MATLAB files). Assume the RRSS mechanism length dimensions in Example 10.1 are given in meters.

18. Figure P.10.1 illustrates an RRSS mechanism used to guide a solar panel. Table P.10.4 includes the planned solar panel positions to be reached and the corresponding RRSS crank displacement angles. Determine the actual solar panel positions reached at these crank displacement angles (using the Appendix F.1 or K.1 MATLAB files).

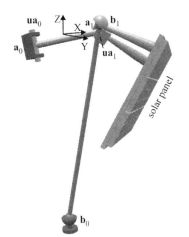

$\mathbf{a}_0 = (0.0164, 0.0015, 2.3579)$
$\mathbf{a}_1 = (-2.0705, 0.5585, 2.363)$
$\mathbf{ua}_0 = \mathbf{ua}_1 = (0.0025, 0.0003, 0.9999)$
$\mathbf{b}_0 = (-0.3058, 0.1265, -2.172)$
$\mathbf{b}_1 = (0.9, -0.9, 0)$

$\mathbf{p}_1 = (1.7321, 0, -1)$
$\mathbf{q}_1 = (1.2321, 0, -1.866)$
$\mathbf{r}_1 = (1.9486, 0, -1.125)$

FIGURE P.10.1
RRSS solar panel mechanism.

TABLE P.10.4

Planned Solar Panel Positions for RRSS Mechanism

θ (°)	p	q	r
50	1.2247, 1.2247, −1	0.8712, 0.8712, −1.8660	1.3778, 1.3778, −1.1250
90	0, 1.7321, −1	0, 1.2321, −1.8660	−0.0631, 1.9416, −1.1210

19. For the RRSS mechanism in Figure P.10.1, calculate the accelerations of coupler points p_1, q_1, and r_1 at 50° and 90° crank rotations considering an initial rotational velocity of 1.9 rad/s and rotational acceleration of −1.1 rad/s^2 (using the Appendix F.1 or K.1 MATLAB files). Assume the RRSS mechanism length dimensions in Example 10.1 are given in meters.

20. If it was suggested that the solar panel be guided by the follower link of the RRSS mechanism in Figure P.10.1 (rather than the coupler link), explain why this option is practical or not.

21. For the RSSR mechanism in Table P.10.5, produce plots of the follower displacement angle versus the crank displacement angle for both mechanism branches over a complete crank rotation (using the Appendix F.2 MATLAB files).

22. For the RSSR mechanism in Table P.10.5, produce plots of the follower angular velocity versus the crank displacement angle for both mechanism branches over a complete crank rotation (using the Appendix F.2 MATLAB files). Consider an initial crank rotational velocity of 1.3 rad/s and a rotational acceleration of 0 rad/s^2.

23. For the RSSR mechanism in Table P.10.5, produce plots of the follower angular acceleration versus the crank displacement angle for both mechanism branches over a complete crank rotation (using the Appendix F.2 MATLAB files). Consider an initial crank rotational velocity of 1.3 rad/s and a rotational acceleration of −0.13 rad/s^2.

24. For the RSSR steering mechanism in Example 10.2, determine the maximum steering-wheel rotation and corresponding tire rotation that can be achieved if an initial angular velocity and angular acceleration of 1.15 rad/s and −0.95 rad/s^2, respectively, were applied to the steering wheel. Consider crank rotation increments of 1°. Use the Appendix F.1 or K.1 MATLAB files.

25. Would the coupler link of the RSSR steering mechanism (see Example 10.2) make a practical alternative for guiding the motion of the tire instead of the follower link?

TABLE P.10.5

RSSR Mechanism Dimensions (m)

Variable	Value
a_0	0, 0, 0
a_1	−0.4125, 0.7145, 0
ua_0	0, 0, 1
b_0	1.65, 0, 0
b_1	1.65, 1.5938, −0.427
ub_0	0, $\sin(\pi/12)$, $\cos(\pi/12)$

26. Using the 4R Spherical mechanism dimensions in Table P.10.6, calculate the path traced by coupler point p_1 for both mechanism branches over a complete crank rotation (using the Appendix F.1 MATLAB files). Consider crank rotation increments of 30°.

27. Using the 4R spherical mechanism dimensions in Table P.10.6, calculate the velocities of coupler point p_1 for both mechanism branches over a complete crank rotation (using the Appendix F.1 MATLAB files). Consider crank rotation increments of 30°, an initial crank rotational velocity of 0 rad/s, and a rotational acceleration of 0.3 rad/s².

28. Figure P.10.2 illustrates a 4R Spherical mechanism used to guide a folding wing. Table P.10.7 includes the planned folding wing positions to be reached and the corresponding 4R Spherical mechanism crank displacement angles. Determine the actual folding wing positions reached at these crank displacement angles (using the Appendix F.1 or K.1 MATLAB files).

29. For the 4R spherical mechanism in Figure P.10.2, calculate the velocities of coupler points p_1, q_1, and r_1 at 29°, 47.5°, and 54.5° crank rotations considering an initial rotational velocity of 1.5 rad/s and a rotational acceleration of −1.18 rad/s² (using the Appendix F.1 or K.1 MATLAB files). Assume the 4R spherical mechanism length dimensions in Figure P.10.2 are given in meters.

TABLE P.10.6

4R Spherical Mechanism Dimensions (mm)

Variable	Value
a_0	0, 0, 29.21
ua_0	0, 0, 1
a_1	0, 11.1787, 26.9872
ua_1	0, 0.3827, 0.9239
b_0	29.21 0, 0
b_1	14.605, −25.2959, 0
p_1	16.5884, −5.2724, 23.4585

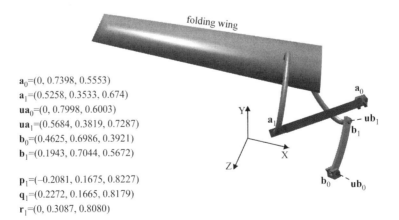

$a_0 = (0, 0.7398, 0.5553)$
$a_1 = (0.5258, 0.3533, 0.674)$
$ua_0 = (0, 0.7998, 0.6003)$
$ua_1 = (0.5684, 0.3819, 0.7287)$
$b_0 = (0.4625, 0.6986, 0.3921)$
$b_1 = (0.1943, 0.7044, 0.5672)$

$p_1 = (−0.2081, 0.1675, 0.8227)$
$q_1 = (0.2272, 0.1665, 0.8179)$
$r_1 = (0, 0.3087, 0.8080)$

FIGURE P.10.2
4R spherical folding-wing mechanism.

TABLE P.10.7

Planned Solar Panel Positions for 4R Spherical Mechanism

θ (°)	p	q	r
29	0.1384, 0.0891, 0.8492	0.4976, 0.2489, 0.6624	0.2570, 0.2970, 0.7707
47.5	0.4964, 0.0163, 0.7082	0.6714, 0.3156, 0.4449	0.5074, 0.2611, 0.6500
54.5	0.7683, −0.0329, 0.3960	0.7575, 0.3608, 0.2106	0.7197, 0.2119, 0.4304

30. For the 4R spherical mechanism in Figure P.10.2, calculate the accelerations of coupler points p_1, q_1, and r_1 at 29°, 47.5°, and 54.5° crank rotations, considering an initial rotational velocity of 0 rad/s and a rotational acceleration of 0.45 rad/s² (using the Appendix F.1 or K.1 MATLAB files). Assume the 4R spherical mechanism length dimensions in Figure P.10.2 are given in meters.

31. Using the 4R Spherical mechanism dimensions in Table P.10.8, produce plots of the follower displacement angle versus the crank displacement angle (for both mechanism branches) over a complete crank rotation (using the Appendix F.2 MATLAB file). Consider crank rotation increments of 1.3°.

32. Using the 4R spherical mechanism dimensions in Table P.10.8, produce plots of the follower angular velocity versus the crank displacement angle (for both mechanism branches) over a complete crank rotation (using the Appendix F.2 MATLAB file). Consider crank rotation increments of 1°, an initial crank rotational velocity of 0 rad/s, and a rotational acceleration of 0.1 rad/s².

33. Using the 4R spherical mechanism dimensions in Table P.10.8, produce plots of the follower angular acceleration versus the crank displacement angle (for both mechanism branches) over a complete crank rotation (using the Appendix F.2 MATLAB files). Consider crank rotation increments of 1.3°, an initial crank rotational velocity of 3 rad/s, and a rotational acceleration of −0.3 rad/s².

34. Using the 4R spherical mechanism in Example 10.4, produce plots of the follower angular velocity versus the crank displacement angle (for the first mechanism branch only) over the given crank rotation (using the Appendix F.2 or K.2 MATLAB files). Consider crank rotation increments of 1°, an initial crank rotational velocity of 0 rad/s, and a rotational acceleration of 0.6 rad/s².

35. Using the 4R Spherical mechanism in Example 10.4, produce plots of the follower angular acceleration versus the crank displacement angle (for the 1st mechanism branch only) over the given crank rotation (using the Appendix F.2 or K.2 MATLAB file). Consider crank rotation increments of 1°, an initial crank rotational velocity of 1 rad/s and rotational acceleration of −0.15 rad/s².

TABLE P.10.8

4R Spherical Mechanism Dimensions

Variable	Value
a_0	0, 0, 29.21
ua_0	0, 0, 1
a_1	0, 11.1787, 26.9872
b_0	29.21, 0, 0
ub_0	1, 0, 0
b_1	14.605, −25.2959, 0

11

Introduction to Robotic Manipulators

CONCEPT OVERVIEW

In this chapter, the reader will gain a central understanding regarding

1. Distinctions and disadvantages of *robotic manipulators* in comparison to classical linkages
2. Matrix-based formulation of displacement equation systems for *Cartesian, cylindrical, spherical, articulated*, and *SCARA* robots
3. *Forward* kinematics of Cartesian, cylindrical, spherical, articulated, and SCARA robots
4. *Inverse* kinematics and workspace of Cartesian, cylindrical, spherical, articulated, and SCARA robots

11.1 Introduction

As explained in Chapter 1 and demonstrated throughout this textbook, a *linkage* (also commonly called a *mechanism*) is an assembly of links and joints where the motion of one link compels the motion of another link in a controlled manner. To enable controlled mechanism motion, they are either initially designed to have a single degree of freedom or ultimately configured (in the case of the geared five-bar mechanism) to have a single degree of freedom. Conventional planar and spatial linkages include the four-bar, slider-crank, geared five-bar, Watt, Stephenson, RRSS, RSSR, and 4R spherical linkages presented in Chapters 4 and 10.

Like the linkages presented in Chapters 4 and 10, a *robotic manipulator* (commonly called a *robot*) also includes an assembly of links and joints and is designed to produce a controlled output motion. In addition to links and joints, however, a robotic manipulator also includes *electronic circuitry, computer-controlled actuators* to compel link motion, and is guided by a *computer program*. Because robotic manipulators include both mechanical and electronic components, they are classified as *electro-mechanical systems*.

To achieve a controlled motion, each joint in a robotic manipulator can be controlled independently. As a result, there is no degree of freedom limit that robotic manipulators can be theoretically designed to have.*

Another common distinction between linkages and robotic manipulators is in their overall design. All of the linkages presented in Chapters 4 and 10 have *closed-loop* designs. With this design, at least two joints in a linkage are connected to ground (thus forming

---

* Robotic manipulators are generally limited to six DOFs because a spatial body has a maximum mobility of six.

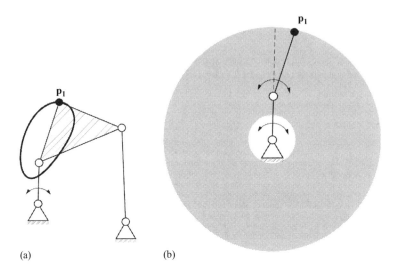

FIGURE 11.1
(a) Four-bar mechanism and coupler curve and (b) robotic manipulator and workspace.

a closed loop). While robotic manipulators can have closed-loop designs, they often have *open loops*, where only one joint is connected to ground.

Figure 11.1 includes a planar four-bar mechanism and a planar, 2-DOF, open-loop robotic manipulator. While coupler Point $\mathbf{p}_1$ on the four-bar mechanism (in the given location on the coupler link) can only trace the curve illustrated, Point $\mathbf{p}_1$ on the robotic manipulator can trace *any* path within the shaded annular area or *workspace*.* Because of the open-loop construction of the robotic manipulator and its mobility, this single manipulator can trace a greater variety of distinct paths than any number of planar four-bar mechanisms.

So, when compared to linkages, robotic manipulators offer advantages such as greater variability—specifically for motion-specific and path-specific tasks. Being computer controlled, robotic manipulators also offer advantages regarding greater precision, accuracy, and repeatability. Lastly, robotic manipulators have the capacity for *remote operation* as well as *autonomous operation* since they can be guided by computer programs (as opposed to mechanisms, which often require a degree of manual operation).

It is becoming increasingly difficult to find an industry where robotic manipulators are not employed, either directly or indirectly. Common industries where robotic manipulators are widely employed (both in product manufacturing and operation) include *automotive, aerospace, defense, electronics,* and *medicine.* The number of applications for robotic systems is rapidly on the increase, since new robotic manipulator capabilities and more practical robot manipulator designs are continually being developed and produced [1].

11.2 Terminology and Nomenclature

To describe the spatial position and orientation of each link in a robotic manipulator, coordinate systems are rigidly attached to each link. These coordinate systems are called

---

* The workspace is the area or volume (for spatial robots) of space that the robot can reach.

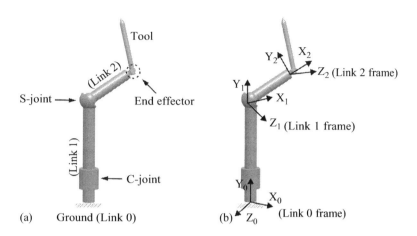

FIGURE 11.2
(a) Robotic manipulator with joint and link descriptions and (b) frames.

frames. In Figure 11.2b, frames are attached to Link 0 (called the *base frame*), Link 1, and Link 2. The positions and orientations between the frames are modeled through transformation matrices (Section 2.5). The general transformation matrix and its application will be further discussed in Sections 11.4–11.6.

A robotic manipulator often includes a component at its free end called an *end effector*. The end effector can serve to handle any tool or component. It can also be the working end of the tool itself if the end effector fully constrains the tool. One common end effector used in robotic manipulators is a *gripper*. In Figure 11.2, the end effector of the robotic manipulator is a gripper that holds the tool. The tool is assumed to be fully constrained by the gripper.*

11.3 Robotic Manipulator Mobility and Types

In Chapter 3, it was noted that Gruebler's Equation is used to determine the mobility of a linkage. These equations can also be used to determine the mobility of robotic manipulators. For planar and spatial robotic manipulators, Gruebler's Equations become

$$DOF_{PLANAR} = 3(L-1) - 2J_1 \tag{11.1}$$

$$DOF_{SPATIAL} = 6(L-1) - 5J_1 - 4J_2 - 3J_3 \tag{11.2}$$

where only 1-DOF joints are used in the planar robotic manipulator and only 1-, 2-, or 3-DOF joints are used in the spatial manipulator.†

Figure 11.3 includes the types of 1-, 2-, and 3-DOF joints used in the robotic manipulators presented in this chapter. Although a robotic manipulator can be designed to include any

---

* If the tool was not fully constrained by the gripper, an additional frame attached to the tool (called a *tool frame*) would have been included.

† As with mechanisms, for robotic manipulators, the ground link should also be counted among the total number of links (*L*) in Gruebler's Equation.

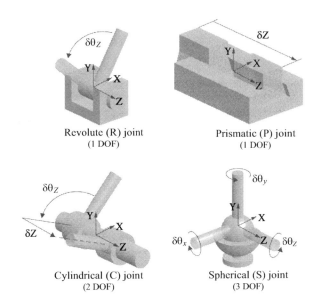

FIGURE 11.3
Example robotic manipulator joint types.

joint type, the revolute, prismatic, cylindrical, and spherical are among those joint types most commonly used in practice.

Figure 11.4 illustrates the five spatial robotic manipulator types considered in this chapter. These particular robotic manipulator configurations are among those commonly

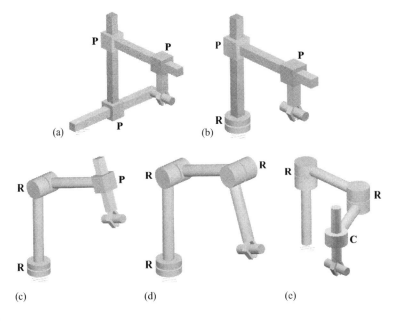

FIGURE 11.4
(a) Cartesian (P-P-P), (b) cylindrical (R-P-P), (c) spherical (R-R-P), (d) articulated (R-R-R), and (e) SCARA (R-R-C) robots.

```
>> L = 4;
>> J1 = 3;
>> DOF_RRR = 6*(L - 1) -5*J1

DOF_RRR =

     3

>> L = 4;
>> J1 = 2;
>> J2 = 1;
>> DOF_RRC = 6*(L - 1) - 5*J1 - 4*J2

DOF_RRC =

     4

>>
```

FIGURE E.11.1
Example 11.1 calculation procedure in MATLAB.

utilized in industrial applications [2]. The designations P-P-P, R-P-P, R-R-P, R-R-R, and R-R-C in Figure 11.4 denote the joint types and joint sequences used in the robotic manipulators.

The P-P-P robotic manipulator is commonly known as a *Cartesian robot* because its degrees of freedom are along the *x*, *y*, and *z*-axes of the Cartesian frame. The R-P-P and R-R-P robotic manipulators are commonly known as *cylindrical* and *spherical robots*, respectively, because their motion is consistent with cylindrical and spherical joints, respectively. The R-R-R and R-R-C robotic manipulators are commonly known as *articulated* and *Selective Compliance Assembly/Articulated Robot Arm* (or *SCARA*) *robots*, respectively.

Example 11.1

Problem Statement: Calculate the mobility values for the R-R-R and R-R-C robotic manipulators.

Known Information: Equation 11.2 and Figure 11.3.

Solution Approach: The R-R-R robotic manipulator is comprised of four links interconnected with three revolute joints. Since there are four links and the revolute joint has a single degree of freedom, $L = 4$ and $J_1 = 3$ in Equation 11.2 for the R-R-R robotic manipulator.

The R-R-C is comprised of four links interconnected with two revolute joints and one cylindrical joint. Since there are four links and the cylindrical joint has two degrees of freedom, $L = 4$, $J_1 = 2$, and $J_2 = 1$ in Equation 11.2 for the R-R-C robotic manipulator.

Figure E.11.1 includes the calculation procedure in the MATLAB® command window.

11.4 The General Transformation Matrix

In Section 2.5, a general spatial transformation matrix was presented. This matrix is used to calculate point coordinates originally established in one link coordinate frame (Frame *j*) in reference to another link coordinate frame (Frame *i*) in a robotic manipulator. If we recall, the general spatial transformation matrix can be expressed as

$$
{}_{j}^{i}[T] = \begin{bmatrix} R_{11} & R_{12} & R_{13} & \Delta_x \\ R_{21} & R_{22} & R_{23} & \Delta_y \\ R_{31} & R_{32} & R_{33} & \Delta_z \\ 0 & 0 & 0 & 1 \end{bmatrix} \tag{11.3}
$$

where:

$R_{11} = \cos \delta_y \cos \delta_z$

$R_{12} = \sin \delta_x \sin \delta_y \cos \delta_z - \cos \delta_x \sin \delta_z$

$R_{13} = \cos \delta_x \sin \delta_y \cos \delta_z + \sin \delta_x \sin \delta_z$

$R_{21} = \cos \delta_y \sin \delta_z$

$R_{22} = \sin \delta_x \sin \delta_y \sin \delta_z + \cos \delta_x \cos \delta_z$

$R_{23} = \cos \delta_x \sin \delta_y \sin \delta_z - \sin \delta_x \cos \delta_z$

$R_{31} = -\sin \delta_y$

$R_{32} = \sin \delta_x \cos \delta_y$

$R_{33} = \cos \delta_x \cos \delta_y$

In this matrix, variables δ_x, δ_y, and δ_z are the angular rotations about a frame's x-, y-, and z-axes, respectively, and variables Δ_x, Δ_y, and Δ_z are the linear translations along the frame's x-, y-, and z-axes, respectively.*

Given $^j\{\mathbf{p}\}$, the spatial coordinates of a point $\mathbf{p}$ in Frame j, the coordinates of this point with respect to Frame i or $^i\{\mathbf{p}\}$ can be calculated as

$$
{}^{i}\{\mathbf{p}\} = {}_{j}^{i}[T]\,{}^{j}\{\mathbf{p}\} \tag{11.4}
$$

where the spatial coordinates of $\mathbf{p}$ in Frame j are $^j\{\mathbf{p}\} = \{p_x\ p_y\ p_z\ 1\}^{T}$.[†]

In Figure 11.5, the robotic manipulator in Figure 11.2b is again considered where Frames 0, 1, and 2 are attached to ground, Link 1 and Link 2, respectively. In this figure, Point $\mathbf{p}_1$ is attached to Link 1 and its coordinates are given with respect to Frame 1. Point $\mathbf{p}_2$ is attached to Link 2 and its coordinates are given with respect to Frame 2.

To calculate the value of Point $\mathbf{p}_1$ with respect to the base frame, Equation 11.4 becomes

$$
{}^{0}\{\mathbf{p}_1\} = {}_{1}^{0}[T]\,{}^{1}[\mathbf{p}_1] \tag{11.5}
$$

where $^0\{\mathbf{p}_1\}$ is the value of $\mathbf{p}_1$ with respect to the base frame (Frame X_0–Y_0–Z_0 in Figure 11.5).[‡] The transformation matrix $_{1}^{0}[T]$ considers Frames 0 and 1. This matrix includes angular displacements δ_{1x}, δ_{1y}, and δ_{1z}, and linear displacements Δ_{1x}, Δ_{1y}, and Δ_{1z}.

To calculate the value of Point $\mathbf{p}_2$ with respect to the base frame, Equation 11.5 becomes

$$
{}^{0}\{\mathbf{p}_2\} = {}_{1}^{0}[T]\,{}_{2}^{1}[T]\,{}^{2}\{\mathbf{p}_2\} = {}_{2}^{0}[T]\,{}^{2}\{\mathbf{p}_2\} \tag{11.6}
$$

---

* In a general *planar* transformation matrix, variables δ_x, δ_y, and Δ_z in the general spatial transformation matrix are all zero.

† The *planar* coordinates of $\mathbf{p}$ in Frame j would be $^j\{\mathbf{p}\} = \{p_x\ p_y\ p_0\ 1\}^{T}$.

‡ A value given with respect to the base frame is also called a *global value*.

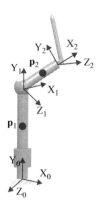

FIGURE 11.5
Robotic manipulator with frames and points.

where $^0\{\mathbf{p}_2\}$ is the value of $\mathbf{p}_2$ with respect to the base frame. Because the transformation matrix $^0_1[T]$ already considers Frames 0 and 1, including the transformation matrix $^1_2[T]$ (which considers Frames 1 and 2) and taking the product of these two matrices produces a single transformation matrix that considers Frames 0 and 2. This matrix includes all of the displacement variables in $^0_1[T]$ as well as the additional angular displacements δ_{2x}, δ_{2y}, and δ_{2z}, and linear displacements Δ_{2x}, Δ_{2y}, and Δ_{2z}.

Let us now assume an additional frame X_3–Y_3–Z_3 is attached to the tool and the coordinates of a point $\mathbf{p}_3$ are given with respect to this frame (Frame 3). To calculate the value of $\mathbf{p}_3$ with respect to the base frame, Equation 11.6 becomes

$$^0\{\mathbf{p}_3\} = {}^0_1[T]\,{}^1_2[T]\,{}^2_3[T]\,{}^3\{\mathbf{p}_3\} = {}^0_2[T]\,{}^2_3[T]\,{}^3\{\mathbf{p}_3\} = {}^0_3[T]\,{}^3\{\mathbf{p}_3\} \tag{11.7}$$

where $^0\{\mathbf{p}_3\}$ is the value of $\mathbf{p}_3$ with respect to the base frame. Because the transformation matrix $^0_2[T]$ already considers Frames 0 and 2, including the transformation matrix $^2_3[T]$ (which considers Frames 2 and 3) and taking the product of these two matrices produces a single transformation matrix that considers Frames 0 and 3. This matrix includes all of the displacement variables in $^0_2[T]$ as well as the additional angular displacements δ_{3x}, δ_{3y}, and δ_{3z}, and linear displacements Δ_{3x}, Δ_{3y}, and Δ_{3z}.

From Equations 11.5 to 11.7, it can be observed that as an additional final frame is introduced, the transformation matrix to calculate the global coordinates of a point in the final frame becomes an increasing product of the transformation matrices for each frame. Knowing this, given a group of N frames, the transformation matrix to calculate the global coordinates of a point in the Nth frame can be expressed as

$$^0_N[T] = {}^0_1[T]\,{}^1_2[T]\,{}^2_3[T]\dots\,{}^{N-1}_N[T]^* \tag{11.8}$$

Therefore, given $^N\{\mathbf{p}_N\}$, the spatial coordinates of a point $\mathbf{p}_N$ in Frame N, the global coordinates of this point with respect to the base frame or $^0\{\mathbf{p}_N\}$ can be calculated as

$$^0\{\mathbf{p}_N\} = {}^0_N[T]^N\{\mathbf{p}_N\} \tag{11.9}$$

---

* When calculating the product of three or more matrices in Equation 11.8, it should be done in right-to-left order.

The coordinate frame x, y, and z rotations and translation variables presented in this section are analogous to *Denavit–Hartenberg* (or *DH*) parameters. Like the δ and Δ variables, DH parameters are also used to attach reference frames to links. In classical DH notation, only rotations about and translations along coordinate frame x- and z-axes are used. Therefore, with the y-axis rotations and translations eliminated, the general transformation matrix becomes

$$
{}^i_j[T] = \begin{bmatrix} \cos\delta_z & -\cos\delta_x\sin\delta_z & \sin\delta_x\sin\delta_z & \Delta_x\cos\delta_z \\ \sin\delta_z & \cos\delta_x\cos\delta_z & -\sin\delta_x\cos\delta_z & \Delta_x\sin\delta_z \\ 0 & \sin\delta_x & \cos\delta_x & \Delta_z \\ 0 & 0 & 0 & 1 \end{bmatrix}^* \tag{11.10}
$$

Example 11.2

Problem Statement: Calculate the elements of transformation matrix ${}^0_3[T]$ for the coordinate frame displacement values given in Table E.11.1.

   Known Information: Matrix 11.3 and Table E.11.1.

   Solution Approach: Because all variables in Matrix 11.3, with the exception of those given in Table E.11.1, are zero, this matrix can be simplified. Figure E.11.2 includes the calculation procedure in MATLAB's command window.

TABLE E.11.1

Frame Displacement Variables (with Unitless Link Lengths)

Frames	δ_x	δ_y	δ_z	Δ_x	Δ_y	Δ_z
1 wrt 0	0	0	55°	0	0	0
2 wrt 1	0	0	0	0	0	1.75
3 wrt 2	0	0	0	3	0	0

```
>> d1z = 55*pi/180;
>> D2z = 1.75;
>> D3x = 3;
>> T01 = [cos(d1z), -sin(d1z), 0, 0; sin(d1z), cos(d1z), 0, 0;...
0, 0, 1, 0; 0, 0, 0, 1];
>> T12 = [1, 0, 0, 0; 0, 1, 0, 0; 0, 0, 1, D2z; 0, 0, 0, 1];
>> T23 = [1, 0, 0, D3x; 0, 1, 0, 0; 0, 0, 1, 0; 0, 0, 0, 1];
>> T03 = T01*T12*T23

T03 =

    0.5736   -0.8192        0   1.7207
    0.8192    0.5736        0   2.4575
         0         0   1.0000   1.7500
         0         0        0   1.0000

>>
```

FIGURE E.11.2
Example 11.2 calculation procedure in MATLAB.

---

* In classical DH notation, the variables θ, α, d, and r are used instead of δ_z, δ_x, Δ_z, and Δ_z, respectively.

11.5 Forward Kinematics

11.5.1 Definition and Application

In *forward kinematics*, the link dimensions and joint motion of a robotic manipulator are known and the corresponding output motion of the links (usually the end effector) is calculated [3]. By formulating an equation system for a robotic manipulator (an equation system to calculate the motion of specific link points) and prescribing the link dimensions and joint motions, the resulting link motion is calculated. The most common application for forward kinematics is for determining end effector motion (e.g., tool paths and orientations).

11.5.2 P-P-P

Figure 11.6 includes the frames specified for the P-P-P robotic manipulator. The transformation matrix ${}^{0}_{3}[T]$ is required to formulate equations to calculate the global position of $\mathbf{p}_3$ on the end effector. To facilitate this procedure, Table 11.1 includes the displacement variables required to align Frame 1 to 0, Frame 2 to 1, and Frame 3 to 2. As this table indicates, a combination of three linear displacements is utilized in the P-P-P robotic manipulator.

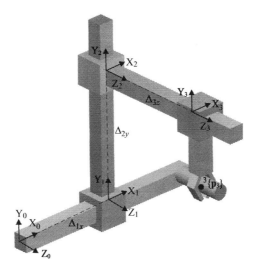

FIGURE 11.6
P-P-P robotic manipulator with frames and point.

TABLE 11.1

Frame Displacement Variables for the P-P-P Robotic Manipulator

Frames	δ_x	δ_y	δ_z	Δ_x	Δ_y	Δ_z
1 wrt 0	0	0	0	Δ_{1x}	0	0
2 wrt 1	0	0	0	0	Δ_{2y}	0
3 wrt 2	0	0	0	0	0	Δ_{3z}

From the information given in Table 11.1, the only nonzero variables are Δ_{1x}, Δ_{2y}, and Δ_{3z}. As a result, Matrix 11.8 becomes

$$
{}^0_3[T] = {}^0_1[T] \, {}^1_2[T] \, {}^2_3[T] =
\begin{bmatrix}
1 & 0 & 0 & \Delta_{1x} \\
0 & 1 & 0 & 0 \\
0 & 0 & 1 & 0 \\
0 & 0 & 0 & 1
\end{bmatrix}
\begin{bmatrix}
1 & 0 & 0 & 0 \\
0 & 1 & 0 & \Delta_{2y} \\
0 & 0 & 1 & 0 \\
0 & 0 & 0 & 1
\end{bmatrix}
\begin{bmatrix}
1 & 0 & 0 & 0 \\
0 & 1 & 0 & 0 \\
0 & 0 & 1 & \Delta_{3z} \\
0 & 0 & 0 & 1
\end{bmatrix}
$$

$$(11.11)$$

Using this transformation matrix and Equation 11.9 produces the following system of equations to calculate the global coordinates of $\mathbf{p}_3$:

$$
\begin{aligned}
{}^0p_{3x} &= {}^3p_{3x} + \Delta_{1x} \\
{}^0p_{3y} &= {}^3p_{3y} + \Delta_{2y} \\
{}^0p_{3z} &= {}^3p_{3z} + \Delta_{3z}
\end{aligned}
$$

$$(11.12)$$

In the P-P-P robotic manipulator, the linear displacement variables Δ_{1x}, Δ_{2y}, and Δ_{3z} are not often assigned constant values because they correspond to the translational displacements of the prismatic joints. More often, displacement ranges are assigned to these variables.

Example 11.3

Problem Statement: Using the joint displacements given in Table E.11.2, calculate the global path points achieved by the end effector of the P-P-P robotic manipulator. In this example, ${}^3\{\mathbf{p}_3\} = [0, -1, 0]^T$.

Known Information: Equation 11.12 and Table E.11.2.

Solution Approach: Table E.11.3 includes the global end effector path point coordinates calculated using Equation 11.12.

TABLE E.11.2

P-P-P Robotic Manipulator Joint Displacements

Point	Δ_{1x}	Δ_{2y}	Δ_{3z}
1	0.5	1.1	−0.1
2	1	1.2	−0.15
3	1.5	1.3	−0.3
4	2	1.2	−0.45
5	2.5	1.1	−0.60

TABLE E.11.3

P-P-P Robotic Manipulator Path Point Coordinates

Point	${}^0p_{3x}$	${}^0p_{3y}$	${}^0p_{3z}$
1	0.5	0.1	−0.1
2	1	0.2	−0.15
3	1.5	0.3	−0.3
4	2	0.2	−0.45
5	2.5	0.1	−0.60

11.5.3 R-P-P

Figure 11.7 includes the frames specified for the R-P-P robotic manipulator. The transformation matrix $_3^0[T]$ is required to formulate equations to calculate the global position of $\mathbf{p}_3$ on the end effector. To facilitate this procedure, Table 11.2 includes the displacement variables required to align Frame 1 to 0, Frame 2 to 1, and Frame 3 to 2. As this table indicates, a combination of a single angular displacement and two linear displacements are utilized in the R-P-P robotic manipulator.

From the information given in Table 11.2, the only nonzero variables are δ_{1z}, Δ_{2z}, and Δ_{3x}. As a result, Matrix 11.8 becomes

$$_3^0[T] = {}_1^0[T]\,{}_2^1[T]\,{}_3^2[T] = \begin{bmatrix} \cos\delta_{1z} & -\sin\delta_{1z} & 0 & 0 \\ \sin\delta_{1z} & \cos\delta_{1z} & 0 & 0 \\ 0 & 0 & 1 & 0 \\ 0 & 0 & 0 & 1 \end{bmatrix} \begin{bmatrix} 1 & 0 & 0 & 0 \\ 0 & 1 & 0 & 0 \\ 0 & 0 & 1 & \Delta_{2z} \\ 0 & 0 & 0 & 1 \end{bmatrix} \begin{bmatrix} 1 & 0 & 0 & \Delta_{3x} \\ 0 & 1 & 0 & 0 \\ 0 & 0 & 1 & 0 \\ 0 & 0 & 0 & 1 \end{bmatrix}$$

$$(11.13)$$

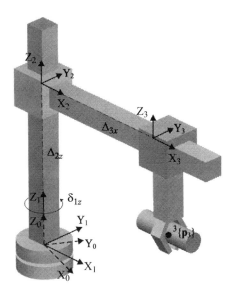

FIGURE 11.7
R-P-P robotic manipulator with frames and point.

TABLE 11.2

Frame Displacement Variables for the R-P-P Robotic Manipulator

Frames	δ_x	δ_y	δ_z	Δ_x	Δ_y	Δ_z
1 wrt 0	0	0	δ_{1z}	0	0	0
2 wrt 1	0	0	0	0	0	Δ_{2z}
3 wrt 2	0	0	0	Δ_{3x}	0	0

Using this transformation matrix and Equation 11.9 produces the following system of equations to calculate the global coordinates of $\mathbf{p}_3$:

$$^0p_{3x} = {}^3p_{3x} \cos \delta_{1z} - {}^3p_{3y} \sin \delta_{1z} + \Delta_{3x} \cos \delta_{1z}$$

$$^0p_{3y} = {}^3p_{3x} \sin \delta_{1z} + {}^3p_{3y} \cos \delta_{1z} + \Delta_{3x} \sin \delta_{1z} \qquad (11.14)$$

$$^0p_{3z} = {}^3p_{3z} + \Delta_{2z}$$

In the R-P-P robotic manipulator, the linear and angular displacement variables Δ_{2z}, Δ_{3x}, and δ_{1z} are not often assigned constant values because they correspond to the translational displacements, the prismatic joints, and the rotational displacements of the revolute joint. More often, displacement ranges are assigned to these variables.

Appendix G.1 includes the MATLAB file user instructions for R-P-P robotic manipulator forward kinematics. In this MATLAB file (which is available for download at www.crcpress.com/product/isbn/9781498724937), Equation 11.14 is used to calculate the global coordinates of $\mathbf{p}_3$.

Example 11.4

Problem Statement: Using the Appendix G.1 MATLAB file with the joint displacements given in Table E.11.4, calculate the global path points achieved by the end effector of the R-P-P robotic manipulator. In this example, ${}^3\{\mathbf{p}_3\} = [0,0,-1]^T$.

Known Information: Table E.11.4 and Appendix G.1 MATLAB file.

Solution Approach: The data for columns δ_{1z}, Δ_{2z}, and Δ_{3x} in Table E.11.4 are first specified in the file *RPP_Input.csv*. Figure E.11.3 includes the input specified (in bold text) in the Appendix G.1 MATLAB file and Table E.11.5 includes the global end effector path point coordinates calculated.

```
p3_3 = [0, 0, -1];
```

FIGURE E.11.3
Specified input (in bold text) in the Appendix G.1 MATLAB file for Example 11.4.

TABLE E.11.4

R-P-P Robotic Manipulator Joint Displacements

Point	δ_{1z} (°)	Δ_{2z}	Δ_{3x}
1	12	1.1	−0.1
2	24	1.2	−0.15
3	36	1.3	−0.3
4	48	1.2	−0.45
5	60	1.1	−0.60

TABLE E.11.5

R-P-P Robotic Manipulator Path Point Coordinates

Point	$^0p_{3x}$	$^0p_{3y}$	$^0p_{3z}$
1	−0.0978	−0.0208	0.1000
2	−0.1370	−0.0610	0.2000
3	−0.2427	−0.1763	0.3000
4	−0.3011	−0.3344	0.2000
5	−0.3000	−0.5196	0.1000

11.5.4 R-R-P

Figure 11.8 includes the frames specified for the R-R-P robotic manipulator. The transformation matrix ${}_3^0[T]$ is required to formulate equations to calculate the global position of $\mathbf{p}_3$ on the end effector. To facilitate this procedure, Table 11.3 includes the displacement variables required to align Frame 1 to 0, Frame 2 to 1, and Frame 3 to 2. As this table indicates, a combination of two angular displacements and three linear displacements are utilized in the R-R-P robotic manipulator.

From the information given in Table 11.3, the only nonzero variables are δ_{1z}, δ_{2x}, and Δ_{3z} and the terms representing Δ_{2z} and Δ_{3y}.* As a result, Matrix 11.8 becomes

$$
{}_3^0[T] = {}_1^0[T]\,{}_2^1[T]\,{}_3^2[T]
$$

$$
=
\begin{bmatrix}
\cos\delta_{1z} & -\sin\delta_{1z} & 0 & 0 \\
\sin\delta_{1z} & \cos\delta_{1z} & 0 & 0 \\
0 & 0 & 1 & 0 \\
0 & 0 & 0 & 1
\end{bmatrix}
\begin{bmatrix}
1 & 0 & 0 & 0 \\
0 & \cos\delta_{2x} & -\sin\delta_{2x} & 0 \\
0 & \sin\delta_{2x} & \cos\delta_{2x} & l_1 \\
0 & 0 & 0 & 1
\end{bmatrix}
\begin{bmatrix}
1 & 0 & 0 & 0 \\
0 & 1 & 0 & l_2 \\
0 & 0 & 1 & \Delta_{3z} \\
0 & 0 & 0 & 1
\end{bmatrix}
$$

$$(11.15)$$

Using the resulting transformation matrix and Equation 11.9 produces the following system of equations to calculate the global coordinates of $\mathbf{p}_3$:

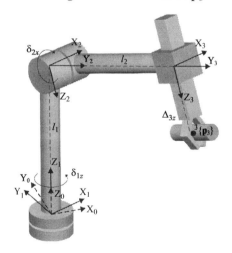

FIGURE 11.8

R-R-P robotic manipulator with frames and point.

TABLE 11.3

Frame Displacement Variables for the R-R-P Robotic Manipulator

Frames	δ_x	δ_y	δ_z	Δ_x	Δ_y	Δ_z
1 wrt 0	0	0	δ_{1z}	0	0	0
2 wrt 1	Δ_{2x}	0	0	0	0	l_1
3 wrt 2	0	0	0	0	l_2	Δ_{3z}

* In Table 11.3, $\Delta_{2z} = l_1$ and $\Delta_{3y} = l_2$.

$$^0p_{3x} = {}^3p_{3x} \cos \delta_{1z} - {}^3p_{3y} \sin \delta_{1z} \cos \delta_{2x} + {}^3p_{3z} \sin \delta_{1z} \sin \delta_{2x} - l_2 \sin \delta_{1z} \cos \delta_{2x}$$

$$+ \Delta_{3z} \sin \delta_{1z} \sin \delta_{2x}$$

$$^0p_{3y} = {}^3p_{3x} \sin \delta_{1z} + {}^3p_{3y} \cos \delta_{1z} \cos \delta_{2x} - {}^3p_{3z} \cos \delta_{1z} \sin \delta_{2x} + l_2 \cos \delta_{1z} \cos \delta_{2x}$$

$$- \Delta_{3z} \cos \delta_{1z} \sin \delta_{2x}$$

$$^0p_{3z} = {}^3p_{3y} \sin \delta_{2x} + {}^3p_{3z} \cos \delta_{2x} + l_2 \sin \delta_{2x} + \Delta_{3z} \cos \delta_{2x} + l_1 \qquad (11.16)$$

In the R-R-P robotic manipulator, variables l_1 and l_2 are assigned constant values because they represent constant link lengths. The angular and linear displacement variables δ_{1z}, δ_{2x}, and Δ_{3z}, however, correspond to the rotational displacements of the revolute joints and the translational displacements of the prismatic joint. More often, displacement ranges are assigned to these variables.

Appendix G.2 includes the MATLAB file user instructions for R-R-P robotic manipulator forward kinematics. In this MATLAB file (which is available for download at www. crcpress.com/product/isbn/9781498724937), Equation 11.16 is used to calculate the global coordinates of $\mathbf{p}_3$.

Example 11.5

Problem Statement: Using the Appendix G.2 MATLAB file with the joint displacements given in Table E.11.6, calculate the global path points achieved by the end effector of the R-R-P robotic manipulator. In this example, $l_1 = l_2 = 0.5$ (unitless link lengths) and $^3\{\mathbf{p}_3\} = [0,0,0]^T$.

Known Information: Table E.11.6 and Appendix G.2 MATLAB file.

Solution Approach: The data for columns δ_{1z}, δ_{2x}, and Δ_{3z} in Table E.11.6 are first specified in the file *RRP_Input.csv*. Figure E.11.4 includes the input specified (in bold text) in the Appendix G.2 MATLAB file and Table E.11.7 includes the global end effector path point coordinates calculated.

```
l1 = 0.5;
l2 = 0.5;
p3_3 = [0, 0, 0];
```

FIGURE E.11.4
Specified input (in bold text) in the Appendix G.2 MATLAB file for Example 11.5.

TABLE E.11.6

R-R-P Robotic Manipulator Joint Displacements

Point	δ_{1z} (°)	δ_{2x} (°)	Δ_{3z}
1	12	10	0.5
2	24	20	1
3	36	30	1.5
4	48	40	2
5	60	50	2.5

TABLE E.11.7

R-R-P Robotic Manipulator Path Point Coordinates

Point	${}^0p_{3x}$	${}^0p_{3y}$	${}^0p_{3z}$
1	−0.0843	0.3967	1.0792
2	−0.0520	0.1168	1.6107
3	0.1863	−0.2564	2.0490
4	0.6707	−0.6039	2.3535
5	1.3802	−0.7969	2.4900

11.5.5 R-R-R

Figure 11.9 includes the frames specified for the R-R-R robotic manipulator. The transformation matrix ${}^0_3[T]$ is required to formulate equations to calculate the global position of $\mathbf{p}_3$ on the tool. To facilitate this procedure, Table 11.4 includes the displacement variables required to align Frame 1 to 0, Frame 2 to 1, and Frame 3 to 2. As this table indicates, a combination of three angular displacements and two linear displacements are utilized in the R-R-R robotic manipulator.

From the information given in Table 11.4, the only nonzero variables are δ_{1z}, δ_{2x}, and δ_{3x}, and the terms representing Δ_{2z} and Δ_{3y}.* As a result, Matrix 11.8 becomes

$$
{}^0_3[T] = {}^0_1[T]\,{}^1_2[T]\,{}^2_3[T]
$$

$$
=
\begin{bmatrix}
\cos\delta_{1z} & -\sin\delta_{1z} & 0 & 0 \\
\sin\delta_{1z} & \cos\delta_{1z} & 0 & 0 \\
0 & 0 & 1 & 0 \\
0 & 0 & 0 & 1
\end{bmatrix}
\begin{bmatrix}
1 & 0 & 0 & 0 \\
0 & \cos\delta_{2x} & -\sin\delta_{2x} & 0 \\
0 & \sin\delta_{2x} & \cos\delta_{2x} & l_1 \\
0 & 0 & 0 & 1
\end{bmatrix}
\begin{bmatrix}
1 & 0 & 0 & 0 \\
0 & \cos\delta_{3x} & -\sin\delta_{3x} & l_2 \\
0 & \sin\delta_{3x} & \cos\delta_{3x} & 0 \\
0 & 0 & 0 & 1
\end{bmatrix}
$$

(11.17)

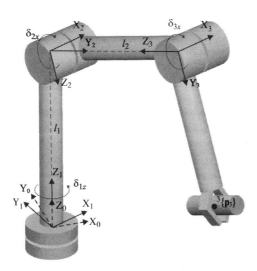

FIGURE 11.9
R-R-R robotic manipulator with frames and point.

* In Table 11.4, $\Delta_{2z} = l_1$ and $\Delta_{3y} = l_2$.

TABLE 11.4

Frame Displacement Variables for the R-R-R Robotic Manipulator

Frames	δ_x	δ_y	δ_z	Δ_x	Δ_y	Δ_z
1 wrt 0	0	0	δ_{1z}	0	0	0
2 wrt 1	δ_{2x}	0	0	0	0	l_1
3 wrt 2	δ_{3x}	0	0	0	l_2	0

Using the resulting transformation matrix and Equation 11.9 produces the following system of equations to calculate the global coordinates of $\mathbf{p}_3$:

$$^0p_{3x} = {^3p_{3x}} \cos \delta_{1z} + {^3p_{3y}} \left(\sin \delta_{1z} \sin \delta_{2x} \sin \delta_{3x} - \sin \delta_{1z} \cos \delta_{2x} \cos \delta_{3x} \right)$$

$$+ {^3p_{3z}} \left(\sin \delta_{1z} \cos \delta_{2x} \sin \delta_{3x} + \sin \delta_{1z} \sin \delta_{2x} \cos \delta_{3x} \right) - l_2 \sin \delta_{1z} \cos \delta_{2x}$$

$$^0p_{3y} = {^3p_{3x}} \sin \delta_{1z} + {^3p_{3y}} \left(\cos \delta_{1z} \cos \delta_{2x} \cos \delta_{3x} - \cos \delta_{1z} \sin \delta_{2x} \sin \delta_{3x} \right)$$

$$+ {^3p_{3z}} \left(-\cos \delta_{1z} \cos \delta_{2x} \sin \delta_{3x} - \cos \delta_{1z} \sin \delta_{2x} \cos \delta_{3x} \right) + l_2 \cos \delta_{1z} \cos \delta_{2x}$$

$$^0p_{3z} = {^3p_{3y}} \left(\sin \delta_{2x} \cos \delta_{3x} + \cos \delta_{2x} \sin \delta_{3x} \right) + {^3p_{3z}} \left(\cos \delta_{2x} \cos \delta_{3x} - \sin \delta_{2x} \sin \delta_{3x} \right)$$

$$+ l_2 \sin \delta_{2x} + l_1 \tag{11.18}$$

In the R-R-R robotic manipulator, variables l_1 and l_2 are assigned constant values because they represent constant link lengths. The angular and linear displacement variables δ_{1z}, δ_{2x}, and δ_{3x}, however, correspond to the rotational displacements of the revolute joints. More often, displacement ranges are assigned to these variables.

Appendix G.3 includes the MATLAB file user instructions for R-R-R robotic manipulator forward kinematics. In this MATLAB file (which is available for download at www.crcpress.com/ product/isbn/9781498724937), Equation 11.18 is used to calculate the global coordinates of $\mathbf{p}_3$.

Example 11.6

Problem Statement: Using the Appendix G.3 MATLAB file with the joint displacements given in Table E.11.8, calculate the global path points achieved by the end effector of the R-R-R robotic manipulator. In this example, $l_1 = l_2 = 0.5$ (unitless link lengths) and $^3\{\mathbf{p}_3\} = [0, 1, 0]^T$.

    Known Information: Table E.11.8 and Appendix G.3 MATLAB file.

    Solution Approach: The data for columns δ_{1z}, δ_{2x}, and δ_{3x} in Table E.11.8 are first specified in the file *RRR_Input.csv*. Figure E.11.5 includes the input specified (in bold text) in

TABLE E.11.8

R-R-R Robotic Manipulator Joint Displacements

Point	δ_{1z} (°)	δ_{2x} (°)	Δ_{3x} (°)
1	12	10	−5
2	24	20	−10
3	36	30	−15
4	48	40	−20
5	60	50	−25

```
l1 = 0.5;
l2 = 0.5;
p3_3 = [0, 1, 0];
```

FIGURE E.11.5
Specified input (in bold text) in the Appendix G.3 MATLAB file for Example 11.3.

TABLE E.11.9

R-R-R Robotic Manipulator Path Point Coordinates

Point	$^0p_{3x}$	$^0p_{3y}$	$^0p_{3z}$
1	−0.3095	1.4561	0.6740
2	−0.5917	1.3289	0.8447
3	−0.8223	1.1318	1.0088
4	−0.9830	0.8851	1.1634
5	−1.0632	0.6139	1.3056

the Appendix G.3 MATLAB file and Table E.11.9 includes the global end effector path point coordinates calculated.

11.5.6 R-R-C

Figure 11.10 includes the frames specified for the R-R-C robotic manipulator. The transformation matrix $^0_3[T]$ is required to formulate equations to calculate the global position of p_3 on the tool. To facilitate this procedure, Table 11.5 includes the displacement variables required to align Frame 1 to 0, Frame 2 to 1, and Frame 3 to 2. As this table indicates, a

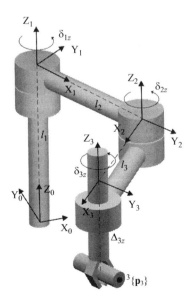

FIGURE 11.10
R-R-C robotic manipulator with frames and point.

TABLE 11.5

Frame Displacement Variables for the R-R-C Robotic Manipulator

Frames	δ_x	δ_y	δ_z	Δ_x	Δ_y	Δ_z
1 wrt 0	0	0	δ_{1z}	0	0	l_1
2 wrt 1	0	0	δ_{2z}	l_2	0	0
3 wrt 2	0	0	δ_{3z}	l_3	0	Δ_{3z}

combination of three angular displacements and four linear displacements are utilized in the R-R-C robotic manipulator.

Using the information given in Table 11.5, the nonzero variables are δ_{1z}, δ_{2z}, δ_{3z}, Δ_{3z}, and the terms representing Δ_{1z}, Δ_{2x}, and Δ_{3x}.* As a result, Matrix 11.8 becomes

$$\begin{aligned}
{}_3^0[T] &= {}_1^0[T]\ {}_2^1[T]\ {}_3^2[T] \\
&= \begin{bmatrix} \cos\delta_{1z} & -\sin\delta_{1z} & 0 & 0 \\ \sin\delta_{1z} & \cos\delta_{1z} & 0 & 0 \\ 0 & 0 & 1 & l_1 \\ 0 & 0 & 0 & 1 \end{bmatrix}
\begin{bmatrix} \cos\delta_{2z} & -\sin\delta_{2z} & 0 & l_2 \\ \sin\delta_{2z} & \cos\delta_{2z} & 1 & 0 \\ 0 & 0 & 0 & 1 \\ 0 & 0 & 0 & 1 \end{bmatrix}
\begin{bmatrix} \cos\delta_{3z} & -\sin\delta_{3z} & 0 & l_3 \\ \sin\delta_{3z} & \cos\delta_{3z} & 0 & 0 \\ 0 & 0 & 1 & \Delta_{3z} \\ 0 & 0 & 0 & 1 \end{bmatrix}
\end{aligned}$$

$$(11.19)$$

Using the resulting transformation matrix and Equation 11.9 produces the following system of equations to calculate the global coordinates of $\mathbf{p}_3$:

$$\begin{aligned}
{}^0p_{3x} &= {}^3p_{3x}\left[\left(\cos\delta_{1z}\cos\delta_{2z} - \sin\delta_{1z}\sin\delta_{2z}\right)\cos\delta_{3z} + \left(-\cos\delta_{1z}\sin\delta_{2z} - \sin\delta_{1z}\cos\delta_{2z}\right)\sin\delta_{3z}\right] \\
&\quad + {}^3p_{3y}\left[-\left(\cos\delta_{1z}\cos\delta_{2z} - \sin\delta_{1z}\sin\delta_{2z}\right)\sin\delta_{3z} + \left(-\cos\delta_{1z}\sin\delta_{2z} - \sin\delta_{1z}\cos\delta_{2z}\right)\cos\delta_{3z}\right] \\
&\quad + l_3\left(\cos\delta_{1z}\cos\delta_{2z} - \sin\delta_{1z}\sin\delta_{2z}\right) + l_2\cos\delta_{1z}
\end{aligned}$$

$$\begin{aligned}
{}^0p_{3y} &= {}^3p_{3x}\left[\left(\sin\delta_{1z}\cos\delta_{2z} + \cos\delta_{1z}\sin\delta_{2z}\right)\cos\delta_{3z} + \left(\cos\delta_{1z}\cos\delta_{2z} - \sin\delta_{1z}\sin\delta_{2z}\right)\sin\delta_{3z}\right] \\
&\quad + {}^3p_{3y}\left[-\left(\sin\delta_{1z}\cos\delta_{2z} + \cos\delta_{1z}\sin\delta_{2z}\right)\sin\delta_{3z} + \left(\cos\delta_{1z}\cos\delta_{2z} - \sin\delta_{1z}\sin\delta_{2z}\right)\cos\delta_{3z}\right] \\
&\quad + l_3\left(\sin\delta_{1z}\cos\delta_{2z} + \cos\delta_{1z}\sin\delta_{2z}\right) + l_2\sin\delta_{1z}
\end{aligned}$$

$$^0p_{3z} = {}^3p_{3z} + l_1 + \Delta_{3z}. \qquad (11.20)$$

In the R-R-C robotic manipulator, variables l_1, l_2, and l_3 are assigned constant values because they represent constant link lengths. The angular and linear displacement variables δ_{1z}, δ_{2z}, δ_{3z}, and Δ_{3y}, however, correspond to the rotational and translational

---

* In Table 11.5, $\Delta_{1z} = l_1$, $\Delta_{2x} = l_2$, and $\Delta_{3x} = l_3$.

displacements of the revolute and prismatic joints, respectively. More often, displacement ranges are assigned to these variables.

Appendix G.4 includes the MATLAB file user instructions for R-R-C robotic manipulator forward kinematics. In this MATLAB file (which is available for download at www.crcpress.com/ product/isbn/9781498724937), Equation 11.20 is used to calculate the global coordinates of $\mathbf{p}_3$.

Example 11.7

Problem Statement: Using the Appendix G.4 MATLAB file with the joint displacements given in Table E.11.10, calculate the global path points achieved by the end effector of the R-R-C robotic manipulator. In this example, $l_1 = 1$, $l_2 = l_3 = 0.5$ (unitless link lengths), and $^3\{\mathbf{p}_3\} = [1, 0, 0]^T$.

Known Information: Table E.11.10 and Appendix G.4 MATLAB file.

Solution Approach: The data for columns δ_{1z}, δ_{2x}, δ_{3z}, and Δ_{3z} in Table E.11.10 are first specified in the file *RRC_Input.csv*. Figure E.11.6 includes the input specified (in bold text) in the Appendix G.4 MATLAB file and Table E.11.11 includes the global end effector path point coordinates calculated.

TABLE E.11.10

R-R-C Robotic Manipulator Joint Displacements

Point	δ_{1z} (°)	δ_{2z} (°)	δ_{3z} (°)	Δ_{3z}
1	12	−5	15	−0.1000
2	24	−10	30	−0.2000
3	36	−15	45	−0.3000
4	48	−20	60	−0.2000
5	60	−25	75	−0.1000

```
l1 = 1;
l2 = 0.5;
l3 = 0.5;
p3_3 = [1, 0, 0];
```

FIGURE E.11.6
Specified input (in bold text) in the Appendix G.4 MATLAB file for Example 11.4.

TABLE E.11.11

R-R-C Robotic Manipulator Path Point Coordinates

Point	$^0p_{3x}$	$^0p_{3y}$	$^0p_{3z}$
1	1.9125	0.5395	0.9000
2	1.6613	1.0190	0.8000
3	1.2780	1.3866	0.7000
4	0.8109	1.6057	0.8000
5	0.3176	1.6595	0.9000

11.6 Inverse Kinematics

11.6.1 Definition and Application

While the robotic manipulator dimensions are known and the end effector motion is calculated in forward kinematics, in *inverse kinematics*, the end effector motion and link dimensions are known and the joint motion required to achieve the end effector motion is calculated [4–7]. Inverse kinematics is often described as the *reverse of forward kinematics*. It is accomplished by calculating joint motion solutions (using the same forward-kinematics-based equation system for a robotic manipulator) given the end effector motion and link lengths. The most common application for inverse kinematics is to determine the joint motion required to achieve the required end effector motion (e.g., the joint motion required to achieve tool paths and orientations).*

11.6.2 P-P-P

Because Equation 11.12 includes three equations each containing a single unknown variable (Δ_{1x}, Δ_{2y}, and Δ_{3z}, respectively), these unknowns can be calculated algebraically for the inverse kinematics of the P-P-P robotic manipulator.

Solving for Δ_{1x}, Δ_{2y}, and Δ_{3z} in Equation 11.12 produces

$$\Delta_{1x} = {}^{0}p_{3x} - {}^{3}p_{3x}$$

$$\Delta_{2y} = {}^{0}p_{3y} - {}^{3}p_{3y} \qquad (11.21)$$

$$\Delta_{3z} = {}^{0}p_{3z} - {}^{3}p_{3z}$$

From Equation 11.21, the user can specify the global end effector coordinates ${}^{0}\{\mathbf{p}_3\}$, the end effector coordinates in Frame 3 ${}^{0}\{\mathbf{p}_3\}$, and calculate the required displacements of each prismatic joint in the P-P-P robotic manipulator (Δ_{1x}, Δ_{2y}, and Δ_{3z}).

The global path point coordinates ${}^{0}\{\mathbf{p}_3\}$ should be prescribed from within the workspace of the P-P-P robotic manipulator. This ensures that the prescribed points will be achieved by the robotic manipulator. As shown in Figure 11.11, the P-P-P robotic manipulator has a cubic (or a rectangular cuboid) workspace with outer x, y, and z dimensions of $\Delta_{1x_{max}}$, $\Delta_{2y_{max}}$ and $\Delta_{3z_{max}}$, respectively (the maximum prismatic joint translations).

Example 11.8

Problem Statement: Calculate the P-P-P joint displacements required to achieve the global path points given in Table E.11.12. In this example, ${}^{3}\{\mathbf{p}_3\} = [0, -1, 0]^{T}$.

Known Information: Equation (11.21) and Table E.11.12.

Solution Approach: Table E.11.13 includes the joint displacements calculated using Equation (11.21). Figure E.11.7 includes the initial and final positions of the P-P-P robotic manipulator over the range of prescribed global end effector points.

---

* Inverse kinematics is similar to motion generation and path generation (see Chapter 5). The key distinction between dimensional synthesis and inverse kinematics is that, with the latter, *joint displacements* are calculated (rather than link dimensions).

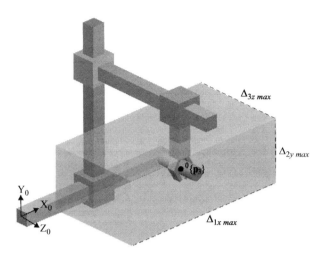

FIGURE 11.11
P-P-P robotic manipulator and workspace.

TABLE E.11.12

P-P-P Robotic Manipulator End Effector Path Point Coordinates

Point	$^0p_{3x}$	$^0p_{3y}$	$^0p_{3z}$
1	0.683	0.375	0.2165
2	0.6764	0.4521	0.2566
3	0.6569	0.5306	0.289
4	0.625	0.6083	0.3125
5	0.5817	0.6826	0.3266
6	0.5283	0.7514	0.3307
7	0.4665	0.8125	0.3248
8	0.3981	0.8641	0.309
9	0.3252	0.9047	0.2838
10	0.25	0.933	0.25

TABLE E.11.13

P-P-P Robotic Manipulator Joint Displacements

Point	Δ_{1x}	Δ_{2y}	Δ_{3z}
1	0.683	1.375	0.2165
2	0.6764	1.4521	0.2566
3	0.6569	1.5306	0.289
4	0.625	1.6083	0.3125
5	0.5817	1.6826	0.3266
6	0.5283	1.7514	0.3307
7	0.4665	1.8125	0.3248
8	0.3981	1.8641	0.309
9	0.3252	1.9047	0.2838
10	0.25	1.933	0.25

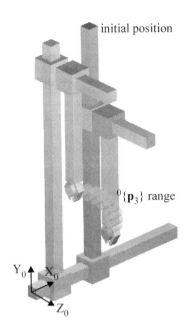

FIGURE E.11.7
P-P-P robotic manipulator in (semitransparent) initial and final positions.

11.6.3 R-P-P

Unlike Equation 11.12, the joint displacement variables δ_{1z}, Δ_{2x}, and Δ_{3x} in Equation 11.14 cannot be calculated algebraically. With unknowns δ_{1z}, Δ_{2x}, and Δ_{3x}, Equation 11.14 becomes a set of three nonlinear simultaneous equations. A root-finding method (see Section 4.2) is required for the inverse kinematics of the R-P-P robotic manipulator. Appendix G.5 includes the user instructions for the MATLAB file to calculate joint displacement solutions for Equation 11.14 given $^3\{\mathbf{p}_3\}$ and a prescribed range of values for $^0\{\mathbf{p}_3\}$.

As shown in Figure 11.12, the R-P-P robotic manipulator has a cylindrical workspace (having a center axis that is collinear with Δ_{2z}) with outer cylinder height and radius dimensions of $\Delta_{2z_{max}}$ and $\Delta_{3x_{max}}$, respectively (the maximum prismatic joint translations).

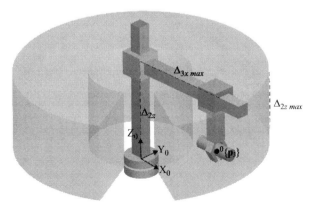

FIGURE 11.12
R-P-P robotic manipulator and workspace.

Example 11.9

Problem Statement: Using the Appendix G.5 MATLAB file, calculate the R-P-P joint displacements required to achieve the global path points given in Table E.11.12. In this example, $^3\{\mathbf{p}_3\} = [0, 0, -1]^T$.

Known Information: Table E.11.12 and Appendix G.5 MATLAB file.

Solution Approach: The data for columns $^0p_{3x}$, $^0p_{3y}$, and $^0p_{3z}$ in Table E.11.12 are first specified in the file *RPP_Input.csv*. Figure E.11.8 includes the input specified (in bold text) in the Appendix G.5 MATLAB file and Table E.11.14 includes the joint displacements calculated. Figure E.11.9 includes the initial and final positions of the R-P-P robotic manipulator over the range of prescribed global end effector points.

```
p3_3 = [0, 0, -1];
```

FIGURE E.11.8

Specified input (in bold text) in the Appendix G.5 MATLAB file for Example 11.9.

TABLE E.11.14

R-P-P Robotic Manipulator Joint Displacements

Point	δ_{1z} (°)	Δ_{2z}	Δ_{3x}
1	28.7689	1.2165	0.7792
2	33.7584	1.2566	0.8136
3	38.929	1.289	0.8444
4	44.2242	1.3125	0.8722
5	49.5629	1.3266	0.8968
6	54.8894	1.3307	0.9185
7	60.1375	1.3248	0.9369
8	65.264	1.309	0.9514
9	70.2287	1.2838	0.9614
10	74.9998	1.25	0.9659

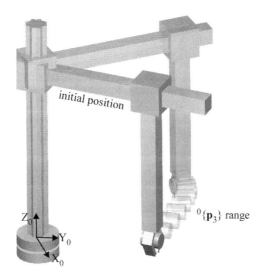

FIGURE E.11.9

R-P-P robotic manipulator in initial and final positions.

11.6.4 R-R-P

The joint displacement variables δ_{1z}, δ_{2x}, δ_{3z} in Equation 11.16 cannot also be calculated algebraically. With unknowns δ_{1z}, δ_{2x}, δ_{3z}, Equation 11.16 becomes a set of three nonlinear simultaneous equations. Like the R-P-P robotic manipulator, a root-finding method is also required for the inverse kinematics of the R-R-P robotic manipulator. Appendix G.6 includes the user instructions for the MATLAB file to calculate joint displacement solutions for Equation 11.16 given $^3\{\mathbf{p}_3\}$ and a prescribed range of values for $^0\{\mathbf{p}_3\}$.

As shown in Figure 11.13, the R-R-P robotic manipulator has a spherical workspace (having center coordinates center = $(0, 0, l_1)$) with an outer radius dimension of

$$r_{\text{outer}} = \sqrt{\left(l_2\right)^2 + \left(\Delta_{3z\,\text{max}}\right)^2}$$

Example 11.10

Problem Statement: Using the Appendix G.6 MATLAB file, calculate the R-R-P joint displacements required to achieve the global path points given in Table E.11.12. In this example, $l_1 = l_2 = 0.5$ (unitless link lengths) and $^3\{\mathbf{p}_3\} = [0,0,0]^T$.

Known Information: Table E.11.12 and Appendix G.6 MATLAB file.

Solution Approach: The data for columns $^0p_{3x}$, $^0p_{3y}$, and $^0p_{3z}$ in Table E.11.12 are first specified in the file *RRP_Input.csv*. Figure E.11.10 includes the input specified (in bold text) in the Appendix G.6 MATLAB file and Table E.11.15 includes the joint displacements calculated. Figure E.11.11 includes the initial and final positions of the R-R-P robotic manipulator over the range of prescribed global end effector points.

```
l1 = 0.5;
l2 = 0.5;
p3_3 = [0, 0, 0];
```

FIGURE E.11.10
Specified input (in bold text) in the Appendix G.6 MATLAB file for Example 11.10.

TABLE E.11.15

R-R-P Robotic Manipulator Joint Displacements

Point	δ_{1z} (°)	δ_{2x} (°)	Δ_{3z}
1	−61.2311	32.9191	0.6614
2	−56.2416	37.2735	−0.6864
3	−51.0710	40.9089	−0.7124
4	−45.7758	43.7775	−0.7388
5	−40.4371	45.8699	−0.7644
6	−35.1105	47.1905	−0.7889
7	−29.8625	47.7679	−0.8115
8	−24.7360	47.6333	−0.8316
9	−19.7713	46.8339	−0.8491
10	−15.0002	45.4141	−0.8634

FIGURE E.11.11
R-R-P robotic manipulator in initial and final positions.

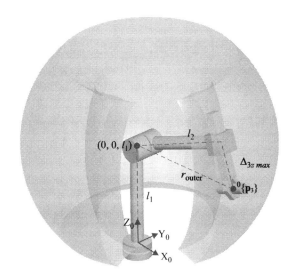

FIGURE 11.13
R-R-P robotic manipulator and workspace.

11.6.5 R-R-R

The joint displacement variables δ_{1z}, δ_{2x}, and δ_{3x} in Equation 11.18 cannot also be calculated algebraically. With unknowns δ_{1z}, δ_{2x}, and δ_{3x}, Equation 11.18 becomes a set of three nonlinear simultaneous equations. Like the R-P-P and R-R-P robotic manipulators, a root-finding method is also required for the inverse kinematics of the R-R-R robotic manipulator. Appendix G.7 includes the user instructions for the MATLAB file to calculate joint displacement solutions for Equation 11.18 given $^3\{\mathbf{p}_3\}$ and a prescribed range of values for $^0\{\mathbf{p}_3\}$.

As shown in Figure 11.14, the R-R-R robotic manipulator has a spherical workspace (having center coordinates center $= (0, 0, l_1)$) with an outer radius dimension of $r_{outer} = l_2 + \|^3\{\mathbf{p}_3\}\|$ where $^3\{\mathbf{p}_3\} = [0, {^0p_{3y}}, {^0p_{3z}}]^T$.

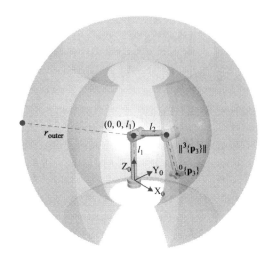

FIGURE 11.14
R-R-R robotic manipulator and workspace.

Example 11.11

Problem Statement: Using the Appendix G.7 MATLAB file, calculate the R-R-R joint displacements required to achieve the global path points given in Table E.11.12. In this example, $l_1 = l_2 = 0.5$ (unitless link lengths) and ${}^3\{\mathbf{p}_3\} = [0,1,0]^T$.

Known Information: Table E.11.12 and Appendix G.7 MATLAB file.

Solution Approach: The data for columns ${}^0p_{3x}$, ${}^0p_{3y}$, and ${}^0p_{3z}$ in Table E.11.12 are first specified in the file *RRR_Input.csv*. Figure E.11.12 includes the input specified (in bold text) in the Appendix G.7 MATLAB file and Table E.11.16 includes the joint displacements calculated. Figure E.11.13 includes the initial and final positions of the R-R-R robotic manipulator.

```
l1 = 0.5;
l2 = 0.5;
p3_3 = [0, 1, 0];
```

FIGURE E.11.12
Specified input (in bold text) in the Appendix G.7 MATLAB file for Example 11.11.

TABLE E.11.16

R-R-R Robotic Manipulator Joint Displacements

Point	δ_{1z} (°)	δ_{2x} (°)	δ_{3x} (°)
1	−61.2311	74.3302	−124.2298
2	−56.2416	75.2908	−121.9275
3	−51.0710	75.4719	−119.5001
4	−45.7758	74.9234	−117.0128
5	−40.4371	73.7565	−114.5580
6	−35.1106	72.0287	−112.1872
7	−29.8625	69.8373	−109.9699
8	−24.7360	67.2585	−107.9608
9	−19.7713	64.3666	−106.2017
10	−15.0002	61.2454	−104.7446

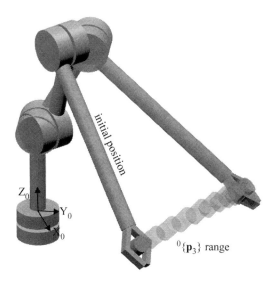

FIGURE E.11.13
R-R-R robotic manipulator in initial and final positions.

11.6.6 R-R-C

The joint displacement variables δ_{1z}, δ_{2x}, δ_{3z}, and Δ_{3y} in Equation 11.20 cannot also be calculated algebraically. With unknowns δ_{1z}, δ_{2x}, δ_{3z}, and Δ_{3y}, Equation 11.20 becomes a set of three nonlinear simultaneous equations. Like the R-P-P, R-R-P, and R-R-R robotic manipulators, a root-finding method is also required for the inverse kinematics of the R-R-C robotic manipulator. Appendix G.8 includes the user instructions for the MATLAB file to calculate joint displacement solutions for Equation 11.20 given ${}^3\{\mathbf{p}^3\}$ and a prescribed range of values for ${}^0\{\mathbf{p}_3\}$.

As shown in Figure 11.15, the R-R-C robotic manipulator has a cylindrical workspace (having a center axis that is collinear with l_1) with outer height and radius dimensions of $\Delta_{3z_{\max}}$ and $r_{\text{outer}} = l_1 + l_2$, respectively.

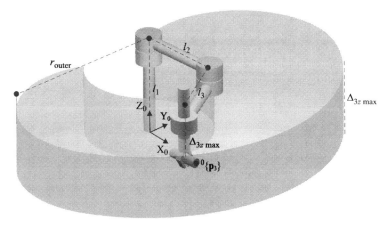

FIGURE 11.15
R-R-C robotic manipulator and workspace.

Example 11.12

Problem Statement: Using the Appendix G.8 MATLAB file, calculate the R-R-C joint displacements required to achieve the global path points given in Table E.11.12. In this example, $l_1 = 1$, $l_2 = l_3 = 0.5$ (unitless link lengths), and $^3\{\mathbf{p}_3\} = [0, 1, -1]^T$.

Known Information: Table E.11.14 and Appendix G.8 MATLAB file.

Solution Approach: The data for columns $^0p_{3x}$, $^0p_{3y}$, and $^0p_{3z}$ in Table E.11.12 are first specified in the file *RRC_Input.csv*. Figure E.11.14 includes the input specified (in bold text) in the Appendix G.8 MATLAB file and Table E.11.17 includes the joint displacements calculated. Figure E.11.15 includes the initial and final positions of the R-R-C robotic manipulator over the range of prescribed global end effector points.

```
l1 = 1;
l2 = 0.5;
l3 = 0.5;
p3_3 = [0, 1, -1];
```

FIGURE E.11.14
Specified input (in bold text) in the Appendix G.8 MATLAB file for Example 11.12.

TABLE E.11.17

R-R-C Robotic Manipulator Joint Displacements

Point	δ_{1z} (°)	δ_{2z} (°)	δ_{3z} (°)	Δ_{3z}
1	−35.4860	−5.7728	46.9980	0.2165
2	−30.0170	−4.5237	44.2350	0.2566
3	−24.5330	−3.1631	41.6230	0.2890
4	−19.0690	−1.7162	39.1470	0.3125
5	−13.6890	−0.21255	36.8220	0.3266
6	−8.4370	1.3269	34.6550	0.3307
7	−3.3822	2.8823	32.6820	0.3248
8	1.4188	4.4435	30.9450	0.3090
9	5.9070	6.0036	29.4950	0.2838
10	10.0190	7.5616	28.3930	0.2500

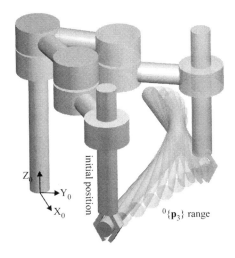

FIGURE E.11.15
R-R-C robotic manipulator in initial and final positions.

11.7 Robotic Manipulator Kinematic Analysis and Modeling in Simmechanics®

As has been noted throughout this chapter, Appendices G.1–G.8 include MATLAB file user instructions for R-P-P, R-R-P, R-R-R, and R-R-C forward and inverse kinematics. In these files, the global end effector coordinates are calculated for prescribed joint displacements (in Appendices G.1–G.4) and the joint displacements are calculated for prescribed global end effector coordinates (in Appendices G.5–G.8).

This textbook also utilizes SimMechanics as an alternate approach for simulation-based kinematic analysis. A library of SimMechanics files is also available for download at www. crcpress. com/product/isbn/9781498724937 for R-P-P, R-R-P, R-R-R, and R-R-C forward kinematics. In addition to calculating the global end effector coordinates for prescribed joint displacements, the motion of the robotic manipulator is also simulated over the joint displacements. The SimMechanics file user instructions for the forward kinematic analysis of the R-P-P, R-R-P, R-R-R, and R-R-C robotic manipulators are given in Appendices L.1–L.4, respectively.

Example 11.13

Problem Statement: Using the Appendix L.4 SimMechanics files, calculate the R-R-C end effector coordinates in Example 11.12.

Known Information: Example 11.12 and Appendix L.4 SimMechanics files.

Solution Approach: The data for columns δ_{1z}, δ_{2x}, δ_{3z}, and Δ_{3z} in Table E.11.17 are first specified in the file *RRC_Input.csv*. Figure E.11.16 includes the input specified (in bold text) in the Appendix L.4 SimMechanics file and Table E.11.18 includes the global R-R-C end effector coordinates calculated. These coordinates are identical to the prescribed end effector coordinates given in Table E.11.12.

```
l1 = 1;
l2 = 0.5;
l3 = 0.5;
p3_3 = [0, 1, -1];
```

FIGURE E.11.16

Specified input (in bold text) in the Appendix L.4 SimMechanics file for Example 11.13.

TABLE E.11.18

Calculated R-R-C Robotic Manipulator End Effector Coordinates

Point	$^0p_{3x}$	$^0p_{3y}$	$^0p_{3z}$
1	0.683	0.375	0.2165
2	0.6764	0.4521	0.2566
3	0.6569	0.5306	0.289
4	0.625	0.6083	0.3125
5	0.5817	0.6826	0.3266
6	0.5283	0.7514	0.3307
7	0.4665	0.8125	0.3248
8	0.3981	0.8641	0.309
9	0.3252	0.9047	0.2838
10	0.25	0.933	0.25

Identical end effector coordinates to those in Table E.11.18 can be replicated for The R-P-P, R-R-P, and R-R-R robotic manipulators by using the joint displacements and dimension input data in Examples 11.9, 11.10, and 11.11, respectively (in the Appendix L.1, L.2 and L.3 SimMechanics files, respectively).

11.8 Summary

Like a linkage, a robotic manipulator (commonly called a robot) includes an assembly of links and joints and is designed to produce a controlled output motion. In addition to links and joints, however, a robotic manipulator also includes electronic circuitry, computer-controlled actuators to compel link motion, and is guided by a computer program. Because robotic manipulators include both mechanical and electronic components, they are classified as electro-mechanical systems.

Another common distinction between linkages and robotic manipulators is in their overall design. Linkages commonly have closed-loop designs. With this design, at least two joints in a linkage are connected to ground (thus forming a closed loop). While robotic manipulators can have closed-loop designs, they often have open loops, where only one joint is connected to ground.

The five spatial robotic manipulator types considered in this chapter are the Cartesian, cylindrical, spherical, articulated, and SCARA robots. They are commonly known as the P-P-P, R-P-P, R-R-P, R-R-R, and R-R-C robotic manipulators, respectively. By prescribing coordinate frames for each link in a robotic manipulator and establishing displacement variables between each frame, equation systems are formulated (using the general spatial transformation matrix) to calculate the motion of any link in the robotic manipulator.

This textbook includes a library of MATLAB files for the forward kinematics (Appendices G.1–G.4) and the inverse kinematics (Appendices G.5–G.8) of the R-P-P, R-R-P, R-R-R, and R-R-C robotic manipulators. In forward kinematics, the link dimensions and joint motion of a robotic manipulator are known and the corresponding output motion of the links (usually the end effector) is calculated. In inverse kinematics, the end effector motion and link dimensions are known and the joint motion required to achieve the end effector motion is calculated. Inverse kinematics is often described as the reverse of forward kinematics.

This textbook also includes a library of MATLAB and SimMechanics files for the forward kinematics (Appendices L.1–L.4) of the R-P-P, R-R-P, R-R-R, and R-R-C robotic manipulators. In addition to calculating the global values for the end effector point $\mathbf{p}_3$, the motion of the robotic manipulator is also simulated in the Appendix L files.

References

1. International Federation of Robotics. Industrial robots: Statistics. http://www.ifr.org. Accessed May 18, 2015.
2. Lewis, F. L., D. M. Dawson, and C. T. Abdallah. 2004. *Robot Manipulator Control: Theory and Practice*. 2nd edn. Section 1.2. New York: Marcel Dekker.

3. Craig, J. J. 2005. *Introduction to Robotics: Mechanics and Control.* 3rd edn. Chapter 3. Upper Saddle River, NJ: Prentice-Hall.

4. Ibid, Chapter 4.

5. Nikiv, S. B. 2001. *Introduction to Robotics: Analysis, Systems and Applications.* pp. 76–82. Upper Saddle River, NJ: Prentice-Hall.

6. Jazar, R. N. 2007. *Theory of Applied Robotics: Kinematics, Dynamics and Control.* Chapter 6. New York: Springer.

7. Kurfess, T. R. 2005. *Robotic and Automation Handbook.* Chapter 3. New York: CRC Press.

Additional Reading

Sandor, G. N. and A. G. Erdman. 1984. *Advanced Mechanism Design: Analysis and Synthesis.* Volume 2. Chapter 6. Englewood Cliffs, NJ: Prentice-Hall.

Wilson, C. E. and J. P. Sadler. 2003. *Kinematics and Dynamics of Machinery.* 3rd edn. Chapter 12. Saddle River, NJ: Prentice Hall.

Problems

1. Explain some of the design similarities and distinctions between robotic manipulators and the linkages presented in Chapters 4 and 10.

2. Explain some of the advantages robotic manipulators have over the linkages presented in Chapters 4 and 10.

3. Calculate the mobility values for the P-P-P, R-P-P, and R-R-P robotic manipulators.

4. Using Matrix (11.3) calculate the transformation matrix $_1^0[T]$ for angular displacements $\delta_{1x} = 12°$ and $\delta_{1z} = -35°$, and linear displacements $\Delta_{1x} = 5.75$ and $\Delta_{1y} = -2.75$ between coordinate frames 0 and 1.

5. Including Matrix (11.3), calculate the transformation matrix $_2^0[T]$ for the displacement data given in Problem 4 and angular displacements $\delta_{2y} = 30°$ and $\delta_{2z} = 15°$, and linear displacements $\Delta_{2x} = 15$, $\Delta_{2y} = -8$, and $\Delta_{2z} = 5$ between coordinate frames 1 and 2.

6. Calculate the transformation matrix $_3^0[T]$ for the coordinate frame displacement values given in Table P.11.1.

TABLE P.11.1

Frame Displacement Variables

Frames	δ_x	δ_y	δ_z	Δ_x	Δ_y	Δ_z
1 wrt 0	20°	0	0	0	2.85	0
2 wrt 1	0	0	15°	0	0	-7.75
3 wrt 2	0	-65°	0	0	7.5	0

7. Explain the distinction between a transformation matrix formulated using *Denavit–Hartenberg* parameters and the transformation matrix given in Matrix 11.3.

8. Calculate the $^0\{\mathbf{p}_3\}$ coordinates produced by the P-P-P robotic manipulator given the joint displacements

$$\Delta_{1x} = 1 + t, \quad \Delta_{2y} = t^{1.45}, \quad \Delta_{3z} = -2t^{0.25}, \quad 0 \le t \le 10$$

   Consider at least 10 equally-spaced curve points. Also $^3\{\mathbf{p}_3\} = [0, -1, 0]^T$.

9. Using the Appendix G.1 or L.1 files, calculate the $^0\{\mathbf{p}_3\}$ coordinates produced by the R-P-P robotic manipulator given the joint displacements

$$\delta_{1z} = t, \quad \Delta_{2z} = t/35°, \quad \Delta_{3x} = (t/40°)^{1.65}, \quad 0 \le t \le 90°$$

   Consider at least 10 equally-spaced curve points. Also $^3\{\mathbf{p}_3\} = [0, 0, -1]^T$.

10. Using the Appendix G.1 or L.1 files, calculate the $^0\{\mathbf{p}_3\}$ coordinates produced by the R-P-P robotic manipulator given the joint displacements

$$\delta_{1z} = (90°)\sin t, \quad \Delta_{2z} = t/55°, \quad \Delta_{3x} = (t/40°)^{1.35}, \quad 0 \le t \le 180°$$

   Consider at least 10 equally-spaced curve points. Also $^3\{\mathbf{p}_3\} = [2, 0, -1.5]^T$.

11. Using the Appendix G.2 or L.2 files, calculate the $^0\{\mathbf{p}_3\}$ coordinates produced by the R-R-P robotic manipulator given the joint displacements

$$\delta_{1z} = t, \quad \delta_{2x} = -2t, \quad \Delta_{3z} = t/55°, \quad 0 \le t \le 45°$$

   Consider at least 10 equally-spaced curve points. Also, $l_1 = l_2 = 2$ and $^3\{\mathbf{p}_3\} = [0, 0, 0]^T$.

12. Using the Appendix G.2 or L.2 files, calculate the $^0\{\mathbf{p}_3\}$ coordinates produced by the R-R-P robotic manipulator given the joint displacements

$$\delta_{1z} = (65°)\sin 1.75t, \quad \delta_{2x} = -t/2.5, \quad \Delta_{3z} = t/65°, \quad 0 \le t \le 90°$$

   Consider at least 10 equally-spaced curve points. Also, $l_1 = 2$, $l_2 = 3.5$, and $^3\{\mathbf{p}_3\} = [0, 0, 0]^T$.

13. Using the Appendix G.3 or L.3 files, calculate the $^0\{\mathbf{p}_3\}$ coordinates produced by the R-R-R robotic manipulator given the joint displacements

$$\delta_{1z} = 2.5t, \quad \delta_{2x} = -1.1t, \quad \delta_{3x} = 3.8t, \quad 0 \le t \le 30°$$

   Consider at least 10 equally-spaced curve points. Also, $l_1 = 2$, $l_2 = 1.5$, and $^3\{\mathbf{p}_3\} = [0, 0.25, 0]^T$.

14. Using the Appendix G.3 or L.3 files, calculate the $^0\{\mathbf{p}_3\}$ coordinates produced by the R-R-R robotic manipulator given the joint displacements

$$\delta_{1z} = (70°)\sin 2t, \quad \delta_{2x} = (35°)\sin 2t, \quad \delta_{3x} = (40°)\sin 2t, \quad 0 \le t \le 90°$$

Consider at least 10 equally-spaced curve points. Also, $l_1 = 2$, $l_2 = 4.5$, and $^3\{\mathbf{p}_3\} = [1, 0.25, 2]^T$.

15. Using the Appendix G.4 or L.4 files, calculate the $^0\{\mathbf{p}_3\}$ coordinates produced by the R-R-C robotic manipulator given the joint displacements

$$\delta_{1z} = 5t, \quad \delta_{2z} = -3t, \quad \delta_{3z} = 2, \quad \Delta_{3z} = -t/100°, \quad 0 \le t \le 30°$$

Consider at least 10 equally-spaced curve points. Also, $l_1 = l_2 = l_3 = 1$ and $^3\{\mathbf{p}_3\} = [0, 1, -0.1]^T$.

16. Using the Appendix G.4 or L.4 files, calculate the $^0\{\mathbf{p}_3\}$ coordinates produced by the R-R-C robotic manipulator given the joint displacements

$$\delta_{1z} = (90°)\sin 4t, \quad \delta_{2z} = (120°)\sin 2t, \quad \delta_{3z} = t/1.75, \quad \Delta_{3z} = -t/150°, \quad 0 \le t \le 90°$$

Consider at least 10 equally-spaced curve points. Also, $l_1 = l_2 = l_3 = 1$ and $^3\{\mathbf{p}_3\} = [2, 1, -2]^T$.

17. In robotic manipulator design, what are the objectives in and applications for *forward* and *inverse* kinematics?

18. Describe the shape and dimensions of the workspace of a P-P-P robotic manipulator with maximum translation distances (in meters) of $\Delta_{1x_{\max}} = 1.65$, $\Delta_{2y_{\max}} = 2.85$, and $\Delta_{3z_{\max}} = 3.2$.

19. Calculate the joint displacements of the P-P-P robotic manipulator required to achieve the curve

$$^0p_{3x} = 1.5 + \cos(t), \quad ^0p_{3y} = 2.75 + \sin(t), \quad ^0p_{3z} = t/2\pi, \quad 0 \le t \le 2\pi$$

Consider at least 10 equally-spaced curve points. Also, $^3\{\mathbf{p}_3\} = [0, -1, 0]^T$.

20. Describe the shape and outer dimensions of the workspace of an R-P-P robotic manipulator with maximum translation distances (in meters) of $\Delta_{2z_{\max}} = 8$ and $\Delta_{3x_{\max}} = 5.75$.

21. Describe the shape and outer dimensions of the workspace of an R-P-P robotic manipulator with maximum translation distances (in meters) of $\Delta_{2z_{\max}} = 7.5$ and $\Delta_{3x_{\max}} = 3.75$.

22. Using the Appendix G.5 files, calculate the joint displacements of the R-P-P robotic manipulator required to achieve the curve

$$^0p_{3x} = 4\cos(t), \quad ^0p_{3y} = 6\sin(t), \quad ^0p_{3z} = t/90°, \quad 0 \le t \le 360°$$

Consider at least 10 equally-spaced curve points. Also, $^3\{\mathbf{p}_3\}=[2,-1,0]^T$.

23. Using the Appendix G.5 files, calculate the joint displacements of the R-P-P robotic manipulator required to achieve the curve

$$^0p_{3x}=t^{1.25}, \quad {}^0p_{3y}=t^2, \quad {}^0p_{3z}=1+t^{1.75}, \quad 0\le t\le 18$$

Consider at least 10 equally-spaced curve points. Also, $^3\{\mathbf{p}_3\}=[-1,3,0]^T$.

24. Describe the shape and outer dimensions of the workspace of an R-R-P robotic manipulator with link lengths (in meters) $l_1=l_2=10$ and a maximum translation distance of $\Delta_{3z_{max}}=8$.

25. Describe the shape and outer dimensions of the workspace of an R-R-P robotic manipulator with link lengths (in meters) $l_1=4.5$ and $l_2=3$, and a maximum translation distance of $\Delta_{3z_{max}}=1.75$.

26. Using the Appendix G.6 files, calculate the joint displacements of the R-R-P robotic manipulator required to achieve the curve

$$^0p_{3x}=1.75\cos(t), \quad {}^0p_{3y}=5.5\sin(t), \quad {}^0p_{3z}=\sin(t), \quad 0\le t\le 360°$$

Consider at least 10 equally-spaced curve points. Also, $l_1=l_2=2.5$ and $^3\{\mathbf{p}_3\}=[0,0,0]^T$.

27. Using the Appendix G.6 files, calculate the joint displacements of the R-R-P robotic manipulator required to achieve the curve

$$^0p_{3x}=7.15+t/360°, \quad {}^0p_{3y}=5.72\sin(t), \quad {}^0p_{3z}=0.52, \quad 0\le t\le 360°$$

Consider at least 10 equally-spaced curve points. Also, $l_1=2.5$, $l_2=2$, and $^3\{\mathbf{p}_3\}=[0,0,0]^T$.

28. Describe the shape and outer dimensions of the workspace of an R-R-R robotic manipulator with link lengths (in meters) $l_1=12.5$ and $l_2=4$, and $^3\{\mathbf{p}_3\}=[0,2,0]^T$.

29. Describe the shape and outer dimensions of the workspace of an R-R-R robotic manipulator with link lengths (in meters) $l_1=8.25$ and $l_2=3.75$, and $^3\{\mathbf{p}_3\}=[0,2,1]^T$.

30. Using the Appendix G.7 files, calculate the joint displacements of the R-R-R robotic manipulator required to achieve the curve

$$^0p_{3x}=3\sin(t), \quad {}^0p_{3y}=2.5\cos(t), \quad {}^0p_{3z}=1.5\cos(t), \quad 0\le t\le 180°$$

Consider at least 10 equally-spaced curve points. Also, $l_1=l_2=3.75$ and $^3\{\mathbf{p}_3\}=[3,1.5,0.75]^T$.

31. Using the Appendix G.7 files, calculate the joint displacements of the R-R-R robotic manipulator required to achieve the curve

$$^0p_{3x} = -2 + t/45°, \quad ^0p_{3y} = \sin(t), \quad ^0p_{3z} = 2t/45°, \quad 0 \le t \le 90°$$

Consider at least 10 equally-spaced curve points. Also, $l_1 = 2.5$, $l_2 = 3$, and $^3\{\mathbf{p}_3\} = [1.5, 0, 1.7]^T$.

32. Describe the shape and outer dimensions of the workspace of an R-R-C robotic manipulator with link lengths (in meters) $l_2 = 2$, $l_3 = 1.25$ and a maximum translation distance of $\Delta_{3z_{max}} = 5$.

33. Describe the shape and outer dimensions of the workspace of an R-R-C robotic manipulator with link lengths (in meters) $l_1 = 1.25$, $l_2 = 2.95$ and a maximum translation distance of $\Delta_{3z_{max}} = 4.5$.

34. Using the Appendix G.8 files, calculate the joint displacements of the R-R-C robotic manipulator required to achieve the curve

$$^0p_{3x} = \frac{t}{45°}\sin(t), \quad ^0p_{3y} = \frac{-t}{45°}\cos(t), \quad ^0p_{3z} = t/30°, \quad 0 \le t \le 180°$$

Consider at least 10 equally-spaced curve points. Also, $l_1 = 3.25$, $l_2 = l_3 = 1.5$, and $^3\{\mathbf{p}_3\} = [0, 1, -0.5]^T$.

35. Using the Appendix G.8 files, calculate the joint displacements of the R-R-C robotic manipulator required to achieve the curve

$$^0p_{3x} = \frac{-t}{120°}\sin(t), \quad ^0p_{3y} = \frac{-t}{60°}\cos(t), \quad ^0p_{3z} = t/45°, \quad 0 \le t \le 180°$$

Consider at least 10 equally-spaced curve points. Also, $l_1 = 2$, $l_2 = l_3 = 1.75$, and $^3\{\mathbf{p}_3\} = [0, 0.5, -0.5]^T$.

Appendix A: User Information and Instructions for MATLAB®

A.1 Required MATLAB Toolkits

To utilize the MATLAB and SimMechanics® files that accompany this textbook (available for download at www.crcpress.com/product/isbn/9781498724937), the following toolkits must be included when installing MATLAB:

- *Symbolic Math Toolbox*: Required for performing symbolic mathematics
- *Optimization Toolbox*: Required for solving equations (e.g., simultaneous equation sets)
- *SimMechanics and Simulink*: Required for running SimMechanics files

To view the installed toolkits, the user can type the command *ver* in the MATLAB command window.

A.2 Description of MATLAB Operators and Functions

For additional descriptions of the functions and operators given in Table A.1, please refer to www. mathworks.com/help/matlab/.

A.3 Preparing and Running Files in MATLAB and Operations in SimMechanics

To execute commands in MATLAB, the user can simply type them line by line in the MATLAB command window (labeled in Figure A.1). This window appears by default in MATLAB. All of the MATLAB commands illustrated in throughout this textbook were implemented using this approach. After typing commands in one line, the user can use the "Enter" button to advance to the text line.

As a set of commands becomes more extensive, the user may prefer to save them as a MATLAB file. To do this, the user can enter the commands in the MATLAB editor window (labeled in Figure A.1) and save them as a *script* (a *.m* file).* The MATLAB editor window

---

* Unlike commands in the MATLAB command line, commands in the MATLAB editor do not include *prompts* (the symbol >>) that precedes every command in the MATLAB command window.

TABLE A.1

MATLAB Functions and Operators Used throughout This Textbook

Function or Operator	Description
$+ - \times \div \wedge$	Add, subtract, multiply, divide, and exponent (respectively)
;	When this operator is used at the end of a command, the results from the command are not displayed
i or j	Complex coefficient $\left(i = j = \sqrt{-1}\right)$
pi	π (3.14159...)
sqrt(x)	Square root of x
atan2(y,x)	The angle between the positive x-axis and the point given by the coordinates (x, y)
abs(x)	Absolute value of a real value or magnitude of a complex value
norm(x)	Vector or matrix norm ($\|x_2\|$)
exp(x)	Exponent (e^x)
syms *var_names* real	Defines symbolic variables (e.g., syms X Y Z real makes variables X, Y, and Z symbolic variables)
syms *name*(t) t real	Defines time-based symbolic variables (e.g., syms X(t) Y(t) Z(t) t real makes variables X, Y and Z time-based symbolic variables)
diff(F)	Calculates the first derivative of function F symbolically
diff(V, N)	Calculates the Nth derivative of function F symbolically
expand(expr)	Symbolic expansion (of polynomials and functions)
simplify(expr)	Algebraic simplification
sin(x)	Sine function
cos(x)	Cosine function
Name = [N1, N2, N3...]	Defines a row vector
Name(cell number)	Calls a vector cell (e.g., Name(1)=N1 in previous vector)
Name[N11, N12, N13...; N21, N22, N23...; N31, N32, N33...]	Defines a matrix
Name(row#, col#)	Calls a matrix cell (e.g., Name(2,3)=N23 in previous matrix)
'	When this operator is used at the end of a matrix or vector, the matrix or vector is transposed
Inv(A)	The inverse of matrix A ($[A]^{-1}$)
A/b	The solution to the matrix system [A]**x**=**b** (**x**=$[A]^{-1}$b)
...	When this operator is used at the end of a command, the command can continue on the next line
f =@(var1, var2...) expression	When this operator is used before a function expression, an anonymous function is defined (e.g., for f=x^2+ 2y, f =@(x, y) x^2+2*y)
[x,fval] = fsolve(fx,x0)	Solves a system of nonlinear equations fx (where fx=[eq1; eq2; eq3...]) for unknown variables x (where x=[var1, va2, var3...]). The variable x0 includes the initial unknown variable values.

can be launched using the "Ctrl+N" buttons or either of the *New Script* buttons encircled in Figure A.1.

To run files in MATLAB, the user can either use the "F5" button or select the *Run* button. Figure A.2 illustrates the MATLAB editor toolbar with the *Run* button encircled. This toolbar appears once a MATLAB file is opened.

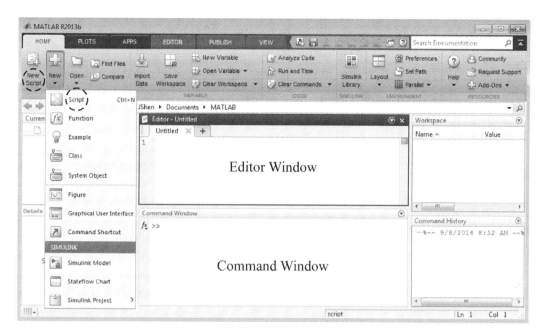

FIGURE A.1
MATLAB command and editor windows and *New Script* buttons (encircled).

FIGURE A.2
MATLAB editor toolbar with run button (encircled).

There are two windows that appear automatically when running SimMechanics files: the *model window* and the *animation window*.* Figure A.3 illustrates the model window for Appendix H.1. In this window, SimMechanics models are constructed. Because the contents in the model window should never be edited in any way, the user should minimize this window whenever it appears.

Figure A.4 illustrates the animation window for Appendix H.1. In this window, the motion of the model (under the user-prescribed dimension and driving link parameters) is simulated. Figure A.5 illustrates the toolbar for the animation window. This toolbar includes options to adjust view settings and play speed, as well as animation recording settings (among others).

---

* The SimMechanics files are associated with Appendices H through L.

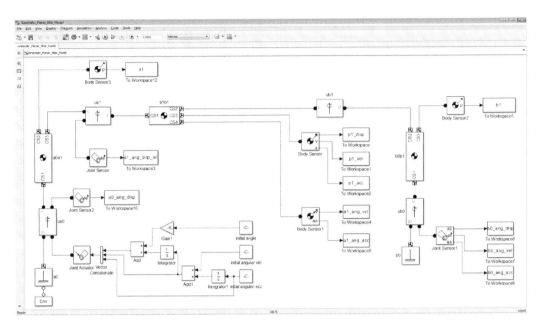

FIGURE A.3
SimMechanics model window (for Appendix H.1).

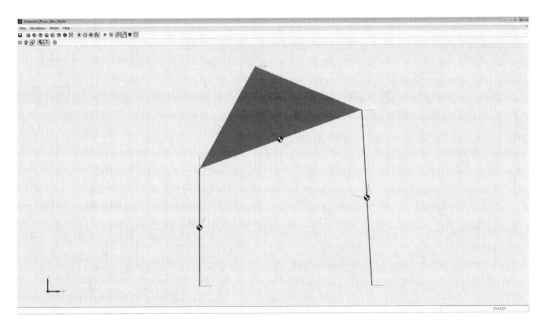

FIGURE A.4
SimMechanics animation window (for Appendix H.1).

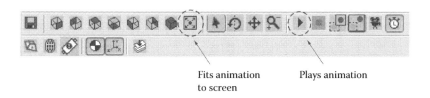

Fits animation
to screen

Plays animation

FIGURE A.5
SimMechanics animation window toolbar.

A.4 Rerunning MATLAB and SimMechanics Files with Existing *.csv Files

When rerunning a MATLAB and SimMechanics file when its *.csv file is open, the user could (1) rename the *.csv file (to prevent it from being overwritten), (2) store the *.csv file in a different folder (also to prevent it from being overwritten), or (3) close the *.csv file in the same folder (so that it may be overwritten). Attempting to rerun a MATLAB and SimMechanics file with its originally named *.csv file open and stored in its original location will prevent the new results from being written (and produce an error message) because a new *.csv file cannot be written when a *.csv file having the same name is in use in the same folder.

A.5 Minimum Precision Requirement for Appendix File User Input

The default precision tolerance for link vector loops in the MATLAB and SimMechanics files is 0.0002. Vector-loop errors beyond this tolerance are not allowed. An error indicating this violation will appear in the command line. When specifying link vectors in any of the MATLAB and SimMechanics files (e.g., vector lengths, vector angles, or vector components), noninteger values should be specified to at least four decimal places.[*]

---

[*] The only exceptions for this requirement are integer values or noninteger values that include a string of zeros before the fourth decimal place (e.g., ½ = 0.5000 = 0.5, ¼ = 0.2500 = 0.25, 5/4 = 1.2500 = 1.25).

Appendix B: User Instructions for Chapter 4 MATLAB® Files

B.1 Planar Four-Bar Mechanism

The Appendix B.1 folder (which is available for download at www.crcpress.com/product/isbn/9781498724937) includes the MATLAB file *Kinematic_ Planar_ 4Bar.m* for the kinematic analysis of planar four-bar mechanisms. To conduct a kinematic analysis, the user specifies the mechanism link dimensions and the crank motion parameters in this file. Values are specified for link variables $\mathbf{W}_1$, $\mathbf{V}_1$, $\mathbf{G}_1$, $\mathbf{U}_1$, and $\mathbf{L}_1$ (Figure B.1a). Values are also specified for the initial crank angle (*start_ ang*), the crank rotation increment (*step_ ang*), and the final crank angle (*stop_ ang*). Lastly, values are specified for the angular velocity (*angular_vel*) and angular acceleration (*angular_ acc*) of the crank.[*] Figure B.2 illustrates the user-input sections of the file *Kinematic_ Planar_ 4Bar.m*, with sample values in bold type.[†] While most of the link vector values in this figure appear in polar exponential form, they can all be specified in any of the rectangular and complex forms given in Equation 2.1.

After specifying the dimensions and driving link parameters in the file *Kinematic_ Planar_4Bar.m*, the next step is to run this file. When running this file, one file (filename *Disp_ Vel_ Acc.csv*) is written to a folder named *Results* (in a format compatible with Microsoft® Excel) that includes the calculated mechanism output at each crank link rotation increment.[‡] The calculated mechanism output is included in Figure B.1b.

B.2 Planar Four-Bar Fixed and Moving Centrode Generation

The Appendix B.2 folder (which is available for download at www.crcpress.com/product/isbn/9781498724937) includes the MATLAB file *Centrodes.m* for planar four-bar (and slider-crank) fixed and moving centrode generation.[§] For centrode generation, the user specifies the mechanism's fixed and moving pivots in this file. Values are specified for pivot variables $\mathbf{a}_0$, $\mathbf{a}_1$, $\mathbf{b}_0$, and $\mathbf{b}_1$ for a planar four-bar mechanism (Figure B.3a) or a slider-crank mechanism (Figure B.3b). Figure B.4 illustrates the user-input section of the file *Centrodes.m*, with sample values in bold type.[¶]

---

[*] If no crank angular velocity or acceleration values are specified, time cannot be calculated and the text *NaN* is written to the time column of the **.csv* file.

[†] The units for crank angular rotation, velocity, and acceleration are degrees, rad/s, and rad/s², respectively.

[‡] At crank rotation increments where circuit defects occur, the text *Inf* is written to the **.csv* file.

[§] The folder *functions* must also accompany the file *Centrodes.m*. This folder includes auxiliary files that are used by *Centrodes.m*.

[¶] The first and second cells for the fixed and moving pivots in *Centrodes.m* (see Figure B.4) correspond to the pivot's *x*- and *y*-components.

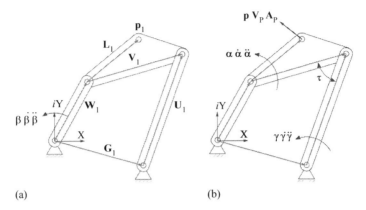

FIGURE B.1
(a) Planar four-bar mechanism and (b) output variables.

```
%----------------------------------------------------------------
% Here, values for the planar 4-bar mechanism variables W1, V1,
% G1, U1 and L1 are assigned.

W1 = 0.5*exp(i*90*pi/180);
V1 = 0.75*exp(i*19.3737*pi/180);
G1 = 0.75 + i*0;
U1 = 0.75*exp(i*93.2461*pi/180);
L1 = 0.5*exp(-i*60.7834*pi/180);
%----------------------------------------------------------------


%----------------------------------------------------------------
% Here, values for the start, step and stop displacement angles
% for the crank link are assigned.

start_ang    = 0;
step_ang     = 1;
stop_ang     = 360;
%----------------------------------------------------------------


%----------------------------------------------------------------
% Here, values for crank link angular velocity and angular
% acceleration are assigned.

angular_vel  = 1.0*ones(N,1);
angular_acc  = 0.1*ones(N,1);
%----------------------------------------------------------------
```

FIGURE B.2
Sections of *Kinematic_ Planar_4Bar.m* with sample values in bold.

After specifying the fixed and moving pivots in the file *Centrodes.m*, the next step is to run this file. When running this file, two files (filenames *Fixed_Centrode.csv* and *Moving_Centrode.csv*) are written to a folder named *Results* (in a format compatible with Microsoft Excel) that include calculated mechanism fixed and moving instant centers at each crank link rotation increment. When considering the slider-crank mechanism for centrode generation, the user should specify a length $\mathbf{b}_0$–$\mathbf{b}_1$ (or the y coordinate of $\mathbf{b}_0$) that produces an acceptable sliding error for $\mathbf{b}_1$. For example, in Example 4.8, using $\mathbf{b}_0 = (2, -100,000)$ produces translations of $\mathbf{b}_1$ that are accurate to five decimal places. Also, when running *Centrodes.m*, a window appears where the four-bar mechanism and its fixed and moving centrodes are displayed.

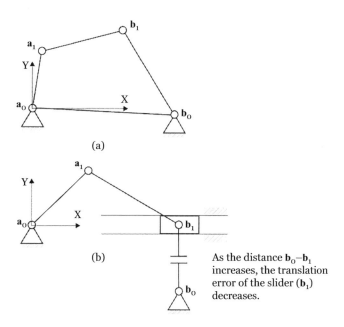

FIGURE B.3
(a) Planar four-bar and (b) slider-crank mechanism fixed and moving pivots.

```
%---------------------------------------------------------------------
% Here, values for the planar 4-bar or slider-crank mechanism
% variables a0, a1, b0 and b1 are assigned.

a0=[0, 0]';
a1=[0, 1]';
b0=[2.5, -100000]';
b1=[2.5, 0]';
%---------------------------------------------------------------------
```

FIGURE B.4
Section of *Centrodes.m* with sample values in bold.

B.3 Slider-Crank Mechanism

The Appendix B.3 folder (which is available for download at www.crcpress.com/product/isbn/9781498724937) includes the MATLAB file *Kinematic_Slider_Crank.m* for the kinematic analysis of slider-crank mechanisms. To conduct a kinematic analysis, the user specifies the mechanism link dimensions and the crank motion parameters in this file. Values are specified for link variables W_1, V_1, and U_1 (Figure B.5a). Values are also specified for the initial crank angle (*start_ang*), the crank rotation increment (*step_ang*), and the final crank angle (*stop_ang*). Lastly, values are specified for the angular velocity (*angular_vel*) and angular acceleration (*angular_acc*) of the crank.[*] Figure B.6 illustrates the user-input sections of the file *Kinematic_Slider_Crank.m*, with sample values in bold type.[†]

---

[*] If no crank angular velocity or acceleration values are specified, time cannot be calculated and the text *NaN* is written to the time column of the **.csv* file.
[†] The units for crank angular rotation, velocity, and acceleration are degrees, rad/s, and rad/s², respectively.

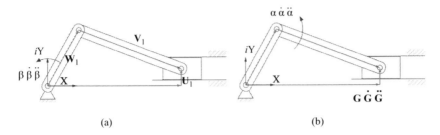

FIGURE B.5
(a) Slider-crank mechanism and (b) output variables.

```
%-----------------------------------------------------------------
% Here, values for the slider-crank mechanism variables W1, U1,
% V1 are assigned and variable G1 calculated.  Variables LW1,
% LU1 and LV1 are the scalar lengths of vectors W1, U1 and V1.

LW1 = 1;
theta = 45*pi/180;
W1 = LW1*exp(i*theta);

LU1 = 0;
U1 = i*LU1;

LV1 = 1.5;
rho = asin((LU1-LW1*sin(theta))/LV1);
V1 = LV1*exp(i*rho);

G1 = W1 + V1 - U1;
%-----------------------------------------------------------------


%-----------------------------------------------------------------
% Here, values for the start, step and stop displacement angles
% for the crank link are assigned.

start_ang   = 0;
step_ang    = 1;
stop_ang    = 360;
%-----------------------------------------------------------------


%-----------------------------------------------------------------
% Here, values for crank link angular velocity and angular
% acceleration are assigned.

angular_vel  = 100*ones(N+1,1);
angular_acc  = 10*ones(N+1,1);
%-----------------------------------------------------------------
```

FIGURE B.6
Sections of *Kinematic_ Slider_Crank.m* with sample values in bold.

After specifying the dimensions and driving link parameters in the file *Kinematic_ Slider_ Crank.m*, the next step is to run this file. When running this file, one file (filename *Disp_Vel_ Acc.csv*) is written to a folder named *Results* (in a format compatible with Microsoft Excel) that includes the calculated mechanism output at each crank link rotation increment.[*] The calculated mechanism output is included in Figure B.5b.

---

[*] At crank rotation increments where circuit defects occur, the text *Inf* is written to the *.csv* file.

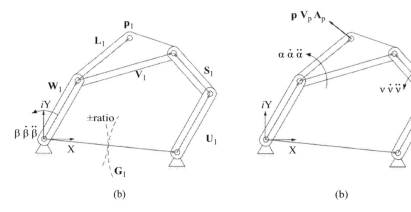

FIGURE B.7
(a) Geared five-bar mechanism and (b) output variables.

B.4 Geared Five-Bar Mechanism (Two Gears)

The Appendix B.4 folder (which is available for download at www.crcpress.com/product/isbn/9781498724937) includes the MATLAB file *Kinematic_5Bar_2Gears.m* for the kinematic analysis of geared five-bar mechanisms having two gears. To conduct a kinematic analysis, the user specifies the mechanism link dimensions and the crank motion parameters in this file. Values are specified for link variables W_1, V_1, G_1, U_1, L_1, S_1, and the gear ratio (Figure B.7a). Values are also specified for the initial crank angle (*start_ang*), the crank rotation increment (*step_ang*), and the final crank angle (*stop_ang*). Lastly, values are specified for the angular velocity (*angular_vel*) and angular acceleration (*angular_acc*) of the crank.[*] Figure B.8 illustrates the user-input sections of the file *Kinematic_Geared_5Bar.m*, with sample values in bold type.[†] While most of the link vector values in this figure appear in polar exponential form, they can all be specified in any of the rectangular and complex forms given in Equation 2.1.

After specifying the dimensions and driving link parameters in the file *Kinematic_Geared_ 5Bar.m*, the next step is to run this file. When running this file, one file (filename *Disp_ Vel_ Acc.csv*) is written to a folder named *Results* (in a format compatible with Microsoft Excel) that includes the calculated mechanism output at each crank link rotation increment.[‡] The calculated mechanism output is included in Figure B.7b.

B.5 Geared Five-Bar Mechanism (Three Gears)

The Appendix B.5 folder (which is available for download at www.crcpress.com/product/isbn/9781498724937) includes the MATLAB file *Kinematic_5Bar_3Gears.m* for the kinematic

---

[*] If no crank angular velocity or acceleration values are specified, time cannot be calculated and the text *NaN* is written to the time column of the *.csv* file.

[†] The units for crank angular rotation, velocity, and acceleration are degrees, rad/s, and rad/s², respectively.

[‡] At crank rotation increments where circuit defects occur, the text *Inf* is written to the *.csv* file.

```
%------------------------------------------------------------------
% Here, values for the geared 5-bar mechanism variables W1, V1,
% G1, U1, L1, S1 and gear ratio are assigned. Only negatiive
% gear ratio values should be specified.


W1 = 0.5*exp(i*90*pi/180);
V1 = 0.75*exp(i*32.7304*pi/180);
G1 = 0.75 + i*0;
U1 = 0.75*exp(i*45*pi/180);
L1 = 0.5*exp(-i*74.1400*pi/180);
S1 = 0.75*exp(i*149.9847*pi/180);
ratio = 2;
%------------------------------------------------------------------


%------------------------------------------------------------------
% Here, values for the start, step and stop displacement angles
% for the crank link are assigned.

start_ang   = 0;
step_ang    = 1;
stop_ang    = 360;
%------------------------------------------------------------------


%------------------------------------------------------------------
% Here, values for crank link angular velocity and angular
% acceleration are assigned.

angular_vel  = 1*ones(N,1);
angular_acc  = 0.1*ones(N,1);
%------------------------------------------------------------------
```

FIGURE B.8
Sections of *Kinematic_Geared_ 5Bar.m* with sample values in bold.

analysis of geared five-bar mechanisms having three gears. The procedure to use this MATLAB file is identical to the procedure given in Section B.4.

B.6 Watt II Mechanism

The Appendix B.6 folder (which is available for download at www.crcpress.com/product/isbn/9781498724937) includes the MATLAB file *Kinematic_Watt_II.m* for the kinematic analysis of Watt II mechanisms. To conduct a kinematic analysis, the user specifies the mechanism link dimensions and the crank motion parameters in this file. Values are specified for link variables W_1, V_1, G_1, U_1, L_1, W_1^*, V_1^*, G_1^*, U_1^*, and L_1^* (Figure B.9a). Values are also specified for the initial crank angle (*start_ang*), the crank rotation increment (*step_ang*), and the final crank angle (*stop_ang*). Lastly, values are specified for the angular velocity (*angular_vel*) and angular acceleration (*angular_acc*) of the crank.* Figure B.10 illustrates the user-input sections of the file *Kinematic_Watt_II.m*, with sample values in bold type.† While most of the link vector values in this figure appear in polar exponential form, they can all be specified in any of the rectangular and complex forms given in Equation 2.1.

---

* If no crank angular velocity or acceleration values are specified, time cannot be calculated and the text *NaN* is written to the time column of the *.csv* file.
† The units for crank angular rotation, velocity, and acceleration are degrees, rad/s, and rad/s^2, respectively.

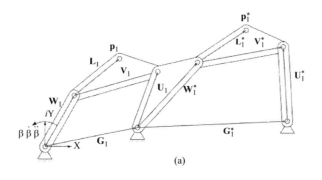

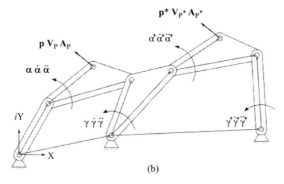

FIGURE B.9

(a) Watt II mechanism and (b) output variables.

```
%----------------------------------------------------------------
% Here, values for the Watt II mechanism variables W1, V1, G1,
% U1, L1, W1s, V1s, G1s, U1s and L1s are assigned.

W1 = 0.5*exp(i*90*pi/180);
V1 = 0.75*exp(i*19.3737*pi/180);
G1 = 0.75 + i*0;
U1 = 0.75*exp(i*93.2461*pi/180);
L1 = 0.5*exp(-i*60.7834*pi/180);

W1s = 0.5*exp(i*45*pi/180);
V1s = 0.75*exp(i*7.9416*pi/180);
G1s = 0.7244 - i*0.1941;
U1s = 0.75*exp(i*60.2717*pi/180);
L1s = 0.5*exp(i*49.3512*pi/180);
%----------------------------------------------------------------


%----------------------------------------------------------------
% Here, values for the start, step and stop displacement angles
% for the crank link are assigned.

start_ang   = 0;
step_ang    = -1;
stop_ang    = -360;
%----------------------------------------------------------------


%----------------------------------------------------------------
% Here, values for crank link angular velocity and angular
% acceleration are assigned.

angular_vel  = -1.5*ones(N,1);
angular_acc  = -0.25*ones(N,1);
%----------------------------------------------------------------
```

FIGURE B.10

Sections of *Kinematic_Watt_ II.m* with sample values in bold.

After specifying the dimensions and driving link parameters in the file *Kinematic_Watt_ II.m*, the next step is to run this file. When running this file, one file (filename *Disp_Vel_ Acc.csv*) is written to a folder named *Results* (in a format compatible with Microsoft Excel) that includes the calculated mechanism output at each crank link rotation increment.[*] The calculated mechanism output is included in Figure B.9b.

B.7 Stephenson III Mechanism

The Appendix B.7 folder (which is available for download at www.crcpress.com/product/ isbn/9781498724937) includes the MATLAB file *Kinematic_ Stephenson_ III.m* for the kinematic analysis of Stephenson III mechanisms. To conduct a kinematic analysis, the user specifies the mechanism link dimensions and the crank motion parameters in this file. Values are specified for link variables $\mathbf{W}_1$, $\mathbf{V}_1$, $\mathbf{G}_1$, $\mathbf{U}_1$, $\mathbf{L}_1$, $\mathbf{W}_1^*$, $\mathbf{V}_1^*$, $\mathbf{G}_1^*$, $\mathbf{U}_1^*$, and $\mathbf{L}_1^*$ (Figure B.11a). Values are also specified for the initial crank angle (*start_ang*), the crank rotation increment

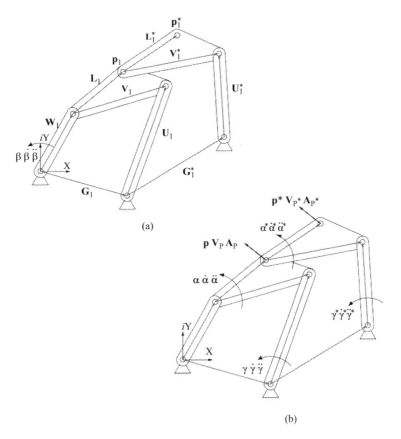

FIGURE B.11
(a) Stephenson III mechanism and (b) output variables.

---

[*] At crank rotation increments where circuit defects occur, the text *Inf* is written to the *.csv* file.

(*step_ang*), and the final crank angle (*stop_ang*). Lastly, values are specified for the angular velocity (*angular_vel*) and angular acceleration (*angular_acc*) of the crank.* Figure B.12 illustrates the user-input sections of the file *Kinematic_ Stephenson_ III.m*, with sample values in bold type.† While most of the link vector values in this figure appear in polar exponential form, they can all be specified in any of the rectangular and complex forms given in Equation 2.1.

After specifying the dimensions and driving link parameters in the file *Kinematic_ Stephenson_ III.m*, the next step is to run this file. When running this file, one file (filename *Disp_Vel_ Acc.csv*) is written to a folder named *Results* (in a format compatible with Microsoft Excel) that includes the calculated mechanism output at each crank link rotation increment.‡ The calculated mechanism output is included in Figure B.11b.

```
%--------------------------------------------------------------------
% Here, values for the Stephenson III mechanism variables W1,
% V1, G1, U1, L1, V1s, G1s, U1s and L1s are assigned.

W1 = 0.5*exp(i*90*pi/180);
V1 = 0.75*exp(i*19.3737*pi/180);
G1 = 0.75 + i*0;
U1 = 0.75*exp(i*93.2461*pi/180);
L1 = 0.5*exp(-i*60.7834*pi/180);

V1s = 1*exp(i*17.1417*pi/180);
G1s = 0.2159 + i*0.2588;
U1s = 1*exp(i*76.4844*pi/180);
L1s = 0.5*exp(i*63.7091*pi/180);
%-------------------------------------------------------------------


%-------------------------------------------------------------------
% Here, values for the start, step and stop displacement angles
% for the crank link are assigned.

start_ang   = 0;
step_ang    = -1;
stop_ang    = -360;
%-------------------------------------------------------------------


%-------------------------------------------------------------------
% Here, values for crank link angular velocity and angular
% acceleration are assigned.

angular_vel   = -1*ones(N,1);
angular_acc   = -0.25*ones(N,1);
%-------------------------------------------------------------------
```

FIGURE B.12
Sections of Kinematic_ *Stephenson_ III.m* with sample values in bold.

---

* If no crank angular velocity or acceleration values are specified, time cannot be calculated and the text *NaN* is written to the time column of the *.csv* file.
† The units for crank angular rotation, velocity, and acceleration are degrees, rad/s, and rad/s², respectively.
‡ At crank rotation increments where circuit defects occur, the text *Inf* is written to the *.csv* file.

Appendix C: User Instructions for Chapter 6 MATLAB® Files

C.1 Planar Four-Bar Mechanism

The Appendix C.1 folder (which is available for download at www.crcpress.com/product/isbn/9781498724937) includes the MATLAB file *Static_Planar_4Bar.m* for the static force analysis of planar four-bar mechanisms. To conduct a static force analysis, the user specifies the mechanism link dimensions, coupler force, gravitational constant, center of mass vectors, link masses, and the crank motion parameters in this file. Values are specified for link variables $\mathbf{W}_1$, $\mathbf{V}_1$, $\mathbf{G}_1$, $\mathbf{U}_1$, $\mathbf{L}_1$, the coupler force $\mathbf{F}_{\mathbf{p}_1}$, the gravitational constant $\mathbf{g}$, the center of mass vectors $\mathbf{R}_1$ through $\mathbf{R}_3$, and link masses m_1 through m_3 (Figure C.1a). Values are also specified for the initial crank angle (*start_ang*), the crank rotation increment (*step_ang*), and the final crank angle (*stop_ang*). Figure C.2 illustrates the user-input sections of the file *Static_Planar_4Bar.m*, with sample values in bold type.* While most of the link vector values in this figure appear in polar exponential form, they can all be specified in any of the rectangular and complex forms given in Equation 2.1.

After specifying the mechanism dimensions and driving link parameters in the file *Static_Planar_4Bar.m*, the next step is to run this file. When running this file, one file (filename *Static_Loads.csv*) is written to a folder named *Results* (in a format compatible with Microsoft® Excel) that includes the calculated mechanism output at each crank link rotation increment.† The calculated mechanism output is included in Figure C.1b.

C.2 Slider-Crank Mechanism

The Appendix C.2 folder (which is available for download at www.crcpress.com/product/isbn/9781498724937) includes the MATLAB file *Static_Slider_Crank.m* for the static force analysis of slider-crank mechanisms. To conduct a static force analysis, the user specifies the mechanism link dimensions, slider force, static friction coefficient, gravitational constant, center of mass vectors, link masses, and the crank motion parameters in this file. Values are specified for link variables $\mathbf{W}_1$, $\mathbf{V}_1$, $\mathbf{U}_1$, the slider force $\mathbf{F}$, the static friction coefficient $\pm \mu$, the gravitational constant $\mathbf{g}$, the center of mass vectors $\mathbf{R}_1$, $\mathbf{R}_2$, and link masses m_1 through m_3 (Figure C.3a). Values are also specified for the initial crank angle (*start_ang*), the crank rotation increment (*step_ang*), and the final crank angle (*stop_ang*). Figure C.4 illustrates the user-input sections of the file *Static_Slider_Crank.m*, with sample values in bold type.*

---

* The unit for crank angular rotation is *degree*.
† At crank rotation increments where circuit defects occur, the text *Inf* is written in the *.csv* file.

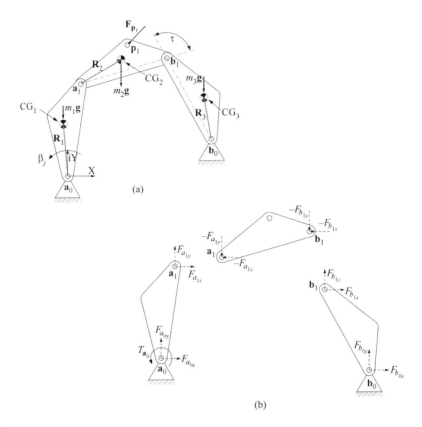

FIGURE C.1
(a) Planar four-bar mechanism and (b) output variables (angle τ shown in a).

```
%---------------------------------------------------------------
% Here, values for the planar 4-bar mechanism variables W1, V1,
% G1, U1 and L1 are assigned. The coupler force Fp1, gravity
% "g", center of mass vectors R1, R2 and R3 and link masses are
% specified here also.

W1 = 0.5*exp(i*90*pi/180);
V1 = 0.75*exp(i*19.3737*pi/180);
G1 = 0.75 + i*0;
U1 = 0.75*exp(i*93.2461*pi/180);
L1 = 0.5*exp(-i*60.7834*pi/180);

Fp1 = [0, 4500]; g = -9.81;

R1 = -0.0932 - i*0.0380;
R2 = 0.0955 + i*0.0159;
R3 = -0.1180 + i*0.1261;

m1 = 8; m2 = 40; m3 = 12;
%---------------------------------------------------------------

%---------------------------------------------------------------
% Here, values for the start, step and stop displacement angles
% for the crank link are assigned.

start_ang    = 0;
step_ang     = -1;
stop_ang     = -60;
%---------------------------------------------------------------
```

FIGURE C.2
Sections of *Static_Planar_4Bar.m* with sample values in bold.

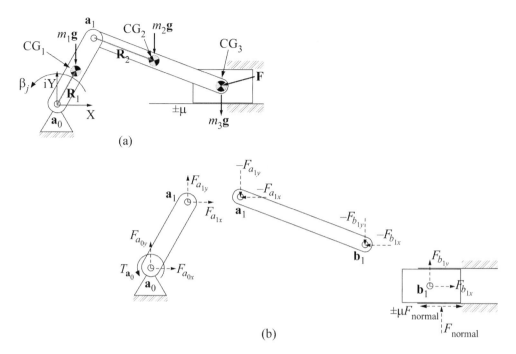

FIGURE C.3

(a) Slider-crank mechanism and (b) output variables.

```
%-------------------------------------------------------
% Here, values for the slider-crank mechanism variables W1, U1
% and V1 are assigned. The slier force F, friction coefficient
% "mu", gravity "g", center of mass vectors R1 and R2 and link
% masses are specified here also.

LW1 = 4;
theta = 45*pi/180;
W1 = LW1*exp(i*theta);

LU1 = 0;
U1 = i*LU1;

LV1 = 6;
rho = asin((LU1-LW1*sin(theta))/LV1);
V1 = LV1*exp(i*rho);

G1 = W1 + V1 - U1;

F = [-100, 0]; mu = 0.1; g = -9.81;

R1 = 0;
R2 = 0.0265 - i*0.0141;

m1 = 0.05; m2 = 0.025; m3 = 0.075;
%-------------------------------------------------------

%-------------------------------------------------------
% Here, values for the start, step and stop displacement angles
% for the crank link are assigned.

start_ang   = 0;
step_ang    = 1;
stop_ang    = 10;
%-------------------------------------------------------
```

FIGURE C.4

Sections of *Static_Slider_Crank.m* with sample values in bold.

After specifying the mechanism dimensions and driving link parameters in the file *Static_Slider_Crank.m*, the next step is to run this file. When running this file, one file (filename *Static_Loads.csv*) is written to a folder named *Results* (in a format compatible with Microsoft Excel) that includes the calculated mechanism output at each crank link rotation increment.* The calculated mechanism output is included in Figure C.3b.

C.3 Geared Five-Bar Mechanism (Two Gears)

The Appendix C.3 folder (which is available for download at www.crcpress.com/product/ isbn/9781498724937) includes the MATLAB file *Static_5Bar_2Gears.m* for the static force analysis of geared five-bar mechanisms having two gears. To conduct a static force analysis, the user specifies the mechanism link dimensions, gear ratio, intermediate link force, gravitational constant, center of mass vectors, link masses, and the crank motion parameters in this

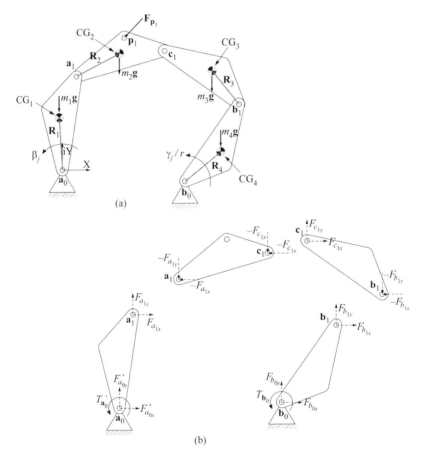

FIGURE C.5
(a) Geared five-bar mechanism and (b) output variables.

---

* At crank rotation increments where circuit defects occur, the text *Inf* is written in the *.csv* file.

```
%-----------------------------------------------------------
% Here, values for the geared 5-bar mechanism variables W1, V1,
% G1, U1, L1 and S1 are assigned. The link force Fp1, gear
% ratio, gravity "g", center of mass vectors R1, R2, R3 and R4
% and link masses are specified here also.
% Only negative gear ratio values should be specified.

W1 = 0.5*exp(i*90*pi/180);
V1 = 0.75*exp(i*32.7304*pi/180);
G1 = 0.75 + i*0;
U1 = 0.75*exp(i*45*pi/180);
L1 = 0.5*exp(-i*74.1400*pi/180);
S1 = 0.75*exp(i*149.9847*pi/180);

Fp1 = [-2500,-3000]; ratio = -2; g = -9.81;

R1 = 0 + i*0.0831;
R2 = 0.2558 + i*0.2955;
R3 = -0.3247 + i*0.1876;
R4 = 0.0356 + i*0.0356;

m1 = 22.54; m2 = 29.785; m3 = 12.075; m4 = 75.67;
%-----------------------------------------------------------

%-----------------------------------------------------------
% Here, values for the start, step and stop displacement angles
% for the crank link are assigned.

start_ang   = 0;
step_ang    = 1;
stop_ang    = 2;
%-----------------------------------------------------------
```

FIGURE C.6
Sections of *Static_5Bar_2Geare.m* with sample values in bold.

file. Values are specified for link variables $\mathbf{W}_1$, $\mathbf{V}_1$, $\mathbf{G}_1$, $\mathbf{U}_1$, $\mathbf{L}_1$, $\mathbf{S}_1$, the intermediate link force $\mathbf{F}_{p_1}$, the gear ratio, the gravitational constant $\mathbf{g}$, the center of mass vectors $\mathbf{R}_1$ through $\mathbf{R}_4$, and link masses m_1 through m_4 (Figure C.5a). Values are also specified for the initial crank angle (*start_ang*), the crank rotation increment (*step_ang*), and the final crank angle (*stop_ang*). Figure C.6 illustrates the user-input sections of the file *Static_5Bar_2Gear.m*, with sample values in bold type.[*] While most of the link vector values in this figure appear in polar exponential form, they can all be specified in any of the rectangular and complex forms given in Equation 2.1.

After specifying the mechanism dimensions and driving link parameters in the file *Static_5Bar_2Gear.m*, the next step is to run this file. When running this file, one file (filename *Static_Loads.csv*) is written to a folder named *Results* (in a format compatible with Microsoft Excel) that includes the calculated mechanism output at each crank link rotation increment.[†] The calculated mechanism output is included in Figure C.5b.

C.4 Geared Five-Bar Mechanism (Three Gears)

The Appendix C.4 folder (which is available for download at www.crcpress.com/product/ isbn/9781498724937) includes the MATLAB file *Static_5Bar_3Gears.m* for the static force analysis of geared five-bar mechanisms having three gears. The procedure to use this MATLAB file is identical to the procedure given in Section C.3.

---

[*] The unit for crank angular rotation is *degree*.
[†] At crank rotation increments where circuit defects occur, the text *Inf* is written in the *.csv* file.

C.5 Watt II Mechanism

The Appendix C.5 folder (which is available for download at www.crcpress.com/product/isbn/9781498724937) includes the MATLAB file *Static_Watt_II.m* for the static force analysis of Watt II mechanisms. To conduct a static force analysis, the user specifies the mechanism link dimensions, coupler forces, gravitational constant, center of mass vectors, link masses, and the crank motion parameters in this file. Values are specified for link variables $\mathbf{W}_1, \mathbf{V}_1, \mathbf{G}_1, \mathbf{U}_1, \mathbf{L}_1, \mathbf{W}_1^*, \mathbf{V}_1^*, \mathbf{G}_1^*, \mathbf{U}_1^*, \mathbf{L}_1^*$, the coupler forces $\mathbf{F}_{\mathbf{p}_1}$ and $\mathbf{F}_{\mathbf{p}_1^*}$, the gravitational constant $\mathbf{g}$, the center of mass vectors $\mathbf{R}_1$ through $\mathbf{R}_5$, and link masses m_1 through m_5 (Figure C.7a). Values are also specified for the initial crank angle (*start_ang*), the crank rotation increment (*step_ang*), and the final crank angle (*stop_ang*). Figure C.8 illustrates the user-input sections

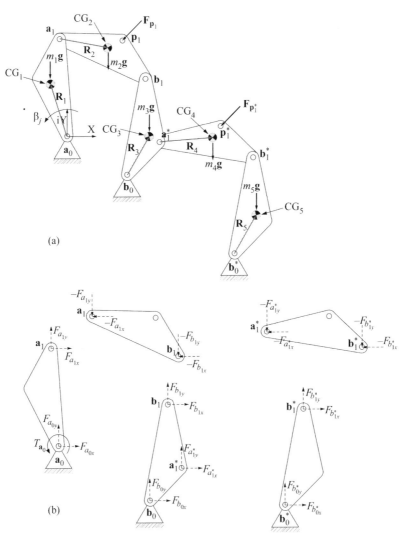

(a)

(b)

FIGURE C.7
(a) Watt II mechanism and (b) output variables.

```
%------------------------------------------------------------
% Here, values for the Watt II mechanism variables W1, V1, G1,
% U1, L1, W1s, V1s, G1s, U1s and L1s are assigned.  The coupler
% forces Fp1 and Fp1s, gravity "g", the center of mass vectors
% R1, R2, R3, R4 and R5 and link masses are specified here also.

W1 = 0.5*exp(i*90*pi/180);
V1 = 0.75*exp(i*19.3737*pi/180);
G1 = 0.75 + i*0;
U1 = 0.75*exp(i*93.2461*pi/180);
L1 = 0.5*exp(-i*60.7834*pi/180);

W1s = 0.5*exp(i*45*pi/180);
V1s = 0.75*exp(i*7.9416*pi/180);
G1s = 0.7244 - i*0.1941;
U1s = 0.75*exp(i*60.2717*pi/180);
L1s = 0.5*exp(i*49.3512*pi/180);

Fp1 = [2500, 3000]; Fp1s = [-1500, 2000]; g = -9.81;

R1 = 0 + i*0.25;
R2 = 0.3172 + i*0.2284;
R3 = 0.1037 + i*0.3675;
R4 = 0.3562 + i*0.161;
R5 = 0.1860 + i*0.3257;

m1 = 8.05; m2 = 29.785; m3 = 33.81; m4 = 29.785; m5 = 12.075;
%------------------------------------------------------------

%------------------------------------------------------------
% Here, values for the start, step and stop displacement angles
% for the crank link are assigned.

start_ang   = 0;
step_ang    = 1;
stop_ang    = 100;
%------------------------------------------------------------
```

FIGURE C.8
Sections of *Static_Watt_II.m* with sample values in bold.

of the file *Static_Wattr_II.m*, with sample values in bold type.* While most of the link vector values in this figure appear in polar exponential form, they can all be specified in any of the rectangular and complex forms given in Equation 2.1.

After specifying the mechanism dimensions and driving link parameters in the file *Static_Watt_II.m*, the next step is to run this file. When running this file, one file (filename *Static_Loads.csv*) is written to a folder named *Results* (in a format compatible with Microsoft Excel) that includes the calculated mechanism output at each crank link rotation increment.† The calculated mechanism output is included in Figure C.7b.

C.6 Stephenson III Mechanism

The Appendix C.6 folder (which is available for download at www.crcpress.com/product/isbn/9781498724937) includes the MATLAB file *Static_Stephenson_III.m* for the static force analysis of Stephenson III mechanisms. To conduct a static force analysis, the user specifies

---

* The unit for crank angular rotation is *degree*.
† At crank rotation increments where circuit defects occur, the text *Inf* is written in the *.csv* file.

the mechanism link dimensions, intermediate link force, gravitational constant, center of mass vectors, link masses, and the crank motion parameters in this file. Values are specified for link variables $\mathbf{W}_1, \mathbf{V}_1, \mathbf{G}_1, \mathbf{U}_1, \mathbf{V}_1^*, \mathbf{G}_1^*, \mathbf{U}_1^*, \mathbf{L}_1^*$ and the intermediate link force $\mathbf{F}_{\mathbf{p}_1^*}$, the gravitational constant $\mathbf{g}$, the center of mass vectors $\mathbf{R}_1$ through $\mathbf{R}_5$, and link masses m_1 through m_5 (Figure C.9a). Values are also specified for the initial crank angle (*start_ang*), the crank rotation increment (*step_ang*), and the final crank angle (*stop_ang*). Figure C.10 illustrates the user-input sections of the file *Static_Stephenson_III.m*, with sample values

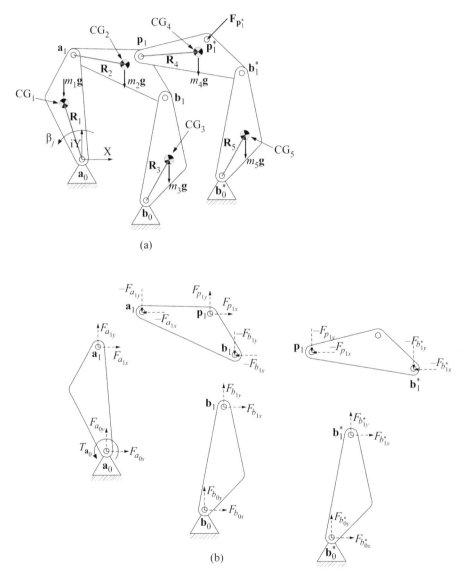

(a)

(b)

FIGURE C.9
(a) Stephenson III mechanism and (b) output variables.

```
%--------------------------------------------------------------
% Here, values for the Stephenson III mechanism variables W1,
% V1, G1, U1, L1, V1s, G1s, U1s and L1s are assigned.  The
% coupler force Fp1s, gravity "g", the center of mass vectors
% R1, R2, R3, R4 and R5 and link masses are specified here also.

W1 = 1.3575*exp(-i*64.4543*pi/180);
V1 = 0.9726*exp(i*57.2740*pi/180);
G1 = 0.9207 - i*2.2989;
U1 = 1.9019*exp(i*84.2513*pi/180);
L1 = 0.6120*exp(-i*143.6057*pi/180);

V1s = 0.5815*exp(-i*125.7782*pi/180);
G1s = -3.3894 + i*2.2487;
U1s = 2.9955*exp(-i*42.1315*pi/180);
L1s = 2.2217*exp(-i*5*pi/180);

Fp1s = [0, -40]; g = -9.81;

R1 = 0.3846 - i*0.3412;
R2 = -0.6343 + i*1.0364;
R3 = -0.2988 - i*0.3639;
R4 = 1.1487 + i*0.0597;
R5 = -0.2384 - i*0.3618;

m1 = 4; m2 = 8; m3 = 4; m4 = 12; m5 = 4;
%--------------------------------------------------------------

%--------------------------------------------------------------
% Here, values for the start, step and stop displacement angles
% for the crank link are assigned.

start_ang   = 0;
step_ang    = 1;
stop_ang    = 40;
%--------------------------------------------------------------
```

FIGURE C.10
Sections of *Static_Stephenson_III.m* with sample values in bold.

in bold type.[*] While most of the link vector values in this figure appear in polar exponential form, they can all be specified in any of the rectangular and complex forms given in Equation 2.1.

After specifying the mechanism dimensions and driving link parameters in the file *Static_Stephenson_III.m*, the next step is to run this file. When running this file, one file (filename *Static_Loads.csv*) is written to a folder named *Results* (in a format compatible with Microsoft Excel) that includes the calculated mechanism output at each crank link rotation increment.[†] The calculated mechanism output is included in Figure C.9b.

---

[*] The unit for crank angular rotation is *degree*.
[†] At crank rotation increments where circuit defects occur, the text *Inf* is written in the *.csv* file.

Appendix D: User Instructions for Chapter 7 MATLAB® Files

D.1 Planar Four-Bar Mechanism

The Appendix D.1 folder (which is available for download at www.crcpress.com/product/isbn/9781498724937) includes the MATLAB file *Dynamic_Planar_4Bar.m* for the dynamic force analysis of planar four-bar mechanisms. To conduct a dynamic force analysis, the user specifies the mechanism link dimensions, coupler force and follower torque, gravitational constant, center of mass vectors, link masses, link mass moments of inertia, and the crank motion parameters in this file. Values are specified for link variables W_1, V_1, G_1, U_1, L_1, the coupler force F_{p1}, the follower torque T_{b0}, the gravitational constant g, the center of mass vectors R_1, R_3, and R_7, link masses m_1 through m_3, and link mass moments of inertia I_1 through I_3 (Figure D.1a). Values are also specified for the initial crank angle (*start_ang*), the crank rotation increment (*step_ang*), and the final crank angle (*stop_ang*). Lastly, values are specified for the angular velocity (*angular_vel*) and angular acceleration (*angular_acc*) of the crank. Figure D.2 illustrates the user-input sections of the file *Dynamic_Planar_4Bar.m*, with sample values in bold type.[*] While most of the link vector values in this figure appear in polar exponential form, they can all be specified in any of the rectangular and complex forms given in Equation 2.1.

After specifying the mechanism dimensions, loads, mass parameters, and driving link parameters in the file *Dynamic_Planar_4Bar.m*, the next step is to run this file. When running this file, one file (filename *Dynamic_Loads.csv*) is written to a folder named *Results* (in a format compatible with Microsoft® Excel) that includes the calculated mechanism output at each crank link rotation increment.[†] The calculated mechanism output is included in Figure D.1b.

D.2 Slider-Crank Mechanism

The Appendix D.2 folder (which is available for download at www.crcpress.com/product/isbn/9781498724937) includes the MATLAB file *Dynamic_Slider_Crank.m* for the dynamic force analysis of slider-crank mechanisms. To conduct a dynamic force analysis, the user specifies the mechanism link dimensions, slider force, dynamic friction coefficient, gravitational constant, center of mass vectors, link masses, link mass moments of inertia, and the crank motion parameters in this file. Values are specified for link variables W_1, V_1, U_1,

---

[*] The units for crank angular rotation, velocity, and acceleration are degree, rad/s, and rad/s², respectively.
[†] At crank rotation increments where circuit defects occur, the text *Inf* is written in the *.csv* file.

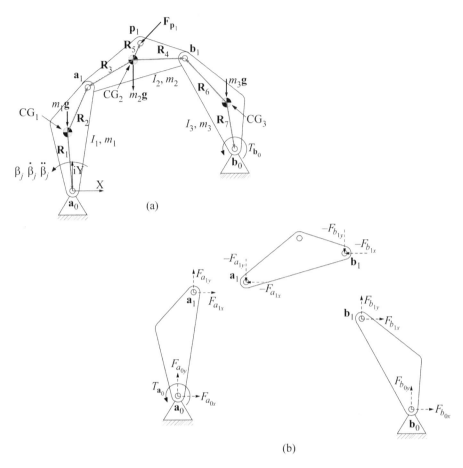

FIGURE D.1
(a) Planar four-bar mechanism and (b) output variables (angle τ not shown).

the slider force **F**, the dynamic friction coefficient $\pm\mu$, the gravitational constant **g**, the center of mass vectors **R**$_1$ and **R**$_3$, link masses m_1 through m_3, and link mass moments of inertia I_1 and I_2 (Figure D.3a). Values are also specified for the initial crank angle (*start_ang*), the crank rotation increment (*step_ang*), and the final crank angle (*stop_ang*). Lastly, values are specified for the angular velocity (*angular_vel*) and angular acceleration (*angular_acc*) of the crank. Figure D.4 illustrates the user-input sections of the file *Dynamic_Slider_Crank.m*, with sample values in bold type.[*] While some the link vector values in this figure appear in rectangular form, they can all be specified in any of the rectangular and complex forms given in Equation 2.1.

After specifying the mechanism dimensions and driving link parameters in the file *Dynamic_Slider_Crank.m*, the next step is to run this file. When running this file, one file (filename *Dynamic_Loads.csv*) is written to a folder named *Results* (in a format compatible with Microsoft Excel) that includes the calculated mechanism output at each crank link rotation increment.[†] The calculated mechanism output is included in Figure D.3b.

---

[*] The units for crank angular rotation, velocity, and acceleration are degree, rad/s, and rad/s², respectively.
[†] At crank rotation increments where circuit defects occur, the text *Inf* is written to the *.csv* file.

```
%------------------------------------------------------------------
% Here, values for the planar 4-bar mechanism variables W1, V1,
% G1, U1 and L1 are assigned. The coupler force Fp1, follower
% torque Tb0, gravity "g", center of mass vectors R1, R3 and
% R7, link masses and link mass moments of inertia are
% specified here also.

W1 = 0.5*exp(i*90*pi/180);
V1 = 0.75*exp(i*19.3737*pi/180);
G1 = 0.75 + i*0;
U1 = 0.75*exp(i*93.2461*pi/180);
L1 = 0.5*exp(-i*60.7834*pi/180);

Fp1 = [0, 0]; Tb0 = 0; g = -9.81;

R1 = 0 - i*0.25;
R3 = -0.3172 - i*0.2284;
R7 = 0.0212 - i*0.3744;

R2 = R1 + W1;  R4 = R3 + V1;  R5 = R3 + L1;  R6 = R7 + U1;

m1 = 8.05;   I1 = 0.805;
m2 = 29.785; I2 = 5.635;
m3 = 12.075; I3 = 2.415;
%------------------------------------------------------------------

%------------------------------------------------------------------
% Here, values for the start, step and stop displacement angles
% for the crank link are assigned.

start_ang   = 0;
step_ang    = 1;
stop_ang    = 360;
%------------------------------------------------------------------

%------------------------------------------------------------------
% Here, values for crank link angular velocity and angular
% acceleration are assigned.

angular_vel   = 1*ones(N,1);
angular_acc   = 0*ones(N,1);
%------------------------------------------------------------------
```

FIGURE D.2
Sections of *Dynamic_Planar_4Bar.m* with sample values in bold.

D.3 Geared Five-Bar Mechanism (Two Gears)

The Appendix D.3 folder (which is available for download at www.crcpress.com/product/isbn/9781498724937) includes the MATLAB file *Dynamic_5Bar_2Gear.m* for the dynamic force analysis and simulation of geared five-bar mechanisms having two gears. To conduct a dynamic force analysis, the user specifies the mechanism link dimensions, gear ratio, intermediate link force, gravitational constant, center of mass vectors, link masses and mass moments of inertia, and the crank motion parameters in this file. Values are specified for link variables $\mathbf{W}_1$, $\mathbf{V}_1$, $\mathbf{G}_1$, $\mathbf{U}_1$, $\mathbf{L}_1$, $\mathbf{S}_1$, the gear ratio, the intermediate link force $\mathbf{F}_{p1}$, the gravitational constant $\mathbf{g}$, the center of mass vectors $\mathbf{R}_1$, $\mathbf{R}_3$, $\mathbf{R}_7$, and $\mathbf{R}_9$, link masses m_1 through m_4, and link mass moments of inertia I_1 through I_4 (Figure D.5a). Values are also specified for the initial crank angle (*start_ang*), the crank rotation increment (*step_ang*), and the final crank angle (*stop_ang*). Lastly, values are specified for the angular velocity (*angular_vel*) and angular acceleration (*angular_acc*) of the crank.

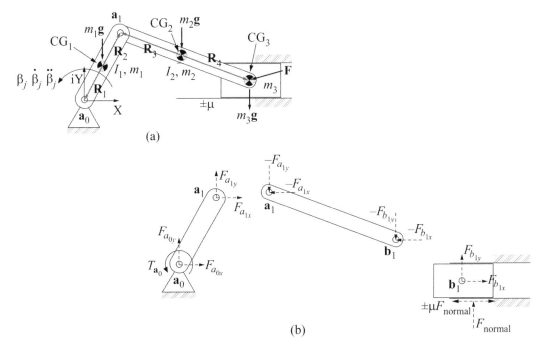

FIGURE D.3
(a) Slider-crank mechanism and (b) output variables.

Figure D.6 illustrates the user-input sections of the file *Dynamic_5Bar_2Gear.m*, with sample values in bold type.* While most of the link vector values in this figure appear in polar exponential form, they can all be specified in any of the rectangular and complex forms given in Equation 2.1.

After specifying the mechanism dimensions, loads, mass parameters, and driving link parameters in the file *Dynamic_5Bar_2Gear.m*, the next step is to run this file. When running this file, one file (filename *Dynamic_Loads.csv*) is written to a folder named *Results* (in a format compatible with Microsoft Excel) that includes the calculated mechanism output at each crank link rotation increment.† The calculated mechanism output is included in Figure D.5b.

D.4 Geared Five-Bar Mechanism (Three Gears)

The Appendix DJ.4 folder (which is available for download at www.crcpress.com/product/isbn/9781498724937) includes the MATLAB file *Dynamic_5Bar_3Gear.m* for the dynamic force analysis and simulation of geared five-bar mechanisms having three gears. The procedure to use this MATLAB file is identical to the procedure given in Section D.3.

---

* The units for crank angular rotation, velocity, and acceleration are degree, rad/s, and rad/s², respectively.
† At crank rotation increments where circuit defects occur, the text *Inf* is written to the *.csv* file.

```
%----------------------------------------------------------------
% Here, values for the planar 4-bar mechanism variables W1, V1,
% G1, U1 and L1 are assigned and/or calculated. Variables LW1,
% LU1 and LV1 are the scalar lengths of vectors W1, U1 and V1.
% The slider force "F", dynamic friction coefficient "mu" and
% gravity "g" are assigned here also. Lastly, the center of mass
% vectors R1 and R3, link masses and link mass moments of
% inertia are specified here.

LW1 = 0.5; theta = 90*pi/180;
W1 = LW1*exp(i*theta);

LU1 = 0;
U1 = i*LU1;

LV1 = 0.9014;
rho = asin((LU1-LW1*sin(theta))/LV1);
V1 = LV1*exp(i*rho);

G1 = W1 + V1 - U1;

F = [0, 0]; mu = 0.5; g = -9.81;

R1 = 0 - i*0.25;
R3 = -0.3750 + i*0.25;

R2 = R1 + W1; R4 = R3 + V1;

m1 = 8.05; I1 = 0.805;
m2 = 14.49; I2 = 4.025; m3 = 30;
%----------------------------------------------------------------

%----------------------------------------------------------------
% Here, values for the start, step and stop displacement angles
% for the crank link are assigned.

start_ang   = 0;
step_ang    = 1;
stop_ang    = 10;
%----------------------------------------------------------------

%----------------------------------------------------------------
% Here, values for crank link angular velocity and angular
% acceleration are assigned.

angular_vel   = 10*ones(N,1);
angular_acc   = 0*ones(N,1);
%----------------------------------------------------------------
```

FIGURE D.4

Sections of *Dynamic_Slider_Crank.m* with sample values in bold.

D.5 Watt II Mechanism

The Appendix D.5 folder (which is available for download at www.crcpress.com/product/ isbn/9781498724937) includes the MATLAB file *Dynamic_Watt_II.m* for the dynamic force analysis and simulation of Watt II mechanisms. To conduct a dynamic force analysis, the user specifies the mechanism link dimensions, coupler link forces and follower torques, gravitational constant, center of mass vectors, link masses and mass moments of inertia, and the crank motion parameters in this file. Values are specified for link variables $\mathbf{W}_1$, $\mathbf{V}_1$, $\mathbf{G}_1$, $\mathbf{U}_1$, $\mathbf{L}_1$, $\mathbf{W}_1^*$, $\mathbf{V}_1^*$, $\mathbf{G}_1^*$, $\mathbf{U}_1^*$, $\mathbf{L}_1^*$, the coupler forces $\mathbf{F}_{p_1}$ and $\mathbf{F}_{p_1^*}$, the follower torques T_{b_0} and $T_{b_0^*}$, the gravitational constant $\mathbf{g}$, the center of mass vectors $\mathbf{R}_1$, $\mathbf{R}_3$, $\mathbf{R}_7$, $\mathbf{R}_9$, and $\mathbf{R}_{13}$, link masses m_1 through m_5, and link mass moments of inertia I_1 through I_5 (Figure D.7a). Values are also specified for the initial crank angle (*start_ang*), the crank rotation increment (*step_ang*), and the final crank angle (*stop_ang*). Lastly, values are specified for the angular velocity (*angular_vel*) and angular acceleration (*angular_acc*) of the crank. Figure D.8 illustrates the

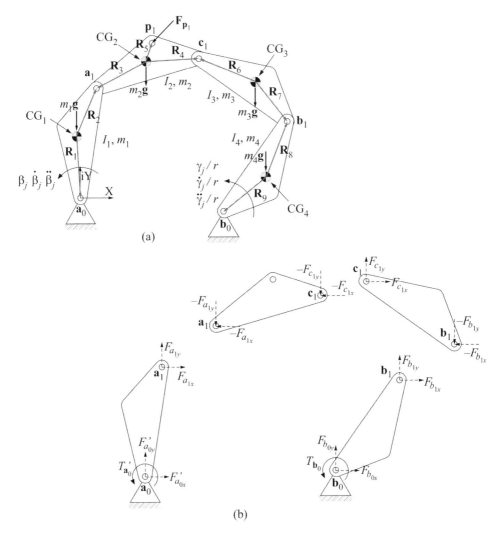

FIGURE D.5
(a) Geared five-bar mechanism and (b) output variables.

user-input sections of the file *Dynamic_Watt_II.m*, with sample values in bold type.* While most of the link vector values in this figure appear in polar exponential form, they can all be specified in any of the rectangular and complex forms given in Equation 2.1.

After specifying the mechanism dimensions, loads, mass parameters, and driving link parameters in the file *Dynamic_Watt_II.m*, the next step is to run this file. When running this file, one file (filename *Dynamic_Loads.csv*) is written to a folder named *Results* (in a format compatible with Microsoft Excel) that includes the calculated mechanism output at each crank link rotation increment.† The calculated mechanism output is included in Figure D.7b.

---

* The units for crank angular rotation, velocity, and acceleration are degree, rad/sec, and rad/sec², respectively.
† At crank rotation increments where circuit defects occur, the text *Inf* is written to the *.csv* file.

```
%-------------------------------------------------------------
% Here, values for the geared 5-bar mechanism variables W1, V1,
% G1, U1, L1 and S1 are assigned. The link force Fp1, gear
% ratio, gravity "g", center of mass vectors R1, R3, R7 and R9,
% link masses and link mass moments of inertia are specified
% here also. Only negative gear ratio values should be
% specified.

W1 = 0.5*exp(i*90*pi/180);
V1 = 0.75*exp(i*32.7304*pi/180);
G1 = 0.75*exp(i*0*pi/180);
U1 = 0.75*exp(i*45*pi/180);
L1 = 0.5*exp(-i*74.1400*pi/180);
S1 = 0.75*exp(i*149.9847*pi/180);

Fp1 = [0,-1000]; ratio = -2; g = -9.81;

R1 = 0 - i*0.0831; R3 = -0.2558 - i*0.2955;
R7 = 0.3247 - i*0.1876; R9 = -0.0356 - i*0.0356;

R2 = R1 + W1; R4 = R3 + V1; R5 = R3 + L1; R6 = R7 + S1;
R8 = R9 + U1;

m1 = 22.54; I1 = 0.505;
m2 = 29.785; I2 = 5.635;
m3 = 12.075; I3 = 2.415;
m4 = 75.67; I4 = 5.635;
%-------------------------------------------------------------

%-------------------------------------------------------------
% Here, values for the start, step and stop displacement angles
% for the crank link are assigned.

start_ang  = 0;
step_ang   = 1;
stop_ang   = 360;
%-------------------------------------------------------------

%-------------------------------------------------------------
% Here, values for crank link angular velocity and angular
% acceleration are assigned.

angular_vel  = 1*ones(N,1);
angular_acc  = 0*ones(N,1);
%-------------------------------------------------------------
```

FIGURE D.6
Sections of *Dynamic_5Bar_2Gear.m* with sample values in bold.

D.6 Stephenson III Mechanism

The Appendix D.6 folder (which is available for download at www.crcpress.com/product/isbn/9781498724937) includes the MATLAB file *Dynamic_Stephenson_III.m* for the dynamic force analysis and simulation of Stephenson III mechanisms. To conduct a dynamic force analysis, the user specifies the mechanism link dimensions, coupler link force and follower torques, gravitational constant, center of mass vectors, link masses and mass moments of inertia, and the crank motion parameters in this file. Values are specified for link variables $\mathbf{W}_1$, $\mathbf{V}_1$, $\mathbf{G}_1$, $\mathbf{U}_1$, $\mathbf{L}_1$, $\mathbf{V}_1^*$, $\mathbf{G}_1^*$, $\mathbf{U}_1^*$, $\mathbf{L}_1^*$ the coupler force $\mathbf{F}_{\mathbf{p}1}$, the follower torques T_{b_0} and $T_{b_0^*}$, the gravitational constant $\mathbf{g}$, the center of mass vectors $\mathbf{R}_1$, $\mathbf{R}_3$, $\mathbf{R}_7$, $\mathbf{R}_8$, and $\mathbf{R}_{12}$, link masses m_1 through m_5, and link mass moments of inertia I_1 through I_5 (Figure D.9a). Values are also specified for the initial crank angle (*start_ang*), the crank rotation increment (*step_ang*), and the final crank angle (*stop_ang*). Lastly, values are specified for

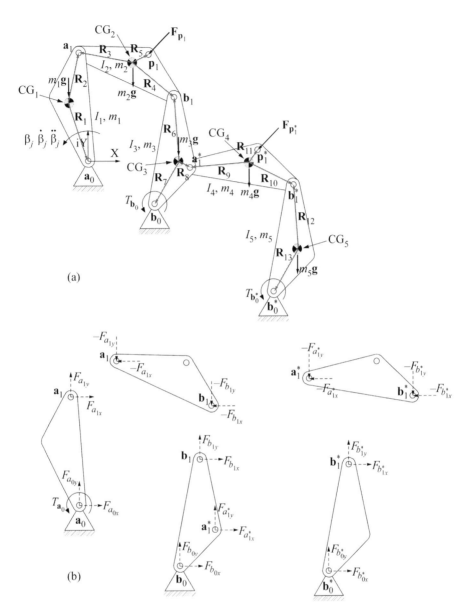

FIGURE D.7
(a) Watt II mechanism and (b) output variables.

the angular velocity (*angular_vel*) and angular acceleration (*angular_acc*) of the crank. Figure D.10 illustrates the user-input sections of the file *Dynamic_Stephenson_III.m*, with sample values in bold type.[*] While most of the link vector values in this figure appear in polar exponential form, they can all be specified in any of the rectangular and complex forms given in Equation 2.1.

_____

[*] The units for crank angular rotation, velocity, and acceleration are degree, rad/s, and rad/s², respectively

```
%-----------------------------------------------------------
% Here, values for the Watt II mechanism variables W1, V1, G1,
% U1, L1, W1s, V1s, G1s, U1s and L1s are assigned. The coupler
% forces Fp1 and Fp1s and follower torques Tb0 and Tb0s, gravity
% "g", the center of mass vectors R1, R3, R7, R9 and R13, link
% masses and link mass moments of inertia are specified here
% also.

W1 = 0.5*exp(i*90*pi/180);
V1 = 0.75*exp(i*19.3737*pi/180);
G1 = 0.75 + i*0;
U1 = 0.75*exp(i*93.2461*pi/180);
L1 = 0.5*exp(-i*60.7834*pi/180);

W1s = 0.5*exp(i*45*pi/180);
V1s = 0.75*exp(i*7.9416*pi/180);
G1s = 0.7244 - i*0.1941;
U1s = 0.75*exp(i*60.2717*pi/180);
L1s = 0.5*exp(i*49.3512*pi/180);

Fp1 = [-500, -500]; Tb0 = 0;
Fp1s = [-1000, 0]; Tb0s = 0; g = -9.81;

R1 = 0 - i*0.25; R3 = -0.3172 - i*0.2284;
R7 = -0.1037 - i*0.3675; R9 = -0.3562 - i*0.161;
R13 = -0.1860 - i*0.3257;

R2 = R1 + W1; R4 = R3 + V1; R5 = R3 + L1; R6 = R7 + U1;
R8 = R7 + W1s; R10 = R9 + V1s; R11 = R9 + L1s; R12 = R13 + U1s;

m1 = 8.05; I1 = 0.805;
m2 = 29.785; I2 = 5.635;
m3 = 33.81; I3 = 5.635;
m4 = 29.785; I4 = 5.635;
m5 = 12.075; I5 = 2.415;
%-----------------------------------------------------------

%-----------------------------------------------------------
% Here, values for the start, step and stop displacement angles
% for the crank link are assigned.

start_ang   = 0;
step_ang    = -1;
stop_ang    = -360;
%-----------------------------------------------------------

%-----------------------------------------------------------
% Here, values for crank link angular velocity and angular
% acceleration are assigned.

angular_vel = -1.5*ones(N,1);
angular_acc = -0.25*ones(N,1);
%-----------------------------------------------------------
```

FIGURE D.8
Sections of *Dynamic_Watt_II.m* with sample values in bold.

After specifying the mechanism dimensions, loads, mass parameters, and driving link parameters in the file *Dynamic_Stephenson_III.m*, the next step is to run this file. When running this file, one file (filename *Dynamic_Loads.csv*) is written to a folder named *Results* (in a format compatible with Microsoft Excel) that includes the calculated mechanism output at each crank link rotation increment.[*] The calculated mechanism output is included in Figure D.9b.

---

[*] At crank rotation increments where circuit defects occur, the text *Inf* is written to the *.csv* file.

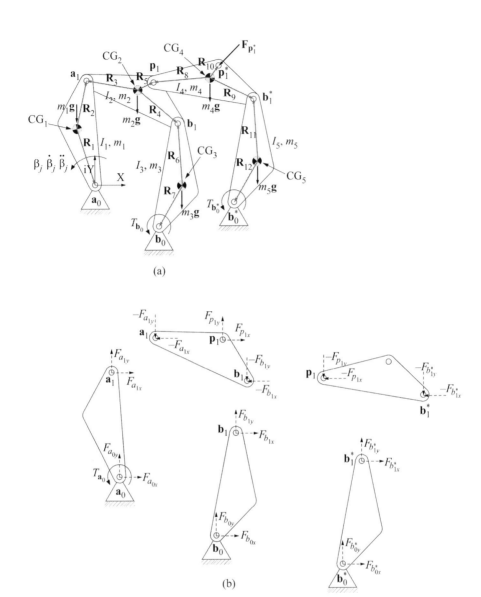

(a)

(b)

FIGURE D.9
(a) Stephenson III mechanism and (b) output variables.

```
%-----------------------------------------------------------------
% Here, values for the Stephenson III mechanism variables W1,
% V1, G1, U1, L1, W1s, V1s, G1s, U1s and L1s are assigned. The
% coupler force Fp1s and follower torques Tb0 and Tb0s,
% gravity "g", the center of mass vectors R1, R3, R7, R9 and
% R13, link masses and link mass moments of inertia are
% specified here also.

W1 = 0.5*exp(i*90*pi/180);
V1 = 0.75*exp(i*19.3737*pi/180);
G1 = 0.75 + i*0;
U1 = 0.75*exp(i*93.2461*pi/180);
L1 = 0.5*exp(-i*60.7834*pi/180);

V1s = 1*exp(i*17.1417*pi/180);
G1s = 0.2159 + i*0.2588;
U1s = 1*exp(i*76.4844*pi/180);
L1s = 0.5*exp(i*63.7091*pi/180);

Fp1s = [-1000, 0]; Tb0 = 0; Tb0s = 0; g = -9.81;

R1 = 0 - i*0.25; R3 = -0.3172 - i*0.2284;
R7 = 0.0212 - i*0.3744; R8 = -0.3923 - i*0.2477;
R12 = -0.1169 - i*0.4862;

R2 = R1 + W1; R4 = R3 + V1; R5 = R3 + L1; R6 = R7 + U1;
R9 = R8 + V1s; R10 = R8 + L1s; R11 = R12 + U1s;

m1 = 8.05; I1 = 0.805;
m2 = 29.785; I2 = 5.635;
m3 = 33.81; I3 = 5.635;
m4 = 29.785; I4 = 5.635;
m5 = 12.075; I5 = 2.415;
%-----------------------------------------------------------------

%-----------------------------------------------------------------
% Here, values for the start, step and stop displacement angles
% for the crank link are assigned.

start_ang   = 0;
step_ang    = -1;
stop_ang    = -360;
%-----------------------------------------------------------------

%-----------------------------------------------------------------
% Here, values for crank link angular velocity and angular
% acceleration are assigned.

angular_vel  = -1*ones(N,1);
angular_acc  = -0.25*ones(N,1);
%-----------------------------------------------------------------
```

FIGURE D.10

Sections of *Dynamic_Stephenson_III.m* with sample values in bold.

Appendix E: User Instructions for Chapter 9 MATLAB® Files

E.1 S, V Profile Generation and Cam Design: Constant Velocity Motion

The Appendix E.1 folder (which is available for download at www.crcpress.com/product/ isbn/9781498724937) includes the MATLAB file *Cam_Const_Vel.m* for the kinematic analysis and design of radial cams for constant velocity motion. To conduct a constant velocity-based kinematic analysis, the user specifies the follower displacement event sequence, the displacement event ranges, the end values for each displacement event, the cam rotation speed, and the cam base circle radius.*

To demonstrate how values are specified in *Cam_Const_Vel.m*, Figure E.1 illustrates an arbitrary follower rise-dwell-fall-dwell displacement event sequence, while Figure E.2 includes the corresponding values (in bold type) specified in *Cam_Const_Vel.m*.†

Rise, fall, and dwell events are specified using R, F, and D (in single quotes in a row matrix), respectively. Because Figure E.1 presents a rise-dwell-fall-dwell sequence, the sequence (R, D, F, D) appears in Figure E.2. Next, the rotation ranges for each displacement event appear in Figure E.2 (representing [β_1, β_2, β_3, β_4]). The rotation ranges should be integers and have a sum of 360°. Next, the displacement value at the end of each displacement event appears. Because the profile in Figure E.1 achieves an arbitrary maximum displacement of 2, the sequence (2, 2, 0, 0) (representing [s_1, s_2, s_3, s_4]) appears in Figure E.2. Lastly, both the cam rotation speed and base radius appear in Figure E.2.

After specifying the follower displacement event sequence parameters, cam rotation speed, and cam base circle radius in the file *Cam_Const_Vel.m*, the next step is to run this file. When running this file, one file (filename *DVAJ.csv*) is written to a folder named *Results* (in a format compatible with Microsoft® Excel) that includes the calculated follower displacement and velocity curve data, and the corresponding cam profile data, at 1° increments.

E.2 S, V, A Profile Generation and Cam Design: Constant Acceleration Motion

The Appendix E.2 folder (which is available for download at www.crcpress.com/product/ isbn/9781498724937) includes the MATLAB file *Cam_Const_Acc.m* for the kinematic

---

* The units for the displacement interval range and cam rotation speed are degrees and rad/s, respectively.
† While this figure illustrates a rise-dwell-fall-dwell displacement event sequence (for demonstration), the user can specify displacement event sequences of any number and any combination of displacement events.

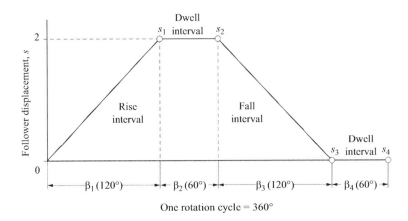

FIGURE E.1
Follower *rise-dwell-fall-dwell* displacement event sequence.

```
%----------------------------------------------------------------
% Here, values for the follower displacement event sequence, the
% corresponding follower displacement event ranges and the end
% value for each follower displacement event are assigned.  The
% cam rotational speed and base circle radius are assigned here
% also.

Event = ['R','D','F','D'];
Beta = [120,60,120,60];
Si = [2,2,0,0];

dTheta = 1.1;
Rbase = 5;
%----------------------------------------------------------------
```

FIGURE E.2
Sections of *Cam_Const_Vel.m* with sample values in bold.

analysis and design of radial cams for constant acceleration motion. To conduct a constant acceleration-based kinematic analysis, the user specifies the follower displacement event sequence, the displacement event ranges, the end values for each displacement event, the cam rotation speed, and the cam base circle radius.[*]

To demonstrate how values are specified in *Cam_Const_Acc.m*, Figure E.3 illustrates an arbitrary follower rise-dwell-fall-dwell displacement event sequence, while Figure E.4 includes the corresponding values (in bold type) specified in *Cam_Const_Acc.m*.[†]

Rise, fall, and dwell events are specified using *R*, *F*, and *D* (in single quotes in a row matrix), respectively. Because Figure E.3 presents a rise-dwell-fall-dwell sequence, the sequence (*R*, *D*, *F*, *D*) appears in Figure E.4. Next, the rotation ranges for each displacement event appear in Figure E.4 (representing [β_1, β_2, β_3, β_4]). The rotation ranges should be integers and have a sum of 360°. Next, the displacement value at the end of each displacement event appears. Because the profile in Figure E.3 achieves an arbitrary maximum displacement of 2, the sequence (2, 2, 0, 0) (representing [s_1, s_2, s_3, s_4]) appears in Figure E.4. Lastly, both the cam rotation speed and base radius appear in Figure E.4.

---

[*] The units for the displacement interval range and cam rotation speed are degrees and rad/s, respectively.
[†] While this figure illustrates a rise-dwell-fall-dwell displacement event sequence (for demonstration), the user can specify displacement event sequences of any number and any combination of displacement events

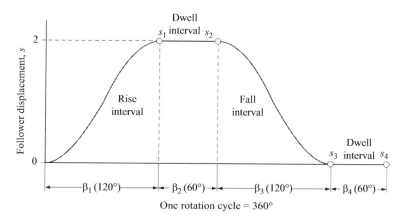

FIGURE E.3
Follower *rise-dwell-fall-dwell* displacement event sequence.

```
%-------------------------------------------------------------------
% Here, values for the follower displacement event sequence, the
% corresponding follower displacement event ranges and the end
% value for each follower displacement event are assigned.  The
% cam rotational speed and base circle radius are assigned here
% also.

Event = ['R','D','F','D'];
Beta = [120,60,120,60];
Si = [2,2,0,0];

dTheta = 1.1;
Rbase = 5;
%-------------------------------------------------------------------
```

FIGURE E.4
Sections of *Cam_Const_Acc.m* with sample values in bold.

After specifying the follower displacement event sequence parameters, cam rotation speed, and cam base circle radius in the file *Cam_Const_Acc.m*, the next step is to run this file. When running this file, one file (filename *DVAJ.csv*) is written to a folder named *Results* (in a format compatible with Microsoft Excel) that includes the calculated follower displacement, velocity and acceleration data, and the corresponding cam profile data, at 1° increments.

E.3 S, V, A, J Profile Generation and Cam Design: Simple Harmonic Motion

The Appendix E.3 folder (which is available for download at www.crcpress.com/product/isbn/9781498724937) includes the MATLAB file *Cam_SHM.m* for the kinematic analysis and design of radial cams for simple harmonic motion. To conduct a simple harmonic-based kinematic analysis, the user specifies the follower displacement event sequence, the displacement event ranges, the end values for each displacement event, the cam rotation speed, and the cam base circle radius.*

To demonstrate how values are specified in *Cam_SHM.m*, Figure E.5 illustrates an arbitrary follower rise-dwell-fall-dwell displacement event sequence, while Figure E.6 includes the corresponding values (in bold type) specified in *Cam_SHM.m*.[†]

Rise, fall, and dwell events are specified using *R*, *F*, and *D* (in single quotes in a row matrix), respectively. Because Figure E.5 presents a rise-dwell-fall-dwell sequence, the sequence (*R*, *D*, *F*, *D*) appears in Figure E.6. Next, the rotation ranges for each displacement event appear in Figure E.6 (representing [$\beta_1, \beta_2, \beta_3, \beta_4$]). The rotation ranges should be integers and have a sum of 360°. Next, the displacement values at the end of each displacement event appear. Because the profile in Figure E.5 achieves an arbitrary maximum displacement of 2, the sequence (2, 2, 0, 0) (representing [s_1, s_2, s_3, s_4]) appear in Figure E.6. Lastly, both the cam rotation speed and base radius appear in Figure E.6.

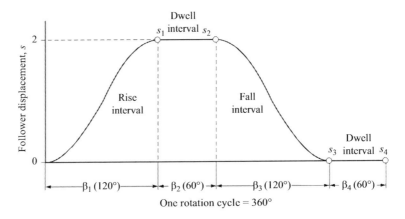

FIGURE E.5
Follower *rise-dwell-fall-dwell* displacement event sequence.

```
%----------------------------------------------------------------
% Here, values for the follower displacement event sequence, the
% corresponding follower displacement event ranges and the end
% value for each follower displacement event are assigned.  The
% cam rotational speed and base circle radius are assigned here
% also.

Event = ['R','D','F','D'];
Beta  = [120,60,120,60];
Si    = [2,2,0,0];

dTheta = 1.1;
Rbase  = 5;
%----------------------------------------------------------------
```

FIGURE E.6
Sections of *Cam_SHM.m* with sample values in bold.

---

* The units for the displacement interval range and cam rotation speed are degrees and rad/s, respectively.
[†] While this figure illustrates a rise-dwell-fall-dwell displacement event sequence (for demonstration), the user can specify displacement event sequences of any number and any combination of displacement events.

After specifying the follower displacement event sequence parameters, cam rotation speed, and cam base circle radius in the file *Cam_SHM.m*, the next step is to run this file. When running this file, one file (filename *DVAJ.csv*) is written to a folder named *Results* (in a format compatible with Microsoft Excel) that includes the calculated follower displacement, velocity, acceleration, and jerk data, and the corresponding cam profile data, at 1° increments.

E.4 S, V, A, J Profile Generation and Cam Design: Cycloidal Motion

The Appendix E.4 folder (which is available for download at www.crcpress.com/product/ isbn/9781498724937) includes the MATLAB file *Cam_Cycloidal.m* for the kinematic analysis and design of radial cams for cycloidal motion. Both the data specified in and calculated from *Cam_Cycloidal.m* are identical to those described in Appendix E.3.

E.5 S, V, A, J Profile Generation and Cam Design: 3-4-5 Polynomial Motion

The Appendix E.5 folder (which is available for download at www.crcpress.com/product/ isbn/9781498724937) includes the MATLAB file *Cam_345_Poly.m* for the kinematic analysis and design of radial cams for 3-4-5 polynomial motion. Both the data specified in and calculated from *Cam_345_Poly.m* are identical to those described in Appendix E.3.

E.6 S, V, A, J Profile Generation and Cam Design: 4-5-6-7 Polynomial Motion

The Appendix E.6 folder (which is available for download at www.crcpress.com/product/ isbn/9781498724937) includes the MATLAB file *Cam_4567_Poly.m* for the kinematic analysis and design of radial cams for 4-5-6-7 polynomial motion. Both the data specified in and calculated from *Cam_4567_Poly.m* are identical to those described in Appendix E.3.

Appendix F: User Instructions for Chapter 10 MATLAB® Files

F.1 RRSS Mechanism

The Appendix F.1 folder (which is available for download at www.crcpress.com/product/isbn/9781498724937) includes the MATLAB file *Kinematic_RRSS.m* for the kinematic analysis of RRSS mechanisms. To conduct a kinematic analysis, the user specifies the mechanism point coordinates and the crank motion parameters in this file. Values are specified for link variables $\mathbf{a}_0$, $\mathbf{ua}_0$, $\mathbf{a}_1$, $\mathbf{ua}_1$, $\mathbf{b}_0$, $\mathbf{b}_1$, $\mathbf{p}_1$, $\mathbf{q}_1$, and $\mathbf{r}_1$ (Figure F.1a). Values are also specified for the initial crank angle (*start_ang*), the crank rotation increment (*step_ang*), and the final crank angle (*stop_ang*). Lastly, values are specified for the angular velocity (*angular_vel*) and angular acceleration (*angular_acc*) of the crank. Figure F.2 illustrates the user-input sections of the file *Kinematic_RRSS.m*, with sample values in bold type.[*]

After specifying the dimensions and driving link parameters in the file *Kinematic_RRSS.m*, the next step is to run this file. When running this file, two files (filenames *RRSS_branch1.csv* and *RRSS_branch2.csv*) are written to a folder named *Results* (in a format compatible with Microsoft® Excel) that includes the calculated mechanism output at each crank link rotation increment for both RRSS branches.[†] The calculated mechanism output is included in Figure F.1b.

The initial RRSS mechanism configuration corresponds to the mechanism dimensions specified in the Appendix F.1 MATLAB file (see Figure F.2). The results for this configuration are included in the file *RRSS_branch1.csv*. The results for the second mechanism configuration are included in the file *RRSS_branch2.csv*.

F.2 RSSR Mechanism

The Appendix F.2 folder (which is available for download at www.crcpress.com/product/isbn/9781498724937) includes the MATLAB file *Kinematic_RSSR.m* for the kinematic analysis of RSSR mechanisms. To conduct a kinematic analysis, the user specifies the mechanism point coordinates and the crank motion parameters in this file. Values are specified for link variables $\mathbf{a}_0$, $\mathbf{ua}_0$, $\mathbf{a}_1$, $\mathbf{b}_0$, $\mathbf{ub}_0$, and $\mathbf{b}_1$ (Figure F.3a). Values are also specified for the initial crank angle (*start_ang*), the crank rotation increment (*step_ang*), and the final crank angle (*stop_ang*). Lastly, values are specified for the angular velocity (*angular_vel*)

---

[*] The units for crank angular rotation, velocity, and acceleration are degrees, rad/s, and rad/s², respectively.

[†] At crank rotation increments where circuit defects occur, the coupler-link motion data written to the *.csv* file appear as complex numbers.

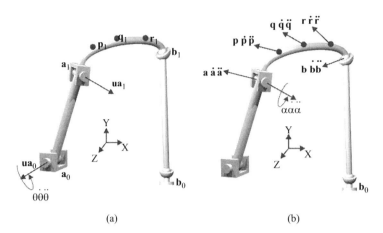

FIGURE F.1
(a) RRSS mechanism and (b) output variables.

```
%--------------------------------------------------------------
% Here, values for the RRSS mechanism variables a0, a1, ua0,
% ua1, b0, b1, p1, q1 and r1 are assigned.  They must all be
% defined as transposed row matrices (as shown).

a0  = [-0.0576, 0.2890, -1.4112]';
a1  = [0.1452, -2.5421, -1.1800]';
ua0 = [-0.0003, 0.0814, 0.9967]';
ua1 = [0.3040, 0.0992, 0.9475]';
b0  = [0.0851, 0.4570, 0.5096]';
b1  = [1.7725, 5.1566, 0.6499]';
p1  = [1.7321, 0, -1]';
q1  = [1.2321, 0, -1.8660]';
r1  = [1.9486, 0, -1.1250]';
%--------------------------------------------------------------


%--------------------------------------------------------------
% Here, values for the start, step and stop displacement angles
% for the crank link are assigned.

start_ang = 0;
step_ang  = 10;
stop_ang  = 200;
%--------------------------------------------------------------


%--------------------------------------------------------------
% Here, values for crank link angular velocity and angular
% acceleration are assigned.

angular_vel = 1.0 * ones(N+1,1);
angular_acc = 0.1 * ones(N+1,1);
%--------------------------------------------------------------
```

FIGURE F.2
Sections of *Kinematic_RRSS.m* with sample values in bold.

and angular acceleration (*angular_acc*) of the crank. Figure F.4 illustrates the user-input sections of the file *Kinematic_RSSR.m*, with sample values in bold type.*

After specifying the dimensions and driving link parameters in the file *Kinematic_RSSR.m*, the next step is to run this file. When running this file, two files (filenames *RSSR_branch1.csv* and *RSSR_branch2.csv*) are written to a folder named *Results* (in a format

---

* The units for crank angular rotation, velocity, and acceleration are degrees, rad/s, and rad/s^2, respectively.

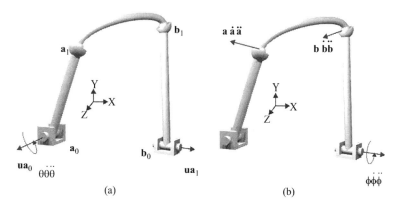

FIGURE F.3
(a) RSSR mechanism and (b) output variables.

```
%--------------------------------------------------------------------
% Here, values for the RSSR mechanism variables a0, a1, ua0, b0,
% b1 and ub0 are assigned.  They must all be defined as
% transposed row matrices as shown).

a0  = [0, 0, -0.4023]';
a1  = [0.3356, -0.0708, -0.4023]';
ua0 = [0, 0, 1]';
b0  = [1, -2.3885, 0]';
b1  = [1, -2.3885, 0.7300]';
ub0 = [0, -1, 0]';
%--------------------------------------------------------------------


%--------------------------------------------------------------------
% Here, values for the start, step and stop displacement angles
% for the crank link are assigned.

start_ang   = 0;
step_ang    = 10;
stop_ang    = 50;
%--------------------------------------------------------------------


%--------------------------------------------------------------------
% Here, values for crank link angular velocity and angular
% acceleration are assigned.

angular_vel  = 0 * ones(N+1,1);
angular_acc  = 0 * ones(N+1,1);
%--------------------------------------------------------------------
```

FIGURE F.4
Sections of *Kinematic_RSSR.m* with sample values in bold.

compatible with Microsoft Excel) that includes the calculated mechanism output at each crank link rotation increment for both RSSR branches.* The calculated mechanism output is included in Figure F.3b.

The initial RSSR mechanism configuration corresponds to the mechanism dimensions specified in the Appendix F.2 MATLAB file (see Figure F.4). The results for this configuration are included in the file *RSSR_branch1.csv*. The results for the second mechanism configuration are included in the file *RSSR_branch2.csv*.

---

* The units for crank angular rotation, velocity, and acceleration are degrees, rad/s, and rad/s², respectively.

Appendix G: User Instructions for Chapter 11 MATLAB® Files

G.1 R-P-P Robotic Manipulator Forward Kinematics

The Appendix G.1 folder (which is available for download at www.crcpress.com/product/isbn/9781498724937) includes two MATLAB files for R-P-P robotic manipulator forward kinematics. These two files are described in Table G.1. To conduct an R-P-P forward kinematics analysis, the user specifies the R-P-P joint displacements in the file *RPP_Input.csv*. This file is compatible with Microsoft® Excel. Also, $\{^3\mathbf{p}_3\}$ is specified in the file *RPP_FK.m*. Figure G.1 illustrates the user-input section of this file, with sample values in bold type.

After specifying the R-P-P joint displacements in *RPP_Input.csv* and $\{^3\mathbf{p}_3\}$ in *RPP_FK.m*, the next step is to run *RPP_FK.m*. When running this file, one file (filename *RPP_p3.csv*) is written to a folder named *Results* (in a format compatible with Microsoft Excel) that includes the $\{^0\mathbf{p}_3\}$ values calculated for the given $\{^3\mathbf{p}_3\}$ value and joint displacement values.

G.2 R-R-P Robotic Manipulator Forward Kinematics

The Appendix G.2 folder (which is available for download at www.crcpress.com/product/isbn/9781498724937) includes two MATLAB files for R-R-P robotic manipulator forward kinematics. These two files are described in Table G.2. To conduct an R-R-P forward kinematics analysis, the user specifies the R-R-P joint displacements in the file *RRP_Input.csv*. This file is compatible with Microsoft Excel. Also, $\{^3\mathbf{p}_3\}$, l_1, and l_2 are specified in the file *RRP_FK.m*. Figure G.2 illustrates the user-input section of this file, with sample values in bold type.

After specifying the R-R-P joint displacements in *RRP_Input.csv* and $\{^3\mathbf{p}_3\}$, l_1, and l_2 in *RRP_FK.m*, the next step is to run *RRP_FK.m*. When running this file, one file (filename *RRP_p3.csv*) is written to a folder named *Results* (in a format compatible with Microsoft Excel) that includes the $\{^0\mathbf{p}_3\}$ values calculated for the given $\{^3\mathbf{p}_3\}$, length, and joint displacement values.

G.3 R-R-R Robotic Manipulator Forward Kinematics

The Appendix G.3 folder (which is available for download at www.crcpress.com/product/isbn/9781498724937) includes two MATLAB files for R-R-R robotic manipulator forward

```
%------------------------------------------------------------------
% Here, the coordinates for the end-effector point p3_3 are
% assigned.

p3_3 = [0, 0, -1];
%------------------------------------------------------------------
```

FIGURE G.1
Section of *RPP_FK.m* with sample values in bold.

TABLE G.1

Appendix G.1 MATLAB Files

Filename	Use of File
RPP_Input.csv	To specify joint displacements
RPP_FK.m	To specify $\{^3\mathbf{p}_3\}$ and write output (in a file compatible with Microsoft Excel)

```
%------------------------------------------------------------------
% Here, the coordinates for the end-effector point p3_3 and the
% lengths l1 and l2 are assigned.

l1 = 0.5;
l2 = 0.5;
p3_3 = [0, 0, 0];
%------------------------------------------------------------------
```

FIGURE G.2
Section of *RRP_FK.m* with sample values in bold.

TABLE G.2

Appendix G.2 MATLAB Files

Filename	Use of File
RRP_Input.csv	To specify joint displacements
RRP_FK.m	To specify $\{^3\mathbf{p}_3\}$, l_1, and l_2, and write output (in a file compatible with Microsoft Excel)

kinematics. These two files are described in Table G.3. To conduct an R-R-R forward kinematics analysis, the user specifies the R-R-R joint displacements in the file *RRR_Input. csv*. This file is compatible with Microsoft Excel. Also, $\{^3\mathbf{p}_3\}$, l_1, and l_2 are specified in the file *RRR_FK.m*. Figure G.3 illustrates the user-input section of this file, with sample values in bold type.

After specifying the R-R-R joint displacements in *RRR_Input.csv* and $\{^3\mathbf{p}_3\}$, l_1, and l_2 in *RRR_FK.m*, the next step is to run *RRR_FK.m*. When running this file, one file (filename *RRR_p3.csv*) is written to a folder named *Results* (in a format compatible with Microsoft Excel) that includes the $\{^0\mathbf{p}_3\}$ values calculated for the given $\{^3\mathbf{p}_3\}$, length, and joint displacement values.

TABLE G.3

Appendix G.3 MATLAB Files

Filename	Use of File
RRR_Input.csv	To specify joint displacements
RRR_FK.m	To specify $\{^3\mathbf{p}_3\}$, l_1, and l_2, and write output (in a file compatible with Microsoft Excel)

```
%------------------------------------------------------------
% Here, the coordinates for the end-effector point p3_3 and the
% lengths l1 and l2 are assigned.

l1 = 0.5;
l2 = 0.5;
p3_3 = [0, 1, 0];
%------------------------------------------------------------
```

FIGURE G.3
Section of *RRR_FK.m* with sample values in bold.

G.4 R-R-C Robotic Manipulator Forward Kinematics

The Appendix G.4 folder (which is available for download at www.crcpress.com/product/isbn/9781498724937) includes two MATLAB files for R-R-C robotic manipulator forward kinematics. These two files are described in Table G.4. To conduct an R-R-C forward kinematics analysis, the user specifies the R-R-C joint displacements in the file *RRC_Input.csv*. This file is compatible with Microsoft Excel. Also, $\{^3\mathbf{p}_3\}$, l_1, l_2 and l_3 are specified in the file *RRC_FK.m*. Figure G.4 illustrates the user-input section of this file with sample values in bold type.

```
%------------------------------------------------------------
% Here, the coordinates for the end-effector point p3_3 and the
% lengths l1, l2 and l3 are assigned.

l1 = 1;
l2 = 0.5;
l3 = 0.5;
p3_3 = [1, 0, 0];
%------------------------------------------------------------
```

FIGURE G.4
Section of *RRC_FK.m* with sample values in bold.

TABLE G.4

Appendix G.4 MATLAB Files

Filename	Use of File
RRC_Input.csv	To specify joint displacements
RRC_FK.m	To specify $\{^3\mathbf{p}_3\}$, l_1, l_2 and l_3, and write output (in a file compatible with Microsoft Excel)

```
%-------------------------------------------------------------
% Here, the coordinates for the end-effector point p3_3 are
% assigned.

p3_3 = [0, 0, -1];
%-------------------------------------------------------------
```

FIGURE G.5
Section of *RPP_IK.m* with sample values in bold.

After specifying the R-R-C joint displacements in *RRC_Input.csv* and $\{^3\mathbf{p}_3\}$, l_1, l_2 and l_3 in *RRC_FK.m*, the next step is to run *RRC_FK.m*. When running this file, one file (filename *RRC_p3.csv*) is written to a folder named *Results* (in a format compatible with Microsoft Excel) that includes the $\{^0\mathbf{p}_3\}$ values calculated for the given $\{^3\mathbf{p}_3\}$, length, and joint displacement values.

G.5 R-P-P Robotic Manipulator Inverse Kinematics

The Appendix G.5 folder (which is available for download at www.crcpress.com/product/isbn/9781498724937) includes two MATLAB files for R-P-P robotic manipulator inverse kinematics. These two files are described in Table G.5. To conduct an R-P-P inverse kinematics analysis, the user specifies the global R-P-P end-effector coordinates ($\{^0\mathbf{p}_3\}$) in the file *RPP_Input.csv*. This file is compatible with Microsoft Excel. Also, $\{^3\mathbf{p}_3\}$ is specified in the file *RPP_IK.m*. Figure G.5 illustrates the user-input section of this file with sample values in bold type.

After specifying the global R-P-P end effector coordinates in *RPP_Input.csv* and $\{^3\mathbf{p}_3\}$ in *RPP_IK.m*, the next step is to run *RPP_IK.m*. When running this file, one file (filename *RPP_Joints.csv*) is written to a folder named *Results* (in a format compatible with Microsoft Excel) that includes the joint displacement values calculated for the given $\{^0\mathbf{p}_3\}$ and $\{^3\mathbf{p}_3\}$ values.

G.6 R-R-P Robotic Manipulator Inverse Kinematics

The Appendix G.6 folder (which is available for download at www.crcpress.com/product/isbn/9781498724937) includes two MATLAB files for R-R-P robotic manipulator inverse

TABLE G.5

Appendix G.5 MATLAB Files

Filename	Use of File
RPP_Input.csv	To specify global end effector coordinates
RPP_IK.m	To specify $\{^3\mathbf{p}_3\}$ and write output (in a file compatible with Microsoft Excel)

```
%-----------------------------------------------------------------
% Here, the coordinates for the end-effector point p3_3 and the
% lengths l1 and l2 are assigned.

l1 = 0.5;
l2 = 0.5;
p3_3 = [0, 0, 0];
%-----------------------------------------------------------------
```

FIGURE G.6
Section of *RRP_IK.m* with sample values in bold.

TABLE G.6

Appendix G.6 MATLAB Files

Filename	Use of File
RRP_Input.csv	To specify global end effector coordinates
RRP_IK.m	To specify $\{^3\mathbf{p}_3\}$, l_1, and l_2, and write output (in a file compatible with Microsoft Excel)

kinematics. These two files are described in Table G.6. To conduct an R-R-P inverse kinematics analysis, the user specifies the global R-R-P end effector coordinates ($\{^0\mathbf{p}_3\}$) in the file *RRP_Input.csv*. This file is compatible with Microsoft Excel. Also, $\{^3\mathbf{p}_3\}$, l_1, and l_2 are specified in the file *RRP_IK.m*. Figure G.6 illustrates the user-input section of this file with sample values in bold type.

After specifying the global R-R-P end effector coordinates in *RRP_Input.csv* and $\{^3\mathbf{p}_3\}$, l_1, and l_2 in *RRP_IK.m*, the next step is to run *RRP_IK.m*. When running this file, one file (filename *RRP_Joints.csv*) is written to a folder named *Results* (in a format compatible with Microsoft Excel) that includes the joint displacement values calculated for the given $\{^0\mathbf{p}_3\}$, $\{^3\mathbf{p}_3\}$, l_1, and l_2 values.

G.7 R-R-R Robotic Manipulator Inverse Kinematics

The Appendix G.7 folder (which is available for download at www.crcpress.com/product/isbn/9781498724937) includes two MATLAB files for R-R-R robotic manipulator inverse kinematics. These two files are described in Table G.7. To conduct an R-R-R inverse kinematics analysis, the user specifies the global R-R-R end-effector coordinates ($\{^0\mathbf{p}_3\}$) in the

TABLE G.7

Appendix G.7 MATLAB Files

Filename	Use of File
RRR_Input.csv	To specify global end effector coordinates
RRR_IK.m	To specify $\{^3\mathbf{p}_3\}$, l_1, and l_2, and write output (in a file compatible with Microsoft Excel)

```
%------------------------------------------------------------
% Here, the coordinates for the end-effector point p3_3 and the
% lengths l1 and l2 are assigned.

l1 = 0.5;
l2 = 0.5;
p3_3 = [0, 1, 0];
%------------------------------------------------------------
```

FIGURE G.7
Section of *RRR_IK.m* with sample values in bold.

file *RRR_Input.csv*. This file is compatible with Microsoft Excel. Also, $\{^3\mathbf{p}_3\}$, l_1, and l_2 are specified in the file *RRR_IK.m*.

Figure G.7 illustrates the user-input section of this file with sample values in bold type. After specifying the global R-R-R end effector coordinates in *RRR_Input.csv* and $\{^3\mathbf{p}_3\}$, l_1, and l_2 in *RRR_IK.m*, the next step is to run *RRR_IK.m*. When running this file, one file (filename *RRR_Joints.csv*) is written to a folder named *Results* (in a format compatible with Microsoft Excel) that includes the joint displacement values calculated for the given $\{^0\mathbf{p}_3\}$, $\{^3\mathbf{p}_3\}$, l_1, and l_2 values.

G.8 R-R-C Robotic Manipulator Inverse Kinematics

The Appendix G.8 folder (which is available for download at www.crcpress.com/product/isbn/9781498724937) includes two MATLAB files for R-R-C robotic manipulator inverse kinematics. These two files are described in Table G.8. To conduct an R-R-C inverse kinematics analysis, the user specifies the global R-R-C end effector coordinates ($\{^0\mathbf{p}_3\}$) in the file *RRC_Input.csv*. This file is compatible with Microsoft Excel. Also, $\{^3\mathbf{p}_3\}$, l_1, l_2, and l_3 are specified in the file *RRC_IK.m*. Figure G.8 illustrates the user-input section of this file with sample values in bold type.

TABLE G.8

Appendix G.8 MATLAB Files

Filename	Use of File
RRC_Input.csv	To specify global end effector coordinates
RRC_IK.m	To specify $\{^3\mathbf{p}_3\}$, l_1, l_2, and l_3, and write output (in a file compatible with Microsoft Excel)

```
%------------------------------------------------------------
% Here, the coordinates for the end-effector point p3_3 and the
% lengths l1, l2 and l3 are assigned.

l1 = 1;
l2 = 0.5;
l3 = 0.5;
p3_3 = [0, 1, -1];
%------------------------------------------------------------
```

FIGURE G.8
Section of *RRC_IK.m* with sample values in bold.

After specifying the global R-R-C end effector coordinates in *RRC_Input.csv* and $\{{}^3\mathbf{p}_3\}$, l_1, l_2, and l_3 in *RRC_IK.m*, the next step is to run *RRC_IK.m*. When running this file, one file (filename *RRC_Joints.csv*) is written to a folder named *Results* (in a format compatible with Microsoft Excel) that includes the joint displacement values calculated for the given $\{{}^0\mathbf{p}_3\}$, $\{{}^3\mathbf{p}_3\}$, l_1, l_2, and l_3 values.

Appendix H: User Instructions for Chapter 4 MATLAB® and SimMechanics® Files

H.1 Planar Four-Bar Mechanism

The Appendix H.1 folder (which is available for download at www.crcpress.com/product/isbn/9781498724937) includes three MATLAB and SimMechanics files for the kinematic analysis and simulation of planar four-bar mechanisms. These three files are described in Table H.1. To conduct a kinematic analysis, the user specifies the mechanism link dimensions and the crank motion parameters in the file *Kinematic_Planar_4Bar_Simulate.m*. Values are specified for link variables W_1, V_1, G_1, U_1, and L_1 (Figure H.1a). Values are also specified for the initial crank angle (*start_ang*), the crank rotation increment (*step_ang*), and the final crank angle (*stop_ang*). Lastly, values are specified for the angular velocity (*angular_vel*) and angular acceleration (*angular_acc*) of the crank. Figure H.2 illustrates the user-input sections of the file *Kinematic_Planar_4Bar_Simulate.m*, with sample values in bold type.* While most of the link vector values in this figure appear in polar exponential form, they can all be specified in any of the rectangular and complex forms given in Equation 2.1.

After specifying the driving link parameters in the file *Kinematic_Planar_4Bar_Simulate.m*, the next step is to run this file. When running this file, one file (filename *Disp_Vel_Acc.csv*) is written to a folder named *Results* (in a format compatible with Microsoft® Excel) that includes the calculated mechanism output at each crank link rotation increment.† The calculated mechanism output is included in Figure H.1b.

Also, when running *Kinematic_Planar_4Bar_Simulate.m*, a graphical user interface appears where the motion of the planar four-bar mechanism is simulated over the defined crank rotation range (see Appendix A.3).

H.2 Slider-Crank Mechanism

The Appendix H.2 folder (which is available for download at www.crcpress.com/product/isbn/9781498724937) includes three MATLAB and SimMechanics files for the kinematic analysis and simulation of slider-crank mechanisms. These three files are described in Table H.2. To conduct a kinematic analysis, the user specifies the mechanism link dimensions and the crank motion parameters in the file *Kinematic_Slider_Crank_Simulate.m*. Values are specified for link variables W_1, V_1, and U_1 (Figure H.3a). Values are also specified for the initial crank angle (*start_ang*), the crank rotation increment (*step_ang*), and the final

---

* The units for crank angular rotation, velocity, and acceleration are *degrees*, *rad/s*, and *rad/s²*, respectively.
† If the mechanism experiences a circuit defect (or if the crank decelerates to a zero velocity state), results beyond this point are not written to the *.csv* file.

TABLE H.1

Appendix H.1 MATLAB and SimMechanics Files

Filename	Use of File
Kinematic_Planar_4Bar_Simulate.m	To specify mechanism link dimensions and crank link controls
Kinematic_Planar_4Bar_Model.slx	To calculate mechanism output and simulate mechanism motion
Post_Simulation_Task.m	To write mechanism output (compatible with Microsoft Excel)

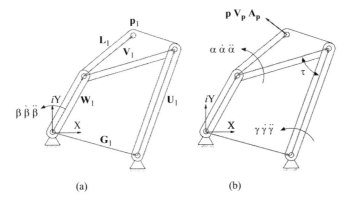

FIGURE H.1
(a) Planar four-bar mechanism and (b) output variables.

```
%------------------------------------------------------------
% Here, values for the planar 4-bar mechanism variables W1, V1,
% G1, U1 and L1 are assigned.

W1 = 0.5*exp(i*90*pi/180);
V1 = 0.75*exp(i*19.3737*pi/180);
G1 = 0.75 + i*0;
U1 = 0.75*exp(i*93.2461*pi/180);
L1 = 0.5*exp(i*60.7834*pi/180);
%------------------------------------------------------------


%------------------------------------------------------------
% Here, values for the start, step and stop displacement angles
% for the crank link are assigned.

start_ang   = 0;
step_ang    = 1;
stop_ang    = 360;
%------------------------------------------------------------


%------------------------------------------------------------
% Here, values for crank link angular velocity and angular
% acceleration are assigned.

angular_vel  = 1.0;
angular_acc  = 0.1;
%------------------------------------------------------------
```

FIGURE H.2
Sections of *Kinematic_Planar_4Bar_Simulate.m* with sample values in bold.

TABLE H.2

Appendix H.2 MATLAB and SimMechanics Files

Filename	Use of File
Kinematic Slider_Crank_Simulate.m	To specify mechanism link dimensions and crank link controls
Kinematic_Slider_Crank_Model.slx	To calculate mechanism output and simulate mechanism motion
Post_Simulation_Task.m	To write mechanism output (compatible with Microsoft Excel)

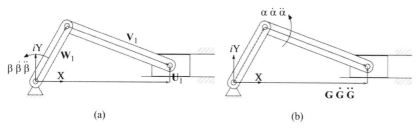

(a) (b)

FIGURE H.3

(a) Slider-crank mechanism and (b) output variables.

crank angle (*stop_ang*). Lastly, values are specified for the angular velocity (*angular_vel*) and angular acceleration (*angular_acc*) of the crank. Figure H.4 illustrates the user-input sections of the file *Kinematic_Slider_Crank_Simulate.m*, with sample values in bold type.*

After specifying the mechanism dimensions and driving link parameters in the file *Kinematic_Slider_Crank_Simulate.m*, the next step is to run this file. When running this file, one File (filename *Disp_Vel_Acc.csv*) is written to a folder named *Results* (in a format compatible with Microsoft Excel) that includes the calculated mechanism output at each crank link rotation increment.† The calculated mechanism output is included in Figure H.3b.

Also, when running *Kinematic_Slider_Crank_Simulate.m*, a graphical user interface appears where the motion of the slider-crank mechanism is simulated over the defined crank rotation range (see Appendix A.3).

H.3 Geared Five-Bar Mechanism (Two Gears)

The Appendix H.3 folder (which is available for download at www.crcpress.com/product/isbn/9781498724937) includes three MATLAB and SimMechanics files for the kinematic analysis and simulation of geared five-bar mechanisms having two gears. These three files are described in Table H.3. To conduct a dynamic force analysis, the user specifies the mechanism link dimensions and the crank motion parameters in the file *Kinematic_5Bar_2Gears_Simulate.m*. Values are specified for link variables W_1, V_1, G_1, U_1, L_1, and S_1 (Figure H.5a). Values are also specified for the initial crank angle (*start_ang*), the crank rotation increment (*step_ang*), and the final crank angle (*stop_ang*). Lastly, values are specified for the angular velocity

---

* The units for crank angular rotation, velocity, and acceleration are *degrees*, *rad/s*, and *rad/s²*, respectively.
† If the mechanism experiences a circuit defect (or if the crank decelerates to a zero velocity state), results beyond this point are not written to the *.csv* file.

```
%------------------------------------------------------------------
% Here, values for the slider-crank mechanism variables W1, U1,
% V1 are assigned and variable G1 calculated.  Variables LW1,
% LU1 and LV1 are the scalar lengths of vectors W1, U1 and V1.

LW1 = 1;
theta = 45*pi/180;
W1 = LW1*exp(i*theta);

LU1 = 0;
U1 = i*LU1;

LV1 = 1.5;
rho = asin((LU1-LW1*sin(theta))/LV1);
V1 = LV1*exp(i*rho);

G1 = W1 + V1 - U1;
%------------------------------------------------------------------


%------------------------------------------------------------------
% Here, values for the start, step and stop displacement angles
% for the crank link are assigned.

start_ang   = 0;
step_ang    = 1;
stop_ang    = 360;
%------------------------------------------------------------------


%------------------------------------------------------------------
% Here, values for crank link angular velocity and angular
% acceleration are assigned.

angular_vel  = 100;
angular_acc  = 10;
%------------------------------------------------------------------
```

FIGURE H.4
Sections of *Kinematic_Slider_Crank_Simulate.m* with sample values in bold.

TABLE H.3

Appendix H.3 MATLAB and SimMechanics Files

Filename	Use of File
Kinematic_5Bar_2Gears_Simulate.m	To specify mechanism link dimensions and crank link controls
Kinematic_5Bar_2Gears_Model.slx	To calculate mechanism output and simulate mechanism motion
Post_Simulation_Task.m	To write mechanism output (compatible with Microsoft Excel)

(*angular_vel*) and angular acceleration (*angular_acc*) of the crank. Figure H.6 illustrates the user-input sections of the file *Kinematic_5Bar_2Gears_Simulate.m*, with sample values in bold type.* While most of the link vector values in this figure appear in polar exponential form, they can all be specified in any of the rectangular and complex forms given in Equation 2.1.

After specifying the mechanism dimensions, loads, mass parameters, and driving link parameters in the file *Kinematic_5Bar_2Gear_Simulate.m*, the next step is to run this file. When running this file, one file (filename *Disp_Vel_Acc.csv*) is written to a folder named *Results* (in a format compatible with Microsoft Excel) that includes the calculated mechanism output at each crank link rotation increment.† The calculated mechanism output is included in Figure H.5b.

---

* The units for crank angular rotation, velocity, and acceleration are *degrees*, *rad/s*, and *rad/s²*, respectively.

† If the mechanism experiences a circuit defect (or if the crank decelerates to a zero velocity state), results beyond this point are not written to the *.csv* file.

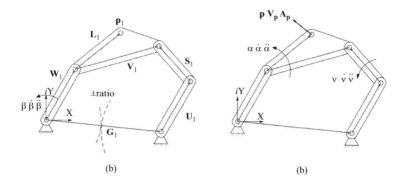

FIGURE H.5
(a) Geared five-bar mechanism and (b) output variables.

```
%-----------------------------------------------------------------
% Here, values for the geared 5-bar mechanism variables W1, V1,
% G1, U1, L1 and S1 are assigned. Only negative gear ratio
% values should be specified.

W1 = 0.5*exp(i*90*pi/180);
V1 = 0.75*exp(i*32.7304*pi/180);
G1 = 0.75*exp(i*0*pi/180);
U1 = 0.75*exp(i*45*pi/180);
L1 = 0.5*exp(i*74.1400*pi/180);
S1 = 0.75*exp(i*149.9847*pi/180);
ratio = -2;
%-----------------------------------------------------------------


%-----------------------------------------------------------------
% Here, values for the start, step and stop displacement angles
% for the crank link are assigned.

start_ang   = 0;
step_ang    = 1;
stop_ang    = 360;
%-----------------------------------------------------------------


%-----------------------------------------------------------------
% Here, values for crank link angular velocity and angular
% acceleration are assigned.

angular_vel   = 1;
angular_acc   = 0;
%-----------------------------------------------------------------
```

FIGURE H.6
Sections of *Kinematic_5Bar_2Gears_Simulate.m* with sample values in bold.

Also, when running *Kinematic_5Bar_2Gear_Simulate.m*, a graphical user interface appears where the motion of the geared five-bar mechanism is simulated over the defined crank rotation range (see Appendix A.3).

H.4 Geared Five-Bar Mechanism (Three Gears)

The Appendix H.4 folder (which is available for download at www.crcpress.com/product/isbn/9781498724937) includes three MATLAB and SimMechanics files for the kinematic

TABLE H.4

Appendix H.4 MATLAB and SimMechanics Files

Filename	Use of File
Kinematic_5Bar_3Gears_Simulate.m	To specify mechanism link dimensions and crank link controls
Kinematic_5Bar_3Gears_Model.slx	To calculate mechanism output and simulate mechanism motion
Post_Simulation_Task.m	To write mechanism output (compatible with Microsoft Excel)

analysis and simulation of geared five-bar mechanisms having three gears. These three files are described in Table H.4. The procedure to use these MATLAB and SimMechanics files is identical to the procedure given in Section H.3.

H.5 Watt II Mechanism

The Appendix H.5 folder (which is available for download at www.crcpress.com/product/ isbn/9781498724937) includes three MATLAB and SimMechanics files for the kinematic analysis and simulation of Watt II mechanisms. These three files are described in Table H.5. To conduct a kinematic analysis, the user specifies the mechanism link dimensions and the crank motion parameters in the file *Kinematic_Watt_II_Simulate.m*. Values are specified for link variables $\mathbf{W}_1$, $\mathbf{V}_1$, $\mathbf{G}_1$, $\mathbf{U}_1$, $\mathbf{L}_1$, $\mathbf{W}_1^*$, $\mathbf{V}_1^*$, $\mathbf{G}_1^*$, $\mathbf{U}_1^*$, and $\mathbf{L}_1^*$ (Figure H.7a). Values are also specified for the initial crank angle (*start_ang*), the crank rotation increment (*step_ang*), and the final crank angle (*stop_ang*). Lastly, values are specified for the angular velocity (*angular_vel*) and angular acceleration (*angular_acc*) of the crank. Figure H.8 illustrates the user-input sections of the file *Kinematic_Watt_II_Simulate.m* with sample values in bold type.[*] While most of the link vector values in this figure appear in polar exponential form, they can all be specified in any of the rectangular and complex forms given in Equation 2.1.

After specifying the mechanism dimensions and driving link parameters in the file *Kinematic_Watt_II_Simulate*, the next step is to run this file. When running this file, one file (filename *Disp_Vel_Acc.csv*) is written to a folder named *Results* (in a format compatible with Microsoft Excel) that includes the calculated mechanism output at each crank link rotation increment.[†] The calculated mechanism output is included in Figure H.7b.

TABLE H.5

Appendix H.5 MATLAB and SimMechanics Files

Filename	Use of File
Kinematic_Watt_II_Simulate.m	To specify mechanism link dimensions and crank link controls
Kinematic_Watt_II_Model.slx	To calculate mechanism output and simulate mechanism motion
Post_Simulation_Task.m	To write mechanism output (compatible with Microsoft Excel)

---

[*] The units for crank angular rotation, velocity, and acceleration are *degrees*, *rad/s*, and *rad/s²*, respectively.

[†] If the mechanism experiences a circuit defect (or if the crank decelerates to a zero velocity state), results beyond this point are not written to the *.csv* file.

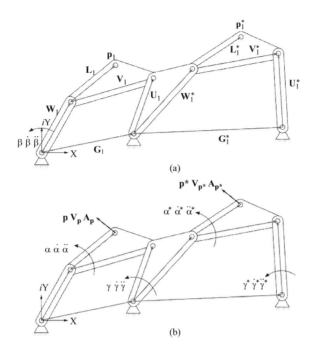

FIGURE H.7
(a) Watt II mechanism and (b) output variables.

```
%-----------------------------------------------------------------
% Here, values for the Watt II mechanism variables W1, V1, G1,
% U1, L1, W1s, V1s, G1s, U1s and L1s are assigned.

W1 = 0.5*exp(i*90*pi/180);
V1 = 0.75*exp(i*19.3737*pi/180);
G1 = 0.75 + i*0;
U1 = 0.75*exp(i*93.2461*pi/180);
L1 = 0.5*exp(i*60.7834*pi/180);

W1s = 0.5*exp(i*45*pi/180);
V1s = 0.75*exp(i*7.9416*pi/180);
G1s = 0.7244 - i*0.1941;
U1s = 0.75*exp(i*60.2717*pi/180);
L1s = 0.5*exp(i*49.3512*pi/180);
%-----------------------------------------------------------------


%-----------------------------------------------------------------
% Here, values for the start, step and stop displacement angles
% for the crank link are assigned.

start_ang   = 0;
step_ang    = -1;
stop_ang    = -360;
%-----------------------------------------------------------------


%-----------------------------------------------------------------
% Here, values for crank link angular velocity and angular
% acceleration are assigned.

angular_vel  = -1.5;
angular_acc  = -0.25;
%-----------------------------------------------------------------
```

FIGURE H.8
Sections of *Kinematic_Watt_II_Simulate.m* with sample values in bold.

Also, when running *Kinematic_Watt_II_Simulate.m*, a graphical user interface appears where the motion of the Watt II mechanism is simulated over the defined crank rotation range (see Appendix A.3).

H.6 Stephenson III Mechanism

The Appendix H.6 folder (which is available for download at www.crcpress.com/product/isbn/9781498724937) includes three MATLAB and SimMechanics files for the kinematic

TABLE H.6

Appendix H.6 MATLAB and SimMechanics Files

Filename	Use of File
Kinematic_Stephenson_III_Simulate.m	To specify mechanism link dimensions and crank link controls
Kinematic_Stephenson_III_Model.slx	To calculate mechanism output and simulate mechanism motion
Post_Simulation_Task.m	To write mechanism output (compatible with Microsoft Excel)

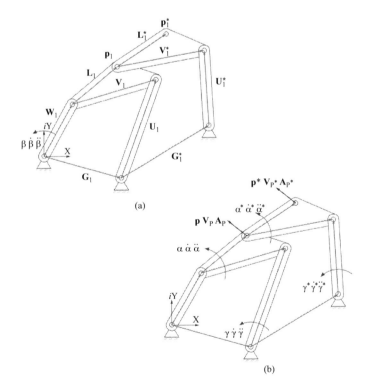

FIGURE H.9
(a) Stephenson III mechanism and (b) output variables.

```
%------------------------------------------------------------------
% Here, values for the Stephenson III mechanism variables W1,
% V1, G1, U1, L1, V1s, G1s, U1s and L1s are assigned.

W1 = 0.5*exp(i*90*pi/180);
V1 = 0.75*exp(i*19.3737*pi/180);
G1 = 0.75 + i*0;
U1 = 0.75*exp(i*93.2461*pi/180);
L1 = 0.5*exp(i*60.7834*pi/180);

V1s = 1*exp(i*17.1417*pi/180);
G1s = 0.2159 + i*0.2588;
U1s = 1*exp(i*76.4844*pi/180);
L1s = 0.5*exp(i*63.7091*pi/180);
%------------------------------------------------------------------


%------------------------------------------------------------------
% Here, values for the start, step and stop displacement angles
% for the crank link are assigned.

start_ang   = 0;
step_ang    = -1;
stop_ang    = -360;
%------------------------------------------------------------------


%------------------------------------------------------------------
% Here, values for crank link angular velocity and angular
% acceleration are assigned.

angular_vel   = -1;
angular_acc   = -0.25;
%------------------------------------------------------------------
```

FIGURE H.10
Sections of *Kinematic_Stephenson_III_Simulate.m* with sample values in bold.

analysis and simulation of Stephenson III mechanisms. These three files are described in Table H.6. To conduct a kinematic analysis, the user specifies the mechanism link dimensions and the crank motion parameters in the file *Kinematic_Stephenson_III_Simulate.m*. Values are specified for link variables W_1, V_1, G_1, U_1, L_1, W_1^*, V_1^*, G_1^*, U_1^*, and L_1^* (Figure H.9a). Values are also specified for the initial crank angle (*start_ang*), the crank rotation increment (*step_ang*), and the final crank angle (*stop_ang*). Lastly, values are specified for the angular velocity (*angular_vel*) and angular acceleration (*angular_acc*) of the crank. Figure H.10 illustrates the user-input sections of the file *Kinematic_Stephenson_III_Simulate.m* with sample values in bold type.[**] While most of the link vector values in this figure appear in polar exponential form, they can all be specified in any of the rectangular and complex forms given in Equation 2.1.

After specifying the mechanism dimensions, loads, mass parameters and driving link parameters in the file *Kinematic_Stephenson_III_Simulate.m*, the next step is to run this file. When running this file, one file (filename *Disp_Vel_Acc.csv*) is written to a folder named *Results* (in a format compatible with Microsoft Excel) that includes the calculated mechanism output at each crank link rotation increment.[†] The calculated mechanism output is included in Figure H.9b.

Also, when running *Kinematic_Stephenson_III_Simulate.m*, a graphical user interface appears where the motion of the Stephenson III mechanism is simulated over the defined crank rotation range (see Appendix A.3).

---

[*] The units for crank angular rotation, velocity, and acceleration are *degrees*, *rad/s*, and *rad/s²*, respectively.
[†] If the mechanism experiences a circuit defect (or if the crank decelerates to a zero velocity state), results beyond this point are not written to the *.csv* file.

Appendix I: User Instructions for Chapter 6 MATLAB® and SimMechanics® Files

I.1 Planar Four-Bar Mechanism

The Appendix I.1 folder (which is available for download at www.crcpress.com/product/isbn/9781498724937) includes three MATLAB and SimMechanics files for the static force analysis and simulation of planar four-bar mechanisms. These three files are described in Table I.1. To conduct a static force analysis, the user first selects the preferred unit system (either Metric or English units) for the written mechanism output.

After selecting the unit system, the user specifies the mechanism link dimensions, coupler force, gravitational constant, center of mass vectors, link masses, and the crank motion parameters in the file *Static_Planar_4Bar_Simulate.m*. Values are specified for link variables $\mathbf{W}_1$, $\mathbf{V}_1$, $\mathbf{G}_1$, $\mathbf{U}_1$, $\mathbf{L}_1$, the coupler force, the gravitational constant $\mathbf{g}$, the center of mass vectors $\mathbf{R}_1$ through $\mathbf{R}_3$, and link masses m_1 through m_3 (Figure I.1a) according to the Metric or English units listed in the file (Figure I.2). Values are also specified for the initial crank angle (*start_ang*), the crank rotation increment (*step_ang*), and the final crank angle (*stop_ang*). Figure I.2 illustrates the user-input sections of the file *Static_Planar_4Bar_Simulate.m*, with sample values in bold type.[*] While most of the link vector values in this figure appear in polar exponential form, they can all be specified in any of the rectangular and complex forms given in Equation 2.1.

After specifying the mechanism dimensions and driving link parameters in the file *Static_Planar_4Bar_Simulate.m*, the next step is to run this file. When running this file, one file (filename *Static_Loads.csv*) is written to a folder named *Results* (in a format compatible with Microsoft® Excel) that includes the calculated mechanism output at each crank link rotation increment.[†] The calculated mechanism output is included in Figure I.1b.

Also, when running *Static_Planar_4Bar_Simulate.m*, a graphical user interface appears where the motion of the planar four-bar mechanism is simulated over the defined crank rotation range (see Appendix A.3).

I.2 Slider-Crank Mechanism

The Appendix I.2 folder (which is available for download at www.crcpress.com/product/isbn/9781498724937) includes three MATLAB and SimMechanics files for the static force analysis and simulation of slider-crank mechanisms. These three files are described in

---

[*] The unit for crank angular rotation is degrees.
[†] When the mechanism experiences a circuit defects occur, results are no longer written to the *.csv* file.

TABLE I.1

Appendix I.1 MATLAB and SimMechanics Files

Filename	Use of File
Static_Planar_4Bar_Simulate.m	To specify mechanism link dimensions, coupler force, and crank link controls
Static_Planar_4Bar_Model.slx	To calculate mechanism output and simulate mechanism motion
Post_Simulation_Task.m	To write mechanism output (compatible with Microsoft Excel)

Table I.2. To conduct a static force analysis, the user first selects the preferred unit system (either Metric or English units) for the written mechanism output.

After selecting the unit system, the user specifies the mechanism link dimensions, slider force, static friction coefficient, gravitational constant, center of mass vectors, link masses, and the crank motion parameters in this file *Static_Slider_Crank_Simulate.m*. Values are specified for link variables $\mathbf{W}_1$, $\mathbf{V}_1$, $\mathbf{U}_1$, the slider force $\mathbf{F}$, the static friction coefficient $\pm\mu$, the gravitational constant $\mathbf{g}$, the center of mass vectors $\mathbf{R}_1$, $\mathbf{R}_2$, and link masses m_1 through m_3 (Figure I.3a) according to the Metric or English units listed in the file (Figure I.4). Values are also specified for the initial crank angle (*start_ang*), the crank rotation increment (*step_ang*), and the final crank angle (*stop_ang*). Figure I.4 illustrates the user-input sections of the file *Static_Slider_Crank_Simulate.m*, with sample values in bold type.*

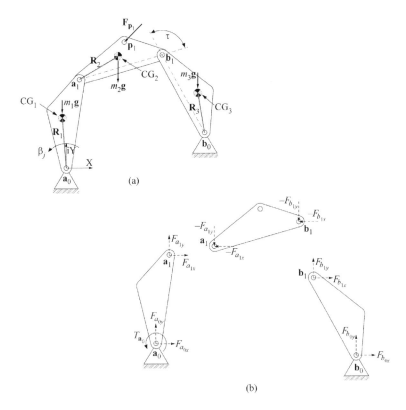

(a)

(b)

FIGURE I.1

(a) Planar four-bar mechanism and (b) output variables (angle τ shown in a).

---

* The unit for crank angular rotation is degrees.

```
%---------------------------------------------------------------
% Here, values for the planar 4-bar mechanism variables W1, V1,
% G1, U1 and L1 are assigned. The coupler force Fp1, gravity
% "g", center of mass vectors R1, R2 and R3 and link masses are
% specified here also.
%
% Available units are the following:
% 'SI' (Metric): Length [meter], Mass [kg] and Force [N]
% 'US' (English): Length [inch], Mass [lbm] and Force [lbf]

unit_select = 'SI';

W1 = 0.5*exp(i*90*pi/180);
V1 = 0.75*exp(i*19.3737*pi/180);
G1 = 0.75 + i*0;
U1 = 0.75*exp(i*93.2461*pi/180);
L1 = 0.5*exp(-i*60.7834*pi/180);

Fp1 = [0, 4500]; g = -9.81;

R1 = -0.0932 - i*0.0380;
R2 = 0.0955 + i*0.0159;
R3 = -0.1180 + i*0.1261;

m1 = 8; m2 = 40; m3 = 12;
%---------------------------------------------------------------

%---------------------------------------------------------------
% Here, values for the start, step and stop displacement angles
% for the crank link are assigned.

start_ang    = 0;
step_ang     = -1;
stop_ang     = -60;
%---------------------------------------------------------------
```

FIGURE I.2

Sections of *Static_Planar_4Bar_Simulate.m* with sample values in bold.

TABLE I.2

Appendix I.2 MATLAB and SimMechanics Files

Filename	Use of File
Static_Slider_Crank_Simulate.m	To specify mechanism link dimensions, coupler force, and crank link controls
Static_Slider_Crank_Model.slx	To calculate mechanism output and simulate mechanism motion
Post_Simulation_Task.m	To write mechanism output (compatible with Microsoft Excel)

After specifying the mechanism dimensions and driving link parameters in the file *Static_Slider_Crank_Simulate.m*, the next step is to run this file. To run the file, the user can use the "F5" button or the *Run Static_Slider_Crank_Simulate.m* button in the toolbar. When running this file, one file (filename *Static_Loads.csv*) is written to a folder named *Results* (in a format compatible with Microsoft Excel) that includes the calculated mechanism output at each crank link rotation increment. The calculated mechanism output is included in Figure I.3b.

Also, when running *Static_Slider_Crank_Simulate.m*, a graphical user interface appears where the motion of the slider-crank mechanism is simulated over the defined crank rotation range (see Appendix A.3).

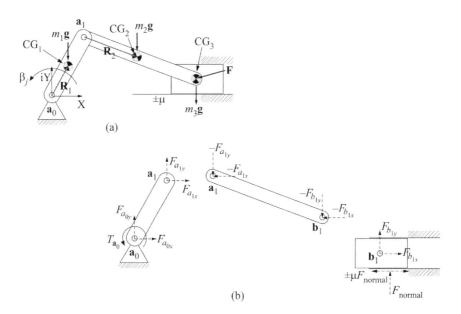

FIGURE I.3

(a) Slider-crank mechanism and (b) output variables.

```
%------------------------------------------------------------------
% Here, values for the slider-crank mechanism variables W1, U1
% and V1 are assigned. The slier force F, friction coefficient
% "mu", gravity "g", center of mass vectors R1 and R2 and link
% masses are specified here also.
%
% Available units are the following:
% 'SI' (Metric): Length [meter], Mass [kg] and Force [N]
% 'US' (English): Length [inch], Mass [lbm] and Force [lbf]

unit_select = 'SI';

LW1 = 4;
theta = 45*pi/180;
W1 = LW1*exp(i*theta);

LU1 = 0;
U1 = i*LU1;

LV1 = 6;
rho = asin((LU1-LW1*sin(theta))/LV1);
V1 = LV1*exp(i*rho);

G1 = W1 + V1 - U1;

F = [-100, 0]; mu = 0.1; g = -9.81;

R1 = 0;
R2 = 0.0265 - i*0.0141;

m1 = 0.05; m2 = 0.025; m3 = 0.075;
%------------------------------------------------------------------

%------------------------------------------------------------------
% Here, values for the start, step and stop displacement angles
% for the crank link are assigned.

start_ang   = 0;
step_ang    = 1;
stop_ang    = 10;
%------------------------------------------------------------------
```

FIGURE I.4

Sections of *Static_Slider_Crank_Simulate.m* with sample values in bold.

TABLE I.3

Appendix I.3 MATLAB and SimMechanics Files

Filename	Use of File
Static_5Bar_2Gear_Simulate.m	To specify mechanism link dimensions, link force, and crank link controls
Static_5Bar_2Gear_Model.slx	To calculate mechanism output and simulate mechanism motion
Post_Simulation_Task.m	To write mechanism output (compatible with Microsoft Excel)

I.3 Geared Five-Bar Mechanism (Two Gears)

The Appendix I.3 folder (which is available for download at www.crcpress.com/product/isbn/9781498724937) includes three MATLAB and SimMechanics files for the static force analysis and simulation of geared five-bar mechanisms having two gears. These three files are described in Table I.3. To conduct a static force analysis, the user first selects the preferred unit system (either Metric or English units) for the written mechanism output.

After selecting the unit system, the user specifies the mechanism link dimensions, intermediate link force, gravitational constant, center of mass vectors, link masses, and the crank motion parameters in the file *Static_5Bar_2Gear_Simulate.m*. Values are specified for link variables $\mathbf{W}_1$, $\mathbf{V}_1$, $\mathbf{G}_1$, $\mathbf{U}_1$, $\mathbf{L}_1$, $\mathbf{S}_1$, the intermediate link force $\mathbf{F}_{p1}$, the gear ratio, the gravitational constant $\mathbf{g}$, the center of mass vectors $\mathbf{R}_1$ through $\mathbf{R}_4$, and link masses m_1 through m_4 (Figure I.5a) according to the Metric or English units listed in the file (Figure I.6). Values are also specified for the initial crank angle (*start_ang*), the crank rotation increment (*step_ang*), and the final crank angle (*stop_ang*). Figure I.6 illustrates the user-input sections of the file *Static_5Bar_2Gear_Simulate.m*, with sample values in bold type.[*] While most of the link vector values in this figure appear in polar exponential form, they can all be specified in any of the rectangular and complex forms given in Equation 2.1.

After specifying the mechanism dimensions and driving link parameters in the file *Static_5Bar_2Gear_Simulate.m*, the next step is to run this file. When running this file, one file (filename *Static_Loads.csv*) is written to a folder named *Results* (in a format compatible with Microsoft Excel) that includes the calculated mechanism output at each crank link rotation increment.[†] The calculated mechanism output is included in Figure I.5b.

Also, when running *Static_5Bar_2Gear_Simulate.m*, a graphical user interface appears where the motion of the geared five-bar mechanism is simulated over the defined crank rotation range (see Appendix A.3).

I.4 Geared Five-Bar Mechanism (Three Gears)

The Appendix I.4 folder (which is available for download at www.crcpress.com/product/isbn/9781498724937) includes three MATLAB and SimMechanics files for the static force analysis and simulation of geared five-bar mechanisms having three gears. These three files are described in Table I.4. The procedure to use these MATLAB and SimMechanics files is identical to the procedure given in Section I.3.

---

[*] The unit for crank angular rotation is degrees.
[†] When the mechanism experiences a circuit defects occur, results are no longer written to the *.csv* file.

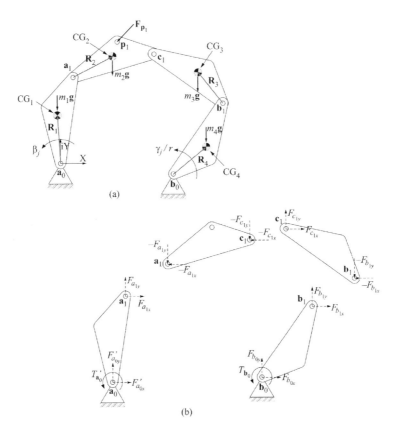

FIGURE I.5
(a) Geared five-bar mechanism and (b) output variables.

I.5 Watt II Mechanism

The Appendix I.5 folder (which is available for download at www.crcpress.com/product/isbn/9781498724937) includes three MATLAB and SimMechanics files for the static force analysis and simulation of Watt II mechanisms. These three files are described in Table I.5. To conduct a static force analysis, the user first selects the preferred unit system (either Metric or English units) for the written mechanism output.

After selecting the unit system, the user specifies the mechanism link dimensions, coupler forces, gravitational constant, center of mass vectors, link masses, and the crank motion parameters in the file *Static_Watt_II_Simulate.m*. Values are specified for link variables $\mathbf{W}_1$, $\mathbf{V}_1$, $\mathbf{G}_1$, $\mathbf{U}_1$, $\mathbf{L}_1$, $\mathbf{W}_1^*$, $\mathbf{V}_1^*$, $\mathbf{G}_1^*$, $\mathbf{U}_1^*$, $\mathbf{L}_1^*$, the coupler forces $\mathbf{F}_{p1}$ and, $\mathbf{F}_{p1^*}$ the gravitational constant $\mathbf{g}$, the center of mass vectors $\mathbf{R}_1$ through $\mathbf{R}_5$, and link masses m_1 through m_5 (Figure I.7a) according to the Metric or English units listed in the file (Figure I.8). Values are also specified for the initial crank angle (*start_ang*), the crank rotation increment (*step_ang*), and the final crank angle (*stop_ang*).

Figure I.8 illustrates the user-input sections of the file *Static_Watt_II_Simulate.m*, with sample values in bold type.* While most of the link vector values in this figure appear in

---

* The unit for crank angular rotation is degrees.

```
%------------------------------------------------------------
% Here, values for the geared 5-bar mechanism variables W1, V1,
% G1, U1, L1 and S1 are assigned. The link force Fp1, gear
% ratio, gravity "g", center of mass vectors R1, R2, R3 and R4
% and link masses are specified here also.
% Only negative gear ratio values should be specified.
%
% Available units are the following:
% 'SI' (Metric) : Length [meter], Mass [kg] and Force [N]
% 'US' (English): Length [inch], Mass [lbm] and Force [lbf]

unit_select = 'SI';

W1 = 0.5*exp(i*90*pi/180);
V1 = 0.75*exp(i*32.7304*pi/180);
G1 = 0.75 + i*0;
U1 = 0.75*exp(i*45*pi/180);
L1 = 0.5*exp(i*74.1400*pi/180);
S1 = 0.75*exp(i*149.9847*pi/180);

Fp1 = [-2500,-3000]; ratio = 2; g = -9.81;

R1 = 0 + i*0.0831;
R2 = 0.2558 + i*0.2955;
R3 = -0.3247 + i*0.1876;
R4 = 0.0356 + i*0.0356;

m1 = 22.54; m2 = 29.785; m3 = 12.075; m4 = 75.67;
%------------------------------------------------------------

%------------------------------------------------------------
% Here, values for the start, step and stop displacement angles
% for the crank link are assigned.

start_ang   = 0;
step_ang    = 1;
stop_ang    = 2;
%------------------------------------------------------------
```

FIGURE I.6

Sections of *Static_5Bar_2Gear_Simulate.m* with sample values in bold.

TABLE I.4

Appendix I.4 MATLAB and SimMechanics Files

Filename	Use of File
Static_5Bar_3Gear_Simulate.m	To specify mechanism link dimensions, link force, and crank link controls
Static_5Bar_3Gear_Model.slx	To calculate mechanism output and simulate mechanism motion
Post_Simulation_Task.m	To write mechanism output (compatible with Microsoft Excel)

polar exponential form, they can all be specified in any of the rectangular and complex forms given in Equation 2.1.

After specifying the mechanism dimensions and driving link parameters in the file *Static_Watt_II_Simulate.m*, the next step is to run this file. When running this file, one file (filename *Static_Loads.csv*) is written to a folder named *Results* (in a format compatible with Microsoft Excel) that includes the calculated mechanism output at each crank link rotation increment.* The calculated mechanism output is included in Figure I.7b.

Also, when running *Static_Watt_II_Simulate.m*, a graphical user interface appears where the motion of the Watt II mechanism is simulated over the defined crank rotation range (see Appendix A.3).

---

* When the mechanism experiences a circuit defects occur, results are no longer written to the *.csv file.

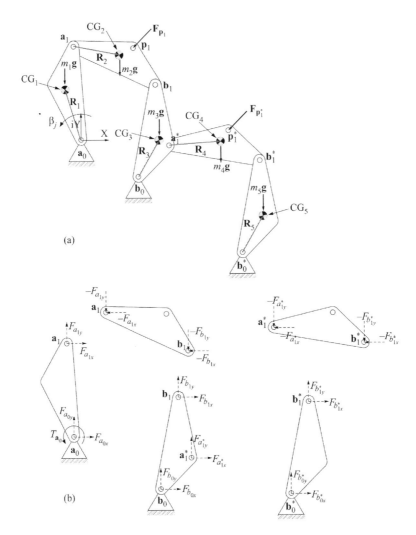

FIGURE I.7
(a) Watt II mechanism and (b) output variables.

I.6 Stephenson III Mechanism

The Appendix I.6 folder (which is available for download at www.crcpress.com/product/ isbn/9781498724937) includes three MATLAB and SimMechanics files for the static force analysis and simulation of Stephenson III mechanisms. These three files are described in Table I.6. To conduct a static force analysis, the user first selects the preferred unit system (either Metric or English units) for the written mechanism output.

After selecting the unit system, the user specifies the mechanism link dimensions, intermediate link force, gravitational constant, center of mass vectors, link masses, and the crank motion parameters in the file *Static_Stephenson_III_Simulate.m*. Values are specified for link variables $\mathbf{W}_1$, $\mathbf{V}_1$, $\mathbf{G}_1$, $\mathbf{U}_1$, $\mathbf{V}_1^*$, $\mathbf{G}_1^*$, $\mathbf{U}_1^*$, $\mathbf{L}_1^*$, and the intermediate link

```
%------------------------------------------------------------
% Here, values for the Watt II mechanism variables W1, V1, G1,
% U1, L1, W1s, V1s, G1s, U1s and L1s are assigned.  The coupler
% forces Fp1 and Fp1s, gravity "g", the center of mass vectors
% R1, R2, R3, R4 and R5 and link masses are specified here also.
%
% Available units are the following:
% 'SI' (Metric) : Length [meter], Mass [kg] and Force [N]
% 'US' (English): Length [inch], Mass [lbm] and Force [lbf]

unit_select = 'SI';

W1 = 0.5*exp(i*90*pi/180);
V1 = 0.75*exp(i*19.3737*pi/180);
G1 = 0.75 + i*0;
U1 = 0.75*exp(i*93.2461*pi/180);
L1 = 0.5*exp(i*60.7834*pi/180);

W1s = 0.5*exp(i*45*pi/180);
V1s = 0.75*exp(i*7.9416*pi/180);
G1s = 0.7244 - i*0.1941;
U1s = 0.75*exp(i*60.2717*pi/180);
L1s = 0.5*exp(i*49.3512*pi/180);

Fp1 = [2500, 3000]; Fp1s = [-1500, 2000]; g = -9.81;

R1 = 0 + i*0.25;
R2 = 0.3172 + i*0.2284;
R3 = 0.1037 + i*0.3675;
R4 = 0.3562 + i*0.161;
R5 = 0.1860 + i*0.3257;

m1 = 8.05; m2 = 29.785; m3 = 33.81; m4 = 29.785; m5 = 12.075;
%------------------------------------------------------------


%------------------------------------------------------------
% Here, values for the start, step and stop displacement angles
% for the crank link are assigned.

start_ang   = 0;
step_ang    = 1;
stop_ang    = 100;
%------------------------------------------------------------
```

FIGURE I.8

Sections of *Static_Watt_II_Simulate.m* with sample values in bold.

TABLE I.5

Appendix I.5 MATLAB and SimMechanics Files

Filename	Use of File
Static_Watt_II_Simulate.m	To specify mechanism link dimensions, coupler force, and crank link controls
Static_Watt_II_Model.slx	To calculate mechanism output and simulate mechanism motion
Post_Simulation_Task.m	To write mechanism output (compatible with Microsoft Excel)

force, $\mathbf{F}_{p1}$ the gravitational constant $\mathbf{g}$, the center of mass vectors $\mathbf{R}_1$ through $\mathbf{R}_5$, and link masses m_1 through m_5 (Figure I.9a) according to the Metric or English units listed in the file (Figure I.10). Values are also specified for the initial crank angle (*start_ang*), the crank rotation increment (*step_ang*), and the final crank angle (*stop_ang*). Figure I.10 illustrates the user-input sections of the file *Static_Stephenson_III_Simulate.m*, with sample values in bold type.* While most of the link vector values in this figure appear in polar exponential

---

* The unit for crank angular rotation is degrees.

TABLE I.6

Appendix I.6 MATLAB and SimMechanics Files

Filename	Use of File
Static_Stephenson_III_Simulate.m	To specify mechanism link dimensions, coupler force, and crank link controls
Static_Stephenson_III_Model.slx	To calculate mechanism output and simulate mechanism motion
Post_Simulation_Task.m	To write mechanism output (compatible with Microsoft Excel)

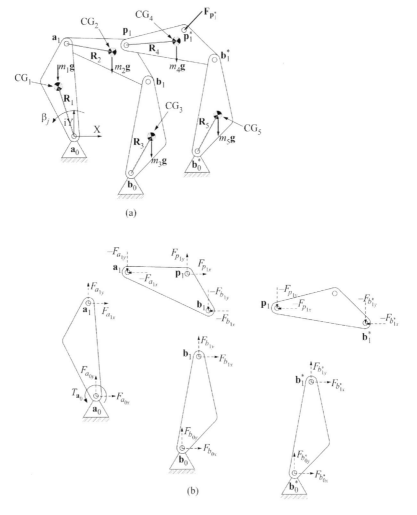

FIGURE I.9

(a) Stephenson III mechanism and (b) output variables.

form, they can all be specified in any of the rectangular and complex forms given in Equation 2.1.

After specifying the mechanism dimensions and driving link parameters in the file *Static_Stephenson_III_Simulate.m*, the next step is to run this file. When running this file, one file (filename *Static_Loads.csv*) is written to a folder named *Results* (in a format compatible

```
%------------------------------------------------------------------
% Here, values for the Stephenson III mechanism variables W1,
% V1, G1, U1, L1, V1s, G1s, U1s and L1s are assigned.  The
% coupler force Fp1s, gravity "g", the center of mass vectors
% R1, R2, R3, R4 and R5 and link masses are specified here also.
%
% Available units are the following:
% 'SI' (Metric) : Length [meter], Mass [kg] and Force [N]
% 'US' (English): Length [inch], Mass [lbm] and Force [lbf]

unit_select = 'SI';

W1 = 1.3575*exp(-i*64.4543*pi/180);
V1 = 0.9726*exp(i*57.2740*pi/180);
G1 = 0.9207 - i*2.2989;
U1 = 1.9019*exp(i*84.2513*pi/180);
L1 = 0.6120*exp(-i*143.6057*pi/180);

V1s = 0.5815*exp(-i*125.7782*pi/180);
G1s = -3.3894 + i*2.2487;
U1s = 2.9955*exp(-i*42.1315*pi/180);
L1s = 2.2217*exp(-i*5*pi/180);

Fp1s = [0, -40]; g = -9.81;

R1 = 0.3846 - i*0.3412;
R2 = -0.6343 + i*1.0364;
R3 = -0.2988 - i*0.3639;
R4 = 1.1487 + i*0.0597;
R5 = -0.2384 - i*0.3618;

m1 = 4; m2 = 8; m3 = 4; m4 = 12; m5 = 4;
%------------------------------------------------------------------

%------------------------------------------------------------------
% Here, values for the start, step and stop displacement angles
% for the crank link are assigned.

start_ang   = 0;
step_ang    = 1;
stop_ang    = 40;
%------------------------------------------------------------------
```

FIGURE I.10
Sections of *Static_Stephenson_III_Simulate.m* with sample values in bold.

with Microsoft Excel) that includes the calculated mechanism output at each crank link rotation increment.[*] The calculated mechanism output is included in Figure I.9b.

Also, when running *Static_Stephenson_III_Simulate.m*, a graphical user interface appears where the motion of the Stephenson III mechanism is simulated over the defined crank rotation range (see Appendix A.3).

---

[*] When the mechanism experiences a circuit defects occur, results are no longer written to the *.csv file.

Appendix J: User Instructions for Chapter 7 MATLAB® and SimMechanics® Files

J.1 Planar Four-Bar Mechanism

The Appendix J.1 folder (which is available for download at www.crcpress.com/product/ isbn/9781498724937) includes three MATLAB and SimMechanics files for the dynamic force analysis and simulation of planar four-bar mechanisms. These three files are described in Table J.1. To conduct a static force analysis, the user first selects the preferred unit system (either Metric or English units) for the written mechanism output.

After selecting the unit system, the user specifies the mechanism link dimensions, coupler force and follower torque, gravitational constant, center of mass vectors, link masses, link mass moments of inertia, and the crank motion parameters in the file *Dynamic_Planar_4Bar_Simulate.m*. Values are specified for link variables $\mathbf{W}_1$, $\mathbf{V}_1$, $\mathbf{G}_1$, $\mathbf{U}_1$, $\mathbf{L}_1$, the coupler force $\mathbf{F}_{\mathbf{p}_1}$, the follower torque $T_{\mathbf{b}_0}$, the gravitational constant $\mathbf{g}$, the center of mass vectors $\mathbf{R}_1$, $\mathbf{R}_3$, and $\mathbf{R}_7$, link masses m_1 through m_3, and link mass moments of inertia I_1 through I_3 (Figure J.1a) according to the Metric or English units listed in the file (Figure J.2). Values are also specified for the initial crank angle (*start_ang*), the crank rotation increment (*step_ang*), and the final crank angle (*stop_ang*). Lastly, values are specified for the angular velocity (*angular_vel*) and angular acceleration (*angular_acc*) of the crank. Figure J.2 illustrates the user-input sections of the file *Dynamic_Planar_4Bar_Simulate.m*, with sample values in bold type.* While most of the link vector values in this figure appear in polar exponential form, they can all be specified in any of the rectangular and complex forms given in Equation 2.1.

After specifying the mechanism dimensions, loads, mass parameters, and driving link parameters in the file *Dynamic_Planar_4Bar_Simulate.m*, the next step is to run this file. When running this file, one file (filename *Dynamic_Loads.csv*) is written to a folder named *Results* (in a format compatible with Microsoft® Excel) that includes the calculated mechanism output at each crank link rotation increment.† The calculated mechanism output is included in Figure J.1b.

Also, when running *Dynamic_Planar_4Bar_Simulate.m*, a graphical user interface appears where the motion of the planar four-bar mechanism is simulated over the defined crank rotation range (see Appendix A.3).

---

* The units for crank angular rotation, velocity, and acceleration are degrees, rad/s, and rad/s², respectively.
† If the mechanism experiences a circuit defect (or if the crank decelerates to a zero velocity state), results beyond this point are not written to the *.csv* file.

TABLE J.1

Appendix J.1 MATLAB and SimMechanics Files

Filename	Use of File
Dynamic_Planar_4Bar_Simulate.m	To specify mechanism link dimensions, mass parameters, coupler force, and crank link controls
Dynamic_Planar_4Bar_Model.slx	To calculate mechanism output and simulate mechanism motion
Post_Simulation_Task.m	To write mechanism output (compatible with Microsoft Excel)

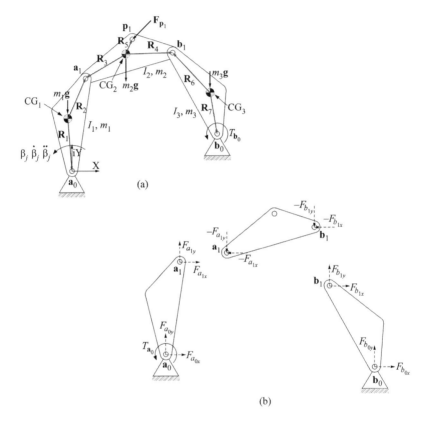

FIGURE J.1

(a) Planar four-bar mechanism and (b) output variables (angle τ shown in Figure I.2a).

J.2 Slider-Crank Mechanism

The Appendix J.2 folder (which is available for download at www.crcpress.com/product/ isbn/9781498724937) includes three MATLAB and SimMechanics files for the dynamic force analysis and simulation of slider-crank mechanisms. These three files are described in Table J.2. To conduct a static force analysis, the user first selects the preferred unit system (either Metric or English units) for the written mechanism output.

After selecting the unit system, the user specifies the mechanism link dimensions, slider force, dynamic friction coefficient, gravitational constant, center of mass vectors, link masses and mass moments of inertia, and the crank motion parameters in the file

```
%--------------------------------------------------------------
% Here, values for the planar 4-bar mechanism variables W1, V1,
% G1, U1 and L1 are assigned. The coupler force Fp1, follower
% torque Tb0, gravity "g", center of mass vectors R1, R3 and R7,
% link masses and link mass moments of inertia are specified
% here also.
%
% Available units are the following:
% 'SI' (Metric) : Length [meter], Mass [kg] and Force [N]
% 'US' (English): Length [inch], Mass [lbm] and Force [lbf]

unit_select = 'SI';

W1 = 0.5*exp(i*90*pi/180);
V1 = 0.75*exp(i*19.3737*pi/180);
G1 = 0.75 + i*0;
U1 = 0.75*exp(i*93.2461*pi/180);
L1 = 0.5*exp(i*60.7834*pi/180);

Fp1 = [0,0]; Tb0 = 0;

R1 = 0 - i*0.25;
R3 = -0.3172 - i*0.2284;
R7 = 0.0212 - i*0.3744;

R2 = R1 + W1; R4 = R3 + V1; R5 = R3 + L1; R6 = R7 + U1;

m1 = 8.05; I1 = 0.805;
m2 = 29.785; I2 = 5.635;
m3 = 12.075; I3 = 2.415;
%--------------------------------------------------------------

%--------------------------------------------------------------
% Here, values for the start, step and stop displacement angles
% for the crank link are assigned.

start_ang  = 0;
step_ang   = 1;
stop_ang   = 360;
%--------------------------------------------------------------

%--------------------------------------------------------------
% Here, values for crank link angular velocity and angular
% acceleration are assigned.

angular_vel  = 1.0;
angular_acc  = 0.1;
%--------------------------------------------------------------
```

FIGURE J.2

Sections of *Dynamic_Planar_4Bar_Simulate.m* with sample values in bold.

TABLE J.2

Appendix J.2 MATLAB and SimMechanics Files

Filename	Use of File
Dynamic_Slider_Crank_Simulate.m	To specify mechanism link dimensions, mass parameters, slider force, and crank link controls
Dynamic_Slider_Crank_Model.slx	To calculate mechanism output and simulate mechanism motion
Post_Simulation_Task.m	To write mechanism output (compatible with Microsoft Excel)

Dynamic_Slider_Crank_Simulate.m. Values are specified for link variables $\mathbf{W}_1$, $\mathbf{V}_1$, $\mathbf{U}_1$, the slider force $\mathbf{F}$, the dynamic friction coefficient $\pm\mu$, the gravitational constant $\mathbf{g}$, the center of mass vectors $\mathbf{R}_1$ and $\mathbf{R}_3$, link masses m_1 through m_3, and link mass moments of inertia I_1 and I_2 (Figure J.3a) according to the Metric or English units listed in the file (Figure J.4).

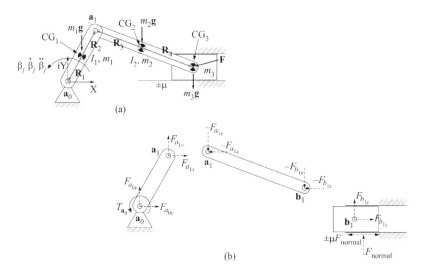

FIGURE J.3
(a) Slider-crank mechanism and (b) output variables.

Values are also specified for the initial crank angle (*start_ang*), the crank rotation increment (*step_ang*), and the final crank angle (*stop_ang*). Lastly, values are specified for the angular velocity (*angular_vel*) and angular acceleration (*angular_acc*) of the crank. Figure J.4 illustrates the user-input sections of the file *Dynamic_Slider_Crank_Simulate.m*, with sample values in bold type.*

After specifying the mechanism dimensions and driving link parameters in the file *Dynamic_Slider_Crank_Simulate.m*, the next step is to run this file. When running this file, one file (filename *Dynamic_Loads.csv*) is written to a folder named *Results* (in a format compatible with Microsoft Excel) that includes the calculated mechanism output at each crank link rotation increment. The calculated mechanism output is included in Figure J.3b.

Also, when running *Dynamic_Slider_Crank_Simulate.m*, a graphical user interface appears where the motion of the slider-crank mechanism is simulated over the defined crank rotation range (see Appendix A.3).

J.3 Geared Five-Bar Mechanism (Two Gears)

The Appendix J.3 folder (which is available for download at www.crcpress.com/product/ isbn/9781498724937) includes three MATLAB and SimMechanics files for the dynamic force analysis and simulation of geared five-bar mechanisms having two gears. These three files are described in Table J.3. To conduct a static force analysis, the user first selects the preferred unit system (either Metric or English units) for the written mechanism output.

After selecting the unit system, the user specifies the mechanism link dimensions, intermediate link force, gravitational constant, center of mass vectors, link masses, mass

---

* The units for crank angular rotation, velocity, and acceleration are degrees, rad/s, and rad/s², respectively.

```
%-----------------------------------------------------------
% Here, values for the slider-crank mechanism variables W1, V1,
% G1 and U1 are assigned and calculated.  Variables LW1, LU1 and
% LV1 are the scalar lengths of vectors W1, U1 and V1. The
% slider force "F", dynamic friction coefficient "mu" and
% gravity "g" are assigned here also. Lastly, the center of mass
% vectors R1 and R3, link masses and link mass moments of
% inertia are specified here.
%
% Available units are the following:
% 'SI' (Metric) : Length [meter], Mass [kg] and Force [N]
% 'US' (English): Length [inch], Mass [lbm] and Force [lbf]

unit_select = 'SI';

LW1 = 0.5; theta = 90*pi/180;
W1 = LW1*exp(i*theta);

LU1 = 0;
U1 = i*LU1;

LV1 = 0.9014;
rho = asin((LU1-LW1*sin(theta))/LV1);
V1 = LV1*exp(i*rho);

G1 = W1 + V1 - U1;

F = [0,0]; mu = 0.5; g = -9.81;

R1 = 0 - i*0.25;
R3 = -0.3750 + i*0.25;

R2 = R1 + W1; R4 = R3 + V1;

m1 = 8.05; I1 = 0.805;
m2 = 14.49; I2 = 4.025; m3 = 30;
%-----------------------------------------------------------

%-----------------------------------------------------------
% Here, values for the start, step and stop displacement angles
% for the crank link are assigned.

start_ang   = 0;
step_ang    = 1;
stop_ang    = 10;
%-----------------------------------------------------------

%-----------------------------------------------------------
% Here, values for crank link angular velocity and angular
% acceleration are assigned.

angular_vel   = 10;
angular_acc   = 0;
%-----------------------------------------------------------
```

FIGURE J.4

Sections of *Dynamic_Slider_Crank_Simulate.m* with sample values in bold.

TABLE J.3

Appendix J.3 MATLAB and SimMechanics Files

Filename	Use of File
Dynamic_5Bar_2Gear_Simulate.m	To specify mechanism link dimensions, mass parameters, coupler force, and crank link controls
Dynamic_5Bar_2Gear_Model.slx	To calculate mechanism output and simulate mechanism motion
Post_Simulation_Task.m	To write mechanism output (compatible with Microsoft Excel)

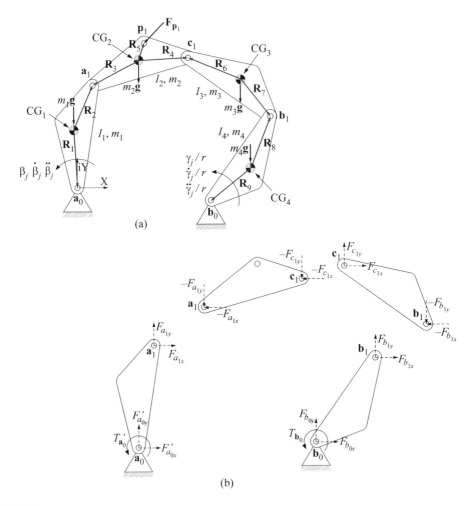

FIGURE J.5
(a) Geared five-bar mechanism and (b) output variables.

moments of inertia, and the crank motion parameters in the file *Dynamic_5Bar_2Gear_Simulate.m*. Values are specified for link variables $\mathbf{W}_1$, $\mathbf{V}_1$, $\mathbf{G}_1$, $\mathbf{U}_1$, $\mathbf{L}_1$, $\mathbf{S}_1$, the intermediate link force $\mathbf{F}_{\mathbf{p}_1}$, the gravitational constant $\mathbf{g}$, the center of mass vectors $\mathbf{R}_1$, $\mathbf{R}_3$, $\mathbf{R}_7$, and $\mathbf{R}_9$, link masses m_1 through m_4, and link mass moments of inertia I_1 through I_4 (Figure J.5a) according to the Metric or English units listed in the file (Figure J.6). Values are also specified for the initial crank angle (*start_ang*), the crank rotation increment (*step_ang*), and the final crank angle (*stop_ang*). Lastly, values are specified for the angular velocity (*angular_vel*) and angular acceleration (*angular_acc*) of the crank. Figure J.6 illustrates the user-input sections of the file *Dynamic_5Bar_2Gear_Simulate.m*, with sample values in bold type.* While most of the link vector values in this figure appear in polar exponential form, they can all be specified in any of the rectangular and complex forms given in Equation 2.1.

After specifying the mechanism dimensions, loads, mass parameters, and driving link parameters in the file *Dynamic_5Bar_2Gear_Simulate.m*, the next step is to run this file.

---

* The units for crank angular rotation, velocity, and acceleration are degrees, rad/s, and rad/s², respectively.

```
%----------------------------------------------------------
% Here, values for the geared 5-bar mechanism variables W1, V1,
% G1, U1, L1 and S1 are assigned. The link force Fp1, gear
% ratio, gravity "g", center of mass vectors R1, R3, R7 and R9,
% link masses and link mass moments of inertia are specified
% here also. Only negative gear ratio values should be
% specified.
%
% Available units are the following:
% 'SI' (Metric) : Length [meter], Mass [kg] and Force [N]
% 'US' (English): Length [inch], Mass [lbm] and Force [lbf]

unit_select = 'SI';

W1 = 0.5*exp(i*90*pi/180);
V1 = 0.75*exp(i*32.7304*pi/180);
G1 = 0.75*exp(i*0*pi/180);
U1 = 0.75*exp(i*45*pi/180);
L1 = 0.5*exp(i*74.1400*pi/180);
S1 = 0.75*exp(i*149.9847*pi/180);

Fp1 = [0,-1000]; ratio = 2; g = -9.81;

R1 = 0 - i*0.0831; R3 = -0.2558 - i*0.2955;
R7 = 0.3247 - i*0.1876; R9 = -0.0356 - i*0.0356;

R2 = R1 + W1; R4 = R3 + V1; R5 = R3 + L1; R6 = R7 + S1;
R8 = R9 + U1;

m1 = 22.54; I1 = 0.505;
m2 = 29.785; I2 = 5.635;
m3 = 12.075; I3 = 2.415;
m4 = 75.67; I4 = 5.635;
%----------------------------------------------------------

%----------------------------------------------------------
% Here, values for the start, step and stop displacement angles
% for the crank link are assigned.

start_ang   = 0;
step_ang    = 1;
stop_ang    = 360;
%----------------------------------------------------------

%----------------------------------------------------------
% Here, values for crank link angular velocity and angular
% acceleration are assigned.

angular_vel  = 1;
angular_acc  = 0;
%----------------------------------------------------------
```

FIGURE J.6

Sections of *Dynamic_5Bar_2Gear_Simulate.m* with sample values in bold.

When running this file, one file (filename *Dynamic_Loads.csv*) is written to a folder named *Results* (in a format compatible with Microsoft Excel) that includes the calculated mechanism output at each crank link rotation increment.* The calculated mechanism output is included in Figure J.5b.

Also, when running *Dynamic_5Bar_2Gear_Simulate.m*, a graphical user interface appears where the motion of the geared five-bar mechanism is simulated over the defined crank rotation range (see Appendix A.3).

---

* If the mechanism experiences a circuit defect (or if the crank decelerates to a zero velocity state), results beyond this point are not written to the *.csv* file.

TABLE J.4

Appendix J.4 MATLAB and SimMechanics Files

Filename	Use of File
Dynamic_5Bar_3Gear_Simulate.m	To specify mechanism link dimensions, mass parameters, coupler force, and crank link controls
Dynamic_5Bar_3Gear_Model.slx	To calculate mechanism output and simulate mechanism motion
Post_Simulation_Task.m	To write mechanism output (compatible with Microsoft Excel)

J.4 Geared Five-Bar Mechanism (Three Gears)

The Appendix J.4 folder (which is available for download at www.crcpress.com/product/isbn/9781498724937) includes three MATLAB and SimMechanics files for the dynamic force analysis and simulation of geared five-bar mechanisms having three gears. These three files are described in Table J.4.

The procedure to calculate dynamic force and torque solutions for geared five-bar mechanism having three gears is identical to the procedure given in Section J.3. In the file *Dynamic_5Bar_3Gear_Simulate.m*, the user also specifies values for variables m_5 and I_5 (the mass and mass moment of inertia of the idler gear, respectively).

J.5 Watt II Mechanism

The Appendix J.5 folder (which is available for download at www.crcpress.com/product/isbn/9781498724937) includes three MATLAB and SimMechanics files for the dynamic force analysis and simulation of Watt II mechanisms. These three files are described in Table J.5. To conduct a static force analysis, the user first selects the preferred unit system (either Metric or English units) for the written mechanism output.

After selecting the unit system, the user specifies the mechanism link dimensions, coupler link forces and follower torques, gravitational constant, center of mass vectors, link masses and mass moments of inertia, and the crank motion parameters in the file *Dynamic_Watt_II_Simulate.m*. Values are specified for link variables $\mathbf{W}_1, \mathbf{V}_1, \mathbf{G}_1, \mathbf{U}_1, \mathbf{L}_1, \mathbf{W}_1^*, \mathbf{V}_1^*, \mathbf{G}_1^*, \mathbf{U}_1^*, \mathbf{L}_1^*$, the coupler forces $\mathbf{F}_{\mathbf{p}_1}$ and $\mathbf{F}_{\mathbf{p}_1}^*$, the follower torques $T_{\mathbf{b}_0}$ and $T_{\mathbf{b}_0}^*$, the gravitational constant $\mathbf{g}$, the center of mass vectors $\mathbf{R}_1, \mathbf{R}_3, \mathbf{R}_7, \mathbf{R}_9$, and $\mathbf{R}_{13}$, link masses m_1 through m_5, and link mass moments of inertia I_1 through I_5 (Figure J.7a) according to the Metric or English units listed in the file (Figure J.8) Values are also specified for the initial crank angle (*start_ang*), the

TABLE J.5

Appendix J.5 MATLAB and SimMechanics Files

Filename	Use of File
Dynamic_Watt_II_Simulate.m	To specify mechanism link dimensions, mass parameters, coupler force, and crank link controls
Dynamic_Watt_II_Model.slx	To calculate mechanism output and simulate mechanism motion
Post_Simulation_Task.m	To write mechanism output (compatible with Microsoft Excel)

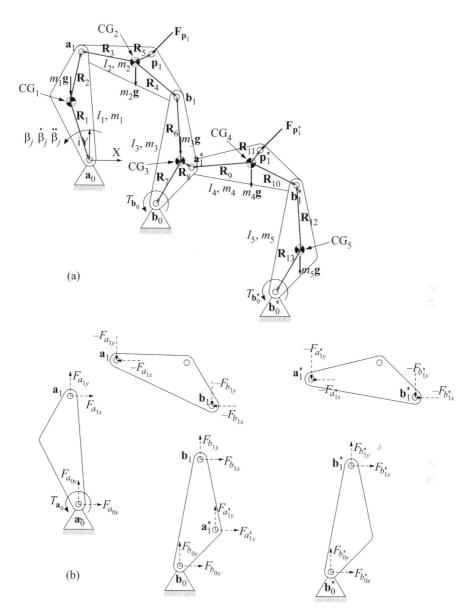

FIGURE J.7
(a) Watt II mechanism and (b) output variables.

crank rotation increment (*step_ang*), and the final crank angle (*stop_ang*). Lastly, values are specified for the angular velocity (*angular_vel*) and angular acceleration (*angular_acc*) of the crank. Figure J.8 illustrates the user-input sections of the file *Dynamic_Watt_II_Simulate.m*, with sample values in bold type.[*] While most of the link vector values in this figure appear in polar exponential form, they can all be specified in any of the rectangular and complex forms given in Equation 2.1.

---

[*] The units for crank angular rotation, velocity, and acceleration are degrees, rad/s, and rad/s², respectively.

```
%----------------------------------------------------------------
% Here, values for the Watt II mechanism variables W1, V1, G1,
% U1, L1, W1s, V1s, G1s, U1s and L1s are assigned.  The coupler
% forces Fp1 and follower torques Tb0 and Tb0s, gravity
% "g", the center of mass vectors R1, R3, R7, R9 and R13, link
% masses and link mass moments of inertia are specified here
% also.
%
% Available units are the following:
% 'SI' (Metric) : Length [meter], Mass [kg] and Force [N]
% 'US' (English): Length [inch], Mass [lbm] and Force [lbf]

unit_select = 'SI';

W1 = 0.5*exp(i*90*pi/180);
V1 = 0.75*exp(i*19.3737*pi/180); G1 = 0.75 + i*0;
U1 = 0.75*exp(i*93.2461*pi/180);
L1 = 0.5*exp(i*60.7834*pi/180);

W1s = 0.5*exp(i*45*pi/180);
V1s = 0.75*exp(i*7.9416*pi/180); G1s = 0.7244 - i*0.1941;
U1s = 0.75*exp(i*60.2717*pi/180);
L1s = 0.5*exp(i*49.3512*pi/180);

Fp1 = [-500, -500]; Tb0 = 0;
Fp1s = [-1000, 0]; Tb0s = 0; g = -9.81;

R1 = 0 - i*0.25; R3 = -0.3172 - i*0.2284;
R7 = -0.1037 - i*0.3675; R9 = -0.3562 - i*0.161;
R13 = -0.1860 - i*0.3257;

R2 = R1 + W1; R4 = R3 + V1; R5 = R3 + L1; R6 = R7 + U1;
R8 = R7 + W1s; R10 = R9 + V1s; R11 = R9 + L1s; R12 = R13 + U1s;

m1 = 8.05; I1 = 0.805; m2 = 29.785; I2 = 5.635; m3 = 33.81;
I3 = 5.635; m4 = 29.785; I4 = 5.635; m5 = 12.075; I5 = 2.415;
%----------------------------------------------------------------

%----------------------------------------------------------------
% Here, values for the start, step and stop displacement angles
% for the crank link are assigned.

start_ang   = 0;
step_ang    = -1;
stop_ang    = -360;
%----------------------------------------------------------------

%----------------------------------------------------------------
% Here, values for crank link angular velocity and angular
% acceleration are assigned.

angular_vel   = -1.5;
angular_acc   = -0.25;
%----------------------------------------------------------------
```

FIGURE J.8
Sections of *Dynamic_Watt_II_Simulate.m* with sample values in bold.

After specifying the mechanism dimensions, loads, mass parameters, and driving link parameters in the file *Dynamic_Watt_II_Simulate*, the next step is to run this file. When running this file, one file (filename *Dynamic_Loads.csv*) is written to a folder named *Results* (in a format compatible with Microsoft Excel) that includes the calculated mechanism output at each crank link rotation increment.[*] The calculated mechanism output is included in Figure J.7b.

---

[*] If the mechanism experiences a circuit defect (or if the crank decelerates to a zero velocity state), results beyond this point are not written to the *.csv* file.

Also, when running *Dynamic_Watt_II_Simulate*, a graphical user interface appears where the motion of the Watt II mechanism is simulated over the defined crank rotation range (see Appendix A.3).

J.6 Stephenson III Mechanism

The Appendix J.6 folder (which is available for download at www.crcpress.com/product/isbn/9781498724937) includes three MATLAB and SimMechanics files for the dynamic force analysis and simulation of Stephenson III mechanisms. These three files are described in Table J.6. To conduct a static force analysis, the user first selects the preferred unit system (either Metric or English units) for the written mechanism output.

After selecting the unit system, the user specifies the mechanism link dimensions, coupler link force and follower torques, gravitational constant, center of mass vectors, link masses and mass moments of inertia, and the crank motion parameters in the file *Dynamic_Stephenson_III_Simulate.m*. Values are specified for link variables $\mathbf{W}_1, \mathbf{V}_1, \mathbf{G}_1, \mathbf{U}_1, \mathbf{L}_1, \mathbf{V}_1^*, \mathbf{G}_1^*, \mathbf{U}_1^*, \mathbf{L}_1^*$, the coupler force $\mathbf{F}_{\mathbf{p}_1^*}$, the follower torques $T_{\mathbf{b}_0}$ and $T_{\mathbf{b}_0^*}$, the gravitational constant $\mathbf{g}$, the center of mass vectors $\mathbf{R}_1, \mathbf{R}_3, \mathbf{R}_7, \mathbf{R}_8$, and $\mathbf{R}_{12}$, link masses m_1 through m_5, and link mass moments of inertia I_1 through I_5 (Figure J.9a) according to the Metric or English units listed in the file (Figure J.10). Values are also specified for the initial crank angle (*start_ang*), the crank rotation increment (*step_ang*), and the final crank angle (*stop_ang*). Lastly, values are specified for the angular velocity (*angular_vel*) and angular acceleration (*angular_acc*) of the crank. Figure J.10 illustrates the user-input sections of the file *Dynamic_Stephenson_III_Simulate.m*, with sample values in bold type.[*] While most of the link vector values in this figure appear in polar exponential form, they can all be specified in any of the rectangular and complex forms given in Equation 2.1.

After specifying the mechanism dimensions, loads, mass parameters, and driving link parameters in the file *Dynamic_Stephenson_III_Simulate.m.*, the next step is to run this file. When running this file, one file (filename *Dynamic_Loads.csv*) is written to a folder named *Results* (in a format compatible with Microsoft Excel) that includes the calculated mechanism output at each crank link rotation increment.[†] The calculated mechanism output is included in Figure J.9b.

TABLE J.6

Appendix J.6 MATLAB and SimMechanics Files

Filename	Use of File
Dynamic_Stephenon_III_Simulate.m	To specify mechanism link dimensions, mass parameters, coupler force, and crank link controls
Dynamic_Stephenson_III_Model.slx	To calculate mechanism output and simulate mechanism motion
Post_Simulation_Task.m	To write mechanism output (compatible with Microsoft Excel)

---

[*] The units for crank angular rotation, velocity, and acceleration are degrees, rad/s, and rad/s², respectively.
[†] If the mechanism experiences a circuit defect (or if the crank decelerates to a zero velocity state), results beyond this point are not written to the *.csv* file.

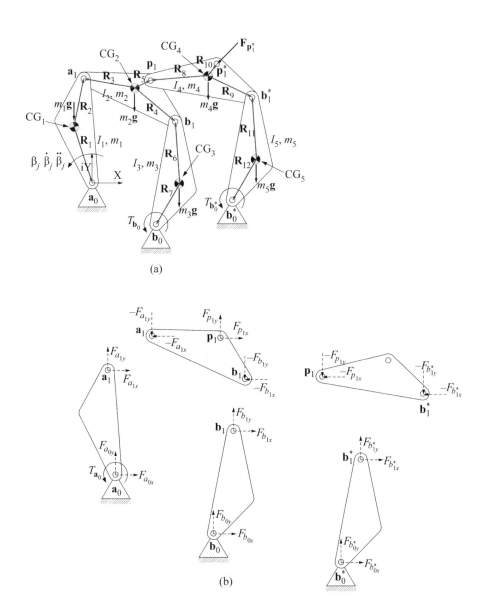

FIGURE J.9
(a) Stephenson III mechanism and (b) output variables.

Also, when running *Dynamic_Stephenson_III_Simulate.m*, a graphical user interface appears where the motion of the Stephenson III mechanism is simulated over the defined crank rotation range (see Appendix A.3).

```
%---------------------------------------------------------------
% Here, values for the Stephenson III mechanism variables W1,
% V1, G1, U1, L1, W1s, V1s, G1s, U1s and L1s are assigned.  The
% coupler forces Fp1 and Fp1s and follower torques Tb0 and Tb0s,
% gravity "g", the center of mass vectors R1, R3, R7, R9 and
% R13, link masses and link mass moments of inertia are
% specified here also.
%
% Available units are the following:
% 'SI' (Metric) : Length [meter], Mass [kg] and Force [N]
% 'US' (English): Length [inch], Mass [lbm] and Force [lbf]

unit_select = 'SI';

W1 = 0.5*exp(i*90*pi/180);
V1 = 0.75*exp(i*19.3737*pi/180);
G1 = 0.75 + i*0;
U1 = 0.75*exp(i*93.2461*pi/180);
L1 = 0.5*exp(i*60.7834*pi/180);

V1s = 1*exp(i*17.1417*pi/180);
G1s = 0.2159 + i*0.2588;
U1s = 1*exp(i*76.4844*pi/180);
L1s = 0.5*exp(i*63.7091*pi/180);

Fp1s = [-1000, 0]; Tb0 = 0; Tb0s = 0; g = -9.81;

R1 = 0 - i*0.25; R3 = -0.3172 - i*0.2284;
R7 = 0.0212 - i*0.3744; R8 = -0.3923 - i*0.2477;
R12 = -0.1169 - i*0.4862;

R2 = R1 + W1; R4 = R3 + V1; R5 = R3 + L1; R6 = R7 + U1;
R9 = R8 + V1s; R10 = R8 + L1s; R11 = R12 + U1s;

m1 = 8.05; I1 = 0.805; m2 = 29.785; I2 = 5.635;
m3 = 33.81; I3 = 5.635; m4 = 29.785; I4 = 5.635;
m5 = 12.075; I5 = 2.415;
%---------------------------------------------------------------

%---------------------------------------------------------------
% Here, values for the start, step and stop displacement angles
% for the crank link are assigned.

start_ang   = 0;
step_ang    = -1;
stop_ang    = -360;
%---------------------------------------------------------------

%---------------------------------------------------------------
% Here, values for crank link angular velocity and angular
% acceleration are assigned.

angular_vel  = -1;
angular_acc  = -0.25;
%---------------------------------------------------------------
```

FIGURE J.10

Sections of *Dynamic_Stephenson_III_Simulate.m* with sample values in bold.

Appendix K: User Instructions for Chapter 10 MATLAB® and SimMechanics® Files

K.1 RRSS Mechanism

The Appendix K.1 folder (which is available for download at www.crcpress.com/prod-uct/isbn/9781498724937) includes the MATLAB and SimMechanics files for the kinematic analysis and simulation of RRSS mechanisms. These three files are described in Table K.1. To conduct a kinematic analysis, the user specifies the mechanism point coordinates and the crank motion parameters in the file *Kinematic_RRSS_Simulate.m*. Values are specified for link variables $\mathbf{a}_0$, $\mathbf{ua}_0$, $\mathbf{a}_1$, $\mathbf{ua}_1$, $\mathbf{b}_0$, $\mathbf{b}_1$, $\mathbf{p}_1$, $\mathbf{q}_1$, and $\mathbf{r}_1$ (Figure K.1a). Values are also specified for the initial crank angle (*start_ang*), the crank rotation increment (*step_ang*), and the final crank angle (*stop_ang*). Lastly, values are specified for the angular velocity (*angular_vel*) and angular acceleration (*angular_acc*) of the crank. Figure K.2 illustrates the user-input sections of the file *Kinematic_RRSS_Simulate.m*, with sample values in bold type.*

After specifying the dimensions and driving link parameters in the file *Kinematic_RRSS_Simulate.m*, the next step is to run this file. When running this file, one file (filename *Disp_Vel_Acc.csv*) is written to a folder named *Results* (in a format compatible with Microsoft® Excel) that includes the calculated mechanism output at each crank link rotation increment.† The calculated mechanism output is included in Figure K.1b.

Also, when running *Kinematic_RRSS_Simulate.m*, a graphical user interface appears where the motion of the RRSS mechanism is simulated over the defined crank rotation range (see Appendix A.3).

K.2 RSSR Mechanism

The Appendix K.2 folder (which is available for download at www.crcpress.com/product/isbn/9781498724937) includes the MATLAB and SimMechanics files for the kinematic analysis and simulation of RSSR mechanisms. These three files are described in Table K.2. To conduct a kinematic analysis, the user specifies the mechanism point coordinates and the crank motion parameters in the file *Kinematic_RSSR_Simulate.m*. Values are specified for link variables $\mathbf{a}_0$, $\mathbf{ua}_0$, $\mathbf{a}_1$, $\mathbf{b}_0$, $\mathbf{ub}_0$, and $\mathbf{b}_1$ (Figure K.3a). Values are also specified for the initial crank angle (*start_ang*), the crank rotation increment (*step_ang*), and the final crank angle (*stop_ang*). Lastly, values are specified for the angular velocity (*angular_vel*)

---

* The units for crank angular rotation, velocity, and acceleration are *degrees*, *rad/s*, and *rad/s²*, respectively.
† If the mechanism experiences a circuit defect (or if the crank decelerates to a zero velocity state), results beyond this point are not written to the *.csv* file.

TABLE K.1

Appendix K.1 MATLAB and SimMechanics Files

Filename	Use of File
Kinematic_RRSS_Simulate.m	To specify mechanism link dimensions and crank link controls
Kinematic_RRSS_Model.slx	To calculate mechanism output and simulate mechanism motion
Post_Simulation_Task.m	To write mechanism output (compatible with Microsoft Excel)

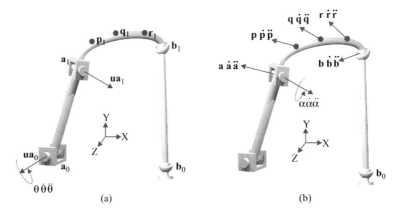

FIGURE K.1
(a) RRSS mechanism and (b) output variables.

```
%-----------------------------------------------------------------
% Here, values for the RRSS mechanism variables a0, a1, ua0,
% ua1, b0, b1, p1, q1 and r1 are assigned.  They must all be
% defined as transposed row matrices (as shown).

a0 = [-0.0576, 0.2890, -1.4112]';
a1 = [0.1452, -2.5421, -1.1800]';
ua0 = [-0.0003, 0.0814, 0.9967]';
ua1 = [0.3040, 0.0992, 0.9475]';
b0 = [0.0851, 0.4570, 0.5096]';
b1 = [1.7725, 5.1566, 0.6499]';
p1 = [1.7321, 0, -1]';
q1 = [1.2321, 0, -1.8660]';
r1 = [1.9486, 0, -1.1250]';
%-----------------------------------------------------------------


%-----------------------------------------------------------------
% Here, values for the start, step and stop displacement angles
% for the crank link are assigned.

start_ang = 0;
step_ang  = 10;
stop_ang  = 200;
%-----------------------------------------------------------------


%-----------------------------------------------------------------
% Here, values for crank link angular velocity and angular
% acceleration are assigned.

angular_vel = 1.0;
angular_acc = 0.1;
%-----------------------------------------------------------------
```

FIGURE K.2
Sections of *Kinematic_RRSS_Simulate.m* with sample values in bold.

and angular acceleration (*angular_acc*) of the crank. Figure K.4 illustrates the user-input sections of the file *Kinematic_RSSR_Simulate.m*, with sample values in bold type.*

After specifying the dimensions and driving link parameters in the file *Kinematic_RSSR_Simulate.m*, the next step is to run this file. When running this file, one file (filenames *Disp_Vel_Acc.csv*) is written to a folder named *Results* (in a format compatible with Microsoft Excel) that includes the calculated mechanism output at each crank link rotation increment.[†]The calculated mechanism output is included in Figure K.3b.

Also, when running *Kinematic_RSSR_Simulate.m*, a graphical user interface appears where the motion of the RSSR mechanism is simulated over the defined crank rotation range (see Appendix A.3).

TABLE K.2

Appendix K.2 MATLAB and SimMechanics Files

Filename	Use of File
Kinematic_RSSR_Simulate.m	To specify mechanism link dimensions and crank link controls
Kinematic_RSSR_Model.slx	To calculate mechanism output and simulate mechanism motion
Post_Simulation_Task.m	To write mechanism output (compatible with Microsoft Excel)

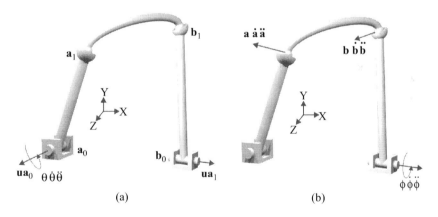

FIGURE K.3
(a) RSSR mechanism and (b) output variables.

---

* The units for crank angular rotation, velocity, and acceleration are *degrees*, *rad/s*, and *rad/s²*, respectively.
[†] If the mechanism experiences a circuit defect (or if the crank decelerates to a zero velocity state), results beyond this pointare not written to the *.csv* file.

```
%------------------------------------------------------------------
% Here, values for the RSSR mechanism variables a0, a1, ua0, b0,
% b1 and ub0 are assigned.  They must all be defined as
% transposed row matrices as shown).

a0 = [0, 0, -0.4023]';
a1 = [0.3356, -0.0708, -0.4023]';
ua0 = [0, 0, 1]';
b0 = [1, -2.3885, 0]';
b1 = [1, -2.3885, 0.7300]';
ub0 = [0, -1, 0]';
%------------------------------------------------------------------


%------------------------------------------------------------------
% Here, values for the start, step and stop displacement angles
% for the crank link are assigned.

start_ang   = 0;
step_ang    = 10;
stop_ang    = 50;
%------------------------------------------------------------------


%------------------------------------------------------------------
% Here, values for crank link angular velocity and angular
% acceleration are assigned.

angular_vel  = 0;
angular_acc  = 0;
%------------------------------------------------------------------
```

FIGURE K.4

Sections of *Kinematic_RSSR_Simulate.m* with sample values in bold.

Appendix L: User Instructions for Chapter 11 MATLAB® and SimMechanics® Files

L.1 R-P-P Robotic Manipulator Forward Kinematics

The Appendix L.1 folder (which is available for download at www.crcpress.com/product/isbn/9781498724937) includes four MATLAB and SimMechanics files for R-P-P robotic manipulator forward kinematics. These four files are described in Table L.1. To conduct an R-P-P forward kinematics analysis, the user specifies the R-P-P joint displacements in the file *RPP_Input.csv*. This file is compatible with Microsoft® Excel. Also, ${}^3\{\mathbf{p}_3\}$ is specified in the file *RPP_FK_Simulate.m*. Figure L.1 illustrates the user-input section of this file, with sample values in bold type.

After specifying the R-P-P joint displacements in *RPP_Input.csv* and ${}^3\{\mathbf{p}_3\}$ in *RPP_FK_Simulate.m*, the next step is to run *RPP_FK_Simulate.m*. When running this file, one file (filename *RPP_p3.csv*) is written to a folder named *Results* (in a format compatible with Microsoft Excel) that includes the ${}^0\{\mathbf{p}_3\}$ values calculated for the given ${}^3\{\mathbf{p}_3\}$ value and joint displacement values.

Also, when running *RPP_FK_Simulate.m*, a graphical user interface appears where the motion of the R-P-P robotic manipulator is simulated over the joint displacements (see Appendix A.3).

L.2 R-R-P Robotic Manipulator Forward Kinematics

The Appendix L.2 folder (which is available for download at www.crcpress.com/product/isbn/9781498724937) includes four MATLAB and SimMechanics files for R-R-P robotic manipulator forward kinematics. These four files are described in Table L.2. To conduct an R-R-P forward kinematics analysis, the user specifies the R-R-P joint displacements in the file *RRP_Input.csv*. This file is compatible with Microsoft Excel.

Also, ${}^3\{\mathbf{p}_3\}$, l_1, and l_2 are specified in the file *RRP_FK_Simulate.m*. Figure L.2 illustrates the user-input section of this file, with sample values in bold type.

After specifying the R-R-P joint displacements in *RRP_Input.csv* and ${}^3\{\mathbf{p}_3\}$, l_1, and l_2 in *RRP_FK_Simulate.m*, the next step is to run *RRP_FK_Simulate.m*. When running this file, one file (file-name *RRP_p3.csv*) is written to a folder named *Results* (in a format compatible with Microsoft Excel) that includes the ${}^3\{\mathbf{p}_3\}$ values calculated for the given ${}^3\{\mathbf{p}_3\}$, length, and joint displacement values.

Also, when running *RRP_FK_Simulate.m*, a graphical user interface appears where the motion of the R-R-P robotic manipulator is simulated over the joint displacements (see Appendix A.3).

TABLE L.1

Appendix L.1 MATLAB Files

Filename	Use of File
RPP_Input.csv	To specify joint displacements
RPP_FK_Simulate.m	To specify $^3\{\mathbf{p}_3\}$
RPP_FK_Model.slx	To calculate manipulator output and simulate manipulator motion
Post_Simulation_Task.m	To write manipulator output (in a file compatible with Microsoft Excel)

```
%--------------------------------------------------------------
% Here, the coordinates for the end-effector point p3_3 are
% assigned.

p3_3 = [0, 0, -1];
%--------------------------------------------------------------
```

FIGURE L.1
Section of *RPP_FK_Simulate.m* with sample values in bold.

TABLE L.2

Appendix L.2 MATLAB Files

Filename	Use of File
RRP_Input.csv	To specify joint displacements
RRP_FK_Simulate.m	To specify $^3\{\mathbf{p}_3\}$, l_1, and l_2
RRP_FK_Model.slx	To calculate manipulator output and simulate manipulator motion
Post_Simulation_Task.m	To write manipulator output (in a file compatible with Microsoft Excel)

```
%--------------------------------------------------------------
% Here, the coordinates for the end-effector point p3_3 and the
% lengths l1 and l2 are assigned.

l1 = 0.5;
l2 = 0.5;
p3_3 = [0, 0, 0];
%--------------------------------------------------------------
```

FIGURE L.2
Section of *RRP_FK_Simulate.m* with sample values in bold.

L.3 R-R-R Robotic Manipulator Forward Kinematics

The Appendix L.3 folder (which is available for download at www.crcpress.com/product/isbn/9781498724937) includes four MATLAB and SimMechanics files for R-R-R robotic manipulator forward kinematics. These four files are described in Table L.3. To conduct an R-R-R forward kinematics analysis, the user specifies the R-R-R joint displacements in the file *RRR_Input.csv*. This file is compatible with Microsoft Excel. Also, $^3\{\mathbf{p}_3\}$, l_1, and l_2 are specified in the file *RRR_FK_Simulate.m*. Figure L.3 illustrates the user-input section of

```
%------------------------------------------------------------------
% Here, the coordinates for the end-effector point p3_3 and the
% lengths l1 and l2 are assigned.

l1 = 0.5;
l2 = 0.5;
p3_3 = [0, 1, 0];
%------------------------------------------------------------------
```

FIGURE L.3
Section of *RRR_FK_Simulate.m* with sample values in bold.

TABLE L.3

Appendix L.3 MATLAB Files

Filename	Use of File
RRR_Input.csv	To specify joint displacements
RRR_FK_Simulate.m	To specify ${}^3\{\mathbf{p}_3\}$, l_1, and l_2
RRR_FK_Model.slx	To calculate manipulator output and simulate manipulator motion
Post_Simulation_Task.m	To write manipulator output (in a file compatible with Microsoft Excel)

this file, with sample values in bold type. After specifying the R-R-R joint displacements in *RRR_Input.csv* and ${}^3\{\mathbf{p}_3\}$, l_1, and l_2 in *RRR_FK_Simulate.m*, the next step is to run *RRR_FK_Simulate.m*. When running this file, one file (filename *RRR_p3.csv*) is written to a folder named *Results* (in a format compatible with Microsoft Excel) that includes the ${}^0\{\mathbf{p}_3\}$ values calculated for the given ${}^3\{\mathbf{p}_3\}$, length, and joint displacement values.

Also, when running *RRR_FK_Simulate.m*, a graphical user interface appears where the motion of the R-R-R robotic manipulator is simulated over the joint displacements (see Appendix A.3).

L.4 R-R-C Robotic Manipulator Forward Kinematics

The Appendix L.4 folder (which is available for download at www.crcpress.com/product/isbn/9781498724937) includes four MATLAB and SimMechanics files for R-R-C robotic manipulator forward kinematics. These four files are described in Table L.4. To conduct an R-R-C forward kinematics analysis, the user specifies the R-R-C joint displacements in the file *RRC_Input.csv*. This file is compatible with Microsoft Excel. Also, ${}^3\{\mathbf{p}_3\}$, l_1, l_2, and l_3 are specified in the file *RRC_FK_Simulate.m*. Figure L.4 illustrates the user-input section of this file, with sample values in bold type.

TABLE L.4

Appendix L.4 MATLAB Files

Filename	Use of File
RRC_Input.csv	To specify joint displacements
RRC_FK_Simulate.m	To specify ${}^3\{\mathbf{p}_3\}$, l_1, l_2, and l_3
RRC_FK_Model.slx	To calculate manipulator output and simulate manipulator motion
Post_Simulation_Task.m	To write manipulator output (in a file compatible with Microsoft Excel)

```
%-----------------------------------------------------------------
% Here, the coordinates for the end-effector point p3_3 and the
% lengths l1, l2 and l3 are assigned.

l1 = 1;
l2 = 0.5;
l3 = 0.5;
p3_3 = [1, 0, 0];
%-----------------------------------------------------------------
```

FIGURE L.4
Section of *RRC_FK_Simulate.m* with sample values in bold.

After specifying the R-R-C joint displacements in *RRC_Iinput.csv* and ${}^3\{\mathbf{p}_3\}$, l_1, l_2, and l_3 in *RRC_FK_Simulate.m*, the next step is to run *RRC_FK_Simulate.m*. When running this file, one file (filename *RRC_p3.csv*) is written to a folder named *Results* (in a format compatible with Microsoft Excel) that includes the ${}^0\{\mathbf{p}_3\}$ values calculated for the given ${}^3\{\mathbf{p}_3\}$, length, and joint displacement values.

Also, when running *RRC_FK_Simulate.m*, a graphical user interface appears where the motion of the R-R-C robotic manipulator is simulated over the joint displacements (see Appendix A.3).

Index

A

Acceleration equations, 63–64, 83–89, 93–95, 104–105, 347–350, 352–355
    coupler angular, 348
    follower angular, 353
    RRSS, 348
    RSSR, 352
Addendum, 275, 278
AGMA, *see* American Gear Manufacturers Association (AGMA)
Aircraft in flight, 4
Aircraft landing-gear example and motion generation, 135
American Gear Manufacturers Association (AGMA), 276, 284
American National Standards Institute (ANSI), 276, 286
Angular acceleration, 65–66, 93, 95, 105, 108
    coupler, 70, 348
    follower, 353
    matrix, 31
Angular velocity, 78, 104, 108, 153–154, 223
    coupler, 68, 82, 347
    crank, 78, 92
    follower, 155, 352
    gear, 271, 288
    link, 64–66, 93, 153
    matrix, 31
    time and driver, 108
ANSI, *see* American National Standards Institute (ANSI)
Aronhold–Kennedy theorem, 76
Articulated robot, 377
    forward kinematics, 387–389, 457–458, 506–507
    inverse kinematics, 397–399, 461
Assembly configurations, 108

B

Backlash, 283
Base circle, 277, 279, 282, 283, 290, 331–332, 334
    diameter, 275, 280
Base frame, 375, 378–379
Base pitch, 280
Bevel gears, 273, 298–301, 305
Binary link, 46

Binding position, 53
Branch defect, 133–135
    elimination, 140–144, 165

C

Cam joint, 46, 56, 59
Cam profile, 331
Cartesian robot, 376, 377
    forward kinematics, 381–382
    inverse kinematics, 392–394
Cayley diagram, 109–111, 118
Center distance, 280–281
Centrode, 78–80, 89–90, 117, 415–416
Change point mechanism, 51
Circuit defect, 53–54, 57, 74, 134, 415
Circular motion, *see* Pure rotation
Circular pitch, 275, 285, 302
Clearance, 275
Closed-loop design, 373
Coefficient matrix, 22
Cognate construction, 109–111
Column matrix, 23
    transpose, addition, subtraction, and multiplication of, 22–26
Compacting mechanism, 7–8
Complex motion, 5, 6, 10, 41, 44–45, 56
Complex number, 37
    addition, 15–16
    forms, 13–15
    multiplication and differentiation, 17–20
Constant acceleration motion, 317–319, 447–449
Constant velocity motion, 315–317, 447
Contact force, 332–334, 332
Contact line, 279
    length, 280
Contact ratio, 280–281
Coriolis acceleration, 66, 87
Coupler, 41, 50, 177, 226, 240, 344
    angular acceleration, 70, 348
    angular velocity, 68, 82, 347
    displacement angles, 75, 138, 152, 158, 345
    locations of interest kinematics, 70–75
    path points, 144
    position, 135
Cramer's rule, 27, 37, 70, 82, 84, 147, 152, 155, 189–199, 228–246

Crank displacement angles, 91, 132, 149, 152, 163, 164
Crank link, 41, 67, 109, 177, 226, 344
Crank-rocker mechanism, 50, 52
Crossed shafts, 284
Cycloid, 279
Cycloidal motion, 322–325, 451
Cylindrical joint, 46, 56, 343, 376, 377
Cylindrical robot, 376, 377
    forward kinematics, 383–384, 457–458, 505–506
    inverse kinematics, 394–335, 460

D

Dedendum, 275
Degrees of freedom (DOF), 3, 46–47, 56, 185, 343
    passive, 344
    and paradoxes, 55–56
Denavit–Hartenberg (DH) parameters, 380
Diametral pitch, 275
Dimensional synthesis, 7, 131
    branch and order defects, 133–135
    elimination, 140–144
    kinematic analysis *vs.*, 132
    mechanism dimensions, 158–163
    path generation *vs.* motion generation, 144–145
    planar four-bar function generation finitely separated positions (FSPs) and multiply separated positions (MSPs), 153–158
    three precision points, 149–153
    planar four-bar motion generation, 135–139
    Stephenson III motion generation, 145–149
    subcategories of, 133
Disk cam, 311
    design and pressure angle, 331–236
    follower motion, 313–331
    constant acceleration motion, 317–319
    constant velocity motion, 315–317
    cycloidal motion, 322–325
    displacement, velocity, acceleration, and jerk, 315
    polynomial motion, 325–331
    rise, fall, and dwell profiles, 313–314
    simple harmonic motion, 320–322
    follower types in, 312–313
    nomenclature, 332
Displacement angles, 113, 149–150, 152, 163
    coupler, 75, 145, 152, 158, 345
    crank, 102, 132, 149, 152, 163, 164
    dyad, 102, 137–138, 140

follower, 132, 149, 152, 163, 164
Displacement equations, 66–67, 80–81, 85, 91–92, 100–102, 344–346, 350–351
    follower, 320, 322
    formulation of, 63
DOF, *see* Degrees of freedom (DOF)
Door linkage, 74
Double-crank mechanism, 50, 53
Double enveloping, 302
Double-rocker mechanism, 50–52
Double thread, 302
Drag-link, *see* Double-crank mechanism
Dynamic force analysis, of planar mechanisms, 223
    dynamic loading in planar space, 224
    four-bar mechanism analysis, 224–229
    geared five-bar mechanism analysis, 233–238
    mass moment of inertia and computer-aided design software, 248–250
    planar mechanism and modeling in SimMechanics, 250–255
    slider-crank mechanism analysis, 230–233
    Stephenson III mechanism analysis, 243–248
    Watt II mechanism analysis, 238–243
Dynamics, meaning of, 1

E

Electro-mechanical systems, 373
End effector, 375
Enveloping worm-gear teeth, 302
Epicycloid, 279
External gear, 272, 280

F

Face width, 275
Filemon's construction, 140–141
Fixed centrode, 78–79, 90, 117, 416
Fixed pivot, 110, 140, 344, 415
Flat-faced cam, *see* Disk cam
Flat-faced follower, 312, 313
Follower angular acceleration equation, 353
Follower displacement angles, 132, 150, 152, 163, 164
Follower link, 41, 50–52, 55–57, 77, 109, 140, 141, 144, 149, 347
Follower motion
    constant acceleration motion, 317–319
    constant velocity motion, 315–317
    cycloidal motion, 322–325

displacement, velocity, acceleration, and jerk, 315
polynomial motion, 325–331
rise, fall, and dwell profiles, 313–314
simple harmonic motion, 320–322
Forward kinematics
definition and application, 381
P-P-P robotic manipulator, 381–382
R-P-P robotic manipulator, 383–384, 457–458, 505–506
R-R-C robotic manipulator, 389–391, 459–460, 507–508
R-R-P robotic manipulator, 385–387, 457, 458, 505, 506
R-R-R robotic manipulator, 387–389, 457–459, 507–508
Frames, 375, 378–380
Free choices, in equations, 137, 147
Friction force, 183, 231, 232, 271, 312
Full joint, *see* Lower pair
Function curve, with precision points, 150
Function generation, 7, 10, 132
planar four-bar mechanism, 158, 163, 164
finitely separated positions (FSPs) and multiply
separated positions (MSPs), 153–158
three precision points, 149–153
Fundamental law of gearing, 279

G

Geared five-bar mechanism, 42–43
acceleration equations, 93–95
displacement equations, 91–92
dynamic force analysis, 233–238
kinematics of intermediate link locations of interest, 95–97
MATLAB, 419–421, 428–429, 437–439, 467–468, 470, 479, 490–491, 494
static force analysis, 185–190
velocity equations, 92–93
Gear joint, 46, 56
Gear mating process, 278
Gear pair, 56, 91, 188–189, 271, 272
Gear-pitch angle, 298
Gear ratio, 91, 188, 292, 305
Gears, 271–272
backlash, 283
center distance and contact ratio, 280–281
gear-tooth interference and undercutting, 282–283
helical-gear nomenclature, 284–287
kinematics, 287–304

bevel gears, 298–301
helical gears, 296–297
planetary gear trains, 291–294
rack and pinion gears, 295
spur gears and gear trains, 287–290
worm gears, 301–304
pressure angle and involute tooth profile, 277–280
spur-gear nomenclature, 274–277
types, 272–274
Gear-tooth interference and undercutting, 282–283
Gear train, 42, 91, 188–189, 271
and spur-gear, 287–291
Global acceleration, 94, 104, 347–348, 352
Global velocity, 92, 103, 346, 352
Grashof's criteria, 50–51, 57
Gripper, 375
Ground link, 2, 43, 47, 140, 245
Gruebler's equation, 46, 55, 56, 66, 80, 375

H

Half joint, *see* Higher pair
Helical gear, 273, 296–297, 301–303, 305
left-hand, 284, 296
nomenclature, 284–287
right-hand, 284, 296
Helix angle, 284–286, 302, 303
Higher pair, 46
Home position, 332
Hypocycloid, 279

I

IC, *see* Instant center (IC)
Identity matrix, 26, 345
Idler gear, 289
Image pole, 143, 144
Imaginary component, 14–16, 67–69 , 117, 136–137
In-line slider crank, 81
Instant center (IC), 76–78, 117, 416
Interference, 282
Intermediate acceleration, 32
Intermediate and total spatial motion, 29–33
Intermediate displacement, 31
Intermediate velocity, 31
Internal gear, 272
International System of Units, 8, 275
Inverse kinematics
definition and application, 392
P-P-P, 392–394

Inverse kinematics (*cont.*)
    R-P-P, 394–395, 460
    R-R-C, 399–400, 462
    R-R-P, 396–397, 460–461
    R-R-R, 397–399, 461
Involute gear tooth, 279

J

Jerk, 315
Joints, 46
    cam, 46, 56
    cylindrical, 46, 56, 343, 376
    full, 46
    gear, 46, 56
    half, 46
    prismatic, 46, 56, 80, 376, 386, 391
    revolute, 46, 66, 76–100, 177–196, 224–244, 376
    spherical, 46, 56, 343, 344, 376

K

Kinematics, 1–2; *see also individual entries*
    chains and mechanisms, 2–3
    dimensional synthesis and, 132, 158–163
    fundamental concepts in, 41–56
    mathematical concepts in, 13–36
    mechanism motion types, 5–6
    mobility, planar, ad spatial mechanisms, 3–5
    planar mechanisms and, 63–117
    software resources, 9
    synthesis, 7–8, 10
    units and conversion, 8–9
Knife-edge follower, 312, 313

L

Level-luffing crane mechanism, 71–73, 144, 176, 359, 361
Linear equation, 22, 37, 152
Linear motion, *see* Pure translation
Linear simultaneous equation systems and matrices, 22–23, 37
Linkage, 373; *see also* Mechanism; Planar mechanism kinematic analysis; Spatial mechanisms
Links
    angular velocity, 64, 93, 153
    binary, 46
    crank, 41, 67, 109, 177, 226, 344
    drag, *see* Double-crank mechanism
    follower, 41, 50–56, 77, 109, 140–149, 347
    ground, 3, 43, 46, 47, 140, 245

    velocity and acceleration components and, 64–66
    rotating, 64
    rotating-sliding, 66
    ternary, 46
Load points, 225, 230, 234
Lower pair, 46

M

Mass moment of inertia and computer-aided design software, 248–249
MATLAB, 9, 10, 14–18, 157–158, 293–297, 380
    calculation procedure, *see individual entries*
    constant acceleration motion, 447–449
    constant velocity motion, 447
    cycloidal motion, 451
    equation formulation in, 88–89
    functions and operators, 410
    geared five-bar mechanism, 419, 428–429, 437–439, 467–469, 479, 490–494
    planar four-bar mechanism, 415–417, 425, 435, 465, 475, 487–488
    3-4-5 polynomial motion, 451
    4-5-6-7 polynomial motion, 451
    precision requirement, 413
    preparing and running files in, 409–412
    rerunning files with existing *.csv files, 413
    R-P-P robotic manipulator forward kinematics, 457–458, 505–506
    inverse kinematics, 460
    R-R-C manipulator forward kinematics, 459–460, 507–508
    inverse kinematics, 462–463
    R-R-P robotic manipulator forward kinematics, 456, 505
    inverse kinematics, 460–461
    R-R-R manipulator forward kinematics, 457, 458, 507
    inverse kinematics, 461–462
    RRSS mechanism, 453, 454, 501–502
    RSSR mechanism, 453, 501–504
    simple harmonic motion, 449–451
    slider-crank mechanism, 417–419, 425–428, 435–437, 465–467, 475–479, 488–490
    Stephenson III mechanism, 422–423, 431–434, 441–445, 472–473, 482–485, 497–499
    vector first-order differentiation procedure in, 19
    Watt II mechanism, 420–422, 430–431, 439–441, 470–472, 480–482, 494–497
Matrix, 22–23, 68, 137
    angular acceleration, 31

angular velocity, 31
coefficient, 22
column, 22–27
combined, 179–193, 227–242, 246
identity, 26, 345
inversion, 26–29
products, 26
row, 23
spatial angular acceleration matrix, 31
spatial angular displacement matrix, 29
spatial angular velocity matrix, 31
square, 24, 26
transformation matrix, 33–37, 377–391
Maverick mechanisms, *see* Paradoxes
    mechanisms
Mechanism inversion, 54–55, 57
Mechanism mobility, 47–47, 56
Mechanisms, 2, 10; *see also* Linkage
Module, 275
Motion generation, 7, 10, 132, 141, 163
analytical, 132, 140
path generation *vs.*, 144–145
planar four-bar mechanism, 135–139, 158
with prescribed timing, 138, 140, 163
Stephenson III mechanism, 145–149, 158
Moving centrode, 79–80, 89–90, 117, 415–417
Moving pivot, 140–143, 344–248, 351–353, 415

N

Newton's first law, 1, 6, 175, 176, 204
Newton's second law, 1, 223, 224, 256
Newton–Raphson method, 64, 117
    flowchart, 64
Nodes, 46–47
Normal acceleration, 68–69, 87–88, 93–84, 104
Normal circular pitch, 285
Normal diametral pitch, 285, 286
Normal direction, 284, 285
Normal force, 183, 231, 289, 293, 296
Normal module, 285
Normal pressure angle, 285
Number of teeth, 274, 292
Number synthesis, 7, 10, 49–50, 56

O

Offset cam system and pressure angle, 332
Open loop design, 373
Order defect, 134, 163–164
    elimination, 140–144
Overhead valve train and actuation
    mechanism, 311, 312

P

Paradox mechanisms, 55
Parking automobile, 3
Passive degree of freedom, 55, 344
Path generation, 131, 133, 163, 144, 163
Periodic functions, 320
Pinion-pitch angle, 298
Pitch circle, 274, 277, 280, 282, 283, 290
Pitch circumference, 302
Pitch curve, 332
Pitch diameter, 274, 280, 285, 298
    worm, 301, 303
Pitch-line velocity, 288
Pitch point, 279
Planar four-bar mechanism, 41, 42, 54, 56, 113
acceleration equations, 68–70
branch defect, 133
with circuit defect, 53
coupler locations of interest kinematics,
    70–75
displacement equations, 66
dyads, 135
dynamic force analysis, 224–229
function generation, 158, 163
    finitely separated positions (FSPs) and
        multiply separated positions (MSPs),
        153–158
    three precision points, 149–153
instant center, centrodes, and centrode
    generation, 76–80
MATLAB, 415–417, 425, 435, 465, 475, 487–488
motion generation, 135–139
RRSS and RSSR kinematic equations and,
    359
static force analysis, 178–181
synthesis and analysis vectors for, 159
vectors and vector expressions in, 160
velocity equations, 67–68
Planar mechanism kinematic analysis, 63
cognate construction, 109–110
four-bar mechanism analysis, 66–80
geared five-bar mechanism analysis, 91–97
link velocity and acceleration components
    and, 64–66
mechanism configurations, 108–109
and modeling in SimMechanics, 112–117
numerical solution method for two
    simultaneous equations, 64
slider-crank mechanism analysis, 80–90
Stephenson III mechanism analysis, 100–108
time and driver angular velocity, 108
Watt II mechanism analysis, 97–100

Planar multiloop six-bar mechanism, 44–45
Planetary gear trains, 291–294
Planet gear, 291, 293, 294
Pliers, 42, 175, 176
    in open and closed positions, 3
    in plane, 5
Point coordinates, 20, 35–37, 377, 383–392
Polar exponential form, 14, 17, 18
Polar forms, of vector, 14
Pole triangle, 143
        3-4-5 Polynomial, 325, 326, 328, 451
        4-5-6-7 Polynomial, 328–331, 335–336, 451
Polynomial motion, 325–331
P-P-P robotic manipulator, *see* Cartesian robot
Precision points, 145, 149–155, 164
Pressure angle, 331
    equation, for disk cam and translating roller
        follower, 333
    and involute tooth profile, 277–280
    and offset cam system, 333
    profile, for 4-5-6-7 polynomial cam and
        translating follower, 335
Pressure line, 277–280, 286, 289, 305
Prime circle, 332
Prismatic joint, 46, 56, 80, 382, 386, 391
Pure rotation, 5, 6, 10, 41, 43, 56
Pure translation, 5, 6, 10, 41, 56

Q

Qualitative kinematic analysis, 63
Quantitative kinematic analysis, 63
Quasi-static state, 175, 223

R

4R, *see* Revolute-revolute-revolute-revolute (4R)
        spherical mechanism
Rack and pinion gear, 272, 295–296, 305
Rack gear, 272, 295, 305
Radial cam, *see* Disk cam
Radial force, 289, 300
Real component, 14, 67–68, 70, 81, 117,
        136–137
Rectangular form, of vector, 14, 16
Revolute joint, 46, 66, 76–100, 177–196, 224–243,
        376
Revolute-revolute-revolute-revolute (4R)
        spherical
    mechanism, 45, 54, 355–359
Revolute-revolute-spherical-spherical (RRSS)
        mechanism, 45, 55–57
    acceleration equations, 347–350

displacement equations, 344–346
MATLAB, 453, 501
velocity equations, 346–347
Revolute-spherical-spherical-revolute (RSSR)
        mechanism, 45, 55–57
    acceleration equations, 352–355
    displacement equations, 350–351
    MATLAB, 453–454, 501–504
    velocity equations, 351–352
Roberts diagram, 110
Robotic manipulators, 5, 373, 374
    forward kinematics
        definition and application, 381
        P-P-P, 381–382
        R-P-P, 383–384
        R-R-C, 389–391
        R-R-P, 385–387
        R-R-R, 387–389
    inverse kinematics
        definition and application, 392
        P-P-P, 392–394
        R-P-P, 394–395
        R-R-C, 399–401
        R-R-P, 396–397
        R-R-R, 397–399
    kinematic analysis, and SimMechanics
        modeling, 401
    mobility and types, 375–377
    terminology and nomenclature, 374–375
    transformation matrix, 377–380
Robotic system, with reference frames, 35
Roller follower, 312–313
    rotating
        and cam system, 334
        and disc cam, 334
Rolling conical wheels, 298
Rolling cylinder pair, 55, 271, 272
Rolling-sliding contact, 46
Root-finding method, 64, 117, 132, 394, 396,
        397–399
Rotating link, 64
Rotating-sliding link, 64
Row matrix, 23
R-P-P robotic manipulator, *see* Cylindrical robot
R-R-C robotic manipulator, *see* Selective
        Compliance Assembly/Articulated
        Robot Arm
(SCARA) robot
R-R-P robotic manipulator, *see* Spherical robot
R-R-R robotic manipulator, *see* Articulated
        robot
RRSS, *see* Revolute-revolute-spherical-spherical
        (RRSS) mechanism

RSSR, *see* Revolute-spherical-spherical-revolute (RSSR) mechanism

S

SCARA, *see* Selective Compliance Assembly/Articulated Robot Arm (SCARA) robot
Screw axis, 4, 9
Screw motion, 4, 5, 9
Selective Compliance Assembly/Articulated Robot Arm (SCARA) robot, 376, 377
　forward kinematics, 389–391, 459–460, 507–508
　inverse kinematics, 399–401, 462
Shaft angle, 298
SimMechanics, 9–10, 112–118, 201–204, 250–255, 362–364, 465–508
　operations in, 411–413
　precision requirement, 413
　rerunning files with existing *.csv files, 413
Simple harmonic motion, 320–322, 449–451
Simultaneous equations
　linear, 22–23, 27, 37, 163, 204, 256
　nonlinear, 67, 92, 102, 117, 394, 396–399
　numerical solution method for, 64
Single thread, 302
Slider-crank mechanism, 41, 42, 54, 77
　acceleration equations, 83–89
　centrode generation, 89–90
　displacement equations, 80–81, 85
　dynamic force analysis, 230–232
　instant centers and circle diagram, 78
　MATLAB, 417–418, 426–427, 435–437, 465–467, 475–479, 487–490
　static force analysis, 182–183
　velocity equations, 81–82, 85
Spatial angular acceleration matrix, 31
Spatial angular displacement matrix, 29
Spatial angular velocity matrix, 31
Spatial four-bar mechanism, 45–46
　types of, 45
Spatial mechanisms, 343
　4R spherical analysis, 355–359
　and modeling in SimMechanics, 362–365
　planar four-bar kinematic analysis using RRSS and RSSR kinematic equations, 359–360
　RRSS mechanism analysis
　　acceleration equations, 347–350
　　displacement equations, 344–346
　　velocity equations, 346–347
　RSSR mechanism analysis
　　acceleration equations, 352–355

　　displacement equations, 350–351
　　velocity equations, 351
Spatial motion, of two-body system, 30
Spatial rotation, of arbitrary body, 30
Speed reduction system, 292
Spherical joint, 46, 56, 343, 344, 376
Spherical robot, 376
　forward kinematics, 385–387, 457, 505
　inverse kinematics, 396–397, 460–461
Spur-gear
　features, 274
　and gear trains, 287–290
　nomenclature, 274–275
　size variation with diametral pitch, 276
　types, 272
Square matrix, 24, 26
Static force analysis, of planar mechanisms, 175–176
　four-bar mechanism analysis, 177–181
　geared five-bar mechanism analysis, 185–190
　planar mechanism and modeling in SimMechanics, 201–204
　slider-crank mechanism analysis, 182–185
　static loading in planar space, 176
　Stephenson III mechanism analysis, 196–200
　Watt II mechanism analysis, 190–195
Statics, meaning of, 1
Stephenson III mechanism, 44–46, 63
　acceleration equations, 68–70
　displacement equations, 100–102
　dynamic force analysis, 243–248
　kinematics of intermediate link locations of interest, 105–108
　MATLAB, 422–423, 431–433, 411–445, 472–473, 482–485, 426–428
　motion generation, 144–149
　static force analysis, 196–200
　synthesis and analysis vectors for, 158
　vectors and vector expressions in, 159
　velocity equations, 103–104
Sun gear, 291, 293
Superposition, 293

T

Tabular method, 291–292
Tangent force, 289, 290
Tangential acceleration, 68, 83, 87–88, 93–94, 104
Tangential velocity, 67, 82, 83, 92, 103
Ternary link, 46
Thread lead, 302
Thread length, 302
Threads, 301

Thrust bearings, 297
Time and driver angular velocity, 108
Tool frame, 375
Torque, *see individual entries*
Total acceleration, 31, 68, 83, 226
Total velocity, 31, 64
Trace point, 332
Train value, 288
Transformation matrix, 33–36, 377–391
Transmission angle, 51–52, 57, 61
    vs. crank displacement angle plot, 113
Transverse direction, 285
Triple-rocker mechanism, 50, 78
    inverted, 78
Triple thread, 302
Type synthesis, 7, 10, 33

U

Undercutting, 283
Unit matrix, *see* Identity matrix
US customary units, 8, 275

V

Valve train assembly, 5–6
Vector; *see also individual entries*
    first-order differentiation procedure in
        MATLAB, 19
    in loop, in two-dimensional complex space,
        15, 16, 18
    and point representation, 20–21
    second-order differentiation procedure in
        MATLAB, 19
    in two-dimensional complex space, 14
Vector-loop, 37, 63, 70, 105, 117
    errors, 413
    for four-bar mechanism, 135, 151, 154, 163,
        164

standard-form, 136
Stephenson III motion generation, 145–146
sum, 18, 81, 85, 102
Velocity equations, 67–68, 81–73, 92–93, 103–104,
    346–347, 351–352
Velocity ratio (VR), 271, 288, 293, 298, 303
Vibration analysis, 229
VR, *see* Velocity ratio (VR)

W

Waldron's construction, 141, 143
Watt II mechanism, 45, 63, 97–100
    dynamic force analysis, 238–243
        MATLAB, 419–422, 430–431, 438–441,
            469–471, 480–482, 494–497
    static force analysis, 190–195
Whole depth, 275
Workspace, 374, 392, 396
Worm gears, 273, 301–304
Worm helix angle, 302
Worm lead angle, 302
Worm pitch, 302
Worm-pitch diameter, 302, 304

X

x-Direction, 176, 226–240, 245–246, 351, 392
x-y Plane, 3, 176, 343, 359
x-y-z Plane, 4, 33–34

Y

y-Direction, 176, 183, 226–245, 334, 343,
    377–380

Z

z-Direction, 175, 343, 351, 359, 377, 392